Lecture Notes in Artificial Intelligence   5625

Edited by R. Goebel, J. Siekmann, and W. Wahlster

Subseries of Lecture Notes in Computer Science

Jacques Carette   Lucas Dixon
Claudio Sacerdoti Coen   Stephen M. Watt (Eds.)

Intelligent Computer Mathematics

16th Symposium, Calculemus 2009
8th International Conference, MKM 2009
Held as Part of CICM 2009
Grand Bend, Canada, July 6-12, 2009
Proceedings

 Springer

Series Editors

Randy Goebel, University of Alberta, Edmonton, Canada
Jörg Siekmann, University of Saarland, Saarbrücken, Germany
Wolfgang Wahlster, DFKI and University of Saarland, Saarbrücken, Germany

Volume Editors

Jacques Carette
McMaster University, Department of Computing and Software
1280 Main Street West, Hamilton, ON L8S 4K1, Canada
E-mail: carette@mcmaster.ca

Lucas Dixon
University of Edinburgh, Informatics, 2.02 Informatics Forum
10 Crichton street, Edinburgh EH8 9AB, UK
E-mail: l.dixon@inf.ed.ac.uk

Claudio Sacerdoti Coen
University of Bologna, Department of Computer Science
via Mura Anteo Zamboni, 7, 40127 Bologna, Italy
E-mail: sacerdot@cs.unibo.it

Stephen M. Watt
University of Western Ontario, Department of Computer Science
London, ON N6A 5B7, Canada
E-mail: stephen.watt@uwo.ca

Library of Congress Control Number: 2009929069

CR Subject Classification (1998): I.2.2, I.1-2, H.3, H.2.8, F.4.1, C.2.4, G.2, G.4

LNCS Sublibrary: SL 7 – Artificial Intelligence

ISSN 0302-9743

ISBN 978-3-642-02613-3 Springer Berlin Heidelberg New York

springer.com

© Springer-Verlag Berlin Heidelberg 2009

Typesetting: Camera-ready by author, data conversion by Scientific Publishing Services, Chennai, India
Printed on acid-free paper   SPIN: 12701551   06/3180   5 4 3 2 1 0

Preface

As computers and communications technology advance, greater opportunities arise for intelligent mathematical computation. While computer algebra, automated deduction and mathematical publishing each have long and successful histories, we are now seeing increasing opportunities for synergy among them. The Conferences on Intelligent Computer Mathematics (CICM 2009) is a collection of co-located meetings, allowing researchers and practitioners active in these related areas to share recent results and identify the next challenges. The specific areas of the CICM conferences and workshops are described below, but the unifying theme is the computerized handling of mathematical knowledge.

The successful formalization of much of mathematics, as well as a better understanding of its internal structure, makes mathematical knowledge in many ways more tractable than general knowledge, as traditionally treated in artificial intelligence. Similarly, we can also expect the problem of effectively using mathematical knowledge in automated ways to be much more tractable. This is the goal of the work in the CICM conferences and workshops. In the long view, solving the problems addressed by CICM is an important milestone in formulating the next generation of mathematical software.

The first CICM was held in Birmingham, UK, in 2008. Although combinations of the constituent meetings had been held together previously, this was the first time this set of conferences and workshops were held together under the CICM name. In some sense this was a symbolic step, recognizing that these areas shared common challenges that should be addressed together. The anchor meetings were the Artificial Intelligence and Symbolic Computation (AISC) conference, the 15th Symposium on the Integration of Symbolic Computation and Mechanized Reasoning (Calculemus 2008) and the 7th International Conference on Mathematical Knowledge Management (MKM 2008). A number of related workshops also joined the meeting.

Those participating in CICM 2008 felt the meeting worked well and wished to hold a federated event again in 2009. The Ontario Research Centre for Computer Algebra (ORCCA) at the University of Western Ontario offered to host the meeting on the shore of Lake Huron in Grand Bend, Ontario. The governing bodies of both Calculemus and MKM agreed to co-locate at CICM 2009. AISC could not participate because it was held only every second year. Two of the workshops of CICM 2008, Mathematical User Interfaces and Towards a Digital Mathematics Library, as well as a number of additional workshops also decided to hold their next event at CICM 2009. Thus, the CICM 2009 meeting included two long-standing international conferences:

- 16th Symposium on the Integration of Symbolic Computation & Mechanized Reasoning (Calculemus 2009)
- 8th International Conference on Mathematical Knowledge Management (MKM 2009)

as well as the following inter-related workshops:

- Second Compact Computer Algebra Workshop (CCA 2009)
- Second Workshop Towards a Digital Mathematics Library (DML 2009)
- W3C Workshop on Ink in Multimodal Applications (InkMMI 2009)
- 4th Mathematical User Interfaces Workshop (MathUI 2009)
- 22nd OpenMath Workshop (OpenMath 2009)
- Third Pen-Based Mathematical Computation (PenMath 2009)

Each of these conferences and workshops had its own successful predecessors, but for each of them it was the first time to be held in North America.

CICM 2009 featured a range of distinguished plenary speakers, representing the interests of the participants. These invited speakers and their hosting events were:

- Rob Arthan (Lemma 1 Ltd and Queen Mary, University of London, UK), "Computational Logic and Continuous Mathematics, Pure and Applied," Calculemus.
- Dorothea Blostein (Queen's University, Canada), "Math-Literate Computers," MKM and PenMath.
- Jacques Calmet (U. Karlsruhe, Germany), "Abstraction-Based Information Technology: A Framework for Open Mechanized Reasoning," Calculemus.
- John Fitch (University of Bath, UK), "CAMAL 40 Years On — Is Small Still Beautiful?" CCA.
- Georges Gonthier (Microsoft Research Cambridge, UK), "Software Engineering for Mathematics," MKM.
- Patrick Ion (Mathematical Reviews, AMS, USA), "Some Traditional Mathematical Knowledge Management," MKM and OpenMath.
- Marko Panić (Microsoft Development, Serbia), "Math Handwriting Recognition in Windows 7 and Its Benefits," MathUI and PenMath.
- David Ruddy (Cornell University, USA), "Assembling the Digital Mathematics Library," DML.

This volume represents the formal proceedings of CICM 2009. It includes a record of the invited talks and the conference papers accepted for the proceedings. Work presented at the workshops and in-progress work presented at the conferences was made available in informal proceedings.

We now describe in more detail the goals and objectives of the constituent meetings of CICM 2009, and the process by which papers were selected for these proceedings.

16th Symposium on the Integration of Symbolic Computation and Mechanized Reasoning (Calculemus 2009)

Calculemus is a series of conferences dedicated to the integration of computer algebra systems and systems for mechanized reasoning, interactive theorem provers or proof assistants and the automated theorem provers. Currently, symbolic computation is divided into several more or less independent branches: traditional ones (e.g., computer algebra and mechanized reasoning) as well as newly emerging ones (on user interfaces, knowledge management, theory exploration, etc.) The main concern of the Calculemus community has been to bring these developments together in order to facilitate the theory, design and implementation of integrated systems for computer mathematics that will routinely be used by mathematicians, computer scientists and engineers in their everyday business. The scope of Calculemus covers all aspects of the interplay of mechanized reasoning and computer algebra, including cross-fertilization between those two research areas, as well as the development of integrated systems that transcend both computer algebra and theorem proving.

Since 1999, to ensure interaction with both the deduction and computer algebra communities, Calculemus has co-located with closely related conferences: Federated Logics Conference 1999 (Trento, Italy), ISSAC 2000 (St. Andrews, UK), IJCAR 2001 (Siena, Italy), AISC 2002 (Marseilles, France), TPHOLs and TABLEAUX 2003 (Rome, Italy), IJCAR 2004 (Cork, Ireland), Formal Methods 2005 (Newcastle upon Tyne, UK), ISSAC 2006 (Genoa, Italy), MKM 2007 (Hagenberg, Austria), and with AISC and MKM within CICM 2008 (Birmingham, UK).

There were 17 full papers submitted to Calculemus 2009. Each of these received at least three reviews, followed by an author response phase. Of these submissions, 10 were accepted for full presentation at the conference and publication in this volume. In addition to these papers, extended abstracts were also solicited to provide a venue for discussion of work in progress. A supplementary proceedings for the work in progress is available at the Calculemus website http://www.calculemus.net .

9th International Conference on Mathematical Knowledge Management (MKM 2009)

The Mathematical Knowledge Management conferences arose similarly from common requirements at the boundaries of neighbouring fields. MKM lies at the intersection of mathematics and computer science with the goal of developing effective techniques, based on formal mathematics and software technology, to take advantage of the enormous knowledge available in current mathematical sources and to organize mathematical knowledge in new ways. Dually, due its very nature, the realm of mathematical information is an attractive candidate for testing innovative theoretical and technological solutions for content-based systems, interoperability, management of machine understandable information, and the Semantic Web. This led to a series of conferences spanning the decade, with meetings held in Hagenberg, Austria (2001), Bertinoro, Italy (2003), Białowieża, Poland (2004), Bremen, Germany (2005), Wokingham, UK (2006), Hagenberg, Austria (2007), Birmingham, UK (2008) and Grand Bend, Ontario, Canada (2009).

MKM 2009 solicited research contributions of two forms: longer papers of about 15 pages and short communications. There were 28 long papers and 6 short communications submitted. Each paper received between 2 and 5 anonymous reviews, for a total of 100 reports. Long paper submissions were also considered for the short communication category. In the end, 16 submissions were accepted as long papers and 6 as short communications for these proceedings. In addition, seven more preliminary submissions were accepted for oral presentation and electronic publication.

Second Compact Computer Algebra Workshop (CCA 2009). The art of compact computer algebra is experiencing a resurgence in relevance and importance. New directions for symbolic computing include the migration from workstations to hand-held devices and the changing role from stand-alone applications to lightweight services within integrated systems. Whether running on a graphing calculator or as support of a client-side web application, certain applications of computer algebra require compact data representation, space-efficient algorithms and effective memory management. The purpose of this workshop was to communicate efforts in research, design, development and applications of compact computer algebra.

Second Workshop Towards a Digital Mathematics Library (DML 2009). Mathematicians dream of a digital archive containing all peer-reviewed mathematical literature ever published, properly linked and validated/verified. It is estimated that the entire corpus of mathematical knowledge published over the centuries does not exceed 100,000,000 pages, an amount easily manageable by current information technologies. The workshop's objectives were to formulate the strategy and goals of a global mathematical digital library and to summarize the current successes and failures of ongoing technologies and related projects.

W3C Workshop on Ink in Multimodal Applications (InkMMI 2009). The goal of this workshop was to identify and prioritize requirements for changes, extensions and additions to digital ink standards, especially in multimodal applications developed based on the W3C's MMI Architecture and as a means of making InkML more useful in current and emerging contexts.

4th Mathematical User Interfaces Workshop (MathUI 2009). This workshop was intended to bring together researchers working on MKM but from the perspective of mathematics manipulated by end users. Accordingly, an emphasis was on providing users with interfaces and software systems that enhance their mathematical working experience. The topics of the workshop centered around presentation and manipulations of mathematical knowledge, workflows induced by mathematical knowledge representations, human communication of mathematical content, user studies with MKM tools or other mathematical interfaces and other novel interfaces to mathematics software.

22nd OpenMath (OpenMath 2009). With the development of MathML 3, OpenMath entered a new phase of its evolution. Topics to be discussed at the workshop included convergence of OpenMath and MathML 3, reasoning with OpenMath, software using or processing OpenMath, as well as new OpenMath Content Dictionaries.

Third Pen-Based Mathematical Computation (PenMath 2009). The use of the pen to enter, edit, and manipulate mathematical expressions can lead to a qualitative improvement in the ease of use of mathematical software. The purpose of this workshop was to explore this area, including pen-based mathematical interfaces for computer algebra and document processing, expression entry editing and manipulation, data collection and analysis, structural analysis, semantic methods, on-line and off-line mathematical handwriting recognition and to receive reports on implementations and experiments. The first workshop in this series was held as a special session of the 2005 Applications of Computer Algebra conference in Nara, Japan, and the second workshop was held as a special session of the conference Communicating Mathematics in the Digital Era conference in Aveiro, Portugal.

Numerous people contributed to making CICM 2009 happen. A list of organizers is to be found on the following pages. We thank them for their very substantial collective effort. To make the meeting as accessible as possible, a number of organizations were approached for financial contributions. We are most grateful for the generosity of the Fields Institute for Research in Mathematical Sciences, our principal sponsor. We also wish to thank McMaster University, the University of Waterloo, the University of Western Ontario (Faculty of Science and Research Western), Wilfrid Laurier University and Maplesoft for financial support. We thank the Ontario Research Centre for Computer Algebra and its members for their assistance and ACM SIGSAM for recommending *in cooperation* status for CICM 2009.

We are at a special point in the development of mathematical software, where systems in each of their individual niches have grown extremely powerful. In continuing to expand their capabilities, they have invariably reached the boundaries of their domains of origin and have started expanding into adjoining areas. A clear understanding of what should happen at these boundaries is essential to lay the foundation for future generations of versatile, integrated and intelligent systems for mathematics. It has been our hope that the discussions at CICM are a fruitful step in this direction.

April 2009   Jacques Carette
                                                            Lucas Dixon
                                                    Claudio Sacerdoti Coen
                                                        Stephen M. Watt

Organization

CICM has an Organizing Committee and the constituent events have their own Program Committees. The CICM Organizing Committee comprises the local organizers, the Program Committee Chairs of the constituent meetings, and past organizers as advisors.

CICM Organizing Committee

Jacques Carette	McMaster University, Canada
James H. Davenport	University of Bath, UK
Michael Doob	University of Manitoba, Canada
William Farmer	McMaster University, Canada
Juergen Gerhard	Maplesoft, Canada
Tetsuo Ida	University of Tsukuba, Japan
Patrick Ion	Mathematical Reviews, AMS, USA
Michael Kohlhase	Jacobs University, Germany
Ilias Kotsireas	Wilfrid Laurier University, Canada
George Labahn	University of Waterloo, Canada
Paul Libbrecht	DFKI, Germany
Robert Miner	Design Science, USA
Claudio Sacerdoti Coen	University of Bologna, Italy
Elena Smirnova	Texas Instruments, USA
Petr Sojka	Masaryk University, Czech Republic
Volker Sorge	University of Birmingham, UK
Masakazu Suzuki	Kyushu University, Japan
Stephen M. Watt (General Chair)	University of Western Ontario, Canada

Calculemus 2009

Program Committee

Markus Aderhold	TU Darmstadt, Germany
Serge Autexier	DFKI, Germany
John Campbell	University College London, UK
Jacques Carette (Co-chair)	McMaster University, Canada
James H. Davenport	University of Bath, UK
Louise Dennis	University of Liverpool, UK
Lucas Dixon (Co-chair)	University of Edinburgh, UK
William Farmer	McMaster University, Canada
Jacques Fleuriot	University of Edinburgh, UK
Herman Geuvers	Radboud University Nijmegen, The Netherlands
Michael Kohlhase	DFKI, Germany
Steve Linton	University of St. Andrews, UK
Tomas Recio	Universidad de Cantabria, Spain
Tom Ridge	University of Cambridge, UK
Julio Rubio	Universidad de La Rioja, Spain
Volker Sorge	University of Birmingham, UK

Additional Reviewers

Jose-Antonio Alonso	Christoph Lüth
F.J. Castro-Jiménez	Grant Olney Passmore
Thierry Coquand	Phil Scott
Josep M. Fortuny	David Sevilla
Chris Heunen	Alan Smaill
Feryal Fulya Horozal	Thomas Sturm
Pouya Larjani	Freek Wiedijk

Mathematical Knowledge Management 2009

Program Committee

Laurent Bernardin	Maplesoft, Canada
Olga Caprotti	University of Helsinki, Finland
Simon Colton	Imperial College, UK
Tetsuo Ida	Tsukuba University, Japan
Mateja Jamnik	University of Cambridge, UK
Tudor Jebelean	RISC Linz, Austria
Alejandro Jofre	University of Chile, Chile
Michael Kohlhase	Jakobs University, Germany
Azzeddine Lazrek	University of Marakech, Morocco
Paul Libbrecht	DFKI, Germany
Bruce Miller	NIST, USA
Robert Miner	Design Science, USA
Laurence Rideau	INRIA, France
Claudio Sacerdoti Coen (Co-chair)	University of Bologna, Italy
Elena Smirnova	Texas Instruments, USA
Volker Sorge	University of Birmingham, UK
Masakazu Suzuki	University of Kyushu, Japan
Joseph Urban	Charles University, Czech Republic
Stephen M. Watt (Co-chair)	University of Western Ontario, Canada
Freek Wiedijk	Rabdoub University, The Netherlands

Additional Reviewers

Romeo Anghelache	Maria Emilia Maietti
Serge Autexier	Roy McCasland
Yves Bertot	Christine Müller
Dominique Duval	Luca Padovani
Herman Geuvers	Matti Pauna
Jeremy Gow	Loïc Pottier
Mikolas Janota	Femke van Raamsdonk
Andrea Kohlhase	Florian Rabe
Temur Kutsia	Matthew Ridsdale
Christoph Lange	Piotr Rudnicki
Yuri Lebedev	Jordi Saludes
Jos Lehmann	Andreas Strotmann
Assia Mahboubi	

Workshops

Compact Computer Algebra 2009 Workshop
Organizers

Elena Smirnova	Texas Instruments, USA
Stephen M. Watt	University of Western Ontario, Canada

Towards a Digital Mathematics Library 2009 Workshop
Program Committee

José Borbinha	Technical University of Lisbon, Portugal
Thierry Bouche	University Grenoble, France
Michael Doob	University of Manitoba, Canada
Thomas Fischer	Goettingen University, Germany
Yannis Haralambous	Télécom Bretagne, France
Václav Hlaváč	Czech Technical University, Czech Republic
Janka Chlebíková	Comenius University, Slovakia
Enrique Maciás-Virgós	University of Santiago de Compostela, Spain
Jiří Rákosník	Academy of Sciences, Czech Republic
Eugenio Rocha	University of Aveiro, Portugal
David Ruddy	Cornell University, USA
Volker Sorge	University of Birmingham, UK
Petr Sojka (Organizer)	Masaryk University, Czech Republic
Masakazu Suzuki	Kyushu University, Japan
Bernd Wegner	Zentralblatt MATH, Germany

Mathematical User Interfaces 2009 Workshop
Program Committee

David Aspinall	University of Edinburgh, UK
Paul Cairns	University of York, UK
Olga Caprotti	University of Helsinki, Finland
Richard Fateman	University of California at Berkeley, USA
Anthony Jameson	DFKI and International University, Germany
Paul Libbrecht (Organizer)	DFKI, Germany
Robert Miner	Design Science, USA
Elena Smirnova	Texas Instruments, USA

OpenMath 2009 Workshop Program Committee

David Carlisle	NAG Ltd., UK
Olga Caprotti	University of Helsinki, Finland
James H. Davenport (Organizer)	University of Bath, UK
Michael Kohlhase	Jacobs University, Germany
Paul Libbrecht	DFKI, Germany
Chris Rowley	Open University, UK
Stephen M. Watt	University of Western Ontario, Canada

PenMath 2009 Workshop Organizers

Oleg Golubitsky	University of Western Ontario, Canada
George Labahn	University of Waterloo, Canada
Edward Lank	University of Waterloo, Canada
Stephen M. Watt	University of Western Ontario, Canada

W3C Workshop on Ink in Multimodal Applications Program Committee

Kazuyuki Ashimura (Co-organizer)	W3C/Keio, Japan
Deborah Dahl	Conversational Technologies, USA
Sriganesh Madhvanath	Hewlett Packard, India
Muthuselvam Selvaraj	Hewlett Packard, India
Raj Tumuluri	Openstream, USA
Tom Underhill	Microsoft, USA
Stephen M. Watt (Co-organizer)	University of Western Ontario, Canada

Table of Contents

1. Joint Invited Talks

2. Calculemus Talks

3. MKM Talks

Computational Logic and
Continuous Mathematics, Pure and Applied

Rob Arthan

Lemma 1 Ltd.
2nd Floor, 31A Chain Street, Reading RG1 2HX, UK
and
QMUL, School of Electronic Engineering and Computer Science,
Queen Mary, University of London, London E1 4NS, UK
rda@lemma-one.com

Continuous problem domains are of ever-increasing importance in the application of computational logic to problems in systems engineering and to problems in mathematics and theoretical computer science. I will outline some recent work both "pure" and "applied" with issues for mechanized reasoning and computer algebra in mind.

For some years, I have been involved with tools used for formally specifying and verifying digital subsystems of avionics control systems. The models used in this work typically have discrete time and continuous data. These discrete models emerge only at the end of a chain of refinements starting from a purely continuous top-level model of the overall system. I will describe a strand of work on methods for dealing with linear continuous systems and discuss issues for mechanized reasoning that are highlighted by this methodological research.

I will sketch some joint work with Robert M. Solovay and John Harrison into decidability and undecidability for various theories of normed spaces and inner product spaces. These are theories that occur naturally both in engineering applications and in applying mechanized reasoning to pure mathematics. On the positive side, we have decision procedures for inner product spaces and for the some fragments of the theory of normed spaces. On the negative side, we can prove undecidability for the theory of normed spaces in general. However, there are still interesting open problem areas to investigate. It is noteworthy that one of the constructions used in the undecidability results provides an interesting challenge problem for mechanized proof and computer algebra.

J. Carette et al. (Eds.): Calculemus/MKM 2009, LNAI 5625, p. 1, 2009.

Math-Literate Computers

Dorothea Blostein

School of Computing
Queen's University, Kingston, Ontario, Canada K7L 3N6
blostein@cs.queensu.ca

Abstract. Math notation is a familiar, everyday tool widely used in society. Computers need math literacy – the ability to read and write math notation – in order to assist people with accessing mathematical documents and carrying out mathematical investigations. In this paper, we discuss issues in making computers math-literate. Software for generating math notation is widely used. Software for recognition of math notation is not as widely used: to avoid the intrusiveness and unpredictability of recognition errors, people often prefer to enter and edit math expressions using a computer-oriented representation, such as LaTeX or a structure-based editor. However, computer recognition of math notation is essential in large-scale recognition of mathematical documents; as well, it offers the ability to create people-centric user interfaces focused on math notation rather than computer-centric user interfaces focused on computer-oriented representations. Issues that arise in computer math literacy include the diversity of math notation, the challenges in designing effective user interfaces, and the difficulty of defining and assessing performance.

1 Introduction

Math notation is a widely-used two-dimensional language for expressing and reasoning about mathematics. This notation developed over centuries, with many variants and dialects. Math notation is fluid, with users creating new forms of math notation as the need arises. Historically, math notation was written and read by people. The recent invention of the computer has lead to widespread use of electronic representations of mathematical expressions. Electronic representations support services such as typesetting, search, and automated reasoning. We need math-literate computers in order to best combine the convenience of paper-based math notation with the power of computer-based math representations.

Currently, computer generation of math notation is common, but recognition is less commonly used: the task of translating math notation into a computer-processable form is often done manually. With continuing advances in math recognition software, the need for manual entry of computer-oriented math formats will decrease.

Input and output of math notation is carried out in various contexts. Here is an informal description of a few scenarios, with and without math-literate computers.

- A person creates a document containing math expressions:

  - With a math-literate computer, this can be done via handwritten entry. The computer software must be able to cope with the variability of handwriting.

J. Carette et al. (Eds.): Calculemus/MKM 2009, LNAI 5625, pp. 2–13, 2009.

Handwritten expressions can be scanned, or the user can write directly on a data tablet; the tablet has the advantage of making stroke-timing information available for use in the recognition process. The user receives feedback about the recognition result, and is thus available to correct recognition errors.

- With manual entry, a person directly enters the structure of a math expression by typing an ASCII form of the expression (as in LaTeX), or by issuing a sequence of commands to a structure-based math editor.

• Paper documents are converted to electronic form:

- With a math-literate computer, scanned documents are interpreted by document-recognition software. Layout analysis is used to separate the document into text regions, math expressions, and figures. Text regions are interpreted by optical character recognition (OCR), math regions are interpreted by math recognition, and figures are interpreted by graphics recognition software [32][34]. When a small number of documents are converted, a person can perform checking and correction of the results. When a large document collection is involved, people perform only very limited checking and correction. In that case, subsequent software must make allowance for the possibility of recognition errors in the electronic documents.

- With manual entry, a person directly enters the structure of the math expressions in the documents. This is can be done for small-scale applications, but is infeasible for large document collections. If manual entry is infeasible and automatic interpretation of math expressions is unavailable, then the math expressions can be left uninterpreted: this leaves math expressions as image regions that can be displayed, but cannot be queried by word-based or symbol-based searches.

• A math document is converted from a notation-oriented electronic form to an information-oriented electronic form (for example, from LaTeX to a symbolic algebra format such as Maple):

- With a math-literate computer, this conversion is done automatically. External information is required, for example to distinguish function names from variable names (see Section 3.1).

- With manual entry, a person directly enters the symbolic algebra form of the math expressions.

• A math document is converted from an electronic form to paper (or to display on a computer monitor):

- This is typically done by computer software. Present-day computers are math-literate in the *writing* direction.

Computer math-literacy is a practically-important subject that offers fascinating research opportunities. In this paper, I present my opinions about challenges and issues that arise in this area. Please note that I am not up-to-date in all the latest publications, but am supplying a retrospective view of the developments in computer processing of math notation over the past twenty years.

2 Diagrams and Notational Conventions

A *diagram* expresses information using a two-dimensional layout of symbols. *Notational conventions* are the constraints that define the mapping between information and two-dimensional layout. A math expression is an example of a diagram, expressed using the notational conventions of math notation. A page of sheet music is another example of a diagram, expressed using the notational conventions of music notation. As illustrated in Figure 1, knowledge of notational conventions is needed both for diagram recognition (*reading* the notation) and for diagram generation (*writing* the notation).

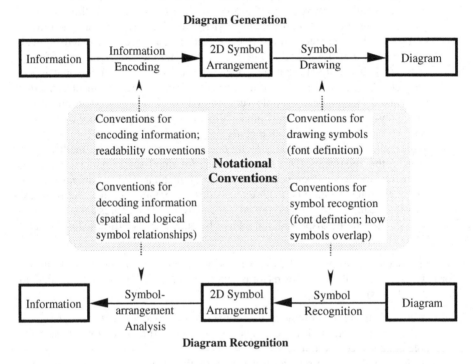

Fig. 1. Notational conventions are used for diagram generation and recognition [4]. When a diagram is generated, notational conventions guide the creation of an aesthetically pleasing diagram that encodes the given information. When a diagram is recognized, notational conventions guide the recognition of symbols and their logical relationships, and dictate how to infer information from this symbol arrangement. This figure illustrates sequential processing steps, but symbol recognition and symbol-arrangement analysis can be concurrent, allowing the use of contextual information to improve symbol-recognition results.

The mapping between information and diagram is not one to one. Many diagrams represent the same information using different layouts. A diagram with *good* layout is easy to read, and is aesthetically appealing. Ideally, diagram generation software automatically chooses a good layout, while diagram recognition software recognizes the information conveyed by the diagram, no matter what the layout of the diagram.

There is significant overlap in the notational conventions used for generation and recognition. However, notational conventions are treated differently due to differences in the two diagram processing tasks. Aesthetic considerations are central in diagram generation: users want nice-looking, readable diagrams. In contrast, diagram recognition systems pay less attention to aesthetics: they are trying to recover the information conveyed by the diagram, and are not trying to judge how nice the diagram layout is. Noise and uncertainty are central in diagram recognition: there is uncertainty about symbol segmentation, symbol recognition, interpretation of the relative placement of symbols, and so on. These problems do not arise in generation. It is interesting to generalize this line of thought, to consider the relationship between the fields of computer vision and computer graphics [21].

The acquisition, representation and exploitation of notational conventions are central to computer math literacy. Notational conventions are equally important in processing other types of two-dimensional notation. The research community as a whole is gradually developing general computational approaches to diagram recognition, which are useful in interpreting diagrams of various types. Much has been published about recognition of various types of diagrams, such as math expressions, engineering drawings, maps, music notation, and bar charts [20][31][32][34]. Publications about diagram generation include graph drawing [36] and visual languages [33]. Many aspects of diagrams are discussed in the Diagrams conferences [35].

2.1 Hard and Soft Notational Conventions

We call a notational convention *hard* if it is used consistently, and *soft* if its use is optional. Hard conventions specify how information is encoded in the two-dimensional notation, and soft conventions specify how to make the diagram readable. Here are some examples, informally expressed. In graph drawing, a hard convention is that "an edge drawn between two nodes represents a relation between the nodes". A soft convention is "choose a graph layout that minimizes the number of edge crossings". In math notation, a hard convention is "division can be encoded by drawing a horizontal line with the dividend expression placed above the line and the divisor expression placed below the line". A soft convention is "when breaking an expression into multiple lines, put the break before a major operator". Petre provides related observations about hard and soft conventions, using the term *secondary notation* for the layout aspects of a diagram [28].

Soft conventions can be applied to a greater or lesser degree, and can be ignored in exceptional circumstances. For example, a soft convention in music notation is "leave more space after long notes than after short notes" [2]. However, this convention is sometimes ignored when music notation is printed very densely, in order to end the page at a pause that gives the performer time to turn the page.

The same information can be represented by a large set of diagrams. These diagrams differ in readability and aesthetic appeal. As an example, the layout of a circuit diagram can be changed; this affects the appearance of the diagram without changing the information being conveyed.

Most diagram recognizers ignore soft conventions, relying wholly on the hard conventions. Poor diagram layout is not noticed, and good diagram layout is not exploited. If diagram recognizers could be expanded to make greater use of soft conventions, this could increase the robustness of recognition: the recognizer can make use of layout and spacing cues. To achieve this, the recognition software needs the ability to reason with constraints that hold "most of the time".

Since diagram generators already use soft conventions, the question arises whether diagram recognizers can be improved by exploiting the knowledge and experience embodied in diagram generators [6]. Specifically, LaTeX software is in widespread use and encodes sophisticated knowledge about the formatting of math notation. Can this generation-oriented knowledge be exploited to improve math-notation recognizers? Possible approaches include reusing generator code to proofread and correct recognizer output, building a model of the generation process into the recognition software, and using a generator to construct cases for a recognizer that uses case-based reasoning [6].

2.2 Sources of Information about the Definition of Math Notation

The design of a math-literate computer system should begin with a definition of math notation: a definition of the syntax and semantics of the two-dimensional language used to express mathematics. Unfortunately, math notation, like most diagram notations, is not formally defined. Rather, it is informally established through common usage. Math notation is only semi-standardized, allowing many variations and drawing styles. The same is true of natural languages, such as English. Building a software model of the notational conventions used in math notation is a complex and time-consuming task.

Sources of information about math notation include written descriptions of math notation, sample documents, coded descriptions built into software for recognizing math notation, coded descriptions built into software for generating math notation, and human experts [5]. Most written descriptions are oriented toward generation of the notation rather than recognition of the notation. Descriptions of math notation for people who want to solve typesetting problems include [12][18][40], and descriptions of math notation oriented toward computational typesetting include [22]. Almost 40 years ago, Martin suggested that the first step in automating the processing mathematical notation is to make a study of the notation, and he went on to present a brief list of the notational conventions found in use in technical publications [25].

Many factors influence how mathematical symbols should be grouped during the recognition of math notation. Some grouping factors are defined for math notation in general; these include operator range and operator precedence. Other grouping factors arise within a particular mathematical expression; these include symbol identity (which often determines whether the symbol is an operator or an operand), relative symbol placement, and relative symbol size and case. Further discussion and related references are provided in [5].

2.3 Electronic Representations of Math Notation

A variety of electronic representations of math notation are in active use, including pixel-oriented representations such as JPEG, symbol-oriented representations such as PDF and PostScript, syntax-oriented representations such as LaTeX, and symbolic-algebra representations such as Maple. Research is needed to better understand these representations: how to define the equivalence of documents and the distance between documents, how to mathematically characterize the mapping between document representations, how to characterize the external information needed to carry out these mappings, and how to characterize the differences between the forward and inverse mappings that occur during document analysis and document production [9].

3 Recognition of Math Notation

In developing software for recognition of math notation, much can be learned from existing research into recognition of other types of diagrams [4]. Over time, our collective experience in recognizing various types of diagrams is giving rise to a general technology for diagram recognition. An appealing analogy is provided by compiler technology: the first compilers were difficult to write, but over time the community developed general techniques for parsing and code generation, which greatly simplify the task of constructing compilers for new source and target languages. Diagram-recognition methods are difficult to generalize, due to the great diversity among diagram notations, and due to the complexity of handling noise and uncertainty. However, algorithms can be shared for common subproblems such as symbol recognition [13][37]. Document recognition contests provide standardized task definitions, including training and testing data, as well as evaluation metrics. Contests have been held for such as dashed-line detection, raster to vector conversion, arc segmentation, symbol recognition, page segmentation, handwriting segmentation, and Arabic handwriting recognition [15].

3.1 External Information Needed for Math Recognition

Math expressions are not self-contained. External information is needed in order to fully understand them. Some of this information, such as the definition of symbols, comes from other parts of the source document. Other information is external: for example, knowledge of symbolic algebra can be applied to find errors in a printed expression or its interpretation. Many dialects of math notation are in use, varying by discipline; the choice of dialect is implicit, and must be inferred by some external means, typically involving familiarity with the math notation used in related publications. Other diagram notations have an analogous need for external information: for example, engineering drawings rely on reader's knowledge of disassembly and kinematics [38], and music notation relies on the reader's knowledge of music theory and performance practice.

The acquisition, representation, and use of external knowledge is a broad and interesting topic, one that is important to the future development of math-literate computers. Without external information, a simple expression such as *a(b)* can be interpreted up to a notation-oriented format such as LaTeX, but further interpretation up to the symbolic algebra level is impossible without the knowledge of whether *a* is a function name or a variable name. When subexpressions are used repeatedly in a document, noticing and exploiting this repetition helps increase the robustness of a recognizer.

3.2 Challenges in Recognizing Math Notation

Many challenges arise in the recognition of math notation [5]. Small symbols, such as dots and commas, are commonly used and are critical to the meaning of the notation; these small symbols are difficult to distinguish from noise. Symbol recognition is difficult because there is a large character set (Roman letters, Greek letters, operator symbols) with a variety of typefaces (normal, bold, italic), and a range of font sizes. A few common symbols in math notation have several possible roles: a dot can represent a decimal point, a multiplication operator, a symbol annotation such as $\dot{x}$, or noise; a horizontal line can indicate a fraction line or a minus sign. The meaning of such symbols must be determined through the contextual information provided by surrounding symbols.

A major challenge in math recognition is identification of the logical meaning of spatial relationships. Implicit mathematical operators are defined entirely by spatial relationships, with no explicit operator symbol; these include superscripts, subscripts, implied multiplication, and matrix structure. Examples of difficult cases are given in many publications from [25] onward. In handwritten notation, the ambiguity of spatial relationships is greatly increased, due to free placement and alignment of symbols. Many researchers (e.g. [39]), have studied the problem of distinguishing horizontal adjacency from superscripts and subscripts: the continuous range of possible symbol

placements $2x \quad 2x \quad 2^x \quad 2^x \quad 2^x$ makes this difficult.

Offsetting these challenges, two characteristics of math notation make it relatively easier to process than many other types of diagram notations. Firstly, most symbols in a math expression are surrounded by white space, which greatly simplifies symbol segmentation. (An exception is handwritten mathematical notation, which may contain overlapping symbols; these can be difficult to segment, particularly if off-line data is used.) Secondly, math notation has a relatively regular and recursive syntax, which makes it well-suited for processing using grammar-based and compiler-like techniques.

3.3 Finding the Math Expressions in a Document

Computer math literacy depends on having automated ways of finding math expressions, or having convenient user interfaces for the user to indicate the location of math expressions of interest. This is not a problem in on-line recognition systems, where a person writes input on a data tablet: in this case, text and math expressions are generally not mixed. The situation is different for paper documents, where math expressions are typically mixed with text, either as offset expressions, or embedded directly into a line of text. The first step in math recognition is to identify where expressions are located on the page. This topic, and document layout analysis in general, has been subject of much research e.g. [34][41].

3.4 Computational Methods for Recognizing Math Notation

Many approaches to math recognition have been explored, including syntactic methods, graph transformation, projection-profile cutting, and procedurally-coded rules [5][10][11]. Noise and uncertainty can be handled by producing lists of alternatives that are passed from one recognition stage to the next, or by executing recognition stages concurrently, using contextual feedback to compensate for noisy input or to reject erroneous input [4]. The many proposals for how to organize a math recognition system are fascinating; each has its own merits and its own advocates. It is difficult to judge which organization is best for a given application.

I have had long-standing interest in development of a general software technology for diagram recognition. My first informal comments, at a workshop in 1990 [1], described a goal of a diagram-recognition technology, analogous to the existing technology available for compilers. Compiler technology makes it (relatively) easy to create compilers with new source and target languages, and similarly a diagram-recognition technology would make it (relatively) easy to create recognizers for new types of diagrams. However, diagram recognition faces added problems due to noise, due to the huge range of types of diagram notations, and due to the natural-language aspects of diagram notations.

Many compiler techniques can be adapted to pattern recognition [7]. Techniques that have been imported include the use of grammars (array grammars, tree grammars, set grammars, graph grammars), and parsing technologies (for example, CYK and Early algorithms for context free grammars, and linear-time LR and LL parsing algorithms for more restricted languages). We illustrate the use of two additional compiler techniques in a math-notation recognition system: use of trees and tree transformation, and a multi-pass control structure, with a clear separation between layout, lexical, syntactic, and semantic analysis [7][42]. The main steps are to (1) find linear structures in the input, and use these as a basis for finding secondary linear structures; (2) organize the linear structures into a tree; (3) divide processing into passes for layout, lexical analysis, syntax, and semantics; (4) use a simple, fixed control structure, such as a sequence of passes; and (5) use tree transformation technology, which provides highly efficient techniques for manipulating trees, and notations for expressing manipulations in a concise, readable form.

3.5   Users Reaction to Math Recognition: The User Interface

Math-recognition software is far less widely used than math generation software. Some of this is due to availability, and to the maturity of the technology. But there are additional, social factors that work against recognition systems. Here are some speculative comments of these factors, with the aim of stimulating discussion and new directions for development of recognition systems.

It seems that the recognition errors made by a computer are quite intrusive, because they are different from the interpretation problems that a person has when reading messy or noisy text. If a person is struggling to read your document, her or she will ask you about semantics: what do you mean here? In contrast, the computer asks about marks on the page: is this a *w*, is that item over there one symbol or two symbols? The computer does not state these questions directly, but a user who is proofreading recognition output – to correct symbol recognition and symbol segmentation errors – is implicitly answering questions like this. Users would find recognition software more inviting if it could move in the direction of allowing users to think more about the meaning of the notation, and less about the marks on the paper. Computer-Human Interaction is a heavily-researched subject that has many ideas to offer [29][30].

Predictability and blame-assignment are two reasons why recognition software isn't as popular as it could be. Consider the case of a user who types a big LaTeX expression, and gets an error because of unbalanced parentheses. Is this user upset at the LaTeX software? No, instead the user blames himself or herself: that was my stupid mistake, I forgot a parenthesis, next time I will do better. Consider instead a user of a pen-based math entry system who gets an error because of a misrecognized symbol. This user is likely to blame the recognition software: what stupid software, even my four-year-old son is better at symbol recognition than that, I hope the software will do better next time -- but it is hard for me to figure out how to help it do better.

A basic question is: do people really want automated recognition of math expressions (assuming that the recognition rate is suitably high)? The answer is certainly *yes* for the case of document recognition, but for manual entry of math expressions, the answer probably depends on the person. I will begin by discussing entry of text, and then move on to entry of math expressions. For entering text, I personally would rather type on a keyboard than handwrite on a data tablet – this is because I can type much faster than I can write by hand. On the other hand, a slow typist may prefer writing on a data tablet (with application of OCR software) over typing on a keyboard. In the case of math expressions, my personal preference is to have an entry method that is focused on the 2D math notation. However, it is possible that some people prefer to use LaTeX because they are so practiced that they can type expressions faster than they can handwrite them.

Before designing a user interface for math recognition, it is worth reviewing the attractive properties of paper: ergonomics, contrast, resolution, weight, viewing angle, durability, cost, life expectancy, and editorial quality [19]. Paper also has limitations: erasing is difficult, it is not possible to "make extra room" to expand parts of a diagram, it is hard to find information in a large stack of paper, and so on. A goal for future computer interfaces is to retain the advantages of paper while also providing the editing and search capabilities lacking in paper. Designers reject the use of computers in the early, conceptual, creative phases of designing, preferring to use paper and pencil, which permits ambiguity, imprecision, and incremental formalization of ideas [16]. Computer based tools force designers into premature commitment, demand inappropriate precision, and are often tedious to use when compared with pencil and paper. For further discussion, see [8].

4 Generation of Math Notation

Less is published on the generation of diagram notations than on the recognition of diagram notations. Diagram generation technology is mature and often proprietary. Prominent among publications on diagram generation is Knuth's work on math notation [22]. We extend ideas from Knuth's text spacing algorithm [23] to music spacing [17]. Also, much has been published on algorithms for graph drawing [36]; many diagram notations are based on graphs.

Interesting issues arise in providing a user interface to generated notation. The user should be allowed to modify the generated notation, without the need to repeat all the modifications when the notation is regenerated [3].

5 Performance Evaluation Issues

Although performance evaluation is an active research area in document image analysis, there are few formally defined performance metrics for diagram generation or recognition. Informally, a generator is successful if it generates information-bearing images that the user finds aesthetically pleasing. There are no ground-truth models of ideal generator output. A generator is debugged on a test suite of diagrams, and in response to user feedback. Evaluating the performance of diagram recognition systems involves defining requirements, characterizing the system's range of inputs and outputs, interpreting published performance evaluation results, reproducing performance evaluation experiments, choosing training and test data, and selecting performance metrics [24].

The user interface is critical to the success of a diagram recognition system. It is difficult to define precise goals for a user interface, and even more difficult to quantify performance of a user interface [8]. Separating user-interface performance from recognition performance is difficult: the time that a user spends correcting recognition errors depends both on the number of recognition errors and on the qualities of the user-interface facilities for finding and correcting errors. The performance of different types of visual feedback in a math recognition system is studied in [43]. One possible performance measure is to compare the time and accuracy of automated and unautomated entry of diagrams, as discussed in [8]. Neilsen defines usability attributes: learnability, efficiency, memorability, errors and satisfaction [26]. Measurable usability parameters include subjective user preference measures, which assess how much the users like the system, and objective performance measures, which measure the speed and accuracy with which users perform tasks on the system [27].

6 Conclusion

Computers should serve people, assisting them in their work. Making computers math literate is an important step in this direction, allowing people to work using the familiar math notation, avoiding the need for them to learn new notations for the convenience of the computers. Computer math literacy provides a smooth transition between paper documents and electronic documents, combining the best properties of paper with the advanced search and evaluation capabilities offered by electronic documents. This paper has summarized some of the issues involved in creating math literate computers. Much progress has been made, and many interesting problems remain to be addressed.

Acknowledgments

Financial support from the Natural Sciences and Engineering Research Council of Canada and from the Xerox Foundation is gratefully acknowledged.

References

[1] Blostein, D.: Structural Analysis of Music Notation. In: Proc. IAPR Workshop on Syntactic and Structural Pattern Recognition, Murray Hill, NJ, p. 481 (1990)

[2] Blostein, D., Haken, L.: Justification of Printed Music. Communications of the ACM 34(3), 88–99 (1991)

[3] Blostein, D., Haken, L.: The Lime Music Editor: A Diagram Editor Involving Complex Translations. Software – Practice and Experience 24(3), 289–306 (1994)

[4] Blostein, D.: General Diagram-Recognition Methodologies. In: Kasturi, R., Tombre, K. (eds.) Graphics Recognition 1995. LNCS, vol. 1072, pp. 106–122. Springer, Heidelberg (1996)

[5] Blostein, D., Grbavec, A.: Recognition of Mathematical Notation. In: Bunke, H., Wang, P. (eds.) Handbook of Character Recognition and Document Image Analysis, pp. 557–582. World Scientific, Singapore (1997)

[6] Blostein, D., Haken, L.: Using Diagram Generation Software to Improve Diagram Recognition: A Case Study of Music Notation. IEEE Trans. Pattern Analysis and Machine Intelligence 21(11), 1121–1136 (1999)

[7] Blostein, D., Cordy, J., Zanibbi, R.: Applying Compiler Techniques to Diagram Recognition. In: Proc. 16th Intl. Conf. on Pattern Recognition, Quebec City, Canada, August 2002, vol. III, pp. 123–126 (2002)

[8] Blostein, D., Lank, E., Rose, A., Zanibbi, R.: User Interfaces for Online Diagram Recognition. In: Blostein, D., Kwon, Y.-B. (eds.) GREC 2001. LNCS, vol. 2390, pp. 92–103. Springer, Heidelberg (2002)

[9] Blostein, D., Zanibbi, R., Nagy, G., Harrap, H.: Document Representations. In: Proc. Fifth IAPR Int'l Workshop on Graphics Recognition (GREC 2003), Barcelona, Spain, July 2003, pp. 3–12 (2003)

[10] Blostein, D.: Graph Transformation in Document Image Analysis: Approaches and Challenges. In: Brun, L., Vento, M. (eds.) GbRPR 2005. LNCS, vol. 3434, pp. 23–34. Springer, Heidelberg (2005)

[11] Chan, K., Yeung, D.: Mathematics Expression Recognition: a Survey. Int'l Journal on Document Analysis and Recognition 3(1), 3–15 (2000)

[12] Chaundy, T., Barrett, P., Batey, C.: The Printing of Mathematics. Oxford University Press, Oxford (1957)

[13] Chhabra, A.: Graphic Symbol Recognition: An Overview. In: Chhabra, A.K., Tombre, K. (eds.) GREC 1997. LNCS, vol. 1389, pp. 68–79. Springer, Heidelberg (1998)

[14] Cushman, W., Ojha, P., Daniels, C.: Usable OCR: What are the Minimum Performance Requirements? In: Proc. ACM SIGCHI 1990 Conference on Human Factors in Computing Systems, Seattle, Washington, April 1990, pp. 145–151 (1990)

[15] Document Recognition Contests, http://www.icdar2007.org/competition.html

[16] Gross, M., Do, E.: Ambiguous Intentions: a Paper-like Interface for Creative Design. In: Proc. Ninth Annual Symposium on User Interface Software and Technology (UIST 1996), Seattle, Washington, November 1996, pp. 183–192 (1996)

[17] Haken, L., Blostein, D.: A New Algorithm for Horizontal Spacing of Printed Music. In: International Computer Music Conference, Banff, September 1995, pp. 118–119 (1995)

[18] Higham, N.: Handbook of Writing for the Mathematical Sciences. SIAM, Philadelphia (1993)

[19] Hsu, R., Mitchell, W.: After 400 Years, Print is Still Superior. CACM 40(10), 27–28 (1997)

[20] Int'l Journal on Document Analysis and Recognition. Springer Verlag (since 1998)

[21] (Int'l Conf.) Computer Vision/ Computer Graphics Collaboration Techniques and Applications, INRIA Rocquencourt, France, May 4-6 (2009)

[22] Knuth, D.: Mathematical Typography. Bulletin of the American Mathematical Society 1(2), 337–372 (1979)

[23] Knuth, D., Plass, M.: Breaking Paragraphs into Lines. Software – Practice and Experience 11, 1119–1184 (1981)

[24] Lapointe, A., Blostein, D.: Issues in Performance Evaluation: A Case Study of Math Recognition. In: Int'l Conf. Document Analysis and Recognition (ICDAR 2009), Barcelona (July 2009) (to appear)

[25] Martin, W.: Computer Input/Output of Mathematical Expressions. In: Proc. 2nd Symposium on Symbolic and Algebraic Manipulations, pp. 78–87. ACM, New York (1971)

[26] Nielsen, J.: Usability Engineering. Academic Press, San Diego (1993)

[27] Nielsen, J., Levy, J.: Measuring Usability: Preference vs. Performance. Communications of the ACM 37(4), 66–75 (1994)

[28] Petre, M.: Why Looking Isn't Always Seeing: Readership Skills and Graphical Programming. Communications of the ACM 38(6), 33–44 (1995)

[29] Proc. ACM Conference on Human Factors in Computing Systems (CHI). ACM Press, New York (annual since 1982)

[30] Proc. ACM Symposium on User Interface Software and Technology (UIST). ACM Press, New York (annual since 1988)

[31] Proc. IAPR Int'l Workshop on Document Analysis Systems (biennial since 1994)

[32] Proc. IAPR Int'l Workshop on Graphics Recognition (biennial since 1995)

[33] Proc. IEEE Symposium on Visual Languages and Human-Centric Computing (annual since 1988) (earlier name: Proc. IEEE Symposium on Visual Languages)

[34] Proc. Int'l Conf. on Document Analysis and Recognition (biennial since 1991) (sponsored by IAPR and IEEE)

[35] Proc. Int'l Conf. on the Theory and Application of Diagrams (biennial since 2000)

[36] Proc. Int'l Symposium on Graph Drawing (held annually since 1993)

[37] Tombre, K., Tabbone, S., Dosch, P.: Musings on Symbol Recognition. In: Liu, W., Lladós, J. (eds.) GREC 2005. LNCS, vol. 3926, pp. 23–34. Springer, Heidelberg (2006)

[38] Vaxivière, P., Tombre, K.: Knowledge Organization and Interpretation Process in Engineering Drawing Interpretation. In: Proc. IAPR Workshop on Document Analysis Systems, Kaiserslautern, Germany, October 1994, pp. 313–321 (1994)

[39] Wang, Z., Faure, C.: Structural Analysis of Handwritten Mathematical Expressions. In: Proc. Ninth Int'l Conf. on Pattern Recognition, Rome, Italy, November 1988, pp. 32–34 (1988)

[40] Wick, K.: Rules for Typesetting Mathematics, translated by V. Boublik and M. Hejlova, The Hague, Mouton (1965)

[41] Workshop on Document Layout Interpretation and its Applications (DLIA) (1999 and 2001)

[42] Zanibbi, R., Blostein, D., Cordy, J.: Recognizing Handwritten Mathematical Expressions Using Tree Transformation. IEEE Trans. Pattern Analysis and Machine Intelligence 24(11), 1455–1467 (2002)

[43] Zanibbi, R., Novins, K., Arvo, J., Zanibbi, K.: Aiding Manipulation of Handwritten Mathematical Expressions through Style-Preserving Morphs. In: Proc. Graphics Interface 2001, Ottawa, Ontario, June 2001, pp. 127–134 (2001)

Abstraction-Based Information Technology:
A Framework for Open Mechanized Reasoning

Jacques Calmet

University of Karlsruhe (TH)
Am Fasanengarten 5, 76131 Karlsruhe, Germany
calmet@ira.uka.de

Abstract. OMRS (Open Mechanized Reasoning Systems) was designed for Automated Theorem Proving and then extended to Computer Algebra. These are the two domains at the heart of the Calculemus approach. An obvious question is to assess whether such an approach can be extended to new domains either within AI or outside of AI. There have been several attempts to turn the world into a computational system. This talk stays away from such general attempts and introduces a framework that is fully set within AI. It extends the basic concepts of OMRS to diverse fields ranging from information technology to sociology through law as illustrated by examples. The main motivation is to claim that whatever the selected approach, Artificial Intelligence is gaining enough strength and power to reach new frontiers and to turn challenges that are not a priori of a purely computational nature into AI domains.

Keywords: mechanized reasoning, abstraction, computational modeling,knowledge, agent.

1 Introduction

OMRS (Open Mechanized Reasoning Systems) [10] were designed for Automated Theorem Proving (ATP) and then extended to Computer Algebra (CA) [2] .These are the two domains at the heart of the Calculemus approach. An obvious question is to assess whether such an approach can be extended to new domains either within AI or outside of AI. There have been several attempts to turn the world into a computational system (model of everything [7], Fredkin's view of the universe [8] as a global cellular automaton, Stephen Wolfram's definition of computing [18] or simply diverse models of the philosophy of sciences [9]). Within AI the very early work of John McCarthy and Patrick Hayes [13] did investigate philosophical problems from the standpoint of AI. Our approach stays away from such general attempts and introduces a framework that is fully set within AI. It extends the basic concepts of OMRS to diverse fields ranging from information technology to sociology through law as illustrated by selected examples. The main motivation is to claim that whatever the selected approach, Artificial Intelligence is gaining enough strength and power to reach new frontiers and to turn challenges that are not a priori of a purely computational nature into AI domains.

J. Carette et al. (Eds.): Calculemus/MKM 2009, LNAI 5625, pp. 14–26, 2009.

The relationship between Calculemus and mathematics cannot be overlooked. We outline some mathematical concepts, such as works of Jacobi and Herbrandt or the problems for the Millenium of the Clay institute, in the framework of ATP and CA. We will survey briefly their links to mechanized reasoning.

Another motivation for our work has its roots in information technology. We are entering an area which will see methodologies based upon artificial intelligent surpassing human capabilities. At the same time knowledge is becoming the building stone of our society. The amount of available knowledge is blowing up. This implies that knowledge management must be modeled along new abstraction paradigms. In other words, artificial artifacts will be substituted to humans in constructive ways enabling an "average" person to mechanize several arts of reasoning in an AI fashion. It will take several years before mathematics delivers constructive methods applicable to ATP or CA in many domains. AI offers ways to design in the near future approaches that can be implemented and used in real world applications. To this goal, we rely more bottom-up (as opposed to top-down) knowledge methodologies. Indeed, the increase in available knowledge leads to huge knowledge whare-houses which are not prone to an easy handling. Using the concept of virtual enterprises (and virtual knowledge community) that is gaining importance in today globalized economy, we try to illustrate that we can approach problems that where never attempted within AI. For instance, we propose an approach to mechanize cultural reasoning that is fully disconnected from today solutions to resolve intercultural differences. These are becoming sensitive issues in today globalized and virtual economy. We extend thus the concept of open mechanized reasoning beyond computer science to cognitive sciences in general.

It is worth noticing that the attempts to introduce abstraction mechanism in Artificial Intelligence (AI) were found mainly in reasoning. Abstraction supposes here a generic process that is proven to lead to the right processing of a problem. This goes beyond the selection of a specific logic since the abstraction concept must be goal oriented, not method oriented. This requires some sort of abstraction mechanism. Abstraction is a paradigm with very many different meanings, usually specialized to a field or subfield of a discipline. For instance, computer science favors abstraction described in the language of category theory but it must compete with different ones (type, semantics, specification, algorithms ...). The paper is structured as follows. The next section reminds classical concepts and facts from mathematics or philosophy which have been investigated along the years and that we do not use as a framework of our work. In section 3 we introduce the open mechanized reasoning framework and the basic results obtained in the domains of theorem proving and symbolic computation. Open means that the methodology is generic and not specific to a particular methodology. The original framework for this approach lies at the heart of AI. Indeed, to mechanize mathematics was indeed a goal of the founding fathers of the field at the Dartmouth meeting in 1956. The achieved results show that successful ones have a three level structure: a theory, a control on this theory and a well-understood interaction with the computing environment. The following section points out that this structure can be extended to almost any domain of knowledge, including law. This leads to the concept of Abstraction Based Information Technology (ABIT). The next section illustrates and discusses this concept in different fields of science and humanities including mechanized legal reasoning. The next section illustrates how a possible implementation can be easily designed in the framework of multiagent systems and virtual knowledge communities. The last section is devoted to some concluding remarks.

2   Mathematics, Philosophy and AI

In a recent paper, the philosopher of sciences, with an education in physics, Michel Ghins [9] asks whether we need a metaphysics to understand the laws of nature. He describes a scientific theory as a set of models together with a set of propositions some of which are laws. This is in fact a study in philosophy in the line of Descartes. There is, as often in this domain of philosophy, a link to logic reasoning. This is not an isolated research track. For instance, the books of Bruno Latour entitled "Science in Action" and "Politics of Nature" are attempts to discover exactly how science works. They are suggested as a worthwhile reading by philosophers and sociologists to computer scientists. A possible reason is the link of Latour to the "actor theory network" and the sociology of knowledge. These are two references, among very many available, towards a possible framework to set a universal model of computation. But, it does not suits the idea of mechanization well.

Conversely, McCarthy and Hayes write in [13] "A computer program capable of acting intelligently in the world must have a general representation of the world in terms of which its inputs are interpreted. Designing such a program requires commitments about what knowledge is and how it is obtained. Thus, some of the major traditional problems of philosophy arise in artificial intelligence". They go on mentioning that the philosophical problems to be solved become clearer with so-called reasoning programs. Their choice for a world-model was the automaton. This view that the universe is a computer or more specifically a cellular automaton was expressed recently by several computer scientists adopting the point of view that there is computational model of everything [7]. Cellular automaton and Turing machines have often be proposed as a generic computational model. The suggestion of Wolfram is in the framework of complex systems but looks to fall in this class . Another very recent and interesting approach to understand the evolution of scientific ideas is found in [11]. The authors propose a network linking the different fields as a way to learn this evolution. For our purposes, it is enough to state that these models are controlled theories without an interaction to an application universe and thus not suitable for our purposes.

Central to logic and ATP is the concept of universe introduced by Herbrandt. Jacques Herbrand wrote his doctoral thesis in 1930 at the age of 22 years. He died in an accident the year after. He set the framework for mechanical reasoning but no one could guess that at that time. As pointed out in [17], he was attending the famous Hadamard's seminar with André Weil, Jean Dieudonné and Claude Chevalley who are legends in French mathematics. This means that mechanized reasoning or a theory for demonstration did not take a very long time to mature. Another student of this Hadamard seminar was Albert Laut-man who was working on mathematical structures [12]. It is nowadays probably better known as a mathematical philosopher. However, Lautman was among the founder of the Bourbaki saga. All these mathematicians are central to CA since they do cover algebra and geometry. It may be that the constructive approach of Bourbaki did inspire CA where solutions to problems must be constructive and complete. A side remark is that it looks like Hadamard was a boring lec-turer but his legacy is enormous. Some other famous names are quoted in [17], namely Vessiot, Frechet and Picard. They are a prototypical link to on-going research problems of interest to the Calculemus community: the investigation of symbolic solutions to systems of partial differential equations (PDE). When the work in logic of Herbrandt took only a few years to demonstrate its use-fulness, the PDE question is around for a few centuries. In [5] it is shown that articles of Jacobi, never previously translated from Latin, have inspired modern mathematics. Computation of normal forms using a sequence of derivations and eliminations, change of orderings, resolvents, characterization of possible normal forms by the rank of Jacobian matrices, a priori bounds on the order of a sys-tem, . . . these posthumous papers of Jacobi develop many themes quite familiar to contemporay research in differential computer algebra. Picard's and Vessiot's works are still acknowledged when attempting to design symbolic solutions of PDEs. Furthermore, it is known that physics and part of biology are PDE-based. This is thus an area where the challenges are high and no solution can be ex-pected in the near future. Some of the problems for the Millenium listed in the web site of the Clay institute, such as the Riemann Hypothesis or the Poincaré Conjecture, need to be solved before CA can propose a constructive solution to this problem. One may think that the abstracted view of algebra, geometry and topology proposed by Grothendiek must be adopted before any significant chance of getting a solution does exist. This is a domain of mathematics where computation is not primarily the main issue which is to understand the suitable mathematical theory and to select a meaningful representation in which to set the problem. We did already mention that the proposed abstraction framework includes a theory, a control and an environment. Therefore, PDEs illustrate that a pure mathematical approach is still a dream and thus, it is not yet suitable for open mechanized reasoning.

3   The Open Mechanized Reasoning Framework

In [10], the Open Mechanized Reasoning System (OMRS) architecture was in-troduced as a mean to specify and implement reasoning systems (e.g., theorem provers) as logical services.

$$Reasoning\ Theory = Sequents + Rules$$
$$Reasoning\ System = Reasoning\ Theory + Control$$
$$Logical\ Service = Reasoning\ System + Interaction$$

In [2], a similar approach was designed for symbolic computer algebra systems under the name of Open Mechanized Computational System architecture.

$$Computation\ Theory = Objects + Algorithms$$
$$Computation\ System = Computation\ Theory + Control$$
$$Algorithmic\ Service = Computation\ System + Interaction$$

In [1], an unified description of both classes of systems was derived and called Open Mechanized Symbolic Computation Systems (OMSCS). It synthesizes the previous definitions into that of Symbolic Mathematical Service. It is based upon definitions of symbolic entities and operations which include the previous definitions of sequents and objects, and of rules and algorithms respectively.

$$Symbolic\ Computation\ Theory = Symbolic\ Entities + Operations$$
$$Symbolic\ Computation\ System = Symbolic\ Computation\ Theory + Control$$
$$Symbolic\ Mathematical\ Service = Symbolic\ Computation\ System + Interaction$$

A key feature is that this is the only example where such a generic approach (e.g., open) is clearly available. Indeed, for logical services and symbolic computation it is possible to prove that a computation always exists and always terminates. In plain words, we have a theory as the initial level, we exercise some control on this theory in the second level while the third one describes how the controlled theory is linked to any environment. In these specific cases, the environment is a computer universe. In [3] an extension to scientific computing was investigated. In this case there is no longer a generic approach but, one may either rely on the standardized arithmetic operations in floating-point arithmetic (IEEE-754 standard) or on specific arithmetics (interval arithmetic for instance) or on routines specific to a given available software. In terms of the previous analysis, this means that instead of considering a generic theory, one may consider various possible theories. For each selected theory, one has to specify the control imposed upon it and the way the interaction with the environment is managed. A general remark is that even for the domains where the problem may be seen as solved, to identify a theory and to analyze its control are rather simple tasks while to formalize the interaction with the environment is always a challenging task. This remark is fully relevant for computational systems where a solution does exist but is often challenging to exhibit it.

4   Abstraction Based Information Technology

A first motivation to extend this work is coming from the state-of-the-art of artificial intelligence (AI) resulting from the extraordinary progresses achieved in recent years by computer technology, both at the hardware and software levels. The claim that within a few years computers will be more efficient than (most) human brains can no longer be disregarded, although this may sound very unpleasant for many scientists or sociologists or humanists. In addition, the hypothesis that brains have mainly a virtual perception of the world enables to investigate a concept of abstraction in AI. A reason is that the images transmitted by the retina to the brain are possibly virtual images. A second motivation is the need to investigate inter/multi/trans-disciplinary problems. The meaning of these concepts have been thoroughly discussed by the French philosopher Edgar Morin, sometimes in collaboration with computer scientists in the framework of complex systems. This implies that we need a common abstraction framework valid for as many disciplines as possible. An underlying assumption is that AI is not simply a subfield of CS but a paradigm to mechanize reasoning processes when dealing with the real world. A second assumption is that we interact with the world through computers and thus we need to model information technology through abstract models. We call this abstraction ABIT (Abstraction Based Information Technology). It is summarized as follows.

- A theory,
- a control on this theory,
- an embedding environment.

It turns out that the three levels of the open mechanized reasoning framework can be used as a basis to design such an abstraction. A further step is to recognize that this ABIT approach is suitable to introduce a concept of abstraction into domains as diverse as physics, philosophy, sociology or even culture [6]. A possible exception is Mathematics, a domain where only theories look to be meaningful since control and link to the environment belong to applications rather than to Mathematics itself.

5   Examples of Abstraction in Various Fields

We present now some simple examples of how ABIT can be defined in various fields of science and humanities.

5.1   Computer Algebra

This is a direct application of the previous sections.

- A theory is a module of algorithms,
- the control consists of a programming language,
- the environment is the computing environment.

5.2   Legal Reasoning

Legal reasoning is thoroughly introduced in [15] that we adopt for reference purposes. The second part of the volume, entitled Legal Logic, provides a large collection of possible theories. Each section or even sub-section can be selected to be a theory. The choice is not limited however to this second part since already in the first part several facets of legal reasoning are presented. The domain is known to be very complex and fragmented. It is thus not surprising to be faced with a large number of optional theories. Some notions, such as for instance doxification [15] , could be used to describe a possible control on theories as well as the embedding into an environment. However, we select a different scheme to introduce legal reasoning. The three following facets define mechanized legal reasoning in ABIT.

- A theory is a set of laws (as voted by legislators),
- the control consists in application decrees,
- the environment is defined by jurisprudence and litigation procedures.

A first limitation is that such a scheme is not universal. For instance, the second step does not look to exist per se in the UK. In fact, laws are valid in a country or in a cluster of countries as anyone working on ciphering for instance quickly notice. The attempt to define, implement and enforce a so-called European law is a good illustration of the complexity of the system. The above mentioned scheme may appear as an over simplified view of legal reasoning but it is very often the view non-experts have. Also, it is suitable to be extended to less trivial approaches. The control may be extended to rules and regulations also as shown for plagiarism or copyright protection. In fact the theory and the control are defined by the administrative, political and judicial instances. They are also responsible for the third level. Law is a domain where electronic agents are investigated for many years and is offering high level conferences such as Jurix or ICAIL or LEA for instance. Thus, there is a proximity to distributed AI. As a summary, straightforward applications amount to select one of the methodologies described in [15] and to verify that the three level architecture is indeed technically suitable. We need to select a specific application to check the third level of ABIT. A natural choice for us is publication rights and plagiarism. The following comments are obviously from a non legal expert. In fact some of the questions we raised were put to lawyers and always got as generic answer: there is a solution since there are laws governing this domain (lawyers without computer culture) or this is under investigation (lawyers with computer culture). It turns out that either as the editor-in-chief of a journal, or as researcher investigating security of mobile systems or as a professor having to fire a student from the university because of a stupid case of plagiarism, some disturbing features and consequences of fully relevant laws have been encountered. This is a motivation to introduce a heuristic tool at the third level: a legal social network. Indeed, plagiarism in projects or thesis is becoming increasingly bothering in the education world with the facilities provided by internet. Tools exist, such as the Turnitin or Copy Tracker or Plagium or Nopliagia or many others software, to check whether part of a report is extracted from a previous publication. Plagiarism is reaching enormous growth since it has been estimated that 20 % of all German master thesis suffer from some sort of plagiarism. We need fast an informative tool to make the students aware of the existence of plagiarism detectors. Since a student signs a form stating that the thesis work presented is his/her own only, plagiarism has legal consequences and a guilty student is thrown out of the university. A legal social network could be very helpful but in any case, mechanized legal reasoning is a requirement.

5.3   Mathematics

Throughout the paper abstraction means solely the abstraction we did introduce. There are obviously many meanings to this word and the next sentences might sound controversial in any other context. A theory is not necessarily an abstraction. Mathematics is about building theories, such as differential algebra for instance, but cares less to set a control on these theories and investigates the link to the environment when moving to applied mathematics. This means that we adopt the old-fashioned distinction between pure and applied mathematics. Thus, differential algebra is a theory. and will become an abstraction when it can be labelled computer algebra. This happens when modules of algorithms are designed. They are controlled by a programming language and the links to the computing universe is fully mastered and understood. Other similar examples are scientific computing or ATP. Topology is another fascinating case. It is obviously a theory but seldom constructive. Although we do not request an abstraction to be constructive, it must be since to master the embedding in a relevant universe, one must master all possible links. The Kenzo specialized system of Francis Sergeraert, an expert on constructive homological algebra, is available from his home page. It is however very difficult to use by non-expert users meaning that the control is not optimal. Thus, we cannot rate the state-of-the-art in this domain to be an abstraction. This is, surprisingly, the only branch of science or humanities that is not concerned with our concept of abstraction although it is the branch where the most abstract knowledge, in the usual meaning, is available.

5.4   Physics

Physics underlines the fact that an abstraction can be time dependent since it depends on our understanding of a domain. Most parts of physics are without doubt abstractions. This statement is also valid for nanotechnology or quantum optics (lasers). But, this is not always true since theoretical physics is still under development. An example taken from history was the computation of the anomalous magnetic moment of leptons to prove that quantum electrodynamics is a theory. This amounted to validate the existence of a renormalization group model. By the way, such computations have certified computer algebra systems as reliable computational tools. When the integration into the theory of the weak and strong interactions became necessary, the model evolved. For instance, one way to extend the renormalization group model to gravity, leads to the so-called string theory. This may be seen as successive theories with some control but, in the case of string theory, with little understanding of the link to environment since it is virtually impossible to probe this model experimentally. Quantum computing is a fascinating area since it is still an open question to know whether an efficient quantum computer will ever be available. Without entering a detailed discussion, one may say that we have an abstraction based on quantum mechanics as a theory, quantum computing provides the control on this theory while the link to the universe is the assumption that quantum computers will be built when technology is adequate.

5.5   Sociology and Political Sciences

Sociology is an area attracting lot of research activities nowadays and where some sort of mechanization is looked for. We will outline in the next section an abstraction in the domain of multiagent systems (MAS) which could be duplicated here. But, on a very general level, we adopt the point of view that sociology is concerned with many possible theories expressing what is the role of agents (human or artificial). The control on the agents with their given roles defines a society. The abstraction is defined as:

- A theory is a set of agents with well defined actions,
- The control consists in defining a society based upon the theory,
- The environment is defined by how this society is governed.

Theories and their control in this abstraction constitute what is usually defined as sociology while the step society to government is the domain of political science. History tends to show however that the form of society implies some constraints on governance. Thus, it is possible to introduce as third level a simulation of the society activities for instance.

5.6   Culture

This is a topic with facets in, at least, philosophy, sociology, psychology, economy, geography, education or business. Consulting companies are training employees of international companies to solve problems arising from intercultural differences. An international transport company such as Hamburg-Sud equips its employees with a booklet listing some trivial cultural differences in the countries they do business with. Numerous examples of cultural problems result from setting up international companies or exchanging international students. Three simple examples are listed here.

- A French-German company hires engineers. They have a health insurance coverage: meaning?
- Exchange of students: What is the aim of education in different cultures?
- Process of decision making in international enterprises.

We want to make such problem solving part of information technology. This means that understanding intercultural differences ought to be an assigned goal of AI. This is only possible when an abstract view of the problem is available. Mechanized cultural reasoning is better described in [6]. A simple introduction is to state that the theory is an ontology, the control is to infer facts from this ontology. Finally, the environment consists in specializing these facts to a specific culture. Technically, this requires that we can abstract cognitive problems.

5.7   From an Ontology to a Methodology

We have so far outlined a possible paradigm for abstraction that is fairly straightforward. Since the goal is to design and implement a mechanized reasoner, we have to define the methods and techniques required to reach this goal. They are based upon the following assumptions. First, we adopt the framework of distributed AI and of multiagent systems. Second, the basic ingredients consist of knowledge. Third, we adopt a bottom-up approach. This means that we want to tailor the solution of a problem to the needs and means of potential users. An overall goal is to define a framework where mechanized reasoning, as understood in AI, is available. A second goal is to design systems that can be used easily by customers. We investigate whether it is possible to use a constructive approach for cognitive systems or complex systems. Consider as example the first cultural problem above. When a company is facing intercultural troubles, it usually calls on a consulting company that will propose a solution. This is very similar to calling on a psychiatrist to solve psychological disorders. We want to design systems using AI mechanized reasoning as a substitute to consulting companies. By the way, health insurance is a generic word with very different. specific meanings in different social systems. Similarly, it is astonishing to set up a double diploma open to top students of two countries, Germany and France specifically, and to notice that after one year of studies these top students do not really understand what is the goal of studying in the partner country. These students will be labelled to have an international experience and probably hired as manager in international enterprises. Then, the third intercultural trouble will likely surface. This is a new goal of AI that we wish to derive from mechanized reasoning and establish as an item in the agenda of AI. We may simply state that culture belongs to the corporate knowledge of a country. This is likely to be received as an horrible point of view by most humanists or cultural experts.

6   Knowledge Methodologies

This section outlines very briefly some methods that have been designed and do enable to propose a meaningful solution for the ABIT framework describe in the previous sections.

6.1   Agent Methodology

We set our approach in the framework of multiagent systems (MAS). They were originally designed as a tool in distributed artificial intelligence. In [4] an Agent Oriented Abstraction (AOA) was proposed to abstract a multiagent system and also to specify what is a society of agents along the line of Weber's fundamental work in Sociology. AOA is based upon six definitions itemized as follows.

- An agent is an entity made of annotated knowledge coupled to a decision mechanism.
- The decision mechanism of an agent is the process by which an agent can reach its assigned goals. It is based upon the contents of the knowledge component. A decision mechanism is characterized by its utility.
- Knowledge annotations are classes or types structuring the knowledge possessed by or associated to agents.

- The utility of a decision mechanism is a measure of the efficiency of this mechanism. It is structured into utility classes.
- A society of agents is the societal organization arising from the actions performed by individual agents in the agent world assigned to a problem.
- A specialization is an implementation of the abstract classes for knowledge or utility.

In our abstraction framework, definitions 1 to 3 define a theory, definitions 4 and 5 define the control on this theory while the last definition acknowledges the link to the environment. A facet of the Agent Oriented Abstraction (AOA) mentioned previously is to enable to select virtual knowledge communities (VKC) as a methodology to represent knowledge.

6.2   Virtual Knowledge Communities

Virtual organizations (VO) are emerging in the information society as a requirement for a new information distribution scheme. They can be defined as a collection of individuals, companies or organizations which have agreed to work together to achieve a purpose. The concept of virtual knowledge communities (VKC) is an specific aspect of knowledge management in the frame of VO. A reasonable denition for VKC can be: Groups of people or agents with similar interests and purposes communicating and interacting by means of information technologies. VKCs are built upon three ground concepts: different entities (the members), similar interests of the members, and electronic communication channels. Since VOs are based on interaction between different entities, it appears natural to consider the concept of virtual knowledge communities within an agent-based abstraction. Indeed, MAS are based on the model of autonomous entities (agents) interacting with each others. AOA considers that agents are composed of two entities: A knowledge component and a decision mechanism. There are several ways to define the knowledge component. VKC is one of them. It is then possible to design the basic operations that can be performed on VKCs and to implement them. Several VKC's implementations have been performed including one for portable devices. Relevant references can be found in [14] and [16].

7   Conclusion

Our goal is to demonstrate that the very basic ideas that are at the source of Calculemus can be extended to provide an abstract framework for most fields of science and humanities. We have outlined a string of ideas and methods to assess that it is possible to switch from logical frameworks to cognitive systems to identify abstraction concepts.

An important motivation lies in the belief that artificial intelligence will be more effective than human actors shortly. At a time when a car is able to run 1,000 km without a driver, we may expect that managing and exploiting knowledge bases may be partly mechanized. Here knowledge is assumed to come in many different arts. We also acknowledge that our technical view of the world is becoming more and more virtual. This concept is changing the way enterprises do collaborate. It is worth designing methodologies that will enable to think of collaborative international management and cultural systems. Although the features of decision making and trust are not considered in this paper, they are central to any knowledge management methodology. Our approach suits this concern.

The first part of the paper is attempting to remind that the search for an universal computational model as always been a goal of humanity. We tried very hard to assert that we do not set our approach within philosophy or any such attempts. A second remark was that although we do believe strongly that pure mathematics will enable breakthroughs towards mechanizing reasoning, we do not set our approach in the framework of mathematics. We have mentioned briefly in the last section bits and pieces that are being shaped up as tools to design abstraction based information technology. A testbed is probably to demonstrate that cultural mechanized reasoning does exist.

References

1. Bertoli, P.G., Calmet, J., Giunchiglia, F., Homann, K.: Specification and integration of theorem provers and computer algebra systems. Fundamenta Informaticae 39(1-2), 39–57 (1999)
2. Calmet, J., Homann, K.: Structures for symbolic mathematical reasoning and computation. In: Limongelli, C., Calmet, J. (eds.) DISCO 1996. LNCS, vol. 1128, pp. 216–227. Springer, Heidelberg (1996)
3. Calmet, J., Lefevre, V.: Toward the Integration of Numerical Computations into the OMSCS Framework. In: Proceedings of the 7th International Workshop on Computer Algebra in Scientific Computing, Saint Petersbourg, Russia, pp. 71–79 (2004)
4. Calmet, J., Maret, P., Endsuleit, R.: Agent-Oriented Abstraction, Revista Real Academia de Ciencias (Madrid). Special Volume on Symbolic Computing and Artificial Intelligence 98(1), 77–83 (2004)
5. Calmet, J., Ollivier, F. (Guest eds.): Jacobi's Legacy. AAECC Special issue 20(1) (April 2009)
6. Calmet, J., Maret, P., Schneider, M.: Cultural Differences as a Tool to Assess Trust in Virtual Enterprises (May 2009) (forthcoming)
7. Carriero, N., Gelernter, D.: A Computational Model of Everything. Communications of the ACM 44(119), 77–81 (2001)
8. Fredkin, E.: Digital Mechanics. Physica D 45, 254–270 (1990)
9. Ghins, M.: Laws of Nature: Do we Need a Metaphysics? Principia 11(2), 127–149 (2007)
10. Giunchiglia, F., Pecchiari, P., Talcott, C.: Reasoning Theories: Towards an Architecture for Open Mechanized Reasoning Systems. Tech. Rep. 9409-15, IRST, Trento, Italy (1994) Short version in: Proc. of the First International Workshop on Frontiers of Combining Systems (FroCoS 1996), Munich, Germany (1996)
11. Herrera, M., Roberts, D.C., Gulbahce, N.: Mapping the Evolution of Scientific Ideas (April 7, 2009) (preprint) arXiv:0904.1234v1

12. Lautman, A.: Essai sur les Notions de Structure et d'Existence en Mathématiques. J. Symbolic Logic 5(1), 20–22 (1940)
13. McCarthy, J., Hayes, P.J.: Some Philosophical Problems from the Standpoint of Artificial Intelligence. Machine Intelligence 4, 463–502 (1969)
14. Maret, P., Calmet, J.: Agent-Based Knowledge Communities. IJCSA 6(2), 3–21 (2009)
15. Sartor, G.: A Treatise of Legal Philosophy and General Jurisprudence. Legal Reasoning, vol. 5. Springer, Heidelberg (2005) (Volume not available separately)
16. Subercaze, J., Maret, P., Calmet, J., Pawar, P.: A service oriented framework for mobile business virtual communities. In: Camarinha-Matos, L.M., Picard, W. (eds.) Pervasive Collaborative Networks. IFIP International Federation for Information Processing, vol. 283, pp. 493–500. Springer, Boston (2008)
17. Wirth, C.-P., Siekmann, J., Benzmueller, C., Autexier, S.: Lectures on Jacques Herbrand as a Logician. SEKI Report SR200901, DFKI Saarbruecken (2009)
18. Wolfram, S.: A New Kind of Science. Wolfram Media (2002)

Software Engineering for Mathematics

Georges Gonthier

Microsoft Research Cambridge
gonthier@microsoft.com

Despite its mathematical origins, progress in computer assisted reasoning has mostly been driven by applications in computer science, like hardware or protocol security verification. Paradoxically, it has yet to gain widespread acceptance in its original domain of application, mathematics; this is commonly attributed to a "lack of libraries": attempts to formalize advanced mathematics get bogged down into the formalization of an unwieldly large set of basic resuts.

This problem is actually a symptom of a deeper issue: the main function of computer proof systems, checking proofs down to their finest details, is at odds with mathematical practice, which ignores or defers details in order to apply and combine abstractions in creative and elegant ways. Mathematical texts commonly leave logically important parts of proofs as "exercises to the reader", and are rife with "abuses of notation that make mathematics tractable" (according to Bourbaki). This (essential) flexibility cannot be readily accomodated by the narrow concept of "proof library" used by most proof assistants and based on 19th century first-order logic: a collection of constants, definitions, and lemmas.

This mismatch is familiar to software engineers, who have been struggling for the past 50 years to reconcile the flexibility needed to produce sensible user requirements with the precision needed to implement them correctly with computer code. Over the last 20 years *object* and *components* have replaced traditional data and procedure libraries, partly bridging this gap and making it possible to build significantly larger computer systems.

These techniques can be implemented in compuer proof systems by exploiting advances in mathematical logic. *Higher-order logics* allow the direct manipulation of functions; this can be used to assign *behaviour*, such as simplification rules, to symbols, similarly to objects. Advanced *type systems* can assign a secondary, contextual meaning to expressions, using mechanisms such as type classes, similarly to the *metadata* in software components. The two can be combined to perform *reflection*, where an entire statement gets quoted as metadata and then proved algorithmically by some decision procedure.

We propose to use a more modest, *small-scale* form of reflection, to implement *mathematical components*. We use the type-derived metadata to indicate *how* symbols, definitions and lemmas should be used in other theories, and functions to implement this usage — roughly, formalizing some of the *exercize* section of a textbook. We have applied successfully this more engineered approch to computer proofs in our past work on the Four Color Theorem, the Cayley-Hamilton Theorem, and our ongoing long-term effort on the Odd Order Theorem, which is the starting point of the proof of the Classification of Finite Simple Groups (the famous "monster theorem" whose proof spans 10,000 pages in 400 articles).

J. Carette et al. (Eds.): Calculemus/MKM 2009, LNAI 5625, p. 27, 2009.

Some Traditional Mathematical Knowledge Management

Patrick D.F. Ion

Mathematical Reviews, AMS
ion@ams.org

What is mathematical knowledge and how can it be managed? There are not only differing views around on the management aspect but there is no real clarity or consensus on what mathematical knowledge is; indeed there are questions as to what knowledge is and what mathematics is. For the sake of definiteness I will adopt a particular stance from which to work, namely that aspect of organizing the knowledge of mathematics represented by Mathematical Reviews, for which I have worked since 1980. From that platform we can explore and speculate both historically and prospectively. Some new results of bibliometric and other machine-enabled examination of the mathematical literature will also be discussed.

J. Carette et al. (Eds.): Calculemus/MKM 2009, LNAI 5625, p. 28, 2009.

Math Handwriting Recognition in Windows 7 and Its Benefits

Marko Panic

Microsoft Development Center Serbia
markop@microsoft.com

Nowadays, writing a math paper in a word processing application or performing calculations in a computational engine often requires spending a considerable amount of time creating math expressions using either a complex UI model with a multitude of drop-down buttons or a complicated and difficult to remember linear format input. As of Windows 7, Microsoft provides users with the most natural and efficient way of inputting math - handwriting recognition, as part of its operating system. Microsoft has taken a completely new approach to this problem and raised math handwriting recognition to a whole new level in terms of functionality, performance and area coverage.

A key power of the math handwriting recognizer in Windows 7 lies in the fact that it outputs the recognition result in MathML format, a standardized mathematical markup language. Any expression written and recognized reaches destination applications in a completely editable form - the output can be inserted and edited just like any other text.

Due to considerable time spent researching and identifying as many areas of math as possible and practically endless different math notations, the final result is a great coverage of high school and college level math, and of even more advanced areas.

The recognizer, developed at Microsoft Development Center Serbia, is exposed through two UI components: Math Input Panel and Math Input Control.

Math Input Panel is a standalone Windows 7 accessory that enables inputting math using handwriting into any MathML-aware Windows application via clipboard interaction.

Math Input Panel

J. Carette et al. (Eds.): Calculemus/MKM 2009, LNAI 5625, pp. 29–30, 2009.

Math Input Control is an ActiveX UI control designed for tighter integration into Windows applications, such as those for word processing, computation and note taking. For applications the added benefit of integrating Math Input Control, as opposed to using Math Input Panel, is better discoverability, process control and a degree of customizability.

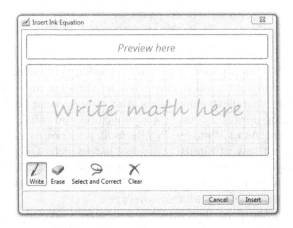

Math Input Control

Math Input Panel and Math Input Control are designed to be used with a tablet pen on a Tablet PC, but they can be used with any input device such as a touchscreen, external digitizer or even a mouse. The user interaction model is designed to be very easy and straightforward. Math expressions are handwritten just like with pencil and paper and the recognizer takes care of the rest.

As no recognizer is perfect, another key power of Windows 7 math handwriting recognition UI is in its ability to provide a great correction experience. In case handwritten math is misrecognized, any part of it (individual symbols or whole sub-structures) can be corrected either by selecting alternates or by rewriting part of the expression.

Therefore, Math Input Panel and Math Input Control bring great benefits to Windows 7 users by providing a painless method of inputting math into their desired applications, using handwriting recognition.

Assembling the Digital Mathematics Library

David Ruddy

Project Euclid, Cornell University
dwr4@cornell.edu

Since discussions of a Digital Mathematics Library (DML) were first formalized, it has been recognized that such a collection would be federated, consisting of a "network of institutions." Implicit in this conception, and explicit in much of the early DML planning documents, is the assumption that this network would be organized in some manner–coordinated and held together by formally accepted policies and practices regarding collection, management, access, and preservation.

This DML, so conceived, has not come to pass. I suggest this is not because the early vision of a DML was particularly flawed, but because, for one, it was enormously more complex than we thought, and two, the approach taken was beyond our capabilities. However inevitable such an approach was, it was unrealistic given our capacity and understanding. We did not, and still do not, possess the technical understanding, the organizational capabilities, or the institutional and political willingness to implement such a grand vision in the manner proposed.

I will argue that the way forward need not abandon the larger vision but rather set aside many of the constructs upon which we assumed it needed to be realized. Chief among these is the notion of a coordinated, planned, or even sensible approach to building the DML–the idea that a central organizing network of institutions will establish, through some formal process, a plan to accomplish the goals of the DML. The future DML, if we can even call it a library, will not be "organized" (in any conventional sense of the word), at least for many years, if ever.

This is not, in my opinion, bad news or even pessimistic. It is rather a natural and expected evolution in our progress. And there has been progress, most notably in our thinking about large scale document networks. There are, further, I will argue, constructive areas of work ahead. For one, we can encourage and promote low-barrier local practices that we are increasingly confident will contribute to a large scale federated digital collection. Such local efforts include digitization methods, local data management practices, and adopting less fearful and more constructive procedures for exposing content. Second, we can engage in more exploration of how to operate in a messy information space, not with the goal of curating or exerting control over a disparate set of data, but aimed at connecting the dots. We should recognize that it is the relationships among exposed content that deserve our attention in this effort. What we may find frustrating about these relationships, their dynamic, shifting, multitudinous nature, is in fact the living nature of our future information environment and the source of its richness.

J. Carette et al. (Eds.): Calculemus/MKM 2009, LNAI 5625, p. 31, 2009.

CAMAL 40 Years on – Is Small Still Beautiful?

John Fitch

Department of Computer Science, University of Bath, UK

Abstract. Over forty years ago an algebra system was written in Cambridge, UK, designed to assist in a number of calculations in celestial mechanics and later in relativity. I present the hardware environment and the main design decisions that led this system, later dubbed CAMAL, to be used in many applications for twenty years. Its performance is investigated, both in its own era, and more recently. It is argued that a compact data representation as in CAMAL has real benefits even in today's larger memory world.

1 Introduction

This paper considers a period in my early academic career, when not only was computer algebra a novel idea but the idea of having a computer at all was close to fantasy. In 1964 I started as an undergraduate at the University of Cambridge, and my first encounter in the first week with the teaching staff was with David Barton, a research student, who was to supervise me in applied mathematics for that term. I still remember that hour with a mixture of amazement and embarrassment at my stupidity. Four years later, having survived a degree and a postgraduate qualification in Computing, I started working for and later with David Barton on his algebra system, which was at that time unnamed.

I wish to impart some of the flavour of that system, why it was written and how it worked. The system survived for over twenty years in some form or other, but I wish to concentrate on the original three systems, and to give some benchmark figures for the performance of the system.

2 The Original Problem and System

David Barton was a research student at the Institute of Theoretical Astronomy, University of Cambridge. His area of interest was celestial mechanics, and the algebra system came into existence in order to solve one problem, to determine the orbit of the moon[2].

Charles Delaunay[7] published the methodology and results of a major hand-calculation. The two-volume book produced an algebraic expression for the orbit of the Moon round the Earth, as perturbed by the Sun in its orbit. The basic methodology was to consider the orbit as an instantaneous ellipse as explained by Newton's laws of motion, but the ellipse will evolve under the influence of the distant Sun (see figure 1).

J. Carette et al. (Eds.): Calculemus/MKM 2009, LNAI 5625, pp. 32–44, 2009.
© Springer-Verlag Berlin Heidelberg 2009

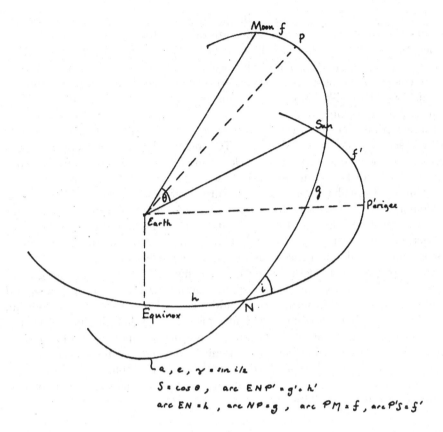

Fig. 1. The Lunar Theory of Delaunay

The problem is posed in terms of 6 variables and 6 angles, and all the variables are "small"; that is the solution is developed as an approximation[1] expanded to a certain level in variables like the eccentricity of the orbits, the ratio of the distances Earth-Moon to the Earth-Sun, and so on. This problem leads to a number of design decisions that at the time were obvious.

All expressions were Fourier series whose coefficients are polynomials in a small fixed number of variables over the rational numbers. The arguments to the sine and cosine functions are linear sums of the six angles. The general form of all expressions is thus

$$\sum P(a, b, c, d, e, f, g, h) \, \frac{\sin}{\cos} \left(\lambda u + \mu v + .. + \gamma z \right)$$

2.1 Hardware Base

The University of Cambridge at that time had an Atlas2 computer, also known as Titan[15]. This computer had a 48 bit <u>word</u> which could hold a floating point

---

[1] After all the 3-body problem is not solvable in gravitation.

number, an instruction, or two 24 bit <u>half-words</u>. The memory was addressed in words by the top 21 bits of a halfword, or in half-words by the top 22 bits. Conceptually a word could also hold 8 six-bit characters, but there was little hardware support for this. The design would probably nowadays be described as three-address, as most instructions had two registers and an address, and the machine was RISC-like, having 128 registers (mainly 24bit); there was also a floating point accumulator.

Titan initially had 64K words of memory (that is 1.5 megabits), although this was soon extended to 128K words. The memory had a cycle time of about 4 μsecs[2], and the second bank was a little faster. There was in addition a 32 word "slave store" which was an instruction cache, working at 300 nanosecond cycle time, although this hardware was often turned off as it was unreliable.

Memory was divided into blocks of 512 words, and this was the basic unit of allocation. Programs had to be in contiguous memory, and there were other restrictions due to the use of an OR rather than a ADD in the base register system, but they do not concern us regarding the algebra system.

To the programmers at the time the real delight of Titan was its B-registers; there where 88 general purpose registers, together with zero in register 0, and register 90 was the subroutine return address. Registers 119-122 were special, and the user's program counter was in register 127. What this means is that programming in assembler was fun and offered many opportunities for optimisation. This was good, but the programming languages available were less so. There was Titan Autocode, a version of Fortran, and a promise of CPL.

The original algebra system was written in the commonest assembler, IIT (Intermediate Input for Titan).

2.2   A Polynomial

The polynomials of the system were held as packed structures. There are only 8 variables, and as the application was approximation, there is no need for exponents to be larger than 31 – there was no hope of that accurate an approximation for the lunar orbit, 10 or 12 being the aim. This remark shows that all variables would be held in one 48 bit word, allowing 5 bits for the exponent and a guard bit to check for exponent overflow. The allocation is fixed, and every term in the polynomial will have a representation for all 8 variables. The advantage in algorithmic terms was that the multiplication of two terms could be achieved using two halfword additions[3], with a mask to check overflow.

The coefficients were held as two integers, in a word each. This was extended later as described below (section 3). Perhaps the only innovation in this section was that the rational was not reduced to lowest terms unless it was being printed or it was about to overflow.

The last part of polynomial design is the list of terms, ordered in increasing total order, that is, the sum of the exponents. By maintaining polynomials in

---

[2] The manual says 2.25 but that was not achieved.

[3] Integers were always in halfwords, and words were only for instructions and floating point.

increasing maximum order it is possible to truncate the multiplication of two polynomials when the terms get too small for the degree of approximation. Assume that we are multiplying the two polynomials

$$A = a_0 + a_1 + a_2 + \ldots$$
$$B = b_0 + b_1 + b_2 + \ldots$$

If we are generating the partial answer

$$a_i(b_0 + b_1 + b_2 + \ldots)$$

then if for some j the product $a_i b_j$ vanishes, then so will all products $a_i b_k$ for $k > j$. This means that the later terms need not be generated. In the product of $1 + x + x^2 + x^3 + \ldots + x^{10}$ and $1 + y + y^2 + y^3 + \ldots + y^{10}$ to a total order of 10 instead of generating 100 term products only 55 are needed. The ordering can also make the merging of the new terms into the answer easier.

2.3   A Fourier Series

The Fourier series part was similarly held as a packed structure with all 6 angles having multipliers packed into the structure. With 8-bit fields there is space for a signed number and a guard bit. The rest of the structure is made from two half-word pointers, one to the polynomial coefficient and one for the primary sum chain. The difference between a sine and cosine term was coded in a single bit on the polynomial pointer; remember that on Titan the bottom 2 bits are ignored on memory access. This is shown in figure 2.

The algorithms for performing the algebra are straightforward for addition, multiplication, differentiation and restricted integration. The product of the Fourier terms are subjected to the linearisation rules.

$$\cos\theta\cos\phi \Rightarrow (\cos(\theta + \phi) + \cos(\theta - \phi))/2,$$

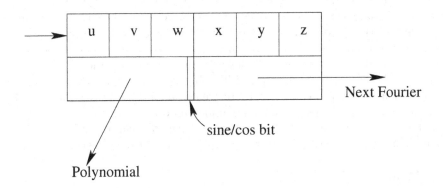

Fig. 2. The Fourier Representation

$$\cos\theta\sin\phi \Rightarrow (\sin(\theta+\phi) - \sin(\theta-\phi))/2,$$
$$\sin\theta\sin\phi \Rightarrow (\cos(\theta-\phi) - \cos(\theta+\phi))/2,$$
$$\cos^2\theta \Rightarrow (1 + \cos(2\theta))/2,$$
$$\sin^2\theta \Rightarrow (1 - \cos(2\theta))/2.$$

Substitution of one Fourier series into another is a more complex operation, but it relies on the approximation mechanism:

$$\sin(\theta+A) \Rightarrow \sin(\theta)\{1 - A^2/2! + A^4/4!\ldots\} +$$
$$\cos(\theta)\{A - A^3/3! + A^5/5!\ldots\}$$
$$\cos(\theta+A) \Rightarrow \cos(\theta)\{1 - A^2/2! + A^4/4!\ldots\} -$$
$$\sin(\theta)\{A - A^3/3! + A^5/5!\ldots\}$$

The actual coding of the operation was not as expressed above, but by the use of Taylor's theorem. It should be noted that the differentiation of a harmonic series is particularly easy.

3   The Bourne-Again CAMAL

Steve Bourne's PhD work was on the Hill formulation of the Lunar theory[3]. This is similar to the Delaunay theory except that it uses a Cartesian coordinate system, and needs more variables and also complex coefficients. In addition some of the coefficients got larger. In order to accommodate these changes Bourne rewrote the system, still in assembler, to add an extended rational coefficient with numerator and denominator up to $2^{76} - 1$, a possibility for overflow to floating point, and four more variables, i through l. The variable i was the complex number and replacement of i^2 by -1 was managed internally always. The final polynomial structure is shown in figure 3.

The other innovation for this second incarnation was the production of a programming language. Prior to this programs were sketched in an informal language and then coded in assembler. The new language[1] was very similar to Titan Autocode, except that the lowercase letters were algebraic variables and angles, and programming variables were single uppercase letters, with A-H and U-Z being algebraic type and I-T being integer. Multiplication was by juxtaposition as in Autocode and there was a looping construct and jumps. There were no subroutines as commonly understood now, but the form $\rightarrow 60 \rightarrow$ could be used to jump to label 60, and store a return link on a stack. The compiler was implemented in a syntax-directed compiler system called Psyco[16,18]

3.1   Memory

The other innovation that was incorporated in the compiler was related to memory allocation. As is appropriate for list processing on a small memory

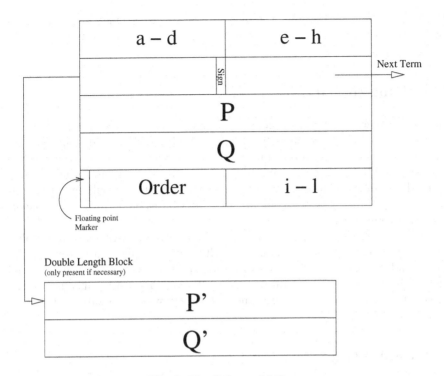

Fig. 3. The Polynomial Term

machine[10] the system worked with an explicit return system. With two sizes for allocation the mechanism was simply two free-chains. However the important feature was the division of operations into those that "naturally" destroyed their arguments and those that did not.

Consider the addition of two polynomials; this is a merge of two chains, and so is a simple example of the first kind of operation. Others include polynomial differentiation and integration. The operation of polynomial multiplication on the other hand does not consume its arguments. The compiler ensured that the necessary copying of structures was made before destructive transformations, *if necessary*. If the operation was then the first argument must be copied but

$$A = B + 5abs$$

not the second. More importantly the user language allowed a colon to follow a variable to indicate that it was not required after use, and the compiler could either not copy or insert a lose call as appropriate. This mechanism, used carefully, could save much memory and allow calculations to complete despite the memory limitations of Titan.

Another feature of the language scheme that was developed over time was to compile into a half-word encoding which was interpreted at run time. The argument at the time was that the basic operations took much longer that the time in the user program. Later a mixed system was used, especially in the third version.

4 Getting the Hump

The work of my PhD was largely in relativity, gravity waves and Einstein space[12]. For this neither the polynomials nor Fourier series were sufficient and so the third algebra system in Cambridge was constructed. The system handled sums and products of elementary functions, with possibly nested arguments, but used the polynomial component of the earlier system as a subsystem. In this way it inherited many of the compact store attributes while providing a faster implementation base.

This code, later called CAMAL(H) or Hump, was not written in assembler, but in a language like the user language, which was compiled into assembler for integer and pointer operations, and into interpreter half-words for polynomial algebra. The compiler was again written in Psyco. This system also introduced use-counts for memory structures and a hashing system to save memory on repeated polynomials.

5 Overall Structure

At this stage in the development of the algebra system we had a collection of systems, polynomial, Fourier, elementary functions and a tensor package (written in the same way as Hump, see [14]), and these could be used together in various ways. A small (1024 word) control program handled space, stacks, and the interpreter, and each component supplied a *jump table* to the functions, in a pre-defined order. In some ways this pre-echoes the internal structure of Scratchpad/2 much later. It was this system that was first named The Cambridge Algebra System, abbreviated to CAMAL[4,5]. The sizes of the code for various configurations is given in table 1.

Table 1. Sizes of Various CAMAL Systems in Blocks(512 words)

Polynomial System	2 + 5	= 7
Fourier System	2 + 7	= 9
Elementary Functions System	2 + 4 + 12	= 18
Fourier Tensors System	2 + 7 + 7	= 16
Elementary Function Tensor System	2 + 4 + 12 + 7 = 25	

The last system described there used three levels of code, linked via interpretation. There are few extant examples of this particular scheme but it was operational.

6   Subsequent History

The CAMAL system, as it was now called, continued for about twenty years as an application tool. Titan was turned off in 1973, and replaced by an IBM370/165. CAMAL was rewritten, first in a dialect of ALGOL68 and then in BCPL[19]. The assembler sections were hand recoded, and the components written in the CAMAL language were compiled to BCPL, again with Psyco. The algorithms and data structures were largely the same, except that the opportunity was taken to allow arbitrary numbers of polynomial variables, fixed for any run, and to allow similar licence for the maximum exponents. The masks and shifts were computed at initialisation time, but the fundamental algorithms were the same. Also arbitrary (unfixed) precision coefficients were introduced, and some hand adjustment of the automatic BCPL code made.

This incarnation was later ported to other architectures, including SUN workstations and VAXen. It fell out of maintenence in the late 1980s, although there was a working version reconstructed in 1999 using an automated BCPL to C translator, which when compared to REDUCE on the same computer showed remarkable speed.

7   Performance

In this paper so far the concentration has been on the design and data structures. This is only of interest if we also can see how this translates into performance.

In the 1970s there was a flurry of small benchmarks that were used in comparing the variety of algebra systems then in existence. The first of these was to calculate the f and g series.

7.1   f and g

The f and g series arise in the solution of orbits, and were first proposed as an algebra benchmark by Sconzo *et al.*[20] in FORMAC. It was widely used for a few years. The series are defined by simple recurrence relations:

$$\dot{\mu} = -3\mu\sigma$$
$$\dot{\sigma} = \epsilon - 2\sigma$$
$$\dot{\epsilon} = -\sigma(\mu + 2\epsilon)$$
$$f_n = \dot{f}_{n-1} - \mu g_{n-1}$$
$$g_n = f_{n-1} + \dot{g}_{n-1}$$
$$f_1 = 1$$
$$g_1 = 0$$

A CAMAL program to calculate this is shown in figure 2.

A table of reported timings is shown in table 3 from about 1973. It is hard to get complete comparisons, but note that the storage requirement of CAMAL is small and the time is short. CAMAL was noted at the time for its speed, but that was never the design. We held to the mantra that memory was finite

Table 2. CAMAL Program for f & g series

```
F[19]; G[19]

F[0] = 1; G[0] = 0; U = -3ab; V = c-2bb; W = -b(a+2c)

FOR N=1:1:19
        F[N] = UdF[N-1]/da + VdF[N-1]/db + WdF[N-1]/dc - aG[N-1]
        G[N] = UdG[N-1]/da + VdG[N-1]/db + WdG[N-1]/dc + F[N-1]
REPEAT

PRINT[F[19]]; PRINT[G[19]]
PRINT[TIME]
STOP
END
```

Table 3. 1973 Comparisons for f & g series

System	Computer	Word	Cycle	× time	Order	Time	Memory
ALTRAN	GE 625/635	36	1.5	6	19	158	51K
CAMAL	Titan	48	4	7	19	6.4	3.8K
CLAM	CDC 6600	60	0.8	1	15	10.6	30K
FORMAC	IBM 7094	32	2	10	12	58.2	??
Korsvald	IBM 7094	32	2	10	12	178.2	??
MATHLAB	PDP 10	36	1	10	12	20	??
PM	IBM 7094	32	2	10	27	105	??
REDUCE	PDP 10	36	1	10	10	68	38K
SAC-1	CDC 1604	48	4.8	36	12	75.9	21
SAC-1/Asm	CDC 1604	48	4.8	36	12	38.5	21

but time infinite[4] and we were pleased by the Memory column of this table. To assist in interpretation the graph of figure 4 shows the growth graphs in time in seconds and memory on 100s of words.

At the 1971 SYMSAC/2 conference there were attempts to run this program on all systems being demonstrated; unfortunately some crashed or apparently looped[5].

For comparison, a recent investigation of the f and g series using contemporary algebra systems and hardware can be found in [6].

7.2 Legendre Polynomials

Another simple benchmark is the calculation of the Legendre Polynomials, which can be defined by the relations

---

[4] although on Titan the time between failures was quite short!

[5] The same was true of a much hyped system in the 1980s.

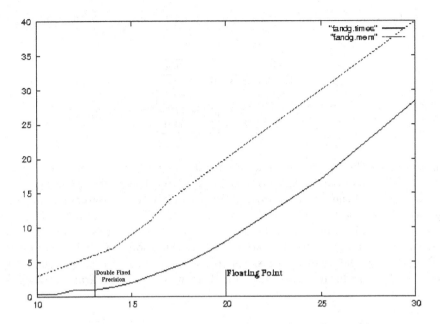

Fig. 4. Time and Memory for f & g series in CAMAL

$$nP_n = (2n - 1)xP_{n-1} - (n - 1)P_{n-2}$$
$$P_0 = 1$$
$$P_1 = x$$

The same polynomials can be calculated by Rodrigues' formula

$$P_n(x) = \frac{1}{2^n n!} \frac{d^n}{dx^n} (x - 1)^n$$

and the simple benchmark is to check that these give the same answers. This is one of the tests made on the 1998 Camal-in-C translation[6]. To order 30 this verification and printing took 0.380s, while the same program in REDUCE on the same computer took 1.071s. Unfortunately my records do not give the memory used.

7.3   Other Benchmarks

Amongst the other benchmarks there is a series published in the SIGSAM Bulletin together with solutions and timings. The f and g series was re-designated as Problem #1, and there were a total of 11 problems proposed. Those interested can find papers reporting timings from March 1972 until the end of the decade. I want here to consider just problem #3, the reversion of a double series[17]. Solutions were proposed by Fitch[8] and by Hall[13] in ALTRAN. Hall's method

---

[6] Unpublished work of Arthur Norman and myself.

Table 4. Comparisons of two algorithms for Problem #3 on PDP10

	Fitch		Hall	
n	Time	Store	time	Store
1	0.224	203	0.250	143
2	0.993	561	0.712	491
3	3.526	1781	2.147	1723
4	12.376	5431	6.759	5401

was to reduce the problem to one of calculating a recurrence problem, while Fitch used repeated approximation, to which CAMAL is well suited. The timing superiority of the CAMAL program led to an inaccurate statement that the algorithm was superior. Later implementations of Hall's method on CAMAL[11] showed that Hall's program used less memory and ran faster (see table 4, time in seconds, memory in words). A compact representation can be deceiving.

But do note the memory sizes. As this is an explicit return system these memory sizes are invariant; in addition there was 14K of program and 3K of fixed data.

7.4   Lunar Disturbing Function

The last comparison I wish to make is in the calculation of the Delaunay Lunar Disturbing Function, which is the first stage of the Delaunay Lunar theory. The details can be found elsewhere(for example [9]) but this is a Fourier series calculation which was bread and butter to the original CAMAL. In 1993 I wrote a module for REDUCE that used as close as I could the CAMAL data structure for Fourier series, while using the standard REDUCE polynomials for coefficients. This system runs significantly faster than the same problem in normal REDUCE(see table 5)

This showed that even within a general purpose algebra system there is scope for a compact subcomponent. For the whole calculation, using both lists and balanced trees, the figures were interesting:

Order of DDF	Reduce	Camal Linear	Camal Tree
2	23.68	11.22	12.9
4	429.44	213.56	260.64
6	>7500	3084.62	3445.54

What came as a surprise at the time was the direct comparison with CAMAL. The CAMAL Disturbing function program could calculate the tenth order with a maximum of 32K words (about 192Kbytes) whereas this system failed to calculate the eighth order in 4Mbytes (taking 2000s before failing). I also have in my archives the output from the standard CAMAL test suite, which includes a sixth order DDF on an IBM 370/165 run on 2 June 1978, taking 22.50s and

Table 5. Solving Kepler's Equation in Reduce and CAMAL-module

Solving Kepler's Equation

Order	REDUCE	Fourier Module
5	9.16	2.48
6	17.40	4.56
7	33.48	8.06
8	62.76	13.54
9	116.06	21.84
10	212.12	34.54
11	381.78	53.94
12	692.56	82.96
13	1247.54	125.86
14	2298.08	187.20
15	4176.04	275.60
16	7504.80	398.62
17	13459.80	569.26
18	***	800.00
19	***	1116.92
20	***	1536.40

using a maximum of 15459 words of memory for heap — or about 62Kbytes. One is tempted to ask if we have made any progress...

8 Conclusions

This paper has presented the design, performance and experience with an early computer algebra system that was designed to run on a small memory computer. Throughout its various incarnations the need to keep memory usage as low as possible was always present. Data structures were packed, memory allocation was simple and explicit, and we always ensured that all memory use was accounted. The polynomial subsystem was especially tuned to approximation techniques, with algorithms that took care of components that were too small to form part of an answer. There were actually many other features that are not mentioned here to keep the focus on the compact data structures and associated algorithms. Clearly the system was designed at a time when computers were much smaller memoried and much slower at execution than that to which we are used today; but there are still applications where memory is at a premium, or battery power considerations indicate slow processing. While not suggesting that CAMAL is necessarily the answer to modern problems, I assert that it still has many things to teach us.

I would like to express my thanks to David Barton (who taught me not to comment code), Steve Bourne from whom I learnt much about assembler, and many others in the CAMAL and algebra communities. I would also give my special thanks to Arthur Norman for the many suggestions he has made over many years that have served to improve CAMAL.

References

1. Barton, D., Bourne, S.R., Burgess, C.J.: A Simple Algebra System. Computer Journal 11, 293–298 (1968)
2. Barton, D.: A New Approach to the Lunar Theory. PhD thesis, University of Cambridge, 196
3. Bourne, S.R.: Automatic Algebraic Manipulation and its Applications to the Lunar Theory. PhD thesis, University of Cambridge (1970)
4. Bourne, S.R., Horton, J.R.: The CAMAL System Manual. The Computer Laboratory, University of Cambridge (1971)
5. Bourne, S.R., Horton, J.R.: The Design of the Cambridge Algebra System. In: Proceedings of SYMSAM/2, pp. 134–143. SIGSAM/ACM (1971)
6. Carette, J., Davenport, J.H., Fitch, J.: Barton and Fitch revisited (2009), http://opus.bath.ac.uk/14083
7. Delaunay, C.: Théorie du Mouvement de la Lune (Extraits des Mém. Acad. Sci.). Mallet-Bachelier, Paris (1860)
8. Fitch, J.P.: A solution to Problem #3. SIGSAM Bulletin 26, 24–27 (1973)
9. Fitch, J.P.: REDUCE meets CAMAL. In: Fitch, J. (ed.) DISCO 1992. LNCS, vol. 721, pp. 104–115. Springer, Heidelberg (1993)
10. Fitch, J.P., Garnett, D.J.: Measurements on the Cambridge Algebra System. In: Proceedings of the International Computing Symposium, Venice, pp. 139–147 (1972)
11. Fitch, J.: A solution of problem #3 using camal. SIGSAM Bulletin 32, 14 (1975)
12. Fitch, J.P.: An Algebraic Manipulator. PhD thesis, University of Cambridge (1971)
13. Hall, A.D.: Solving a problem in eigenvalue approximation with a symbolic algebra system. SIGSAM Bulletin 26, 145–223 (1975)
14. Horton, J.R.: A System for the Manipulation of Tensors on an Automatic Computer. University of Cambridge, Dissertation for the Diploma in Computer Science (1969)
15. I.C.T. Ltd and Cambridge University. Atlas 2 System Programmers Manual, E.P. 59 (April 1964)
16. Irons, E.T.: A Syntax Directed Compiler for ALGOL 60. Comm. ACM 4(1), 51–55 (1961)
17. Lew, J.S.: Problem #3 – reversion of a double series. SIGSAM Bulletin 23, 6–7 (1972)
18. Matthewman, J.R.: Syntax Directed Compilers. PhD thesis, University of Cambridge (1966)
19. Richards, M.: BCPL — a tool for compiler writing and systems programming. In: Proceedings of the Spring Joint Computer Conference, vol. 34, pp. 557–566 (1969)
20. Sconzo, P., LeSchack, A.R., Tobey, R.G.: Symbolic Computation of f and g Series by Computer. Astronomical Journal 70(4), 269–271 (1965)

Conservative Retractions of Propositional Logic Theories by Means of Boolean Derivatives: Theoretical Foundations*

Gonzalo A. Aranda-Corral[1], Joaquín Borrego-Díaz[1], and M. Magdalena Fernández-Lebrón[2]

[1] Departamento de Ciencias de la Computación e Inteligencia Artificial
[2] Departamento de Matemática Aplicada I
E.T.S. Ingeniería Informática, Universidad de Sevilla,
Avda. Reina Mercedes s.n. 41012-Sevilla, Spain
{garanda,jborrego,lebron}@us.es

Abstract. We present a specialised (polynomial-based) rule for the propositional logic called the *Independence Rule*, which is useful to compute the conservative retractions of propositional logic theories. In this paper we show the soundness and completeness of the logical calculus based on this rule, as well as other applications. The rule is defined by means of a new kind of operator on propositional formulae. It is based on the *boolean derivatives* on the polynomial ring $\mathbb{F}_2[\mathbf{x}]$.

Keywords: Conservative retraction, Independence Rule, boolean derivatives.

1 Introduction

A theory T is a conservative extension of a theory T' (or T' is a *conservative retraction*) if every consequence of T in the language of T' is a consequence of T' already. Conservative extensions have been deeply investigated in Mathematical Logic, and they allow to formalize several notions concerning refinements and modularity in Computer Science (for example, in formal verification [20,1,21]). In this paper we investigate how to compute a conservative retraction of a theory. In particular, we are interested in the following problem:

Conservative Retraction Problem (CRP):
- Input: A finite theory T in a language L, and $L' \subseteq L$.
- Output: A conservative retraction of T to the language L'.

Given a sublanguage L' of the language of T, a conservative retraction on L' has two basic properties:

---

* Partially supported by Minerva -Services in Mobility Platform- Project *WeTeVe* (2C/040) and *Ayudas a grupos de investigación, Junta de Andalucía* (TIC 137).

J. Carette et al. (Eds.): Calculemus/MKM 2009, LNAI 5625, pp. 45–58, 2009.

– There always exists a conservative retraction of T. For example, such a theory is

$$\{F \in \text{Form}(L') \;:\; T \models F\} \qquad\qquad (\dagger)$$

– Any two conservative retractions of T in the same sublanguage are equivalent theories.

We will denote by $[T, L']$ a conservative retraction of T to the sublanguage L' throughout the paper. This paper is concerned with the problem of computing (finitely axiomatized) conservative retractions. The importance of the computing of conservative retractions, in any logic, is based on its potential applications. For example:

– *Location principle for Knowledge Based Systems (KBS) reasoning:* Suppose that KB is a knowledge base, and let F be a formula. Suppose also that the language of F is L'. The question

$$KB \overset{?}{\models} F$$

can be solved in two steps:

- A conservative retraction $[KB, L']$ has to be computed
- We have to decide whether $[KB, L'] \models F$

Note that the second question usually has lower complexity than the original one, due to relatively small size of L'. This observation is extremely interesting when KB is a huge ontology.

– It is usual to approach the retraction by means of syntactic analysis, in order to locate the reasoning on certain axioms ([23]). In these cases, the conservative retraction would be very useful.

For example, let us consider the following ontology, in Propositional Description Logic (see [3] and [16] for details):

$$\Sigma = \left\{ \begin{array}{l} \text{Virus} \sqsubseteq \text{Animal} \sqcup \text{MobileEntity} \\ \text{Mammals} \sqsubseteq \text{Animal} \sqcap \text{MobileEntity} \\ \text{Animal} \sqsubseteq \neg\text{Plant} \end{array} \right.$$

Suppose that we want to specialize the reasoning in the concepts {Virus, Mammals, Plants, Animal} (because the concept MobileEntity is not contained in the superconcept LivingBeing), but we do not want to lose any knowledge about these concepts originally entailed by Σ. Note that it could be very hard, or not possible, to transform the ontology to obtain the conservative retraction by a syntactic analysis. Since this case is a propositional description logic ontology, it is possible to apply the method presented in this paper, obtaining thus the conservative retraction

$$\partial_{\text{MobileEntity}}(\Sigma) = \left\{ \begin{array}{l} \text{Animal} \sqsubseteq \neg\text{Plant} \\ \text{Mammals} \sqsubseteq \text{Animal} \end{array} \right.$$

At higher levels of expressivity, one can observe that existing tools provide syntactic modularity, but no semantic modularity.

- *Contextual reasoning.* In a similar way as in the above example, the conservative retraction $[KB, L']$ ensures the maximality of context knowledge with respect to the ontology source.

  A similar problem, in the complex case of ontological reasoning in OWL, is the use of partitioning methods by means $\mathcal{E}$-connections ([15]). Indeed the partitioning to an $\mathcal{E}$-connection provides modularity benefits; it typically contains several "free-standing" components, that is, sub-KBs which do not "use" information from any other components (observation also made in [15]).

- In SAT-based planning, the number of propositional variables is bigger than the size of any formula. Since the formulas without variables of L' are not used for the computing of a conservative retraction (as we will see in this paper), then computing conservative retractions may be a good strategy to synthetize partial plans. Similar ideas can be applied to obtain better partition-based reasoning algorithms for propositional logics ([2]).

- With regard to the specific case of the use of Computer Algebra Systems (CAS) for reasoning with knowledge-based systems in real problems (see e.g. [19]), the rule presented in this paper has interesting features. The use of CAS is based on a faithful translation of logical formulas into polynomials on finite fields. The algebraic counterpart of the Independence Rule, in algebraic geometry terms, is a tool for projecting varieties in positive characteristics. This interpretation is very useful to design new applications of Gröbner basis to Knowledge Based Systems.

On the one hand, to the best of our knowledge, there is no calculus specifically focused on the computing of conservative retractions. The main reason for this is that the notion of *conservative extension* is more interesting (for example in incremental specification/verification of systems). For instance, the Isabelle and ACL2 theorem provers adopt this methodology by providing a language for conservative extensions by definition (even for the specification and verification of the logic itself, see e.g. [1]). Another example are the formal approaches to Ontological reasoning and extending (see e.g. the conservative extensions generated by definitional methodologies [6]). And finally, weaker notions than conservative extensions are used in methods for ontological extensions assisted by automated reasoning systems (see [8,9]).

On the other hand, although the conservative retraction of theories can be interesting itself, in expressive logics (like first order logic) the retraction may not be finitely axiomatized (for example, in first order theories of arithmetic). It is even possible that it involves undecidable problems. In the concrete case of propositional logics, computing conservative retractions are feasible. One can, for example, translate the theory to clauses and then select a conservative retraction from the saturation by resolution of the clausal translation.

The main contribution of this paper is a new propositional rule, called *Independence Rule*, specifically designed to compute (and to deal with) conservative retractions. This is the first tool designed for effectively computing conservative retractions. The Independence Rule allows the systematic elimination of propositional variables outside the sublanguage preserving, at the same time, the logical consequences in the sublanguage. Moreover, the rule is also useful to deal with other propositional logical problems, as it will be described.

Finally, it is necessary to note that the theoretical existence shown in (†) does not illustrate how to obtain a finite axiomatization of the conservative retraction. The method presented in this paper outputs a finite axiomatization of $[T, L']$.

The paper is organized as follows. The next section reviews the relationship between propositional logic and the ring $\mathbb{F}_2[\mathbf{x}]$. In the third section the *boolean derivatives* are introduced. Section 4 shows the soundness and completeness of a complete calculus based on them. Section 5 presents basic properties of the rule which are useful to simplify the computing. In section 6 we formalize the location principle as a basis for the computing of the conservative retraction. Section 7 is devoted to show other interesting applications of the Independence Rule, such as theory merging and conservative extensions built by hierarchical merging. We conclude with some remarks about future work.

2   Propositional Logic and the Ring $\mathbb{F}_2[\mathbf{x}]$

The algebraic translation of Propositional Logic into Polynomial Algebra is based on a well known translation of propositional logic in this kind of algebras (see [18], and also [12]). There exist several approaches and applications of this traslation, which allow the use of algebraic tools (as Gröbner Basis) for solving logical problems (see e.g. [5,13,10] and the application given in [19]). This section is devoted to review the main features.

We fix a propositional language $PV = \{p_1, \ldots, p_n\}$, $PForm$ denotes the set of propositional formulas in this language, and $var(F)$ denotes the set of variables of F.

The ring we work on is $\mathbb{F}_2[\mathbf{x}]$ (where $\mathbf{x} = x_1, \ldots, x_n$). A key ideal is $\mathbb{I}_2 := (x_1 + x_1^2, \ldots, x_n + x_n^2)$. To clarify the reasoning, we fix an identification $p_i \mapsto x_i$ (or $p \mapsto x_p$) between PV and the set of indeterminates.

Given $\alpha = (\alpha_1, \ldots, \alpha_n) \in \mathbb{N}^n$, let us define $|\alpha| := \max\{\alpha_1, \ldots, \alpha_n\}$, and $sg(\alpha) := (\delta_1, \ldots, \delta_n)$, where δ_i is 0 if $\alpha_i = 0$ and 1 otherwise. If $a(\mathbf{x}) \in \mathbb{F}_2[\mathbf{x}]$,

$$\deg_\infty(a(\mathbf{x})) := \max\{|\alpha| \ : \ \mathbf{x}^\alpha \text{ is a monomial of } a\},$$

and $\deg_i(a(\mathbf{x}))$ is the degree w.r.t. x_i. If $\deg_\infty(a(\mathbf{x})) \leq 1$, $a(\mathbf{x})$ is called a *polynomial formula*.

Three maps represent the standard starting point for the translation from propositional logic into $\mathbb{F}_2[\mathbf{x}]$:

- The *flatening* map $\Phi : \mathbb{F}_2[\mathbf{x}] \to \mathbb{F}_2[\mathbf{x}]$ is defined by

$$\Phi\left(\sum_{\alpha \in I} \mathbf{x}^\alpha\right) := \sum_{\alpha \in I} \mathbf{x}^{sg(\alpha)}$$

Note that Φ satisfies

$$\Phi(\mathbb{I}_2) = (0) \text{ and } \Phi(a \cdot b) = \Phi(\Phi(a) \cdot \Phi(b))$$

- The *polynomial interpretation* $P : PForm \to \mathbb{F}_2[\mathbf{x}]$ assigns a polynomial to each logical formula. This is achieved by assigning to each propositional variable p_i a monomial x_i and defining, for each connective, the function as follows:
  - $P(\bot) = 0, P(p_i) = x_i, P(\neg F) = 1 + P(F)$
  - $P(F_1 \wedge F_2) = P(F_1) \cdot P(F_2)$
  - $P(F_1 \vee F_2) = P(F_1) + P(F_2) + P(F_1)P(F_2)$
  - $P(F_1 \to F_2) = 1 + P(F_1) + P(F_1)P(F_2)$
  - $P(F_1 \leftrightarrow F_2) = 1 + P(F_1) + P(F_2)$
- The *propositional interpretation* $\Theta : \mathbb{F}_2[\mathbf{x}] \to PForm$ is defined by:
  - $\Theta(0) = \bot$, $\Theta(1) = \top$, $\Theta(x_i) = p_i$,
  - $\Theta(a \cdot b) = \Theta(a) \wedge \Theta(b)$, and
  - $\Theta(a + b) = \neg(\Theta(a) \leftrightarrow \Theta(b))$.

We have that

$$\Theta(P(F)) \equiv F \text{ and } P(\Theta(a)) = a$$

Since we shall be frequently applying $\Phi \circ P$, we take $\pi := \Phi \circ P$, called the *polynomial projection*.

Next we list some basic results that we will use later on.

Lemma 1. *Let* $v : PV \to \{0, 1\}$ *be a valuation with* $v(p_i) = \delta_i$. *Then for every* $F \in PForm$, $v(F) = P(F)(\delta_1, \ldots \delta_n)$.

From any subset X of $\mathbb{F}^n$ we can cook up an ideal $I(X)$, the ideal of polynomials vanishing on X. From any subset I of $\mathbb{F}_2[\mathbf{x}]$ we can cook up an algebraic set $V(I)$, the "vanishing set" of the ideal. The behaviour of the ideals of $\mathbb{F}_2[\mathbf{x}]$ is well known:

- If $A \subseteq (\mathbb{F}_2)^n$, then $V(I(A)) = A$,
- For every $\mathfrak{I} \in Ideals(\mathbb{F}_2[\mathbf{x}])$, it holds that $I(V(\mathfrak{I})) = \mathfrak{I} + \mathbb{I}_2$.

Therefore $F \equiv F'$ if and only if $P(F) = P(F') \pmod{\mathbb{I}_2}$ which is also equivalent to $\Phi \circ P(F) = \Phi \circ P(F')$.

The following theorem states the main relationship between propositional logic and $\mathbb{F}_2[\mathbf{x}]$:

Theorem 1. *The following conditions are equivalent:*

1. $\{F_1, \ldots, F_m\} \models G$.
2. $1 + P(G) \in (1 + P(F_1), \ldots, 1 + P(F_m)) + \mathbb{I}_2$.
3. $\mathtt{NF}(1 + P(G), \mathtt{GB}\,[(1 + P(F_1), \ldots, 1 + P(F_m)) + \mathbb{I}_2]) = 0$. *(where* $\mathtt{GB}$ *denotes Gröbner basis) and* $\mathtt{NF}$ *denotes normal form.*

3   Boolean Derivatives and Non-clausal Theorem Proving

Boolean derivative is a well known tool in Boolean Function Calculus (cf. [22]). We introduce here the operator on propositional formulas as a translation of the usual derivation on $\mathbb{F}_2[\mathbf{x}]$. Recall that a derivation on a ring R is a map $d : R \to R$ verifying:

1. $d(a + b) = d(a) + d(b)$
2. $d(a \cdot b) = d(a) \cdot b + a \cdot d(b)$

Definition 1. *A map $\partial : PForm \rightarrow PForm$ is a boolean derivation if there exists a derivation d on $\mathbb{F}_2[\mathbf{x}]$ such that the following diagram commutes:*

$$
\begin{array}{ccc}
PForm & \overset{\partial}{\rightarrow} & PForm \\
\pi \downarrow & \# & \uparrow \Theta \\
\mathbb{F}_2[\mathbf{x}] & \overset{d}{\rightarrow} & \mathbb{F}_2[\mathbf{x}]
\end{array}
$$

That is,

$$\partial = \Theta \circ d \circ \pi$$

If the derivation on $\mathbb{F}_2[\mathbf{x}]$ is $d = \frac{\partial}{\partial x_p}$, we denote ∂ as $\frac{\partial}{\partial p}$. This derivation has an interesting property : The formula $\frac{\partial}{\partial p}(F)$ represents the change of truth value of F if the truth value of p is changed (recall that $F\{p/G\}$ denotes the formula obtained by substitution of p by the formula G in F).

Proposition 1. $\frac{\partial}{\partial p}F \equiv \neg(F\{p/\neg p\} \leftrightarrow F)$.

Proof. It is easy to see that

$$\pi(F\{p/\neg p\})(\mathbf{x}) = \pi(F)(x_1, \ldots, x_p + 1, \ldots, x_n).$$

Since $\frac{\partial}{\partial x}a(x) = a(x + 1) + a(x)$ holds for polynomial formulas, one has

$$P(\frac{\partial}{\partial p}F) = \frac{\partial}{\partial x_p} \circ \pi(F)(\mathbf{x}) = \pi(F)(x_1, \ldots, x_p + 1, \ldots, x_n) + \pi(F)(\mathbf{x})$$

hence

$$\frac{\partial}{\partial x_p} \circ \pi(F)(\mathbf{x}) = \Phi(P(F\{p/\neg p\}) + P(F)) = \pi(\neg(F\{p/\neg p\} \leftrightarrow F)).$$

By application of Θ we have that $\frac{\partial}{\partial p}F \equiv \neg(F \leftrightarrow F\{p/\neg p\})$.

An important feature of the boolean derivative above defined is that the value of $\frac{\partial}{\partial p}F$ with respect to a valuation does not depend on p. Thus, we can apply any valuation on $PV \setminus \{p\}$ to this formula. That is, since for polynomial formulas

$$\Theta(\frac{\partial}{\partial x}a) \equiv \frac{\partial}{\partial p}\Theta(a)$$

we can assume that

$$\frac{\partial}{\partial p}F := \Theta(\frac{\partial}{\partial x_p}\pi(F))$$

so $p \notin var(\frac{\partial}{\partial p}F)$.

Definition 2. *The Independence Rule (or ∂-rule) on polynomial formulas a_1, a_2 $\in \mathbb{F}_2[\mathbf{x}]$ is defined as:*

$$\partial_x(a_1, a_2) := \frac{a_1, \ a_2}{1 + \Phi\left[(1 + a_1 \cdot a_2)(1 + a_1 \cdot \frac{\partial}{\partial x}a_2 + a_2 \cdot \frac{\partial}{\partial x}a_1 + \frac{\partial}{\partial x}a_1 \cdot \frac{\partial}{\partial x}a_2)\right]}$$

In terms of polynomial coefficents, if we write $a_i = b_i + x_p \cdot c_i$, with $\deg_{x_p}(b_i) = \deg_{x_p}(c_i) = 0$ $(i = 1, 2)$, then

$$\partial_{x_p}(a_1, a_2) := \frac{b_1 + x_p \cdot c_1, \ b_2 + x_p \cdot c_2}{\Phi\left[1 + (1 + b_1 \cdot b_2)[1 + (b_1 + c_1)(b_2 + c_2)]\right]}$$

Note that the rule is symmetric. The Independence Rule on formulas is defined by translating the above rule to formulas:

$$\partial_p(F_1, F_2) := \Theta(\partial_{x_p}(\pi(F_1), \pi(F_2))).$$

This is the propositional interpretation of the result of applying the (polynomial) independence rule to the polynomial projection of the formulas.

For example,

$$\partial_b(c \to a \vee b, d \to a \wedge b) = \Theta[\partial_b(1 + c(1 + a)(1 + b), \ 1 + d(1 + ab))]$$
$$= \neg c \vee a \vee \neg d$$

Lemma 2. *Let $F \in PForm$ and p be a propositional variable. There exists $F_0 \in PForm$, such that $p \notin var(F_0)$ and*

$$F \equiv \neg(F_0 \leftrightarrow p \wedge \frac{\partial}{\partial p}F)$$

Proof. Consider the polynomial formula $a = \pi(F)$. Since $\deg_{x_p}(a) \leq 1$, there exists $b \in \mathbb{F}_2[\mathbf{x}]$ such that $\deg_{x_p}(b) = 0$ and $a = b + x_p \cdot \frac{\partial}{\partial x_p}a$.

By applying Θ, we conclude that

$$F \equiv \Theta(b + x_p \cdot \frac{\partial}{\partial x_p}a) \equiv \neg(\Theta(b) \leftrightarrow p \wedge \frac{\partial}{\partial p}F)$$

4 Soundness and Completeness of Independence Rule

The Independence Rule induces a concept of proof in the standard way, that we denote as $\vdash_\partial$.

Proposition 2. *The Independence Rule is sound.*

Proof. It is sufficent to see that $\{F_1, F_2\} \models \partial_p(F_1, F_2)$. If $\pi(F_i) = a_i$, with $a = b_i + x_p \cdot c_i$ and $\deg_{x_p}(b_i) = \deg_{x_p}(c_i) = 0$, then $F_i \equiv \neg[\Theta(b_i) \leftrightarrow p \wedge \Theta(c_i)]$ $(i = 1, 2)$.

Assume that v satisfies $v(F_1) = v(F_2) = 1$. If $v(p) = 1$ then

$$v(\neg[\Theta(b_i) \leftrightarrow \Theta(c_i)]) = 1;$$

if $v(p) = 0$ then $v(\Theta(b_i)) = 1$. Both cases imply that

$$\{F_1, F_2\} \models \bigwedge_{i=1,2} \neg[\Theta(b_i) \leftrightarrow \Theta(c_i)] \vee \bigwedge_{i=1,2} \Theta(b_i)$$

By application of π to the right-hand formula, and knowing that $c_i = \frac{\partial}{\partial x_p} \pi(F_i)$, one obtains the result by rewriting.

As seen in the above proof, $\deg_i(\partial_{x_i}(a_1, a_2)) = 0$, and the valuations are considered with respect to every possible value on p. Therefore, it is straightforward to prove the following property:

Corollary 1. *Let* $v : PV \backslash \{p\} \to \{0, 1\}$. *The following conditions are equivalent:*

1. $v \models \partial_p(F_1, F_2)$.
2. *Some extension of* v *to* PV *is a model of* $\{F_1, F_2\}$.

For example, consider the propositional formula $p_1 \wedge \neg p_2$. It has that

$$\pi(p_1 \wedge \neg p_2) = x_1(1 + x_2)$$

We have that

$$\partial_{x_1}(x_1(1 + x_2), x_1(1 + x_2)) = 1 + x_2,$$

so the valuation v such that $v(\neg p_2) = 1$ is the only one that we can extend to a model of $p_1 \wedge \neg p_2$. In the case of that $\partial_p(\pi(F_1), \pi(F_2)) = 1$, every partial valuation is extendable to a model of $\{F_1, F_2\}$. Analogously, if $\partial_p(\pi(F_1), \pi(F_2)) = 0$, then there is no valuation extendable to a model of both formulas.

The refutation procedure can be applied to formulas or their equivalent polynomials formulas. Let us see an example. An ∂-refutation for the set $\pi[\{p \to q, q \vee r \to s, \neg(p \to s)\}]$ is

1. $1 + x_1 + x_1 x_2$ $[\![\pi(p \to q)]\!]$
2. $1 + (x_2 + x_3 + x_2 x_3)(1 + x_4)$ $[\![\pi(q \vee r \to s)]\!]$
3. $x_1(1 + x_4)$ $[\![\pi(\neg(p \to s))]\!]$
4. $1 + x_1 + x_3 + x_1 x_4 + x_3 x_4 + x_1 x_3 + x_1 x_3 x_4$ $[\![\partial_{x_2} \text{ to } (1), (2)]\!]$
5. 0 $[\![\partial_{x_1} \text{ to } (3), (4)]\!]$

The following theorem states the refutational completeness of ∂-rule:

Theorem 2. *If* Γ *is inconsistent then* $\Gamma \vdash_\partial \bot$.

Proof. Let $\partial_k[\Gamma]$ ($k \leq n$) be the set of formulas defined by recursion as follows: $\partial_0[\Gamma] := \Gamma$ and, if $k \geq 1$,

$$\partial_k[\Gamma] := \{\partial_{p_k}(F_1, F_2) \ : \ F_1, F_2 \in \partial_{k-1}[\Gamma]\}$$

Note that if $F \in \partial_k[\Gamma]$, then $var(F) \subseteq \{p_{k+1}, \ldots p_n\}$. Thus $\partial_n[\Gamma] \subseteq \{\top, \bot\}$. Therefore it is sufficent to prove that Γ is inconsistent if and only if $\bot \in \partial_n(\Gamma)$.

Since the rule is sound, if $\bot \in \partial_n[\Gamma]$, the set Γ has no models.

Assume now that $\partial_n[\Gamma] = \{\top\}$. Then the constant valuation 1 is a model of $\partial_n[\Gamma]$. By applying induction on k up to 0, it is sufficent to prove that one can extend a model of $\partial_k[\Gamma]$ to a model of $\partial_{k-1}[\Gamma]$.

Let $v : \{p_{k+1}, \ldots, p_n\} \to \{0, 1\}$ be a model of $\partial_k[\Gamma]$, and assume that v can not be extended to a model of $\partial_{k-1}[\Gamma]$. That is, if $v_i = v \cup \{(p_k, i)\}$, then there exists $F^i \in \partial_{k-1}[\Gamma]$ such that $v_i(F^i) = 0$ $(i = 0, 1)$. Note that $v_i(\partial_{p_k}(F^0, F^1)) = 1$ $(i = 1, 2)$.

By rewriting F^i as in lemma 2,

$$F^i \equiv \neg(F_0^i \leftrightarrow p_k \wedge \frac{\partial}{\partial p_k} F^i).$$

We conclude then that $v_0(F_0^0) = 0$, and hence $v(F_0^0) = 0$. Furthermore,

$$v(\neg(F_0^1 \leftrightarrow \frac{\partial}{\partial p_k} F^1)) = v_1(\neg(F_0^1 \leftrightarrow p \wedge \frac{\partial}{\partial p_k} F^1)) = 0.$$

Both facts imply that $v(\partial_{p_k}(F^0, F^1)) = 0$, leading to a contradiction, because $v \models \partial_k[\Gamma]$.

Applying induction, a model of $\partial_0[\Gamma] = \Gamma$ can be found.

The above proof suggests how to find models of Γ (when it is consistent). The decision procedure sketched in the proof is based on the partial saturation of Γ by the ∂-rule. Therefore the method can have a high cost, $O(|\Gamma|^{2^n})$.

5 Properties of the Independence Rule

The following result lists some basic properties that facilitate the computations:

Proposition 3. *Let F, G be propositional formulas*

1. $\partial_p(p, F) \equiv F\{p/\top\}$
2. *If $p \notin var(F)$ then $\partial_p(F, G) \equiv F \wedge \partial_p(G, G)$*
3. *If $p \notin var(F) \cup var(G)$ then $\partial_p(F, G) \equiv F \wedge G$*
4. $\partial_p(G, G) \equiv G\{p/\bot\} \vee \neg(G\{p/\top\} \leftrightarrow G\{p/\bot\})$
5. $\partial(F_1 \wedge F_2, F_3) \equiv \partial_p(F_1, F_2 \wedge F_3)$
6. $\partial_p(F_1 \vee F_2, F_3) \equiv \partial_p(F_1, F_3) \vee \partial_p(F_2, F_3)$
7. $\partial_p(F_1, F_2) \equiv \partial_p(F_2, F_1)$

Proof. The proofs are based on algebraic manipulation of polynomial translation, except property (4), which follows from corollary 1.

Entailment can also be reduced by means of the Independence Rule:

Proposition 4.
$$\Gamma \models G \quad \Longrightarrow \quad \partial_p[\Gamma] \models \partial_p(G)$$

6   Location Principle as Conservative Retraction of Theories

Given $Q = \{q_1, \ldots, q_k\} \subseteq PV$ the operator $\partial_Q := \partial_{q_1} \circ \cdots \circ \partial_{q_k}$ is well defined modulo logical equivalence. This follows from corollary 1, because for every $p, q \in PV$,

$$\partial_p \circ \partial_q[\Gamma] \equiv \partial_q \circ \partial_p[\Gamma]$$

A consequence of corollary 1 and theorem 2 (its proof) is that entailment problem can be reduced to another one where only appears the variables of the goal:

Corollary 2. *(Location principle)* $\Gamma \models F \quad \Longleftrightarrow \quad \partial_{PV \setminus var(F)}[\Gamma] \models F$

Proof. If $\Gamma \models F$ then $\partial_{PV \setminus var(F)}[\Gamma] \cup \{F\}$ is inconsistent (if not, a model of this set can be extended to a model of $\Gamma \cup \{F\}$ by corollary 1). The other implication is true because $\Gamma \models \partial_{PV \setminus var(F)}[\Gamma]$.

The corollary states that $\partial_{PV \setminus L'}[\Gamma]$ is an conservative retraction of Γ to L' (an instance of $[\Gamma, L']$). Thus, CRP problem is solved in this way for propositional logic.

From here, to simplify the notation, we identify $[\Gamma, L']$ with $\partial_{PV \setminus L'}[\Gamma]$.

7   Theory Merging and Hierarchical Theory Merging

In this section we describe how the Independence Rule can be used for theory merging. The following theorem can be considered a version of Craig's Interpolation Lemma for conservative retractions:

Theorem 3. *Let T_1 and T_2 be consistent theories with languages L_1 and L_2 respectively. The following conditions are equivalent:*

1. *$T_1 \cup T_2$ is consistent.*
2. *$[T_1, L_1 \cap L_2] \cup [T_2, L_1 \cap L_2]$ is consistent.*

Proof. (1) $\Longrightarrow$ (2) follows from the soundness of the Independence Rule, because a model of $T_1 \cup T_2$ is model of both retractions.

(2) $\Longrightarrow$ (1) follows from the completeness of the Independence Rule: if

$$v \models [T_1, L_1 \cap L_2] \cup [T_2, L_1 \cap L_2]$$

then there exists two extensions of v, v_1 and v_2, such that $v_1 \models T_1$ and $v_2 \models T_2$. Since the common variables to L_1 and L_2 are in the domain of v, we have that $v_1 \cup v_2$ is a well defined valuation which models $T_1 \cup T_2$.

The above theorem establishes a necessary and suffcient condition for theory merging. However, there are some situations where the merging is inconsistent but it would be interesting to extend one of the theories with consistent knowledge entailed by the other one. For example, when we aim to merge ontologies which have uncertain concepts.

Consider the ontology

$$\Sigma' = \left\{ \begin{array}{r} \mathsf{Bacteria} \sqsubseteq \mathsf{Animal} \sqcup \mathsf{MobileEntity} \\ \mathsf{Fish} \sqsubseteq \mathsf{Animal} \sqcap \mathsf{MobileEntity} \\ \mathsf{MobileEntity} \sqsubseteq \neg\mathsf{Mammals} \end{array} \right.$$

It has that $\Sigma \cup \Sigma'$ entails $\mathsf{Mammals} \equiv \bot$, thus the union is inconsistent. However, it is feasible to extend Σ with knowledge from Σ'. The idea is to retract the second theory to interesting concept symbols, for example to the set

$$\{\mathsf{Bacteria}, \mathsf{Fish}, \mathsf{Animal}, \mathsf{Mammals}\}$$

In this case, the resultant ontology is consistent:

$$\Sigma \cup \partial_{\mathsf{MobileEntity}}(\Sigma') = \left\{ \begin{array}{r} \mathsf{Virus} \sqsubseteq \mathsf{Animal} \sqcup \mathsf{MobileEntity} \\ \mathsf{Mammals} \sqsubseteq \mathsf{Animal} \sqcap \mathsf{MobileEntity} \\ \mathsf{Animal} \sqsubseteq \neg\mathsf{Plant} \\ \mathsf{Bacteria} \sqsubseteq \mathsf{Animal} \\ \mathsf{Fish} \sqsubseteq \mathsf{Animal} \end{array} \right.$$

It is also possible for the ontology obtained in this way to be inconsistent. The following result shows a case in which the extension of the ontology source is consistent:

Lemma 3. *Let T_1 and T_2 be consistent theories in the languages L_1 and L_2, respectively. The theory $T_1 \cup \partial_{L_1 \cap L_2}(T_2)$ is consistent*

In order to formalize the above ideas, we introduce the notion of *hierarchical merging*.

Definition 3. *Let T_1 and T_2 be consistent theories in the languages L_1 and L_2, respectively. A* hierarchical merging *of T_1 and T_2, is a theory T such that:*

1. *T is a conservative extension of T_1.*
2. *For any formula F in the language of $L_2 \setminus L_1$,*

$$T \models F \iff T_2 \models F$$

3. *Whenever theory T' satisfies (1) and (2), $T' \models T$ is verified.*

Thus, the Independence Rule is useful to show that the hierarchical merging of two theories is unique modulo equivalence (when it exists). The result is straightforward from the properties of ∂:

Theorem 4. *Under the conditions of the above definition, $T_1 \cup \partial_{L_1 \cap L_2}(T_2)$ is a hierarchical merging of T_1 and T_2.*

8   Related Work, Conclusions and Future Work

A related rule is the *general resolution* (cf. [4]):

$$Res_p(F,G) : \qquad \frac{F,\ G}{F\{p/\top\} \vee G\{p/\bot\}}$$

(although it is expressed with respect to propositional variables, the original rule allows substitution of any subformula). For polynomial formulas $a_1, a_2 \in \mathbb{F}_2[\mathbf{x}]$ the rule is translated as follows:

$$Res_x(a_1, a_2) : \qquad \frac{a_1,\ a_2}{\Phi(1 + (1 + a_1 + (x+1)\frac{\partial}{\partial x}a_1)(1 + a_2 + x\frac{\partial}{\partial x}a_2))}$$

The general resolution is sound and refutationally complete. It is easy to see that

$$\models \partial_x(F,G) \to Res_x(F,G)$$

but in general it is not an equivalence[1].

Throughout the paper we pointed out related work using similar tools to the used here. To the best of our knowledge, no work on algebraic methods applied to conservative retraction was ever done. However, it is possible to use the elimination theorem on Gröbner basis in order to obtain a conservative retraction (see [14]). However, the elimination of polynomial variables depends on the selected lex ordering on variables for computing the Gröbner basis.

The future work may follow two lines. The first one is the extension to many-valued logics and their applications (see e.g. [19]). For this, a careful generalization of boolean derivatives, with nice logical meaning, seems necessary (in that case, it seems interesting to use another kind of derivations on polynomials on finite fields, as for example the Hasse-Schmidt derivations, see [17]). In the short term we are working on the extensions of ∂_p-rule to certain Description Logics with limited expressivity (as EL logic, [20], and some members of the $DL - lite$ family of Description Logics, see [11]), as well as the use of this rule for solving problems about definability in these logics.

References

1. Alonso, J.-A., Borrego-Díaz, J., Hidalgo, M.-J., Martín-Mateos, F.-J., Ruiz-Reina, J.-L.: A Formally Verified Prover for the ALC Description Logic. In: Schneider, K., Brandt, J. (eds.) TPHOLs 2007. LNCS, vol. 4732, pp. 135–150. Springer, Heidelberg (2007)
2. Amir, E., McIlraith, S.: Partition-based logical reasoning for first-order and propositional theories. Artificial Intelligence 162(1-2), 49–88 (2005)
3. Baader, F., Calvanese, D., McGuinnes, D.L., Nardi, P., Patel-Schneider, P.F.: The Description Logics Handbook. Theory, Implementations and Applications. Cambridge University Press, Cambridge (2003)

---

[1] This fact also implies refutational completeness for ∂-rule. We showed the proof of th. 2 to remark how to find models and to explain the role of the operator $\partial_p[.]$.

4. Bachmair, L., Ganzinger, H.: A theory of resolution. In: Robinson, J.A., Voronkov, A. (eds.) Handbook of Automated Reasoning, vol. I, pp. 19–99. Elsevier Science Pub., Amsterdam (1998)

5. Beame, P., Impagliazzo, R., Krajícek, J., Pitassi, T., Pudlák, P.: Lower Bounds on Hilvert's Nullstellensatz and propositional proofs. Proc. of London Mathematical Society 73, 1–26 (1996)

6. Bennett, B.: Relative Definability in Formal Ontologies. In: Proc 3rd Int. Conf. Formal Ontology in Information Systems (FOIS 2004), pp. 107–118. IOS Press, Amsterdam (2004)

7. Bochmann, D., Posthoff, C.: Binäre dynamishe systeme. Akademieverlag, Berlin (1981)

8. Borrego-Díaz, J., Chávez-González, A.M.: Extension of ontologies assisted by automated reasoning systems. In: Moreno Díaz, R., Pichler, F., Quesada Arencibia, A. (eds.) EUROCAST 2005. LNCS, vol. 3643, pp. 247–253. Springer, Heidelberg (2005)

9. Borrego-Díaz, J., Chávez-González, A.M.: Controlling ontology extension by uncertain concepts through cognitive entropy. In: Proc. Workshop ISWC05 Uncertainty Reasoning on the Semantic Web URSW 2005, pp. 56–66 (2005), http://ftp.informatik.rwth-aachen.de/Publications/CEUR-WS/

10. Buresh-Oppenheim, J., Clegg, M., Imppagliazzo, R., Pitassi, T.: Homogeneization and the Polynomial Calculus. Computational Complexity 11, 91–108 (2003)

11. Calvanese, D., De Giacomo, G., Lombo, D., Lenserini, M., Rosati, R.: Tractable Reasoning and Efficient Query Answering in Description Logics: The $DL - Lite$ Family. J. Automated Reasoning 39, 385–429 (2007)

12. Chazarain, J., Alonso-Jiménez, J.A., Briales-Morales, E., Riscos-Fernández, A.: Multi-valued logic and Gröbner bases with applications to modal logic. Journal Symbolic Computation 11, 181–194 (1991)

13. Clegg, M., Edmonds, J., Impagliazzo, R.: Using Gröbner Basis algorithm to find proofs of unsatisfiability. In: Proc. ACM Symposium of Computing, pp. 174–183 (1996)

14. Cox, D., Little, J., O'Shea, D.: Ideals, Varieties and Algorithms: An Introduction to Computational Algebraic Geometry and Commutative Algebra. Springer, Heidelberg (2005)

15. Cuenca-Grau, B., Parsia, B., Sirin, E., Kalyanpur, A.: Automatic Partitioning of OWL Ontologies Using E-Connections. In: Proc. 2005 Int. Workshop on Description Logics (DL2005) (2005), http://sunsite.informatik.rwth-aachen.de/Publications/CEUR-WS/ Vol-147/21-Grau.pdf

16. Giunchiglia, F., Yatskevitch, M., Shvaiko, P.: Semantic Matching: Algorithms and Implementations. J. Data Semantics 9, 1–38 (2007)

17. Fernández-Lebrón, M., Narváez-Macarro, L.: Hasse-Schmidt Derivations and Coefficient Fields in Positive Characteristics. J. of Algebra 265(1), 200–210 (2003)

18. Kapur, D., Narendran, P.: An equational approach to theorem proving in first-order predicate calculus. In: Proc. 9 Int. Joint Conf. on Artificial Intelligence (IJCAI 1985), pp. 1146–1153 (1985)

19. Laita, L.M., Roanes-Lozano, E., de Ledesma, L., Alonso-Jiménez, J.A.: A computer algebra approach to verification and deduction in many-valued knowledge systems. Soft Computing 3, 7–19 (1999)

20. Lutz, C., Wolter, F.: Conservative extensions in the lightweight description logic EL. In: Pfenning, F. (ed.) CADE 2007. LNCS, vol. 4603, pp. 84–99. Springer, Heidelberg (2007)
21. Martín-Mateos, F.J., Alonso, J.A., Hidalgo, M.J., Ruiz-Reina, J.L.: Formal Verification of a Generic Framework to Synthetize SAT-Provers. J Aut. Reasoning 32(4), 287–313 (2004)
22. Thayse, A.: Boolean Calculus of Differences. Springer, Berlin (1981)
23. Tsarkov, D., Horrocks, I.: Optimised Classification for Taxonomic Knowledge Bases. In: Proc. 2005 Int. Workshop on Description Logics (DL 2005) (2005), http://sunsite.informatik.rwth-aachen.de/Publications/CEUR-WS/Vol-147/39-TsarHorr.pdf

Combining Coq and Gappa
for Certifying Floating-Point Programs*

Sylvie Boldo[1,2], Jean-Christophe Filliâtre[2,1], and Guillaume Melquiond[1,2]

[1] INRIA Saclay - Île-de-France, ProVal, Orsay, F-91893
[2] LRI, Université Paris-Sud, CNRS, Orsay, F-91405
Sylvie.Boldo@inria.fr
Jean-Christophe.Filliatre@lri.fr
Guillaume.Melquiond@inria.fr

Abstract. Formal verification of numerical programs is notoriously difficult. On the one hand, there exist automatic tools specialized in floating-point arithmetic, such as Gappa, but they target very restrictive logics. On the other hand, there are interactive theorem provers based on the LCF approach, such as Coq, that handle a general-purpose logic but that lack proof automation for floating-point properties. To alleviate these issues, we have implemented a mechanism for calling Gappa from a Coq interactive proof. This paper presents this combination and shows on several examples how this approach offers a significant speedup in the process of verifying floating-point programs.

1   Introduction

Numerical programs typically use floating-point arithmetic [1]. Due to their limited precision and range, floating-point numbers are only an approximation of real numbers. Each operation may introduce an inaccuracy and their total contribution is called the *rounding error*. Moreover, some real operations may not be available as sequences of floating-point operations, *e.g.*, infinite sums or integrals. This introduces another inaccuracy called the *method error*. Both errors make it somehow complicated to know what floating-point programs actually compute with respect to the initial algorithms on real numbers.

One way to proceed is to give a program a precise specification of its accuracy. Generally speaking, a specification explains what can be expected from the result given facts about the inputs. Typically, it bounds the sum of both rounding and method errors. For example, the specification for a function `float_cos` defined on the `double` type of floating-point numbers may be the following:[1]

$$\forall x : \texttt{double}, \quad |x| \leq 2\pi \;\Rightarrow\; \left| \frac{\cos(x) - \texttt{float\_cos}(x)}{\cos(x)} \right| \leq 2^{-53}.$$

---

* This research was partially supported by projects ANR-05-BLAN-0281-04 "CerPAN" and ANR-08-BLAN-0246-01 "F∮ST".
[1] Note that $\pi/2$ cannot be represented by a floating-point number, therefore $\cos(x)$ cannot be zero.

J. Carette et al. (Eds.): Calculemus/MKM 2009, LNAI 5625, pp. 59–74, 2009.
© Springer-Verlag Berlin Heidelberg 2009

Such an inequality is typically proved using pen and paper. It can be deduced from the specification of each floating-point operation and the mathematical properties of the cosine function. Such proofs are notoriously difficult and error-prone.

Ideally, the code of a software would be analyzed by a certification tool, which would answer whether it is correct or not. In order to increase the confidence in the answer (and therefore in the software), the tool may rely on formal methods. Obviously, such a tool is not conceivable, but we aim at making this process as automatized as possible.

In this paper, we will describe how we have combined existing tools to help in the process of formally certifying numerical codes. From a user point of view, the first step is to annotate the source code with the specifications of the software. This annotated code is sent to a first tool that produces proof obligations corresponding to the correctness of the software. The examples of this paper are C programs and we are using the Caduceus tool (Section 2.1). By computing weakest preconditions based on the code and the annotations, it generates theorem statements to be verified by automated theorem provers or proof assistants.

Some theorems, hopefully most of them, will be automatically discharged. For instance, numerical properties may be discharged by the Gappa tool (Section 2.3), which is very efficient at proving bounds, especially on rounding errors. But Gappa only tackles the floating-point fragment of a program, so properties that involve more than just floating-point arithmetic may not be handled.

The remaining theorems will have to be manually handled by the user in a proof assistant with a suitable floating-point formalization. The paper focuses on the use of Coq for this task (Section 2.2). When using a proof assistant, the user issues tactics to split the goal into simpler subgoals. Examples of such tactics are logical cut, case analysis, or induction. This will become tedious if the user has to repeat this process until all the subgoals are discharged, which may require a high number of explicit proof steps. Especially frustrating is the fact that, once simplified, the subgoals may fit into specific logic fragments, which some tools, such as Gappa, could handle automatically outside Coq.

In order to benefit from Gappa inside Coq, we have implemented a mechanism for calling the tool from an interactive proof. From a technical point of view, Gappa is called as an external prover. This does not weaken the confidence in Coq formal proofs since Gappa produces a proof trace that is checked by Coq (Section 4).

This combination of Coq and Gappa does not radically change the way to tackle rounding and method errors. It simply eases the use of traditional approaches in a formal setting. C programs illustrating this point are given in Section 3. The combination of all these tools (Caduceus, Coq, Gappa) makes it possible to formally verify a source code while benefiting from automation.

There have been previous work on formally proving numerical components (especially hardware ones) while relying on automated tools. Among them, the certification of the IEEE-compliance [1] of a gate-level design for the Pentium Pro processor used Forte, a combination of two model checkers and a lightweight theorem prover [2]. Another work made the ACL2 theorem prover interact with a VHDL verification tool in order to prove the correctness of a hardware multiplier [3]. While not at the source-code level, two other proof assistants have been thoroughly used for verification of floating-point properties: both PVS and HOL Light provide some automation for performing error analysis [4,5].

2   Caduceus, Coq, and Gappa

2.1   Verification of C Programs

The Why platform[2] is a set of tools for deductive verification of Java and C programs [6]. In this paper, we only focus on verification of C programs but the results would apply to Java programs as well. The verification of a given C program proceeds as follows. First, the user specifies requirements as annotations in the source code, in a special style of comments. Then, the annotated program is fed to the tool Caduceus [7], which is part of the Why platform, and verification conditions (VCs for short) are produced. These are logical formulas whose validity implies the soundness of the code with respect to the given specification. Finally, the VCs are discharged using one or several theorem provers, which range from interactive proof assistants such as Coq to purely automatic theorem provers such as Alt-Ergo [8]. The workflow is illustrated on Figure 1.

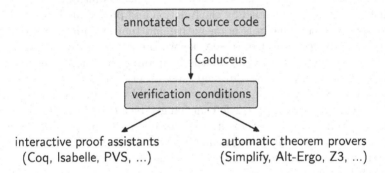

Fig. 1. The Caduceus tool

Annotations are inserted in C source code using comments with a leading @. They are written in first-order logic and re-use the syntax of side-effect free C expressions. For instance, here is a code excerpt where an array t is searched for a zero value.

```
//@ invariant 0 <= i
for (i = 0; i < n; i++) {
  if (t[i] == 0) break;
}
//@ assert i < n => t[i] == 0
```

The **for** loop is given a loop invariant, as in traditional Hoare logic [9]. (In that case, the invariant could be found automatically.) A loop invariant typically

---

[2] Available at http://why.lri.fr/

generates two VCs: one to show that it holds right before the loop is entered; and one to show that it is preserved by the loop body. In this example, an assertion is also manually inserted right after the loop, which results in a VC for this program point. Additional VCs are produced to establish the safe execution of the code, *i.e.*, that the program does not perform any division by zero or any array access out of bounds. In this example, a VC requires to show that t[i] is a legal array access, which may or may not be provable depending on hypotheses regarding t and n.

Verification with Caduceus is modular: each function is given a *contract* and proved correct with respect to the contracts of the functions it calls.[3] For instance, a partial contract for a function sorting an array of integer could be

```
/*@ requires
  @   0 <= n && \valid_range (t, 0, n-1)
  @ assigns
  @   t[0..n-1]
  @ ensures
  @   \forall int i,j; 0 <= i <= j < n => t[i] <= t[j] */
void sort(int *t, int n);
```

The contract contains three clauses. Keyword **requires** introduces a precondition, that is a property assumed by the function and proved at the caller site. In this example, it states that n is nonnegative and that all indices from 0 to n-1 in t can be safely accessed. Conversely, keyword **ensures** introduces a postcondition, that is a property provided by the function, right before it returns. Here it states that the array is sorted in increasing order.[4] Finally, keyword **assigns** introduces the memory locations possibly modified by the function, which means that any other memory location is left unchanged by a call to this function. Here, it states that only the array elements t[i] for $0 \leq i < n$ are possibly assigned.

Caduceus handles a large fragment of ANSI C, with the notable exception of pointer casts and unions. It handles floating-point arithmetic, using a model where each floating-point number is seen as a triple of real numbers [10]. The first component is the floating-point number itself, as it is computed. The second component is the real number that would have been computed if roundings were not performed. The third component is a ghost variable attached to the floating-point number and which represents the ideal value that the programmer intended to compute. Annotations are written using real numbers only, and the three components of a floating-point variable x can be referred to within annotations: x itself stands for the first component; \exact(x) for the second one; and \model(x) for the third one. Thus the user can refer to the rounding error as the difference between the first two, and to method error as the difference between the last two. Examples are given in Section 3.

---

[3] That means we only establish *partial correctness* of recursive functions.

[4] For the specification to be complete, the postcondition should also state that the array is a permutation of its initial value. It can be done, but is omitted here for the sake of simplicity.

Since the general-purpose automatic provers do not support this model of floating-point arithmetic, we have formalized it in the Coq proof assistant.

2.2   The Coq Proof Assistant

The Coq proof checker [11,12] is a proof assistant based on higher-order logic. One may express properties such as "there exists a function which has such and such properties" or "every relation that verifies such hypothesis has a certain property" and check proofs about these. Proofs are built using tactics (such as applying a theorem, rewriting, computing, etc.). A Coq file contains the statement of lemmas and their proofs as a sequence of tactics in the Coq language.

The Coq standard library contains an axiomatization of real numbers [13]. Few automation is provided to reason about real numbers. As a consequence, the proof of a typical lemma such as $0 < 1 - 2^{-52}$ is already a few lines long:

```
1  Lemma OneMinusUlpPos:  (0 < 1 - powerRZ 2 (-52))%R.
2  Proof.
3    apply Rlt_Rminus.
4    unfold powerRZ.
5    rewrite <- Rinv_1 at 3.
6    apply Rinv_1_lt_contravar ; auto with real.
7  Qed.
```

The proof is done backward, by transforming the conclusion until it trivially derives from the hypotheses. This proof starts by applying the theorem Rlt_Rminus (line 3) since $0 < 1 - 2^{-52}$ is a consequence of $2^{-52} < 1$. The definition of powerRZ is then unfolded (line 4) so that 2^{-52} is converted to $(2^{52})^{-1}$. We replace 1 by 1^{-1} (theorem Rinv_1, line 5). At this point, the goal has become $(2^{52})^{-1} < 1^{-1}$. After applying theorem Rinv_1_lt_contravar (line 6), the remaining goals are $1 < 2^{52}$ and $1 \leq 1$, which are solved automatically by the tactic auto (line 6).

A high-level formalization of floating-point arithmetic [14,15] is also available in Coq. A floating-point number is a pair of integers (m, e) which represents the real number $m \times 2^e$. The value of the mantissa m and the exponent e are bounded according to the floating-point format. For example, in IEEE-754 double-precision format [1], the pair verifies $|m| < 2^{53}$ and $-1074 \leq e$. This library[5] contains a large number of floating-point definitions and theorems and has been used to prove many old and new properties [16].

Most floating-point proofs rely on computations on real numbers, such as deciding $0 < 1 - 2^{-52}$ or bounding method error. Such goals can be addressed using the interval tactic [17]. This reflexive tactic, based on interval arithmetic, decides inequalities by bounding real expressions thanks to guaranteed floating-point arithmetic. Once done with method error, the user is left with VCs related to rounding errors, which Gappa is typically designed for.

---

[5] Available at http://lipforge.ens-lyon.fr/www/pff/

2.3   The Gappa Tool

Gappa[6] is a tool dedicated to proving arithmetic properties on numerical programs [18,19]. Given a logical proposition expressing bounds on mathematical real-valued expressions, Gappa checks that it holds. The following is such a proposition and below is its transcription in Gappa's input language.

$$\forall x, y \in \mathbb{R}, \quad |x| \le 2 \wedge y \in [1, 9] \Rightarrow x \times x + \sqrt{y} \in [1, 7]$$

```
{ |x| <= 2 /\ y in [1,9] -> x * x + sqrt(y) in [1,7] }
```

In order to verify the proposition, Gappa first analyzes which expressions may be of interest. Then it tries to enclose them in intervals by performing a saturation over its library of theorems on interval arithmetic, forward error analysis, and algebraical identities. Gappa stops when it reaches enclosures small enough to be compatible with the right-hand side of the proposition or when the saturation does no longer improve the enclosures.

Once Gappa has verified the proposition, it generates a formal proof. To increase confidence, this proof script can then be mechanically checked by an independent proof system, such as Coq or HOL Light.[7]

If Gappa fails to prove the proposition, the user can suggest to the tool that it should perform a bisection—splitting input intervals until the proposition holds on each sub-intervals—or augment the library of theorems with new mathematical identities. Gappa will then assume these equalities hold; they will appear as hypotheses of the generated formal proof.

Expressing Floating-Point Programs. In addition to universally-quantified variables on $\mathbb{R}$, basic arithmetic operators ($+$, $-$, $\times$, $\div$, $\sqrt{\cdot}$), and numerical constants, Gappa expressions can also contain *rounding* operators. The integer-part functions, $\lfloor \cdot \rfloor$ and $\lceil \cdot \rceil$, are instances of such operators. Since the IEEE-754 standard [1] mandates that "a floating-point operator shall behave as if it was first computing the *infinitely-precise* value and then *rounding* it so that it fits in the destination floating-point format", having appropriate rounding operators is sufficient to express the computations of a floating-point program.

The following script is similar to the previous one, but all the expressions are now as if they had been computed in single precision with rounding to nearest (tie-breaking to even mantissa).

```
@rnd = float<ieee_32,ne>;
z = rnd(rnd(x * x) + rnd(sqrt(y)));
{ |x| <= 2 /\ y in [1,9] -> z in [1,7] }
```

Note that Gappa only manipulates expressions on real numbers. As a consequence, infinities and NaNs (Not-a-Numbers) are no part of this formalism:

---

[6] Available at http://lipforge.ens-lyon.fr/www/gappa/

[7] The generated proofs depend on a library of facts written for the target system. Currently, this formalization has been proved for Coq only.

rounding operators return a real value and there is no upper bound on the magnitude of the input numbers. This means that NaNs and infinities will not be generated nor propagated as they would in IEEE-754 arithmetic.

We can, however, use Gappa to prove that a given code does not produce any of these exceptional values. Indeed, if one proves that a Gappa-rounded value is smaller than the biggest floating-point number in the working format, then the actual IEEE-754 computation is guaranteed not to overflow, by definition of overflow. Therefore, in order to check that computations in the previous example are overflow-safe, one can run Gappa on the following script:[8]

```
@rnd = float<ieee_32,ne>;
z = rnd(rnd(x * x) + rnd(sqrt(y)));
{ |x| <= 2 /\ y in [1,9]
  -> z in [1,7] /\ |rnd(x * x)| <= 0x1.FFFFFEp127 /\
     |rnd(sqrt(y))| <= 0x1.FFFFFEp127 }
```

The absence of infinities and NaNs is not a deficiency, as reasoning about them is usually done by case analysis. This can be easily performed using Coq traditional tactics. For the cases without infinities and NaNs, which are the complicated ones, the Gappa tactic applies.

Verifying Accuracy. While the previous examples show that Gappa can bound ranges of floating-point variables, this is only a small part of its purpose. This tool was designed to prove bounds on computation errors, which also happen to be real-valued expressions. Let us assume that the developer actually needed the infinitely-precise result $M_z = x^2 + \sqrt{y}$. Is the computed result z sufficiently close to this ideal value M_z? This can be answered by bounding the absolute error $z - M_z$:[9]

```
@rnd = float<ieee_32,ne>;
Mz = x * x + sqrt(y);
z = rnd(rnd(x * x) + rnd(sqrt(y)));
{ |x| <= 2 /\ y in [1,9] -> |z - Mz| <= 1b-21 }
```

For the sake of simplicity, M_z has the same operations as z, but without rounding. This is not a requirement, as Gappa is also able to bound errors when M_z is a completely different expression. Note also that Gappa is not limited to absolute errors; it can handle relative errors in a similar way, which is especially important when proving floating-point properties.

3   Proving Floating-Point Programs

Before describing the inner working of the Coq-Gappa combination, we illustrate its use on the verification of three typical floating-point programs.

---

[8] The number 0x1.FFFFFEp127 is the biggest finite floating-point numbers for IEEE-754 single-precision format, written with the notation of the standard of the ISO C language (1999).

[9] The number 1b-21 means 2^{-21}, which is almost the optimal upper bound on the specified absolute error.

3.1   Naive Cosine Computation

The first example is an implementation of the cosine function for single-precision floating-point arithmetic. To present a complete Coq proof, we have simplified the function by removing its argument-reduction step. Thus, input x is required to have already been reduced to a value close to zero; only the polynomial evaluation has to be performed. The specification of the function states that, for $|x|$ smaller than 2^{-5}, the computed value \result is equal to $\cos(x)$ up to 2^{-23}.

```
/*@ requires  |x| <= 1./32
  @ ensures   |\result - cos(x)| <=  2^^(-23)
  @ */
float toy_cos(float x) {
    return 1.f - x * x * .5f;
}
```

Note that 2^{-23} is a tight bound on the error of this function. It ensures that the computed result is one of the floating-point numbers close to the mathematical value $\cos(x)$.

Given this annotated C code, Caduceus generates a VC stating the accuracy of the result, which can be formally proved with the Coq script below.[10]

```
1  Proof.
2      intros; why2gappa; unfold cos.
3      assert ( Rabs ((1 - (f*f) * (5/10)) - Rtrigo_def.cos f)
4                      <= 7/134217728 )%R
5          by interval with (i_bisect_diff f).
6      gappa.
7  Qed.
```

The first part of the proof script (line 2) turns the goal into a user-friendly form: the why2gappa tactic cleans the goal by expanding and rewriting some Caduceus-specific notations. At this point, assuming that $\circ(\cdot)$ is the rounding operation from a real number to the nearest single-precision floating-point number, the user has to prove the following goal:

$$\forall x : \texttt{float}, \quad |x| \leq \tfrac{1}{32} \Rightarrow$$
$$|\circ(\circ(1) - \circ(\circ(x \times x) \times \circ(5/10))) - \cos(x)| \leq 2^{-23}.$$

As would be done with a pen-and-paper verification, the formal proof of this goal starts by computing and proving a bound on the method error. Since the polynomial is chosen so that the computed result is close to the cosine, the method error is known beforehand. A typical way to obtain a polynomial approximation and its error is to use a computer algebra system.

---

[10] The Coq script is reproduced *verbatim*. In particular, some terms are obfuscated due to Coq renaming them to prevent conflicts. So f designates in fact the variable x; and Rtrigo_def.cos is the name of the cosine function in Coq's standard library.

Here, the method error is smaller than 7×2^{-27}. So we are asserting this property in Coq (lines 3 and 4) and we prove it (line 5). The assertion is proved by the `interval` tactic [17]. Its option `i_bisect_diff` tells the tactic to recursively perform a bisection on the interval enclosure $[-2^{-5}, 2^{-5}]$ of x, until a first-order interval evaluation of the method error $(1 - (x \times x) \times (5/10)) - \cos(x)$ gives a compatible bound on all the sub-intervals.

Once the assertion is proved and hence available as an hypothesis, the user has to prove the following property:

$$\forall x : \texttt{float}, \quad |x| \leq \tfrac{1}{32} \Rightarrow$$
$$|(1 - (x \times x) \times (5/10)) - \cos(x)| \leq 7 \cdot 2^{-27} \Rightarrow$$
$$|\circ(\circ(1) - \circ(\circ(x \times x) \times \circ(5/10))) - \cos(x)| \leq 2^{-23}.$$

This is achieved by the `gappa` tactic (line 6). It calls Gappa and then uses the Coq script that the tool generates in order to finish the proof. Note that the Gappa tool takes advantage of the inequality proved by the `interval` tactic as it knows nothing about the cosine.

3.2 Discretization of a Partial Differential Equation

The second example is a numerical code about acoustic waves by F. Clément [20]. Given a rope attached at its two ends, a force initiates a wave, which then undulates according to the following mathematical equation:

$$\frac{\partial^2 u(x,t)}{\partial t^2} - c^2 \frac{\partial^2 u(x,t)}{\partial x^2} = 0.$$

The value $u(x,t)$ gives the position of the rope at the abscissa x and the time t. It is discretized both in space and time with steps $(\Delta x, \Delta t)$. The result is a matrix p of size $ni \times nk$ where $p[i][k] = p_i^k$ is the position of the rope at the abscissa $i \times \Delta x$ and the time $k \times \Delta t$. The matrix p is computed by the following piece of code [21], where a is an approximation of an exact constant A derived from c, ni, and nk:

```
/*@ invariant 1 <= k <= nk
        && analytic_error (p,ni,ni,k,a) */
for (k=1; k<nk; k++) {
  p[0][k+1] = 0.;

  /*@ invariant 1 <= i <= ni
          && analytic_error (p,ni,i-1,k+1,a) */
  for (i=1; i<ni; i++) {
    dp = p[i+1][k] - 2.*p[i][k] + p[i-1][k];
    p[i][k+1] = 2.*p[i][k] - p[i][k-1] + a*dp;
  }

  p[ni][k+1] = 0.;
}
```

The predicate `analytic_error` states the exact analytical expression of the rounding error. It also states that the rounding error of a single iteration is smaller than a known value. More precisely, it bounds the absolute value of

$$\varepsilon_i^{k+1} := p_i^{k+1} - (2p_i^k - p_i^{k-1} + A \times (p_{i+1}^k - 2p_i^k + p_{i-1}^k)).$$

Under some hypotheses on A and the ranges of (p_i^k), we could prove that $|\varepsilon_i^{k+1}| \leq 85 \times 2^{-52}$ (and a similar property concerning the initialization of the p_i^1) [21]. The original Coq proof amounts to 735 lines of tactics. Thanks to the gappa tactic, we were able to

- improve the result: we now have the formal proof that $|\varepsilon_i^{k+1}| \leq 80 \times 2^{-52}$;
- drastically cut off the size of the proof script: the 735 lines of tactics reduce to 10.

This is a tremendous improvement. Not only is the new proof script dramatically shorter and simpler to write, but it is also more amenable to future changes and maintenance. Indeed, if the program is to be modified in such a way that the error slightly increases, the initial proof would be completely broken and only a small part could be re-used. Using Gappa, the situation is different: while the statement of the theorem would change, the proof would probably be robust enough to remain valid.

3.3   Preventing Overflows

Another type of proof that greatly benefits from automation is overflow proofs. Typically, one wants to guarantee that no overflows happen. To do so, it is usually sufficient to bound the program inputs. The resulting VCs are especially tedious to prove. As a consequence, the bounds are often over-estimated in order to simplify the demonstrations. This is the case for the following example. This program computes an accurate discriminant using Kahan's algorithm [22]. The accuracy is measured in ulps (unit in the last place), which is the distance between two consecutive floating-point numbers. The discriminant algorithm relies on the `exactmult` function which computes the rounding error of a multiplication.

```
/*@ requires xy==round(x*y) &&
  @    (x*y==0 || 2^^(-969) <= |x*y|) &&
  @    |x| <= 2^^995 && |y| <= 2^^995 && |x*y| <= 2^^1022
  @ ensures \result==x*y-xy
  @ */
double exactmult(double x, double y, double xy);

/*@ requires
  @        (b==0   || 2^^(-916) <= |b*b|) &&
  @        (a*c==0 || 2^^(-916) <= |a*c|) &&
  @        |b| <= 2^^510 && |a| <= 2^^995 && |c| <= 2^^995 &&
  @        |a*c| <= 2^^1021
```

```
@ ensures \result==0 ||
@       |\result-(b*b-a*c)| <= 2*ulp(\result)
@ */

double discriminant(double a, double b, double c) {
  double p,q,d,dp,dq;
  p=b*b;
  q=a*c;

  if (p+q <= 3*fabs(p-q))
    d=p-q;
  else {
    dp=exactmult(b,b,p);
    dq=exactmult(a,c,q);
    d=(p-q)+(dp-dq);
  }
  return d;
}
```

The formal proofs for this program (including overflows) have been presented in [23,24]. Here we only focus on the overflows of the discriminant; we do not care about the `exactmult` function.

All the overflow proofs were first done prior to the **gappa** tactic. For seven proof obligations, it took more than 420 lines of Coq. Using the tactic, the proofs reduce to 35 lines (about 5 lines per theorem). The Coq compilation time, however, is about 5 times greater.[11] Nevertheless, the profit is clear in the verification process as the time for developing the proof overwhelms the time for compiling it.

It is also interesting to note that the specification is also improved. The hypothesis $|a \times c| \leq 2^{1020}$ in the original proof [24] was too strong; we proved instead that $|a \times c| \leq 2^{1021}$ is sufficient to guarantee that no overflows occur. The proof was not modified at all after changing the annotations. This means that the automation is sufficient to use exactly the same proof when modifying slightly the specification. This is really worthwhile for proof maintenance.

4   Implementation Details

The **gappa** tactic is part of the standard V8.2 Coq distribution.[12] It relies on Gappa, which is an external stand-alone tool and comes with its own library of Coq theorems.

Figure 2 describes the process of performing a formal certification of a C program using Coq and Gappa. Starting with an annotated C program, Caduceus generates VCs corresponding to the specification of this C code. Lots of these proof obligations can be discharged by automatic provers. The most complicated

---

[11] Should Coq only check proofs generated by Gappa instead of embedding them, the compilation time would be equivalent. See end of Section 4.

[12] Available at http://coq.inria.fr/

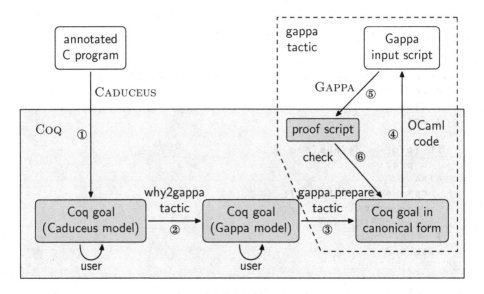

Fig. 2. Dataflow in the Coq and Gappa combination

ones, especially those involving floating-point properties, are left to the user. The Caduceus tool therefore generates template Coq scripts (Step ①), and the user has to fill in the blanks.

At this point, the Coq goals are expressed in the floating-point model of Caduceus. As usual with a proof assistant, the user issues tactics in order to split the goal into simpler subgoals that can be handled automatically. If one subgoal is in the scope of the Gappa tool, the user can proceed as follows.

First of all, the goal has to be translated to the floating-point model of Gappa. Stored in an auxiliary library, some theorems state that both models are equivalent. In particular, if a floating-point number is the closest to a real number in the Caduceus model, then it is the result of a rounding function in the Gappa model, and reciprocally. The why2gappa tactic automatically applies these theorems to rewrite the goal and its context (Step ②). It also unfolds the Caduceus model of floating-point numbers, with its floating-point, exact, and model parts. At this point, the goal is made of inequalities between real-valued expressions potentially containing rounding operators matching Gappa's ones.

Now, the user can launch the Gappa tool to finish the proof, thanks to a single call to the **gappa** tactic (Steps ③, ④, ⑤, and ⑥). During Step ④, some OCaml code embedded into Coq reads the goal and outputs a text file suitable for Gappa. This code then runs Gappa and asks for a Coq script of the result (Step ⑤). Another OCaml code loads this script into Coq, checks it, and generates the corresponding λ-term, and uses it to finish the proof (Step ⑥).

This process will succeed only if the type of the Gappa λ-term matches exactly the Coq goal of the user. Otherwise, Coq would rightfully complain that the overall proof is not well-typed. For instance, the user goal could mention the inverse x^{-1} of a variable, while the generated proof would consider $1/x$ instead. Although equal, these two terms are not convertible, so type-checking would fail. Rather than transforming the generated script afterward, we decided to transform the goal beforehand. The gappa_prepare tactic (Step ③), called internally by the gappa tactic, makes sure that the goal will not leave any margin of interpretation to the OCaml code nor to Gappa.

This subtactic is written in Ltac, the tactic language embedded into Coq and available to user scripts. It transforms all the hypotheses and the goal so that they are enclosures of the form $m_1 \cdot 2^{e_1} \leq expr \leq m_2 \cdot 2^{e_2}$, with m_1, e_1, m_2, and e_2 explicit integers. Moreover, the $expr$ part should only contain the basic arithmetic operators $(+, -, \times, \div, \sqrt{\cdot}, \text{and } |\cdot|)$, rounding operators, identifiers, and constants $(m \cdot 2^e \text{ or } m \cdot 10^e)$. For instance, if a proposition is $|\exp(x) + 5 \times y| \leq 3/8$, the tactic will generalize $\exp(x)$ to a fresh identifier e everywhere. Then it will replace the proposition by the (equivalent yet not convertible) proposition $0 \cdot 2^0 \leq |e + (5 \cdot 2^0) \times y| \leq 3 \cdot 2^{-3}$.

In order to transform the propositions, the tactic could perform some pattern-matching to find all the sub-terms that look unadapted and apply rewriting theorems to them. This method is easy to implement but slow, as a huge number of rewriting operation may be needed, especially for constants. (For instance, the real number 11 is implicitly stored by Coq as $1 + (1+1) \times (1 + (1+1) \times (1+1))$.) Instead, the tactic builds an inductive object that represents the syntax tree of the expressions. Some Coq functions (defined in the logic language, not in the tactic language) then implement the previous transformations. We have proved they generate a syntax tree whose evaluation as a real-valued expression gives the same result as the previous expression. Hence applying this single theorem is enough to get a suitable goal. In other words, the tactic simplifies the goal by convertibility and reflexivity [25,17], which is both time- and space-efficient.

Step ④ is then trivial: the OCaml code just has to select the propositions that are enclosures, to visit the nodes of their simplified syntax trees, and to produce the corresponding Gappa script. If Gappa succeeds in verifying the script (Step ⑤), the OCaml code can then load the produced proof and have Coq check it. It takes only a few seconds for the gappa tactic to reach this point after it is called.

We, however, wanted Coq not only to check the generated proof, but also to embed it into the current user λ-term. Therefore, the gappa tactic, while calling an external prover, does produce a complete Coq proof of the goal. Unfortunately, Coq is unable to deal with two scripts at once. So the tactic first launches a separate Coq session that produces a λ-term in the context of Gappa's libraries. Then it runs another separate session to get a λ-term with fully-qualified names and no notations. This last λ-term can finally be loaded in the original user session, without interfering with user-defined names and notations. This incurs a noticeable slowdown for the user. It could be fixed in two ways: enhance Coq so that other scripts can be checked in the same session, or enhance Gappa so that it directly produces a plain λ-term. Embedding the proof script into the user proof hardly increases the confidence though, since the script has already been checked by Coq. So, Step ⑥ could be reduced to type-checking the Gappa proof, creating an axiom with its type, and applying this axiom to the goal. This would ensure that the gappa tactic takes a few seconds only, without having to modify either Coq or Gappa.

Lastly, note that the gappa tactic accesses only a small part of Gappa's features. Indeed, when using the tool directly, the user can pass hints regarding properties of the problem, such as mathematical identities, to guide it. As long as the goals do not need any particular user hint, the tactic is as powerful as the tool.

5   Conclusion

We have presented an integration of the Gappa automated prover in the Coq proof assistant. This greatly eases the verification process of numerical programs. As shown with realistic examples, the gappa tactic significantly reduces the size of Coq proofs, and improves their maintainability. This tactic is part of the V8.2 Coq standard distribution.

This paper focuses on C programs verified with the Caduceus tool. However, this approach is generic enough to apply to other verification tools, such as Frama-C[13] for C programs or Krakatoa for Java programs [6]. Indeed, our work builds upon the Why platform, which provides a common backend for Caduceus, Frama-C/Jessie, and Krakatoa. Said otherwise, any verification technology using Why to produce Coq verification conditions can benefit from the gappa tactic.

The current gappa tactic does not encompass all the features of the Gappa tool. As explained before, there is no way to pass hints to Gappa, such as interval bisection or equalities. Moreover, while the tool can infer enclosures for variables and expressions, the tactic does not offer a way to query them. This feature would relieve the user from the burden of guessing logical cuts.

The Gappa tool is limited to a small logical fragment dedicated to floating-point arithmetic, and so is the gappa tactic. A more ambitious perspective is to integrate Gappa to a state-of-the-art SMT solver such as Alt-Ergo [8]. This would result in more VCs discharged automatically but also in more automation when invoked from Coq.

References

1. Microprocessor Standards Subcommittee: IEEE Standard for Floating-Point Arithmetic. IEEE Std. 754-2008, pp. 1–58 (August 2008)
2. O'Leary, J., Zhao, X., Gerth, R., Seger, C.J.H.: Formally verifying IEEE compliance of floating-point hardware. Intel Technology Journal (Q1), 1–10 (February 1999)
3. Reeber, E., Sawada, J.: Combining ACL2 and an automated verification tool to verify a multiplier. In: Manolios, P., Wilding, M. (eds.) 6th International Workshop on the ACL2 Theorem Prover and its Applications, August 2006, pp. 63–70 (2006)

---

[13] Available at http://www.frama-c.com/

4. Carreño, V.A., Miner, P.S.: Specification of the IEEE-854 floating-point standard in HOL and PVS. In: HOL 1995: 8th International Workshop on Higher-Order Logic Theorem Proving and Its Applications, Aspen Grove, UT (September 1995)
5. Harrison, J.: A machine-checked theory of floating-point arithmetic. In: Bertot, Y., Dowek, G., Hirschowitz, A., Paulin, C., Théry, L. (eds.) TPHOLs 1999. LNCS, vol. 1690, pp. 113–130. Springer, Heidelberg (1999)
6. Filliâtre, J.-C., Marché, C.: The Why/Krakatoa/Caduceus Platform for Deductive Program Verification. In: Damm, W., Hermanns, H. (eds.) CAV 2007. LNCS, vol. 4590, pp. 173–177. Springer, Heidelberg (2007)
7. Filliâtre, J.C., Marché, C.: Multi-prover verification of C programs. In: Davies, J., Schulte, W., Barnett, M. (eds.) ICFEM 2004. LNCS, vol. 3308, pp. 15–29. Springer, Heidelberg (2004)
8. Conchon, S., Contejean, E., Kanig, J., Lescuyer, S.: CC(X): Semantic Combination of Congruence Closure with Solvable Theories. In: 5th International Workshop on Satisfiability Modulo Theories (SMT 2007). Electronic Notes in Computer Science, vol. 198(2), pp. 51–69. Elsevier Science Publishers, Amsterdam (1985)
9. Hoare, C.A.R.: An axiomatic basis for computer programming. Communications of the ACM 12(10), 576–580, 583 (1969)
10. Boldo, S., Filliâtre, J.C.: Formal Verification of Floating-Point Programs. In: 18th IEEE International Symposium on Computer Arithmetic, Montpellier, France, June 2007, pp. 187–194 (2007)
11. Bertot, Y., Castéran, P.: Interactive Theorem Proving and Program Development. Coq'Art: The Calculus of Inductive Constructions. Texts in Theoretical Computer Science. Springer, Heidelberg (2004)
12. The Coq Development Team: The Coq Proof Assistant Reference Manual – Version V8.2. (2008), http://coq.inria.fr/
13. Mayero, M.: The Three Gap Theorem (Steinhauss conjecture). In: Coquand, T., Nordström, B., Dybjer, P., Smith, J. (eds.) TYPES 1999. LNCS, vol. 1956, pp. 162–173. Springer, Heidelberg (2000)
14. Daumas, M., Rideau, L., Théry, L.: A generic library of floating-point numbers and its application to exact computing. In: 14th International Conference on Theorem Proving in Higher Order Logics, Edinburgh, Scotland, pp. 169–184 (2001)
15. Boldo, S.: Preuves formelles en arithmétiques à virgule flottante. PhD thesis, École Normale Supérieure de Lyon (November 2004)
16. Boldo, S.: Pitfalls of a Full Floating-Point Proof: Example on the Formal Proof of the Veltkamp/Dekker Algorithms. In: 3rd International Joint Conference on Automated Reasoning (IJCAR), Seattle, USA, August 2006, pp. 52–66 (2006)
17. Melquiond, G.: Proving bounds on real-valued functions with computations. In: Armando, A., Baumgartner, P., Dowek, G. (eds.) IJCAR 2008. LNCS, vol. 5195, pp. 2–17. Springer, Heidelberg (2008)
18. de Dinechin, F., Lauter, C., Melquiond, G.: Assisted verification of elementary functions using Gappa. In: ACM Symposium on Applied Computing, Dijon, France, pp. 1318–1322 (2006)
19. Melquiond, G.: De l'arithmétique d'intervalles à la certification de programmes. PhD thesis, École Normale Supérieure de Lyon, Lyon, France (November 2006)
20. Bécache, E.: Étude de schémas numériques pour la résolution de l'équation des ondes. ENSTA (September 2003)
21. Boldo, S.: Floats & Ropes: a case study for formal numerical program verification. In: 36th International Colloquium on Automata, Languages and Programming. LNCS. Springer, Heidelberg (2009)

22. Kahan, W.: On the Cost of Floating-Point Computation Without Extra-Precise Arithmetic (November 2004),
    http://www.cs.berkeley.edu/~wkahan/Qdrtcs.pdf
23. Boldo, S., Daumas, M., Kahan, W., Melquiond, G.: Proof and certification for an accurate discriminant. In: 12th IMACS-GAMM International Symposium on Scientific Computing, Computer Arithmetic and Validated Numerics, Duisburg, Germany (September 2006)
24. Boldo, S.: Kahan's algorithm for a correct discriminant computation at last formally proven. IEEE Transactions on Computers 58(2), 220–225 (2009)
25. Boutin, S.: Using reflection to build efficient and certified decision procedures. Theoretical Aspects of Computer Software, 515–529 (1997)

A Comparison of Equality in Computer Algebra and Correctness in Mathematical Pedagogy

Russell Bradford[1], James H. Davenport[1], and Christopher J. Sangwin[2]

[1] Department of Computer Science
University of Bath, Bath BA2 7AY, United Kingdom
{R,J.Bradford,J.H.Davenport}@bath.ac.uk
[2] Maths Stats & OR Network, School of Mathematics
Birmingham, B15 2TT, United Kingdom
C.J.Sangwin@bham.ac.uk

Abstract. How do we recognize when an answer is "right"? This is a question that has bedevilled the use of computer systems in mathematics (as opposed to arithmetic) ever since their introduction. A computer system can certainly say that some answers are definitely wrong, in the sense that they are provably not an answer to the question posed. However, an answer can be mathematically right without being pedagogically right. Here we explore the differences and show that, despite the apparent distinction, it is possible to make many of the differences amenable to formal treatment, by asking "under which congruence is the pupil's answer equal to the teacher's?".

1   Introduction

The purpose of this paper is to examine current computer aided assessment (CAA) practise from a theoretical computer science point of view. In particular, we envisage a student being asked a mathematical question in an online automated assessment system such as the following, "what is $\frac{\mathrm{d}\sin^2 2x}{\mathrm{d}x}$?". Such online assessments are becoming rather commonplace, and an example from the STACK system [27] is shown in Figure 1. Neverthless, the field is still at the 'craft' stage, and is bedevilled by the various meanings attached to the concept of "right answer". This paper aims to provide a more formal underpinning than has existed hitherto, which we hope will allow better communication, collaboration and understanding.

In this paper we are not concerned with the very real difficulties of mathematical input, see for example [29]. Note in Figure 1, for example, the student has not been diligent in making every multiplication explicit, although the feedback interpreting this one-dimensional string in traditional format has. We are instead interested in automatically establishing equality of two expressions. Henceforth, we assume we have valid parse trees which represent the student's answer.

While a computer algebra system can typically only say "right or wrong" (coded here as T or F), a teacher using CAA normally requires three outcomes. The first is a numerical mark (also called a score), which we will normalise to

J. Carette et al. (Eds.): Calculemus/MKM 2009, LNAI 5625, pp. 75–89, 2009.
© Springer-Verlag Berlin Heidelberg 2009

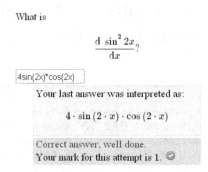

Fig. 1. Typical computer aided assessment

be out of 1. The second outcome is feedback, which is text given to the student. Figure 1 shows these two outcomes. The last outcome is a *note* for the teacher. Typically in CAA questions are randomly generated and the feedback, if given, may contain manipulated expressions which depend on the random parameters, or student's answer. Hence, if the teacher wishes to generate statistics of the outcomes, the feedback and score are not helpful. Instead the note records the logical outcome, regardless of the exact question asked, or precise answer given. Some typical answers to the question in Figure 1 are given in Table 1.

No. 1 represents the correct answer, in the expected *form*. No. 2 evaluates to the correct answer, however in practice we would not allow restatements of the question like this as a valid answer. For this example, such behaviour could be by disabling the the 'derivative' key on the input palette or similar, but in other circumstances this is harder to prevent. When we consider the equality of two expressions, we do so here with the differentiation operator being a *noun* in the student's expression. In no. 3 and no. 4 it is, arguably, clear the student is operating along sensible lines, and so some partial credit has been awarded. Perhaps some feedback "don't forget that differentiating the square gives you a factor of two, *and* the factor of $2x$ gives you a factor of two" might be appropriate in no. 3. Regarding no. 5 as correct assumes we had configured our algebra system to use `trigreduce` or the equivalent, and this emphasises the importance of a relatively sophisticated algebra engine to mark even relatively simple exercises. In no. 7 the student has integrated by mistake, and this is a situation we can anticipate and for which we may provide helpful feedback, though this is outside the main scope of *this* paper.

There are many examples of CAA which evaluate student's answers in a sophisticated mathematical way. An early example using Maple was AiM [20,31]. While this system was, and remains, useful for assessing many questions, it cannot provide some kinds of detailed feedback, particularly at an elementary level. Other Computer Algebra Systems (CAS) have been used, e.g. the STACK system uses Maxima, see [27,28]. It is not necessary to use a *mainstream* CAS to process students' responses in a CAA system. For example, CALM (see [1]), Metric (see [24]) and Aplusix (see [9]) developed their own mathematical libraries. We argue these are "computer algebra" in its broadest sense. Fundamental to all these systems is the need to compare two mathematical expressions using computer algebra of some kind. This is the issue we examine in the remainder of this paper. We note that systems like WebWorK, [16], check for mathematical correctness by evaluating at a number of points. They are therefore essentially testing extensional equivalence and so do not fall within the scope of this paper, although in practice there is probably considerable scope for a hybrid approach.

Table 1. Typical human-marked answers

No.	Student's answer	C.A.	Score
1.	$4\sin 2x \cos 2x$	T	1
2.	$\frac{d \sin^2 2x}{dx}$	T	0
3.	$2\sin 2x \cos 2x$	F	0.7
4.	$2 \times 2\sin 2x \cos 2x$	T	0.8
5.	$2\sin 4x$	T?	1
6.	$2\sin 2x \cos 2x + 2\sin 2x \cos 2x$	T	0.8
7.	$x/4 - \sin(4x)/8$	F	0

2   What Is a 'Right' Answer (Pedagogically)?

Ever since its introduction (probably in [25], see [6, (I), p. 165]), the sign '=' has had several meanings. There are at least six senses in which this synonym is currently used in traditional written notation (and even in computer algebra, equality has many meanings [12]):

(i) assignment of a value to a variable ($x = 1$);
(ii) to denote an equation yet to be solved ($x^2 + 1 = 0$);
(iii) definition of a function ($f(x) = x^2$);
(iv) as notation for λ-reduction, or combinatory reduction, as in "$\mathbf{K}MN = M$" [2, Corollary 2.1.26], and hence informally as in "what is 1+1 equal to?";
(v) as a "variant" of $\in$, as in $f(x) = O(x^2)$ [13, Section 8]; and
(vi) as a Boolean infix operator, returning either TRUE or FALSE.

It is not symmetric in uses (i), (iii) and (v), and not always in (iv).

It is the last sense we wish to examine in detail in this paper, since it is a crucial component in mathematical pedagogy, particularly in the assessment process. But here we are not concerned with = as an operator on statements of predicate logic, but in establishing the equality of mathematical expressions — yet another potential usage for this symbol.

Furthermore, we are concerned with getting, not merely "a correct" answer, but also "the right" answer. As a further example, let us assume a teacher has asked a student to

$$\text{expand out } (x+1)^2 \tag{1}$$

and the response they have from one student is $x^2 + x + x + 1$. This is "correct" in the sense that it is algebraically equivalent to $(x+1)^2$ and is in expanded form (actually two separate mathematical properties) but "incorrect" in the sense that the student has not *gathered like terms* by performing an addition $x + x$. We might say that the student has fallen at one of the two hurdles:

explicit task — do the expansion;
implicit task — do the necessary tidying up afterwards.

What about a response $2x + x^2 + 1$? This is, arguably, better in the sense that the terms are gathered, but the student here has not *ordered* terms to write their expression in the conventional form[1]. We might say that it is:

mathematically correct, in that it is equal to the question posed;
pedagogically correct, in that the student has done the task required;
aesthetically incorrect, in that there are more conventional ways of writing the answer.

We will not go further into aesthetic correctness here except to point out that it is more difficult, and subjective, than it seems — Table 2 shows two different questions with what most people would agree to be the aesthetically correct answers. Note that the aesthetic answers, while different, are *mathematically* the same, and indeed on a deeper level the questions are the same. Nevertheless, we hope the three-fold classification above is useful.

Table 2. Aesthetically correct answers

Question	Aesthetic answer
Simplify $\frac{x^5-1}{x-1}$	$x^4 + x^3 + x^2 + x + 1$
First five terms of Maclaurin series for $\frac{1}{1-x}$	$1 + x + x^2 + x^3 + x^4$

We should note that we have refrained from using the word "simplify" here. The word is ambiguous and indeed it can be used for the opposite mathematical operations. For example, in [33] the word "simplify" is usually taken to mean (e.g. p. 11, Ex 8) "*simplify by removing brackets and collecting like terms*". (Arguably "removing" should be "expanding" here). But, "simplify" is also later used to implicitly mean *factor and cancel like terms*. For example,

$$\text{p. 139, (77) Simplify} \frac{a^4 + a^2b^2 + b^4}{a^3 - b^3}.$$

---

[1] We might use the phrase "canonical form", but this has a technical meaning in computer algebra [14, p. 79].

This is typical[2] of contemporary usage. "Simplify" may mean little more than "do what I've just shown you"[3]: a more refined vocabulary is necessary.

In formal computer science however, for two equivalent expressions A and B, [8] argued that A was simpler than B when *"the length of the description of A is shorter then the length of the description of B"*. Applying his formal definitions to binary encodings of the integers and operations $+$, $\times$, $-$ and exponentiation he argued that 2^7 is more complex than 128, but that 2^8 is simpler than 256. However, given two explicit integers n and m *"it is never the case that the algebraic expression $n + m$ is simpler than the integer q equal to $n + m$."* A restricted version of this, but adequate for our purposes, is implemented in Maple's `simplify(...,size)`.

These issues might, at first, appear utterly trivial. The expert does not worry about such distinctions: a hallmark of their expertise is that they work modulo such "technicalities". But, during elementary mathematical instruction this is *the point of the work*. One application of computer algebra is to automatic computer aided assessment of mathematics and current systems go well beyond multiple choice or similar question types. In particular, students are expected to provide a mathematical expression as their answer and a computer algebra system seeks to establish its properties. On the basis of these properties feedback is provided. In our examples above the teacher might like to say, for example, *"yes, but ... "*. Only if such fine grained distinctions can be made may sufficiently sophisticated feedback be provided. But why is it important to provide such detailed feedback? Is it not sufficient to provide only a binary correct/incorrect outcome, an associated mark and give a student a summary percentage at the end? It is a paradigm in education that "feedback promotes learning". But a closer inspection reveals a much more complex picture. The meta-analysis of [21] examined about 3000 educational studies and found that over one third of feedback interventions *decreased performance*: a counterintuitive and largely ignored outcome. It is not feedback, *per se*, but the nature of the feedback which determines its effectiveness. In particular, feedback which concentrates on specific *task related* features and on how to improve is found to be effective, whereas feedback which focuses on the *self* is detrimental. A low end of test summary mark — hardly a specific form of feedback — may be interpreted as a personal and general comment on the ability of the student, whereas detailed feedback on each task points to where improvement can be made.

---

[2] The quality and variety of exercises in [33] is, in the opinion of the third author, somewhat better than many current algebra textbooks. C. O. Tuckey was a very well respected teacher, president of the Mathematical Association, author of many books and widely circulated reports (e.g. [34]) into effective teaching. This example is not a personal criticism, but rather an example of typical usage.

[3] This was brought out when the second author taught a summer school of teachers in the French "classes préparatoires" — a system that does not fit into the Bologna framework [4], but is part of the higher education system [3]. The teachers eventually admitted that "simplify" (actually "simplifier", but in this case the English and French words seem to be in close correspondence) meant "give me what I expect".

3   Theoretical Models of Computer Algebra

We have seen there is a big difference between pedagogically correct and mathematically correct, so it might be thought that it might be difficult to reconcile the two, but it is our thesis that it is possible to make the differences amenable to formal treatment. To do this we need to set up some formalism.

While computer algebra has a long history (early approaches include [19,22]), it was largely aimed at supporting specific calculations, and theoretical underpinnings were slower to emerge. The most relevant for our point of view is the "universal algebra" approach underpinning Axiom [18] and Magma [7].

This is generally considered via the 'multi-sorted approach' [32], though in fact algebra systems in practice use an 'order-sorted' approach [17], for the reasons given in [15]. Such a typed approach is very relevant for mathematics as, although the Zermelo-Frankel formalisation of mathematics is untyped, most mathematics in practice is typed, and the mathematical operations have a type structure. In this approach we introduce various operators, so a trivial construction of the integers would introduce pred and succ operators to define predecessors and successors of numbers, and we introduce axioms, such as the axiom

$$\text{pred}(\text{succ}(z)) = z. \tag{2}$$

We then let $\equiv$ be the congruential closure of our given axioms, such as (2), i.e. the smallest relation containing the axioms and satisfying

R for all t, $t \equiv t$;
S if $t_1 \equiv t_2$, then $t_2 \equiv t_1$,
T if $t_1 \equiv t_2$ and $t_2 \equiv t_3$, then $t_1 \equiv t_3$;
C if $t_1 \equiv t_2$, then

$$f(u_1, \ldots, u_{k-1}, t_1, u_{k+1}, \ldots, u_n) \equiv f(u_1, \ldots, u_{k-1}, t_2, u_{k+1}, \ldots, u_n),$$

where f is any n-ary operator.

Because of condition **C**, the operators are well-defined on the equivalence classes of $\equiv$. $\equiv$ is then said to be a **congruence**, and the corresponding logical system "equality up to $\equiv$" is said to be **congruential** [12, Definition 1]. Note that we are *not* saying that computer algebra systems *are* implemented this way, merely that one can formalise what they are doing in this structure.

Hence the first question one can ask of a computer algebra system is the following.

Question 1. Which axioms generate the congruence =?

Most algebra systems do not answer this question in full generality, with fully-typed systems such as Axiom and Magma coming the nearest, in the sense that one can inspect the code for that component of = acting on a particular sort.

As far as the authors[4] can determine, for Maple acting explicitly[5] on Laurent polynomial objects built up from the integers and variables (or expressions which behave like variables) with +, -, * and raising to explicit integer[6] powers, the following axioms generate Maple's equality.

1. Associativity of addition (essentially by regarding it as an n-ary operation).
2. Associativity of multiplication (also by regarding it as an n-ary operation).
3. Commutativity of addition.
4. Commutativity of multiplication.
5. Arithmetic evaluations on integer sub-expressions.
6. Collection of $mZ + nZ$ into $(m+n)Z$, where m, n are integers.
7. Replacing $m(Z_1+Z_2)$ by mZ_1+mZ_2 *where*[7] we have precisely a two-element product.
8. Collection of $Z^m Z^n$ into $Z^{(m+n)}$, where m, n are integers (and possibly not explicit if they are 1, though they are stored as such internally).
9. Suppression of $+0$.
10. Suppression of $*1$.
11. Replacing Z^0 by 1.

We note that this does *not* include the general distributive law for multiplication (or its corollary, the expansion of powers). Other systems may vary here.

Where practicable, algebra systems go further, and wish to create a **canonical** representation [14, p. 79], i.e. reduce every element of an equivalence class to one particular representation. In the case just mentioned above, Maple does this by applying rules (5–11) in the left-to-right sense, and storing the components of sums and products in a unique order determined by Maple's internal hash coding system [10]. It is this internal order that causes *apparently* strange results, e.g.

$$\text{simplify}\left(\frac{x^{105} - 1}{x - 1}\right) = 1 + x + x^{88} + x^{89} + x^{104} + x^{90} + \cdots.$$

Even apart from this problem, asking that a student return the canonical representation is normally too strong. For example, only one of these two expressions can be canonical:

$$\sin x \cos x \text{ or } \cos x \sin x, \tag{3}$$

but it would be a rare teacher who marked one right and the other wrong.

---

[4] They are grateful to Jacques Carette for his assistance here, but the authors bear the responsibility for any misconceptions.

[5] That is to say, where every sub-expression in the expression is built up this way, rather than by having more complicated operators 'cancel'.

[6] Including negative integers. Note that, while Maple *prints* x^ (-2) as $\frac{1}{x^2}$, it is in fact stored as x^{-2}. Similarly, x is stored as x^1.

[7] This caveat means that Maple's = relation is not actually a congruence.

4   Theoretical Models of Computer-Assisted Pedagogy

This section contains examples, which commonly occur in pedagogy, of senses in which two expressions are the same. It is rare that when assessing a question the teacher considers a single property. In fact, they make a number of separate judgements and construct feedback on the basis of these multiple outcomes. Hence, in a particular situation, the teacher might wish to consider a number of comparisons to build the appropriate feedback. Exactly what outcomes, i.e. mark, feedback and note, to assign is highly context dependent. For example, successfully establishing equivalence to an incorrect answer known to arise from a common misconception may result in no mark, but helpful feedback. Typically, the first test is to establish that the student's answer is "equal" to the correct answer given by the teacher.

There are many senses in which two expressions are considered "equal". We shall describe some of these now, from the most restrictive to the most liberal senses. We provide examples where establishing this sense of equality is a crucial component in the assessment of a mathematical question. However, we do not comment at this stage of the technical feasibility or the efficiency (i.e. computational cost) of doing so.

$==$ The most restrictive sense of equality is *absolutely identical expressions*. For example, the teacher may want exactly $x^2 + 2x + 1$. The order of the terms here is a key component. This kind of equality is closely related to the equality of the parse trees representing the two expressions.

$=_{AC}$ The next notion of equality is that up to commutativity and associativity of the basic arithmetic operations. However, the basic arithmetical operations are assumed to be *nouns*. This means they *represent* the operation, but do not perform the calculation. Hence, $2x + y =_{AC} y + 2x$ but $x + x + y \neq_{AC} 2x + y$. This is a very useful test for checking that an answer is the "same" but "simplified". Since distribution amounts to *doing* multiplication we have $2(x + 1) \neq_{AC} 2x + 2$.

$=_{\text{ext}}$ Extensional equivalence is perhaps the most common notion of equivalence. Take two expressions ex_1 and ex_2, which might contain multi-variables, be an equation, list, set, matrix, etc. If when values are assigned to the variables (from some agreed sets) they always evaluate identically then ex_1 and ex_2 are extensionally equivalent. This notion of equivalence carries over to equations and inequalities. From a technical point of view, we cannot simply evaluate an expression over an infinite set, such as the real numbers. Hence, the starting point for CAS-supported CAA was to evaluate the difference of two expressions symbolically and look for a zero result. There are significant difficulties in establishing extensional equivalence for complex (inverse) trigonometrical expressions [5], and even the apparently trivial

$$\log\left(\frac{1}{x}\right) = -\log x \tag{4}$$

is true everywhere except on a set of measure zero (the branch cut for log, traditionally $(-\infty, 0)$). It is also not immediately clear how in practice to establish whether two *equations* are extensionally equivalent. Systems of in-equalities are similarly difficult.

$=_\alpha$ Consider the following question.

> A rectangle has length 8cm greater than its width. If it has an area of 33cm^2, write down an equation which relates the side lengths to the area of the rectangle.

The kind of answer the teacher is looking for is $x(x+8) = 33$, or $l(l-8) = 33$, or indeed $l^2 + 8l - 33 = 0$, or .... One might argue that the phrase "use x to denote the length of the shortest side" could be used here to reduce the technical difficulty of automatically assessing the answer. However, this might significantly reduce the level of difficulty of the problem for the student by making a crucial choice in precisely the modeling step the problem is designed to assess. Given two expressions ex_1 and ex_2, we need to establish whether there exists a substitution of the variables of ex_2 into ex_1 which renders ex_1 extensionally (or whatever other kind of equivalence we are asking for) equivalent to ex_2. This is the idea of α-equivalence, denoted $\equiv_\alpha$ in [2, Definition 2.1.11].

This last notion of equality, i.e. modulo variable names used, might be extended to other kinds of equalities. Indeed, we might well want to know whether there exists a substitution of the variables of ex_2 into ex_1 which renders the new parse tree for the substituted version of ex_1 identical to the parse tree for ex_2.

5   Unifying the Approaches

Let us assume that we *are* trying to use a computer algebra system to get close to *intensional* equivalence, which in the pedagogic context could be described as "does it *mean* the right thing?".

Notation 1. *Let us suppose we have a question, to which the teacher has supplied a formula as the answer, f_T, and the pupil has supplied an answer f_P.*

There are various questions we might ask.

1. Is f_P *identical* to f_T, written $f_P == f_T$? We should note that this question is not trivial to answer, even at the level of MathML-Presentation [11], however, we will assume that it is answerable. If so, the answer is presumably mathematically, pedagogically and even aesthetically correct.
2. Is f_P *mathematically equal* to f_T, at least as far as our algebra system can deduce it, written $f_P =_{\text{CAS}} f_T$? Note that $=_{\text{CAS}}$ may well have to be more sophisticated than just $=_{\text{Maxima}}$ or $=_{\text{Maple}}$. To get answer 5 for Table 1 correct, we need to use $=_{\text{Maxima:trigexpand}}$ and so on. If this is the case, the student has produced a mathematically, even if not pedagogically, correct answer. If not, there are then logically two possibilities.

- The algebra system is wrong (or at least inadequate). This is one of those "should not happen" cases, but might, particularly if the problem-setter (e.g. teacher) has allowed a more powerful input syntax than was intended, as might happen if a (troublesome) pupil answered (1) with

$$x^2 + \left(\max_{n \in \mathbf{N}} \exists x, y, z \in \mathbf{N}^* x^n + y^n = z^n \right) x + 1. \tag{5}$$

Of course, it is only since the Wiles–Taylor proof that we have known that this was well-defined, never mind correct. In this category also belong all the "computer algebra is undecidable" paradoxes [26], which in practice do not crop up, and can generally be excluded syntactically — after all what business has a pupil got using syntax like (5) at this level?
- The pupil's answer is definitely wrong. This system can do no more to help, and we may wish to look at "buggy rules" (see [23]) or other techniques to determine how much partial credit to allow.

So we are left with a mathematically correct answer, and the question is "how many marks, if any, should be allocated?" (recalling that, in Table 1, one correct answer got no marks), with a supplementary of "what feedback do I need to give?". To be concrete, consider the example of Figure 1, for which we have to add more rules to (1–11) above, say the following.

12. $\frac{d u^n}{d x} = n u^{n-1} \frac{d u}{d x}$.
13. $\frac{d \sin(g(x))}{d x} = \cos(g(x)) \frac{d g(x)}{d x}$.
14. $\frac{d u v}{d x} = u \frac{d v}{d x} + v \frac{d u}{d x}$.
15. $\frac{d x}{d x} = 1$.
16. $\frac{d n}{d x} = 0$ (n a number).

Furthermore, we will only let these rules act from left to right (matters might be different if we were setting integration problems rather than differentiation ones). Let us classify the rules into three classes.

underlying: those which we believe do not really change the expression in *form* as well as substance, and which "ought" to be part of ==. Call this class $\mathcal{U}$, and the congruence generated by $\mathcal{U} \equiv_{\mathcal{U}}$. In our case, $\mathcal{U}$ would be (1)–(4).
venial: those which the pupil *ought* to have used, and which should not be left in the pupil's answer. Call this class $\mathcal{V}$, and the congruence generated by $\mathcal{U} \cup \mathcal{V} \equiv_{\mathcal{V}}$. In our case, $\mathcal{V}$ would be (5)–(11).
fatal: those which the pupil *had* to apply, and which *must* not be left in the pupil's answer. Call this class $\mathcal{F}$, and the congruence generated by $\mathcal{U} \cup \mathcal{V} \cup \mathcal{F} \equiv_{\mathcal{F}}$. In our case, $\mathcal{F}$ would be (12)–(16).

In fact, $\equiv_{\mathcal{F}}$ should be $=_{\mathrm{CAS}}$, i.e. everything that our algebra system can prove.

Then we can propose the following strategy (Table 3) for our automated marker, depending on the *finest* relation R for which $f_P R f_T$. This leads to the results in Table 4, where we see two differences from Table 1: item 3 is simply marked wrong, rather than being given 0.7, and item 5 is marked wrong, whereas

Table 3. Putative Strategy

Relation	Score	Feedback
$\equiv_{\mathcal{U}}$	1.0	Well done
$\equiv_{\mathcal{V}}$	0.8	OK, but there are better ways of writing it
$\equiv_{\mathcal{F}}$	0.0	You were meant to *do* the differentation
—	0.0	I'm sorry, that's not right

in fact it is right. The first of these is a "buggy rule" issue, as discussed earlier. As regards the second, the problem is that we have not told the system about trigonometric contraction. This involves adding a new rule

17. $\sin x \cos x = \frac{1}{2} \sin 2x$

as well as various rules about fractions, which should pretty certainly be added to class $\mathcal{V}$. Before we can discuss the correct classification of rule 17, we need a digression.

Table 4. Table 1 according to table 3

No.	Student's answer	Score	Feedback
1.	$4 \sin 2x \cos 2x$	1	Well done
2.	$\frac{\mathrm{d} \sin^2 2x}{\mathrm{d}x}$	0	*do* the differentation
3.	$2 \sin 2x \cos 2x$	0	I'm sorry, that's not right
4.	$2 \times 2 \sin 2x \cos 2x$	0.8	better ways of writing it
5.	$2 \sin 4x$	0	I'm sorry, that's not right
6.	$2 \sin 2x \cos 2x + 2 \sin 2x \cos 2x$	0.8	better ways of writing it
7.	$x/4 - \sin(4 * x)/8$	0	I'm sorry, that's not right

6   What Is a 'Right' Answer (Algorithmically)?

This is an important question. If we are following [16] and generating *questions*, we must also generate *answers*, as well as mark schemes on the lines of Table 3. Here we follow the suggestion of [8] and say that a 'right' answer a must be:

(a) equivalent under $\mathcal{U} \cup \mathcal{V} \cup \mathcal{F}$ to the question asked (a is mathematically an answer);

(b) invariant under the application of the rules in $\mathcal{F}$ (a isn't the question restated);

(c) a smallest such member, i.e. there is no a' with $a \equiv_{\mathcal{V}} a'$ and $|a'| < |a|$ for some size measure $|\cdot|$.

Quite what we take as our definition of $|\cdot|$ is not clear, and probably needs further experimentation. For the moment we are taking the number of printed characters in the answer, though a case could certainly be made for including ⁢ and &FunctionApplication; as well. In this context, assuming (17) is not in $\mathcal{F}$, we see that $2 \sin 4x$ is 'a' right answer, and indeed 'the' right answer.

6.1   Trigonometric Contraction Revisited

With this preamble, we can now ask about the classification of (17). If we are concerned merely about differentiation, we could class it with $\mathcal{U}$. If we had already taught trigonometric contraction, we could class it in $\mathcal{V}$. Alternatively, we could be more subtle. Suppose we had taught it before, and wanted to give the students a gentle reminder of it. We could define $\equiv_{\mathcal{U}'}$ to be the congruence generated by $\mathcal{U} \cup \{(17)\}$, and use the score and feedback from Table 5. Alternatively, we may have made the point before, and want to reinforce it. Then could define $\equiv_{\mathcal{V}'}$ to be the congruence generated by $\mathcal{U} \cup \mathcal{V} \cup \{(17)\}$.

Table 5. Putative Strategy Refined

Relation	Score	Feedback
$\equiv_{\mathcal{U}}$	1.0	Well done
$\equiv_{\mathcal{U}}'$	0.9	Well done, but you forgot about trigonometric contraction
$\equiv_{\mathcal{V}}$	0.8	OK, but there are better ways of writing it
$\equiv_{\mathcal{V}'}$	0.6	You *really* should use trigonometric contraction
$\equiv_{\mathcal{F}}$	0.0	You were meant to *do* the differentation
—	0.0	I'm sorry, that's not right

6.2   (1) Revisited

To resolve (1) in this framework, we would let $\mathcal{U}$ be rules (1)–(4), $\mathcal{V}$ be rules (5)–(11), and $\mathcal{F}$ be the following (interpreted as left→right rules).

18. $Z^n = Z \cdot Z^{n-1}$.
19. n-ary distributive law.

This gives us the results in Table 6.

Table 6. Question 1 according to table 3

No.	Student's answer	Congruence	Score	Feedback
1.	$x^2 + 2x + 1$	$\equiv_{\mathcal{U}}$	1	Well done
2.	$(x+1)(x+1)$	$\equiv_{\mathcal{F}}$	0	*do* the expansion
3.	$x^2 + 2x + 2$	—	0	I'm sorry, that's not right
4.	$x^2 + x + x + 1$	$\equiv_{\mathcal{V}}$	0.8	better ways of writing it

6.3   Other Issues

The way systems deal with numbers is also crucial. The representation 0.5 actually means five tenths. Students are apt to write things such as $0.5x^2 + 1/3$, a perfectly accurate representation for $x^2/2 + 1/3$, but one which does not conform to notational conventions. So, the system should establish an equality when floating point numbers are used within expressions. However, 0.33 is often not an acceptable approximation for $1/3$. Whether the teacher will reject all expressions containing floating point numbers, or whether they wish to say "yes you are correct, but we don't normally use floats" is a matter of pedagogy and should not be a work around a technical restriction. Some of these issues can be solved in our framework: for example we could have a rule converting decimals into the corresponding rationals, and place it into $\mathcal{U}$ (no penalty), $\mathcal{V}$, or possibly some $\mathcal{V}''$ with a penalty, and feedback, of its own.

Lastly, particularly for assessment of science, we would like to deal with units. Here is is necessary to establish whether the student has the correct value and correct units, or the "correct value" using different but dimensionally consistent units to that of the teacher. This area requires its own reasoning [30], but which could nevertheless be incorporated into this framework by asking which rules (OpenMath Formal Mathematical Properties) were used, and whether their use incurs a penalty.

We highlight some closely related issues. For example, a naïve set is a collection of objects, without duplication. Given the many senses of equality above, it is necessary to be explicit about how the set construction function decides on the equality of two given expressions. Rarely is extensional equivalence actually used. For example, in Maple 9.5 we have the following session:

```
> S:={x^2-1,(x-1)*(x+1)};
```

$$S := \left\{ (x-1)(x+1), x^2 - 1 \right\}$$

```
> map(simplify,S);
```

$$S := \left\{ x^2 - 1 \right\}$$

When written in different algebraic forms, Maple is happy to tolerate duplicates in sets. The default notion of equality is not that of extensional equivalence. Indeed, for many CAS the notion of equality for the purposes of sets is simply that of identity of internal representations, once any default "simplification" has been done. Here again we would need a set of rules, some underlying (e.g. order of elements in a set doesn't matter), some venial or fatal (e.g. removal of duplicates), depending on the pedagogical point being stressed.

7 Conclusion

This paper has looked at the question "how hard is it to use computer algebra to decide if a 'Calculus 101' answer is correct?", and, we hope, convinced the reader that it is rather harder than it looks. No matter which CAS we use, $=_{\text{CAS}}$ is simultaneously too strong and and too weak for what we want.

Too strong: it may decide that simple restatement of the questions are "correct", because they are algebraically equivalent to the answer.

Too weak: the built-in $=_{\text{CAS}}$ does not apply enough rules, such as trigonometric contraction.

Too coarse: it cannot produce the "mostly right but" answers we have allocated 0.8 to above. Of course, the number of marks is, of course, a matter of taste for each teacher to decide and our somewhat arbitrary allocations should not be taken too seriously.

Too inflexible: the set of rules allowed, and their status within the marking scheme, will vary *during* a single course, never mind between courses.

This is not to say that computer algebra is not useful, and indeed $=_{CAS}$ will probably be an important *component* of any scheme. But such a scheme will need to have different levels of equality. For simplicity, we have illustrated a linear hierarchy, but in practice one would probably have a lattice of various classes of "venial" rules.

References

1. Ashton, H., Beevers, C.E., Koraninski, A.A., Youngson, M.A.: Incorporating partial credit in computer aided assessment of mathematics in secondary education. British Journal of Educational Technology 37, 93–119 (2006)
2. Barendregt, H.P.: The Lambda Calculus: Its Syntax and Semantics. North-Holland, Amsterdam (1984)
3. Belhoste, B.: Historique des classes préparatoires. Exposé au Colloque de l'UPS (2003), ftp://trf.education.gouv.fr/pub/edutel/sup/cpge/historique.pdf
4. Bergen Conference of European Ministers Responsible for Higher Education. The framework of qualifications for the European Higher Education Area (2005), http://www.bologna-bergen2005.no/EN/BASIC/050520_Framework_qualifications.pdf
5. Bradford, R.J., Davenport, J.H.: Towards Better Simplification of Elementary Functions. In: Mora, T. (ed.) Proceedings ISSAC 2002, pp. 15–22 (2002)
6. Cajori, F.: A history of mathematical notations. Open Court (1928)
7. Cannon, J., Playoust, C.: An Introduction to MAGMA. Springer, Heidelberg (1997)
8. Carette, J.: Understanding expression simplification. In: Gutierrez, J. (ed.) Proceedings of ISSAC 2004, pp. 72–79 (2004)
9. Chaachoua, H., Nicaud, J.F., Bronner, A., Bouhineau, D.: APLUSIX, a learning environment for algebra, actual use and benefits. In: Proceedings of the International Congress on Mathematics Education (ICME-10), Copenhagen, Denmark (2004)
10. Char, B.W., Geddes, K.O., Gentleman, M.W., Gonnet, G.H.: The Design of MAPLE: A Compact, Portable and Powerful Computer Algebra System. In: van Hulzen, J.A. (ed.) ISSAC 1983 and EUROCAL 1983. LNCS, vol. 162, pp. 101–115. Springer, Heidelberg (1983)
11. World-Wide Web Consortium. Mathematical Markup Language (MathML) Version 2.0, 2nd edn. (2003), http://www.w3.org/TR/MathML2/
12. Davenport, J.H.: Equality in computer algebra and beyond. J. Symbolic Comp. 34, 259–270 (2002)
13. Davenport, J.H., Libbrecht, P.: The Freedom to Extend OpenMath and its Utility. Mathematics in Computer Science (to appear, 2009)
14. Davenport, J.H., Siret, Y., Tournier, E.: Computer Algebra, 2nd edn. Academic Press, London (1993)
15. Doye, N.J.: Automated Coercion for Axiom. In: Dooley, S. (ed.) Proceedings ISSAC 1999, pp. 229–235 (1999)

16. Gage, M., Pizer, A., Roth, V.: WeBWorK: Generating, delivering, and checking math homework via the Internet. In: Proc. ICTM2 international congress for teaching of mathematics at the undergraduate level (2002), http://www.math.uoc.gr/~ictm2/Proceedings/pap189.pdf
17. Goguen, J.A., Meseguer, J.: Order-sorted Algebra I: Equational deduction for multiple inheritance, polymorphism and partial operations. Theor. Comp. Sci. 105, 217–293 (1992)
18. Jenks, R.D., Sutor, R.S.: AXIOM: The Scientific Computation System. Springer, Heidelberg (1992)
19. Kahrimanian, H.G.: Analytic differentiation by a digital computer. M.A. Thesis, Temple University (1953)
20. Klai, S., Kolokolnikov, T., Van den Bergh, N.: Using Maple and the web to grade mathematics tests. In: Proceedings of the International Workshop on Advanced Learning Technologies, Palmerston North, New Zealand, December 4–6 (2000)
21. Kluger, A.N., DeNisi, A.: Effects of feedback intervention on performance: A historical review, a meta-analysis, and a preliminary feedback intervention theory. Psychological Bulletin 119(2), 254–284 (1996)
22. Nolan, J.: Analytic differentiation on a digital computer. M.A. Thesis, M.I.T. (1953)
23. O'Shea, T.: A self improving quadratic tutor. In: Sleeman, D., Brown, J.S. (eds.) Intelligent Tutoring Systems, ch. 13, pp. 309–336. Kluwer, Dordrecht (1982)
24. Ramsden, P.: Fresh Questions, Free Expressions: METRIC's Web-based Self-test Exercises. Maths Stats and OR Network online CAA series (June 2004), http://www.mathstore.ac.uk/repository/mathscaa_jun2004.pdf
25. Recorde, R.: The Whetstone of Witte. J. Kyngstone, London (1557)
26. Richardson, D.: Some Unsolvable Problems Involving Elementary Functions of a Real Variable. Journal of Symbolic Logic 33, 514–520 (1968)
27. Sangwin, C.J.: STACK: making many fine judgements rapidly. In: CAME (2007)
28. Sangwin, C.J.: What is a Mathematical Question? In: Proceedings of the JEM conference, Lisbon (Feburary 2007)
29. Sangwin, C.J., Ramsden, P.: Linear syntax for communicating elementary mathematics. Journal of Symbolic Computation 42(9), 902–934 (2007)
30. Stratford, J.D., Davenport, J.H.: Unit Knowledge Management. In: Autexier, S., Campbell, J., Rubio, J., Sorge, V., Suzuki, M., Wiedijk, F. (eds.) AISC 2008, Calculemus 2008, and MKM 2008. LNCS, vol. 5144, pp. 382–397. Springer, Heidelberg (2008)
31. Strickland, N.: Alice interactive mathematics. MSOR Connections 2(1), 27–30 (2002), http://ltsn.mathstore.ac.uk/newsletter/feb2002/pdf/aim.pdf
32. Thatcher, J.W., Wagner, E.G., Wright, J.B.: Data Type Specification: Parameterization and the Power of Specification Techniques. ACM TOPLAS 4, 711–732 (1982)
33. Tuckey, C.O.: Examples in Algebra. Bell & Sons, London (1904)
34. Tuckey, C.O.: The teaching of algebra in schools. A Report for the Mathematical Association. G. Bell & Sons (1934)

Exploring a Quantum Theory with Graph Rewriting and Computer Algebra

Aleks Kissinger

Oxford University Computing Laboratory
alexander.kissinger@comlab.ox.ac.uk

Abstract. It can be useful to consider complex matrix expressions as circuits, interpreting matrices as parts of a circuit and composition as the "wiring," or flow of information. This is especially true when describing quantum computation, where graphical languages can vastly reduce the complexity of many calculations [3,9]. However, manual manipulation of graphs describing such systems quickly becomes untenable for large graphs or large numbers of graphs. To combat this issue, we are developing a tool called Quantomatic, which allows automated and semi-automated explorations of graph rewrite systems and their underlying semantics. We emphasise in this paper the features of Quantomatic that interact with a computer algebra system to discover graphical relationships via the unification of matrix equations. Since these equations can grow exponentially with the size of the graph, we use this method to discover small identities and use those identities as graph rewrites to expand the theory.

1 Introduction

Quantomatic[1] is a tool designed to use automated graph rewriting techniques in conjunction with a computer algebra system to expand and enrich graphical theories. The primary contribution of this paper is an exposition of the methods, features and limitations of this tool, as well as a detailed look at its application to a real problem.

Quantomatic was created to explore the theory of complementary classical structures (CCS), which witness in an abstract sense the interaction of non-commuting observables in quantum states and protocols [5]. The theory of CCS is primarily expressed in the language of monoidal categories and builds upon a large body of work concerned with formalising quantum information within category theory (see for example [1,4,18,20,21,22]). Certain kinds of monoidal categories lend themselves well to graphical representations [12,13], which often provide a simpler and more intuitive interpretation of concepts like entanglement

---

[1] Quantomatic is being developed by the author, Lucas Dixon, and Ross Duncan. The source code, including a series of Mathematica[19] notebooks used in this paper, is currently available for Subversion checkout. For details, see http://dream.inf.ed.ac.uk/projects/quantomatic.

J. Carette et al. (Eds.): Calculemus/MKM 2009, LNAI 5625, pp. 90–105, 2009.
© Springer-Verlag Berlin Heidelberg 2009

and generalised information flow. In the interest of providing a minimal introduction to the motivating theory for Quantomatic, many of the results from this
body of work are given in their concrete form, where the monoidal category is
taken to be FdHilb, the category of finite-dimensional Hilbert spaces and linear
maps.

In section 2 we provide a summarised definition of typed graph rewriting, as
defined in [9,14]. In section 3, we provide an algebraic construction of classical
structures and the related notions of unbiased points, complementarity, and the
spider theorem. We then introduce a graphical notation for CCS using graphs
of black and white dots and give a short exposition of the theory in terms of
a graph rewrite system. We also explain how the algebraic constructions from
the previous section provide a concrete semantics for the graphical language. In
section 5, we explain in more detail the current and future methods that
Quantomatic employs to automatically apply rewrites and to communicate with
a computer algebra system to deduce new rules in the theory. In section 6, we
work through a short case study, using Quantomatic to build a graph identity,
then exporting a system of phase equations to Mathematica for reduction. We
then illustrate the solution to these equations being fed back into Quantomatic
to deduce a non-trivial aspect of the behaviour of a genuinely entangled state
called the W state.

2   Graph Rewrite Systems

Definition 2.1. *For a partial order $(T, \leq)$, we define a T-graph as a pair
(G, τ_G), where G is a directed graph and $\tau_G : V_G \to T$ is called the typing
function of G.*

Definition 2.2. *A T-graph homomorphism $f : G \to H$ is a graph homomorphism (f_V, f_E) with an additional component $f_T : T \to T$ that is monotone
with respect to $\leq$ and is consistent with the typing functions of G and H, i.e.
$f_T \circ \tau_G = \tau_H \circ f_V$.*

Remark 2.1. Defining T-graph isomorphisms in the usual way, we have by anti-
symmetry that $f_T = id_T$. Therefore, as in the case of untyped graphs, it is
natural to say that isomorphic T-graphs are "essentially" the same.

Remark 2.2. If we think of the elements of T as formal expressions, then it is
useful to think of $\leq$ as a unifiability or pattern-matching condition.

Definition 2.3. *If T has a bottom element $\perp$, we call this the boundary type.
We call vertices of this type boundary vertices and all other vertices internal
vertices. We say a T-graph G is well-bounded if each of its boundary vertices is
incident to exactly one edge. If the boundary vertex is the source of an edge, it
is called an input, and if it is the destination of an edge, it is called an output.*

Definition 2.4. *For the set of well-bounded T-graphs $\mathcal{G}$, a graph rewrite system
(GRS) is a set S of triples (L, R, ρ), where $L, R \in \mathcal{G}$ and ρ is a bijection on the
boundary vertices of L and R.*

To see how we actually perform rewrites, we need the concept of a matching.

Definition 2.5. *A T-graph homomorphism* $f : G \to H$ *is strict on a set of vertices* $V' \subseteq V_G$ *if for all vertices* $v \in V'$ *and all edges* $e \in H$ *that are incident to* $f_V(v)$, e *is in the image of* f_E.

Definition 2.6. *For* L, G *well-bounded, a T-graph matching* $m : L \to G$ *is a T-graph homomorphism such that* m *is injective on edges and strictly injective on internal vertices.*

For a well-bounded T-graph G, a rewrite rule (L, R, ρ), and a matching $m :$ $L \to G$ we perform a rewrite by replacing the sub-graph matched by L with R, "gluing" on ρ. For an explicit definition, see [14] or [9]. We let the resultant graph be called $G[(L, R, \rho), m]$ and make the following definition.

Definition 2.7. *For a graph rewrite system* $(\mathcal{G}, S)$, *we define the reduction relation* $\to_S$ *as follows.*

$$G \to_S H \Leftrightarrow \exists (L, R, \rho) \in S, m : L \to G. \; G[(L, R, \rho), m] \cong H$$

It is often useful to describe infinite sets of graph rewrites using pattern graphs.

Definition 2.8. *[9] A pattern graph is a well-bounded T-graph* G *with a pairwise disjoint family* $\mathcal{B}$ *of subsets of* V_G *called !-boxes (bang-boxes). We introduce a refinement order* $\preceq$ *on pattern graphs.* $G \preceq H$ *if and only if* H *can be obtained from* G *via the following !-box operations.*

copy: copies a !-box $B \in \mathcal{B}$. For $v \in B$, add a new vertex v' of the same type, as well as a new e' for every edge incident to v', including those connected to vertices outside of the !-box. Form a new !-box B' of all the new vertices.

drop: remove B from $\mathcal{B}$, leaving the vertices of G intact.

kill: remove all $v \in B$ from G and remove B from $\mathcal{B}$.

merge: if the vertices of two !-boxes $B_1, B_2 \in \mathcal{B}$ share no edges, merge them into a new !-box $B_1 \cup B_2$.

We represent !-boxes graphically by drawing a box around sets of vertices. We can extend this definition to rewrites in the obvious way. For a rewrite (L, R, ρ), we can associate !-boxes $\mathcal{B}_L$ and $\mathcal{B}_R$, subject to the following conditions.

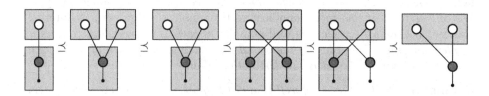

Fig. 1. Operations on !-boxes. Left to right: **copy**, **merge**, **copy**, **drop**, **kill**.

- A bijection $\mu : \mathcal{B}_L \to \mathcal{B}_R$ exists
- For each $B \in \mathcal{B}_L$, the restriction of ρ to B is a bijection from the boundary vertices in B to the boundary vertices in $\mu(B)$.

Let $r = (L, R, \rho, \mathcal{B}_L, \mathcal{B}_R, \mu)$ be a pattern rewrite. For $B \in \mathcal{B}_L$ and some operation **op** from definition 2.8, obtain a new rewrite r' by applying **op**(B) to L and **op**$(\mu(B))$ to R then applying the suitable restriction or extension of ρ. If we define pattern graph homomorphisms as T-graph homomorphisms $f : G \to H$ such that for each $B \in \mathcal{B}_G$, $f(B) \in \mathcal{B}_H$, we recover a suitable definition of pattern graph matching and rewriting.

3   Classical Structures, Algebraically

Within the standard, pure-state theory of quantum mechanics [16], vectors in a Hilbert space are called *states*[2] and self-adjoint linear maps are called *observables* [11]. States represent a physical system, and observables represent measurable, physical quantities of the system, such as position or momentum of a particle. In this paper, we shall focus on *non-degenerate* observables, or those observables O in a Hilbert space $\mathcal{H}$ with $\dim(\mathcal{H})$ distinct eigenvalues. In such a case, the (normalised) eigenvectors of O form an orthonormal basis of $\mathcal{H}$.

The eigenvalues of an observable can be interpreted as the possible outcomes of measuring it, and the associated eigenvectors are the classical physical states that correspond to these outcomes. For this reason, we call the eigenvectors of an observable *classical points*. An arbitrary quantum state is then a superposition of one or more classical points. A property that is unique to classical data (as opposed to quantum data) is that it can be copied and deleted. To witness this, we can define a *classical structure* on an observable [5].

Definition 3.1. *For a non-degenerate observable O and an orthnormal basis $\{\mathbf{u}_j\}_j$ of eigenvectors of O, a* classical structure *is a pair of maps (δ_O, ϵ_O) defined as follows.*

$$\delta_O : \mathcal{H} \to \mathcal{H} \otimes \mathcal{H} :: \mathbf{u}_j \mapsto \mathbf{u}_j \otimes \mathbf{u}_j$$
$$\epsilon_O : \mathcal{H} \to \mathbb{C} :: \mathbf{u}_j \mapsto 1$$

Let $(-)^\dagger$ be the linear adjoint (conjugate-transpose) of a map. For a classical structure $(\mathcal{H}, \delta_O, \epsilon_O)$, any vector ψ induces a map $O_\psi : \mathcal{H} \to \mathcal{H} := \delta_O^\dagger \circ (1 \otimes \psi)$.

Remark 3.1. In general, a *classical structure* is any triple $(A, \delta : A \to A \otimes A, \epsilon : A \to I)$ in a †-symmetric monoidal category $(\mathcal{C}, \otimes, I, (-)^\dagger)$ that induces a *special commutative Frobenius algebra*. We shall omit the details of this construction here (see for example [5,7]) and only consider the case where $\mathcal{C} = $ FdHilb, the category of finite-dimensional Hilbert spaces and linear maps, $\otimes$ is the normal tensor product, and $I = \mathbb{C}$, the complex numbers.

---

[2] In this paper, we shall use the terms *vector*, *state*, and *point* interchangeably.

Definition 3.2. *If O_ψ is unitary, we say ψ is* unbiased *with respect to O. If $\delta_O \circ \psi = \frac{1}{\sqrt{\dim(\mathcal{H})}} (\psi \otimes \psi)$, we say ψ is* classical *with respect to O.*

Remark 3.2. For an observable $O \in \mathbb{C} \otimes \mathbb{C}$ with an eigenbasis $\{\mathbf{u}_j\}_j$, any unbiased point can be represented as $e^{i\beta} (\mathbf{u}_1 + e^{i\alpha}\mathbf{u}_2)$ for $\alpha, \beta \in [0, 2\pi)$. Furthermore, we don't distinguish two points that differ by only a global phase, so we can write all unique unbiased points as simply $p(\alpha) := \mathbf{u}_1 + e^{i\alpha}\mathbf{u}_2$. Note also that the induced map $O_{p(\alpha)}$ is diagonal with respect to the basis $\{\mathbf{u}_j\}_j$, so it commutes with δ_O.

For classical structures, we have a result called the "spider theorem." This theorem exists in various guises in the literature [5,15,18]. Its statement for finite-dimensional Hilbert spaces is as follows, where $\mathbf{u}_j^{\otimes n}$ is the tensor product of n copies of $\mathbf{u}_j$.

Theorem 3.1. *(Spider in FdHilb) For a classical structure $(\mathcal{H}, \delta, \epsilon)$, any map $f : \mathcal{H}^{\otimes m} \to \mathcal{H}^{\otimes n}$ containing only arbitrary compositions and tensor products of $(\delta, \epsilon, \delta^\dagger, \epsilon^\dagger)$ and swaps $\sigma : \psi \otimes \varphi \mapsto \varphi \otimes \psi$ can be expressed as $\bigotimes_k f_k$, where each component $f_k : \mathcal{H}^{\otimes r} \to \mathcal{H}^{\otimes s}$ is of the form:*

$$f_k(\psi) = \begin{cases} \lambda \mathbf{u}_j^{\otimes s} & \text{if } \psi = \lambda \mathbf{u}_j^{\otimes r} \text{ for some } j, \lambda \\ 0 & \text{otherwise} \end{cases}$$

The name "spider" is due to the graphical interpretation of classical structures, which we shall see shortly. We now consider a space particularly important to quantum computing, the two-dimensional space $\mathcal{Q} := \mathbb{C} \otimes \mathbb{C}$. Elements of $\mathcal{Q}$ are referred to as quantum bits, or *qubits*. A common choice for observables in $\mathcal{Q}$ are the Pauli spin matrices. We shall focus on these two:

$$Z := \begin{pmatrix} 1 & 0 \\ 0 & -1 \end{pmatrix} \qquad X := \begin{pmatrix} 0 & 1 \\ 1 & 0 \end{pmatrix}$$

Let $\{|0\rangle, |1\rangle\}$ be an orthonormal basis of eigenvectors of Z. We fix $\{|+\rangle, |-\rangle\}$ as an eigenbasis with respect to X, where

$$|+\rangle := \tfrac{1}{\sqrt{2}} (|0\rangle + |1\rangle) \quad \text{and} \quad |-\rangle := \tfrac{1}{\sqrt{2}} (|0\rangle - |1\rangle).$$

For the two-dimensional Hilbert space $\mathcal{Q}$, we shall define the classical structures $(\mathcal{Q}, \delta_Z, \epsilon_Z)$ and $(\mathcal{Q}, \delta_X, \epsilon_X)$ as above. By remark 3.2, we shall represent the unbiased points

$$z_\alpha = |0\rangle + e^{i\alpha}|1\rangle$$
$$x_\beta = |+\rangle + e^{i\beta}|-\rangle$$

and their associated maps Z_α, X_β as above. Note that $X_0 = \sqrt{2} \cdot |0\rangle$, $X_\pi = \sqrt{2} \cdot |1\rangle$, $Z_0 = \sqrt{2} \cdot |+\rangle$, and $Z_\pi = \sqrt{2} \cdot |-\rangle$. Note that X_0 and X_π are classical points with respect to Z and vice-versa. This means that X and Z induce *complementary* classical structures.

Definition 3.3. *Two classical structures $A = (\mathcal{H}, \delta_A, \epsilon_A)$ and $B = (\mathcal{H}, \delta_B, \epsilon_B)$ are called* complementary *if the classical points of A are unbiased with respect to B and the classical points of B are unbiased with respect to A.*

Complementary classical structures (or CCS) have a rich set of identities, which are most easily explained in a graphical language.

4   Classical Structures, Graphically

We represent classical structures as the following graphs.

$$\delta_Z := \qquad \epsilon_Z := \qquad z_\alpha := \boxed{\alpha} \qquad Z_\alpha := \boxed{\alpha}$$

$$\delta_X := \qquad \epsilon_X := \qquad x_\beta := \boxed{\beta} \qquad X_\beta := \boxed{\beta}$$

Since both δ_Z and δ_X are commutative, we can compose by "gluing" graphs together on the boundary nodes and we can tensor by simple juxtaposition. Taking the adjoint $(-)^\dagger$ of a map flips its graph upside-down and reverses the sign of all the phase angles.

We employ *spiders* as a short-hand for trees of δ_Z, $\delta_Z^\dagger$, ϵ_Z, and $\epsilon_Z^\dagger$. Writing $\circ$ for graph composition, this is defined as follows.

$$sp_Z(m, n) := sp_Z'(n) \circ sp_Z'(m)^\dagger$$
$$sp_Z'(0) := \epsilon_Z$$
$$sp_Z'(1) := 1$$
$$sp_Z'(n) := (\delta_Z \otimes 1) \circ sp_Z'(n-1)$$

We define sp_X similarly, using δ_X, and ϵ_X. Note that sp_Z and sp_X both generate matrices that are sparse in their respective bases. This plays a key role in optimising matrix output from a graph. In short, we make the following definitions.

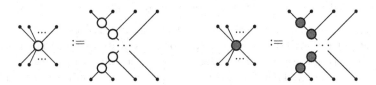

As such, we have an equivalent statement of the spider theorem.

Theorem 4.1. *(Spider, graphical) Any connected graph of a single colour is uniquely determined by the number of inputs and outputs.*

One can see the connection between this theorem and thm. 3.1 if one thinks of each f_k from the previous theorem as analogous to connected graph components. By rem. 3.2, unbiased maps commute through dots of the same colour, so we often write the sum of all the phases contained in a spider on its vertex. Thus, our most general graph components are

We define the types of these vertices as follows. For a set of free variables F, let $\mathcal{LR}[F] \subset \mathbb{Q}[F \cup \{\pi\}]$ be the set of linear polynomials with rational coefficients over $F \cup \{\pi\}$. Then, for $\mathcal{B} = \{X, Z\}$, our set of vertex types is $T = (\mathcal{B} \times \mathcal{LR}[F]) \cup \{\perp\}$. Type subsumption $\leq$ is defined as:

$$(b_1, e_1(\bar{\alpha})) \leq (b_2, e_2(\bar{\beta})) \iff (b_1 = b_2) \wedge (\exists \sigma : F \to \mathcal{LR}[F].\ e_1(\sigma(\bar{\alpha})) = e_2(\bar{\beta}))$$

This means one vertex matches another if it is the same colour and there exists a substitution on F such that e_1 can be unified with e_2. Using this typing, a vertex with m inputs, n outputs, and type (Z, α) has an interpretation as a linear map.

$$sp_Z(\alpha, m, n) := sp_Z(1, n) \circ Z_\alpha \circ sp_Z(m, 1)$$

Unrolling the recursion, we get the map:

$$sp_Z(\alpha, m, n) :: |0\rangle^{\otimes m} \mapsto |0\rangle^{\otimes n},\ |1\rangle^{\otimes m} \mapsto e^{i\alpha}|1\rangle^{\otimes n},\ other \mapsto 0$$

$sp(0, r, s)$ then corresponds to the definition of f_k given in the spider theorem for FdHilb (thm. 3.1).

Every term involving classical structures has an interpretation as a directed acyclic graph. Furthermore, every graph can be represented as a term, including those with cycles. To understand this, we introduce the notion of *map-state duality*.

Proposition 4.1. *For any finite-dimensional Hilbert space $\mathcal{H}$, linear maps from $\mathcal{H}^{\otimes m}$ to $\mathcal{H}^{\otimes n}$ are in bijective correspondence to the elements of $\mathcal{H}^{\otimes(m+n)}$.*

We shall use this to discuss states as linear maps or vice-versa. In quantum computation, it is common to view entangled states as a channel for information flow. This interpretation is the basis of protocols such as quantum teleportation[2].

Remark 4.1. We can change the direction of any internal edge in a graph of classical structures without changing the contents of its matrix representation. This is a non-trivial result of the spider theorem, the compact closure of FdHilb, and map-state duality. Therefore we can take graphs to be undirected or directed acyclic whenever it is convenient. For details, see [8,7].

It is useful to think of the spider theorem as a graph identity that lets us merge adjacent vertices of the same colour. By this philosophy, we can express the spider theorem as the following rewrite patterns.

From the complementarity of X and Z, we can derive several other rewrites [5]. In order to avoid expanding spiders, we use versions of these rules that are locally confluent with "sp." Here are two examples.

5   Quantomatic

Quantomatic was created to explore theories based on graph rewrite systems. It consists of a core written in ML and a GUI based on a Java graph library called JUNG [17]. We also make use of a small library of support functions, written as a Mathematica package, for generating matrices with free variables from Quantomatic output. The most notable functions are T[], the n-ary tensor product, xsp[] and zsp[], which implement $sp_X(\alpha, m, n)$ and $sp_Z(\beta, m, n)$ respectively, and sig[], which implements the generalised tensor swap function $\sigma(p)$ described in eqn. (2).

The features of Quantomatic fall into two operational components: a graphical component that operates on a rewrite theory, and an algebraic component that operates on the semantics of the theory. Using just the graphical component, a theory can be expanded with derived rewrites and completions as follows.

1. A potential LHS is constructed in Quantomatic.
2. Rewrites and converse rewrites are performed to yield a new RHS.
3. The rule "LHS → RHS" can be included back into the theory as a derived rewrite if only rewrites were used and as a completion if rewrites and converse rewrites were used.

We draw the distinction between derived rewrites and completions because the first has no affect on the confluence and termination properties of the rewrite theory, whereas the latter might. We can also use the graphical and algebraic components together to systematically develop a theory. In general, this process is as follows.

1. A graphical identity $G(\bar{\alpha}) = H(\bar{\beta})$ is conjectured, where $\bar{\alpha}$ and $\bar{\beta}$ are lists of free variables such as phase angles.
2. G and H are input into Quantomatic and potentially normalised with respect to a reduction strategy.
3. Quantomatic exports the interpretations of G and H as tensor terms (i.e. terms constructed with $\otimes$ and $\circ$).

4. The CAS is used to search for a substitution $\sigma : \{\bar{\beta}\} \to \mathcal{LR}[\bar{\alpha}]$ and a scalar λ such that $G(\bar{\alpha})=\lambda H(\sigma(\bar{\beta}))$. This amounts to solving a system of equations of the following form, for unknowns $\{\alpha_k\}$, with coefficients $c_{jl}, d_{jkl} \in \mathbb{R}$.

$$\left\{ \sum_j c_{jl} e^{i \sum_k d_{jkl} \alpha_k} = 0 \right\}_l \tag{1}$$

5. If a substitution is found, a new rewrite $G(\bar{\alpha}) \to \lambda H(\sigma(\bar{\beta}))$ is incorporated into the theory.

Remark 5.1. Solving general systems in the form of (1) with a CAS is a hard problem. However, sometimes a certain choice of LU or ILO representative of a graph (see sec.6) separates some or all of the variables. The techniques described here could be greatly improved by finding general methods for doing this, either graphically or algebraically.

5.1   Dag-Ification and Tensor Term Export

By remark 4.1, we can choose any ordering for the edges of a graph G.

Proposition 5.1. *For a finite graph G, we can always form a semantically equivalent directed acyclic graph G', called the dag-ification.*

Proof. Remove all self-loops from G with the "tr" rewrite, then define a strict order $<$ on the vertices of G such that any two connected vertices are comparable. Such an order always exists because, for example, a strict linear order on the vertices of G will work. Form G' from G by directing all edges such that $u \to v$ iff $u < v$.

G' depends heavily on the choice of $<$ and is not unique in general. Once we have dag-ified a graph, we can reconstruct a term using *components*. A component is a triple (i, t, o), where i and o are lists of edges and t is a tensor term generated by sp_X, sp_Z, 1, and a tensor permutation function σ, defined as follows for a permutation p.

$$\sigma(p) :: \psi_1 \otimes \psi_2 \otimes \ldots \otimes \psi_n \mapsto \psi_{p(1)} \otimes \psi_{p(2)} \otimes \ldots \otimes \psi_{p(n)} \tag{2}$$

σ acts as a generalised swap function. We can recover the "normal" swap function as $\sigma((2\ 1))$. We construct components recursively as follows.

- For a single vertex v of type $\bot$, let $[\![v]\!] = ([e], 1, [e])$, where e is the unique edge connected to v. Otherwise, v has type (b, α) with in-edges in_v and out-edges out_v, let $[\![v]\!]$ be a component $(in_v, sp_b(\alpha, \#in_v, \#out_v), out_v)$.
- For components $c_1 = (i_1, t_1, o_1)$ and $c_2 = (i_2, t_2, o_2)$ that share no edges:

$$c_1 \otimes c_2 = (i_1 \cdot i_2, (t_1 \otimes t_2), o_1 \cdot o_2)$$

- For components c_1 and c_2 where o_1 and i_2 share some edges, we can make c_1' and c_2' be such that o_1' and i_2' share all edges by padding out c_1 and c_2

with identity components $([e], 1, [e])$. After finding a permutation p such that $p(o_1') = i_2'$, we form composition as:

$$c_2 \circ c_1 = (i_1', (t_2' \circ \sigma(p) \circ t_1'), o_2')$$

A component is *total* on G if it contains every vertex in G. Any total component will then represent a valid semantic interpretation of G. For a dag-ified graph G', we can always chose a sequence of tensors and compositions that will construct a total component. Take, for instance, a partition of the vertices of G' into dag ranks. Tensoring together the vertices of each rank and composing the ranks yields a total component.

Given suitable definitions for the constructors, composition, and tensor product, we can import the generated tensor term into a computer algebra and evaluate it as a matrix.

5.2 Rewrite Strategies

Quantomatic will implement a variety of different strategies for automatic graph rewrites. We describe two strategies here, both designed to reduce the total number of vertices, edges, and non-zero phase angles.

Strategy: CONV

Method: for $R = \{\text{sp, tr, el, ha, cc}\}$, do $G \downarrow R$

Termination: G is a normal form with respect to R

It was shown in [14] that R is confluent and terminating, so this strategy always terminates with unique normal forms with respect to R.

If we consider each undirected rewrite as a set of directed rewrites, one for each possible ordering, we can make a finer-grained choice of rewrites to apply in a strategy. We'll define an example of this kind of strategy to reduce phase angles.

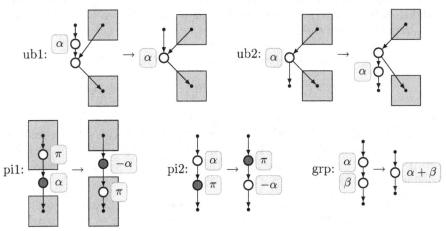

Note that these rewrites move white angles strictly downward. If we take G to be a dag-ification of an undirected graph, normalising with respect to these rewrites will always terminate. Let A_1 be the above rewrites and A_2 be the same with the colours swapped. We define the *directed angle push* strategy as follows.

Strategy: DAP

Method: repeatedly normalise first with respect to A_1, then A_2

Termination: G is a fixed point

Proposition 5.2. *DAP is a terminating strategy.*

Proof. A_1 and A_2 terminate individually for directed acyclic graphs. For each vertex $v \in V_G$, let $w(v)$ be the size of the longest directed walk from v. For all the vertices $V' \subseteq V_G$ that are labeled with a non-zero angle, let $W = \sum_{v \in V'} w(v)$. All of the above rewrites are strictly non-increasing on W. If $G \downarrow A_1 \downarrow A_2$ preserves W, then all the angles are blocked and G is a fixed point, otherwise W decreases. Since $W \geq 0$, this procedure terminates.

6   Example: Exploring Tripartite Entanglement[3]

Often when classifying entangled states, one wants to know which states can be transformed into one another using local operations.

Definition 6.1. *Two states $\psi, \varphi \in \mathcal{H}^n$ are ILO-equivalent if there exist n invertible operators $L_i : \mathcal{H} \to \mathcal{H}$ such that $(L_1 \otimes \ldots \otimes L_n) \circ \psi = \varphi$. If each L_i is unitary, ψ and φ are said to be LU-equivalent.*[4]

We can now use Quantomatic to explore the behaviour and ILO-equivalence classes of 3-qubit states. If any part of the state is separable, the problem reduces to that of 2-qubit states, which is trivial. Therefore, we shall only consider true entangled states. From [10], we know that there are only two ILO-equivalence classes, represented by the following maps:

$$GHZ :: \{|0\rangle \mapsto |0\rangle \otimes |0\rangle, |1\rangle \mapsto |1\rangle \otimes |1\rangle\} \text{ and}$$

$$W :: \{|0\rangle \mapsto (|0\rangle \otimes |1\rangle) + (|1\rangle \otimes |0\rangle), |1\rangle \mapsto (|0\rangle \otimes |0\rangle)\}$$

Note that $GHZ = \delta_Z$. Figure 2 shows the W state, up to a scalar.

For this example, we shall work mainly with the W state. We start by finding a more general form of W-like (ILO-equivalent) states. We shall then identify a rewrite rule for supplementary angles that induces a new kind of graphical behaviour for W-like states.

---

[3] Much of this case study follows notes by Bob Coecke and Bill Edwards, see [6].

[4] LU is commonly referred to in quantum information literature by the (equivalent) condition of local operations with classical communication (LOCC), and ILO is often referred to as stochastic LOCC, or SLOCC. For details, see [10].

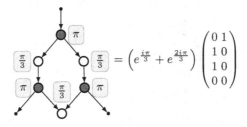

Fig. 2. The W state as a graph

6.1 Finding a Better Representative for W-Like States

We postulate that we can find an ILO-equivalence to the W state for any state of the form given by Fig. 3.

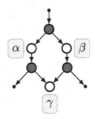

Fig. 3. A general form for the W state

To help with our search, we narrow down the types of linear maps we will look for. Since we're trying to change the unbiased-Z angles, we'll look at various kinds of Z phase shifts. We try the standard, unitary shifts Z_α as well as the two "partial" shifts Zuc_α and Zdc_α.

$$Z_\alpha := \begin{pmatrix} 1 & 0 \\ 0 & e^{i\alpha} \end{pmatrix} \quad Zuc_\alpha := \begin{pmatrix} \cos\alpha & 0 \\ 0 & 1 \end{pmatrix} \quad Zdc_\alpha := \begin{pmatrix} 1 & 0 \\ 0 & \cos\alpha \end{pmatrix}$$

Using Quantomatic and Mathematica, we discovered that conjugation by Zuc_α yields a unification. To do this, we first define Zuc_α in graphical terms (Fig. 4). We then feed the equation in Fig. 5 into Quantomatic and export the matrix terms to Mathematica.

A first call to Reduce[eq] yields the condition $(a + b + c = \pi) \wedge (\ldots)$. So, letting $c = \pi - a - b$, and calling Reduce[] again, we find a substitution

$$\sigma = \left\{ d \mapsto \frac{\pi}{2} - a - b, \; e \mapsto \frac{\pi}{2} - b, \; f \mapsto \frac{\pi}{2} - a \right\}$$

that satisfies the equation in Fig. 5. Therefore any state of the form given by (3) such that the angles sum to π is ILO-equivalent to the W state.

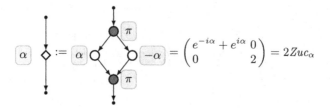

Fig. 4. The definition of Zuc_α

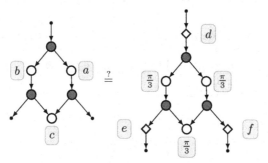

```
eq = (T[zsp[(1/3)Pi,2,0],id2[1],id2[1]].sig[3,2,0,1].T[xsp[Pi,2,1],xsp[Pi,2,1],
id2[2]].sig[1,0,3,2,4,5].T[zsp[e,1,1],zsp[-e,1,1],zsp[-f,1,1],zsp[f,1,1],id2[2]
].sig[1,0,3,2,4,5].T[xsp[Pi,1,2],xsp[Pi,1,2],id2[2]].sig[2,0,1,3].T[xsp[0,1,2],
xsp[0,1,2]].sig[1,0].T[zsp[(1/3)Pi,1,1],zsp[(1/3)Pi,1,1]].xsp[0,1,2].xsp[Pi,2,1
].sig[1,0].T[zsp[d,1,1],zsp[-d,1,1]].sig[1,0].xsp[Pi,1,2].id2[1])==k*(T[zsp[c,2
,0],id2[1],id2[1]].sig[3,0,2,1].T[xsp[0,1,2],xsp[0,1,2]].sig[1,0].T[zsp[a,1,1],
                  zsp[b,1,1]].xsp[0,1,2].id2[1])
```

Fig. 5. Equation, conjugating by Zuc, in Quantomatic, then Mathematica

6.2 Supplementary Angle Condition

To aid in the reduction of entangled states, we'll search for angles that induce the following graph disconnect.

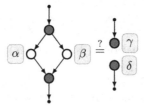

We use Quantomatic to export this identity as a matrix equation. We use Mathematica to solve for phase angles and a scaling factor k.

$$\begin{pmatrix} k = \frac{1}{e^{i\alpha}+e^{i\beta}} \\ \wedge\, \gamma = \quad \pi \\ \wedge\, \delta = \quad \pi \\ \wedge\, \pi = \quad \alpha+\beta \\ \wedge\, \pi \neq \quad \alpha-\beta \end{pmatrix} \ \vee\ \begin{pmatrix} k = \frac{1}{1+e^{i(\alpha+\beta)}} \\ \wedge\, \gamma = \quad 0 \\ \wedge\, \delta = \quad 0 \\ \wedge\, \pi = \quad \alpha-\beta \\ \wedge\, \pi \neq \quad \alpha+\beta \end{pmatrix}$$

From these conditions, we can deduce that the angles α and β on the LHS are precisely those that are both non-zero and their sum or difference is π. Therefore, we call these identities supplementary angle identities. We can now introduce two new rewrite rules to the theory.

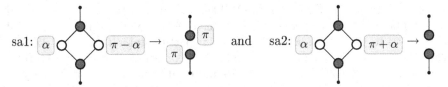

6.3   Emergent Property

Let CONV+SA be the strategy CONV, but with rules $R' = R \cup \{\text{sa1, sa2}\}$. CONV+SA still terminates because all of the rules are strictly reductive on the graph complexity. If we consider the definition of the W state as a map, it behaves like a controlled two-qubit entanglement. That is, an input of $|0\rangle$ yields $(|0\rangle \otimes |1\rangle)+(|1\rangle \otimes |0\rangle)$, which is a fully entangled state called the Einstein-Podolsky-Rosen (EPR) state. An input of $|1\rangle$ yields the separable state $|0\rangle \otimes |0\rangle$. If we note the following representations,

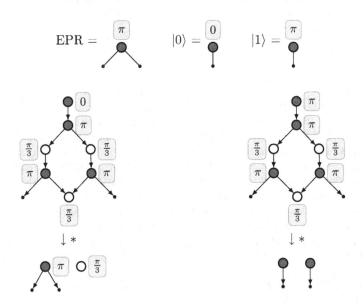

then alternate applications of CONV+SA and DAP, we get the following normalisations.

Taking any W-like state, we can apply the rewrite derived in section 6.1 in conjunction with the method above to prove that a $|1\rangle$ input yields a separable state and a $|0\rangle$ yields a state that is ILO-equivalent to the EPR state.

7   Conclusion and Future Work

We have shown that Quantomatic is already a useful tool for working with a graphical theory. The example in this paper expanded on some important properties of tripartite states that are W-like, namely that they permit a general form and behave like a controlled EPR state. To follow on from this, the natural next step is to give a similar treatment to tripartite states that are GHZ-like. When the angles in Fig. 3 do *not* add up to π, we conjecture that the state is ILO-equivalent to the GHZ state. We hope to find local linear maps that are easy to express in the graphical language to prove this identity. After this, the next step is to explore states involving more qubits or higher-dimensional generalisations of qubits such as qutrits for 3 dimensions and qudits for d dimensions.

Quantomatic itself also could benefit from a more flexible theory engine, better support for strategies and cleaner interaction with the computer algebra system. Also, the use of a general-purpose CAS can be limiting in the class of equations it can solve. A specialised CAS that can better cope with phase equations and periodic unknowns could reduce the amount of manual help that is needed to push systems of equations through a reduction routine. Also, a better implementation of sparse matrices that takes into account the properties of classical structures and tensor products could drastically reduce the resource requirements and increase the effective size limits CAS-based methods.

References

1. Abramsky, S., Coecke, B.: A categorical semantics of quantum protocols. In: Proceedings from LiCS (Feburary 2004) arXiv:quant-ph/0402130v5
2. Bennett, C.H., Brassard, G., Crepeau, C., Jozsa, R.: Teleporting an unknown quantum state via dual classical and EPR channels. Phys. Rev. Lett. (January 1993)
3. Coecke, B.: Kindergarten quantum mechanics (2005) arXiv:quant-ph/0510032v1
4. Coecke, B.: Introducing categories to the practicing physicist. Advanced Studies in Mathematics and Logic 30, 45–74 (2006) arXiv:0808.1032v1 [quant-ph]
5. Coecke, B., Duncan, R.: Interacting quantum observables. In: Aceto, L., Damgård, I., Goldberg, L.A., Halldórsson, M.M., Ingólfsdóttir, A., Walukiewicz, I. (eds.) ICALP 2008, Part II. LNCS, vol. 5126, pp. 298–310. Springer, Heidelberg (2008)
6. Coecke, B., Edwards, B.: Three qubit entanglement analysed with graphical calculus. Technical Report PRG-RR-09-03, Oxford University Computing Laboratory (2009)
7. Coecke, B., Paquette, E.O., Pavlovic, D.: Classical and quantum structuralism. Semantic Techniques for Quantum Computation, p. 43 (October 2008)

8. Coecke, B., Paquette, E.O., Perdrix, S.: Bases in diagrammatic quantum protocols (August 2008) arXiv:0808.1029v1 [quant-ph]
9. Dixon, L., Duncan, R.: Extending graphical representations for compact closed categories with applications to symbolic quantum computation. AISC/MKM/Calculemus, pp. 77–92 (June 2008)
10. Dür, W., Vidal, G., Cirac, J.I.: Three qubits can be entangled in two inequivalent ways. Phys. Rev. A 62(6) (November 2000)
11. Hannabuss, K.: An Introduction to Quantum Theory. Oxford University Press, Oxford (1997)
12. Joyal, A., Street, R.: The geometry of tensor calculus I. Advances in Mathematics 88, 55–113 (1991)
13. Kelly, M., Laplaza, M.L.: Coherence for compact closed categories. Journal of Pure and Applied Algebra 19, 193–213 (1980)
14. Kissinger, A.: Graph rewrite systems for classical structures in dagger-symmetric monoidal categories. Master's thesis, Oxford University (Feburary 2008)
15. Lack, S.: Composing props. Theory and Applications of Categories 13(9), 147–163 (2004)
16. Von Neumann, J., Beyer, R.T.: Mathematical foundations of quantum mechanics. Princeton University Press, Princeton (1996)
17. O'Madadhain, J., Fisher, D., Nelson, T.: JUNG: Java universal network/graph framework, http://jung.sourceforge.net
18. Paquette, E.O.: Categorical quantum computation. PhD thesis, Université de Montréal (Feburary 2008)
19. Wolfram Research. Mathematica (2007)
20. Selinger, P.: Dagger compact closed categories and completely positive maps (extended abstract). Electronic Notes in Theoretical Computer Science 170, 139–163 (2007)
21. Selinger, P.: A survey of graphical languages for monoidal categories (2009), http://www.mscs.dal.ca/~selinger/papers.html
22. Vicary, J.: A categorical framework for the quantum harmonic oscillator (Jun 2007), arXiv:0706.0711v2 [quant-ph]

ACL2 Verification of Simplicial Degeneracy Programs in the Kenzo System*

Francisco-Jesus Martín-Mateos[1], Julio Rubio[2], and Jose-Luis Ruiz-Reina[1]

[1] Computational Logic Group
Dept. of Computer Science and Artificial Intelligence, University of Seville
E.T.S.I. Informática, Avda. Reina Mercedes, s/n. 41012 Sevilla, Spain
{fjesus,jruiz}@us.es
[2] Dept. of Mathematics and Computation, University of La Rioja
Edificio Vives, Luis de Ulloa s/n. 26004 Logroño, Spain
julio.rubio@unirioja.es

Abstract. Kenzo is a Computer Algebra system devoted to Algebraic Topology, and written in the Common Lisp programming language. It is a descendant of a previous system called EAT (for Effective Algebraic Topology). Kenzo shows a much better performance than EAT due, among other reasons, to a smart encoding of degeneracy lists as integers. In this paper, we give a complete automated proof of the correctness of this encoding used in Kenzo. The proof is carried out using ACL2, a system for proving properties of programs written in (a subset of) Common Lisp. The most interesting idea, from a methodological point of view, is our use of EAT to build a model on which the verification is carried out. Thus, EAT, which is logically simpler but less efficient than Kenzo, acts as a mathematical model and then Kenzo is formally verified against it.

1 Introduction

The Kenzo system [8] is a Common Lisp program, developed by F. Sergeraert and devoted to Algebraic Topology. It was written mainly as a research tool and has got relevant results which have not been confirmed nor refuted by any other means. Being a compact program (around 16000 lines of Common Lisp, implementing complicated algorithms), the question of Kenzo reliability (beyond testing) came up in a natural way.

Several approaches based on Formal Methods have been used to undertake this problem, ranging from the Algebraic Specification of its data structures ([12], [7], and recently computer aided with Coq [6]) to the application of Proof Assistants to study the correctness of *algorithms* implemented in Kenzo. In this second line, the most important contributions have been the Isabelle/HOL proof of the Basic Perturbation Lemma [3] and the project by Coquand and Spiwack which is based on Constructive Type Theory and Coq [5]. As it is well-know, Coq

---

* This work has been supported by Ministerio de Educación y Ciencia, project MTM2006-06513.

J. Carette et al. (Eds.): Calculemus/MKM 2009, LNAI 5625, pp. 106–121, 2009.

proofs carry their corresponding programs, and also some work has been done to produce running code from Isabelle/HOL proofs in this context [4]. Nevertheless, the extracted programs are not comparable with the real Kenzo system, both from the efficiency and the programming languages points of view (OCaML or ML code instead of Common Lisp).

Due to this drawback of the approaches based on Isabelle and Coq, a new research line was launched, focused on the ACL2 theorem prover. ACL2 is oriented to prove properties of Common Lisp programs, and thus it could seem, at first sight, very promising to verify Kenzo. Nevertheless, since the ACL2 logic is first-order, the *full* verification of Kenzo is not possible, since it uses intensively higher order functional programming (to encode, in particular, topological spaces of infinite dimension). This observation, however, does not close the possibility of verifying first order *fragments* of Kenzo with ACL2. Some preliminary works in this line have been published in [1] and [2]. It is worth noting that in those papers we undertake the problem of verifying some Common Lisp programs about simplicial topology (in particular, algebraic manipulation and simplicial properties of Kenzo algorithms), but that no *actual* Kenzo fragment was studied.

In this paper we present for the first time the verification of a Kenzo fragment within the ACL2 theorem prover. The verified fragment is small in number of lines, but it is central to the efficiency got by Kenzo. This is compared to the predecessor of Kenzo, another Common Lisp system called EAT [15], based on the same Sergeraert's ideas, but whose performance was much poorer than that of Kenzo. One of the reasons why Kenzo performs better than EAT is because of a smart encoding of *degeneracy lists.* These combinatorial objects are usually presented in the Simplicial Topology literature as decreasing lists of natural numbers, and so they were encoded in EAT. On the contrary, in Kenzo degeneracy lists are encoded as natural numbers. Since to generate and compose degeneracy lists are operations which appear in an *exponential* manner in most Kenzo calculations (through the Eilenberg-Zilber theorem [14]), it is clear that the benefits of having a better way for storing and processing degeneracy lists is very important. But, on the negative side, the algorithms are somehow *obscured* in Kenzo, with respect to the clean and comprehensible approach in EAT. Therefore, to prove the correctness of the implementation of degeneracy algorithms in Kenzo seems to be a good test-bed to apply computer-aided formal methods.

A complete ACL2 proof of the correctness of the degeneracy programs in Kenzo is described in this paper. The main methodological contribution of the proof is, in our opinion, using EAT to build a model with respect to the verification is carried out. Thus, EAT, which is logically simpler (i.e., easier to be verified) but less efficient than Kenzo, acts as a mathematical model and then Kenzo is formally verified against it.

The organization of the rest of the paper is as follows. In Section 2, we introduce briefly both Simplicial Topology and the role of degeneracy operators in it. In Section 3, we give a brief introduction to the ACL2 system. Even if a *first order* fragment of Kenzo (and EAT) has been chosen, the Kenzo functions cannot be directly defined in ACL2 (due to Common Lisp features, like loops or destructive updates, which are not available in ACL2). Thus, in Section 4 we explain how to obtain actual ACL2 functions from Kenzo and EAT degeneracy programs, in a safe and reliable way. Sections 5 and 6 are devoted to the description of the ACL2 proof of correctness and other important properties. Finally we comment some conclusions and point out possible further work.

Due to the lack of space, we will not give here details about the proofs obtained and some function definitions will be omitted. The interested reader may consult [13], where the complete development is available.

2 The Role of Degeneracy Operators in Simplicial Topology

Simplicial Topology [14] is a subarea of Topology devoted to replace topological spaces by combinatorial models, in order to ease their study. The simplest combinatorial model of a topological space is a *simplicial complex*. Let V be a set together with a partial order $<$ on it. A *n-simplex* is a list $[v_0, v_1, \ldots, v_n]$ where $v_0 < v_1 < \ldots < v_n$ are elements of V. For each index i we consider the *i-face* operator ∂_i that given a n-simplex constructs a $(n-1)$-simplex deleting the element at position i. A *simplicial complex* K (over $(V, <)$) is a set of simplices closed with respect to the face operators.

Each n-simplex can be *realized* as an affine geometrical simplex (for instance, a 0-simplex is realized as a point, a 1-simplex as a segment, a 2-simplex as a triangle, a 3-simplex as a tetrahedron and so on). Thus, simplicial complexes are models for *triangulated spaces*, which are a class of topological spaces sufficiently large to develop much of the general and algebraic topology. Nevertheless, simplicial complexes have a severe drawback: one needs many simplices to model relatively simple spaces. For instance, to model a sphere with a tetrahedron we need 4 vertices, 6 edges and 4 triangles. Since the topological notions are quite flexible, we could use a much more efficient way of representing a sphere: by means of a triangle where all the edges and vertices are collapsed to just one point. The problem with this new representation is the "dimension jump": there is one element of dimension 2 (the triangle) and one element of dimension 0 (the point), and then this set of simplices is not closed with respect to the face operators.

The solution to this problem is to move from simplicial *complexes* to simplicial *sets*. In addition to the face operators, new operators of *degeneracy* are considered. These operators create "artificial" simplexes (with no geometrical meaning) but allowing "jumping" among dimensions. To give an idea of this sophisticated instrument let us comment briefly on how a simplicial complex can be viewed as a simplicial set. The trick is to accept simplexes that are ordered but not necessarily strictly ordered; that is, repeated elements are allowed. Then for each index i with $0 \leq i \leq n$, we define the *i-degeneracy* operator η_i that given a n-simplex constructs a $(n+1)$-simplex repeating the element at position i.

Based on this idea, we define a *simplicial set* as a graded set $\{K_q\}_{q \in \mathbb{N}}$ of *abstract* simplexes (i.e. not necessarily lists of elements) with the i-face and i-degeneracy operators, satisfying the following *simplicial identities* (see [14] for details):

$$\begin{array}{ll}
\forall i < j & \partial_i \partial_j = \partial_{j-1} \partial_i \\
\forall i \le j & \eta_i \eta_j = \eta_{j+1} \eta_i \qquad (1) \\
\forall i < j & \partial_i \eta_j = \eta_{j-1} \partial_i \\
\forall i,j & \partial_i \eta_i = Id = \partial_{j+1} \eta_j \\
\forall i > j+1 & \partial_i \eta_j = \eta_j \partial_{i-1}
\end{array}$$

A simplicial set represents a topological space in a much less expensive manner than a simplicial complex. For instance, a sphere of dimension n can be represented with just two non-degenerate simplices: one in dimension n and other in dimension 0 (geometrically, all the faces on the affine n-simplex are collapsed over a unique point, producing a topological sphere; think in a segment where the two extremes are identified, producing a circle, a 1-sphere).

A simplex is *degenerate* if it is obtained as the application of some operator η_i. It could be proved that given a simplex x there exists a unique non-degenerate simplex y and a unique strictly decreasing list of natural numbers $[i_0, i_1, \ldots, i_n]$ such that $\eta_{i_0} \eta_{i_1} \ldots \eta_{i_n}(y) = x$. (This fundamental result of Simplicial Topology has been proved in ACL2 as documented in [2]). We call this list of indices $[i_0, i_1, \ldots, i_n]$ a *degeneracy list* and we say that x is obtained applying the degeneracy list $[i_0, i_1, \ldots, i_n]$ to y.

In general, the application of degeneracy lists to simplexes is a very common operation in Kenzo, even for degenerate simplexes. Let us note that the application of a degeneracy list $[i_0, \ldots, i_n]$ to an degenerate simplex x, that is the result of applying another degeneracy list $[j_0, \ldots, j_m]$ to a non-degenerate simplex y, is the result of applying the composition of the two degeneracy lists, $[i_0, \ldots, i_n] \circ [j_0, \ldots, j_m]$, to y. The *composition* of two degeneracy lists is defined as the composition of the degeneracy operators: $[i_0, \ldots, i_n] \circ [j_0, \ldots, j_m] = \eta_{i_0} \cdots \eta_{i_n} \eta_{j_0} \cdots \eta_{j_m}$; repeatedly applying equation (1) above, this could be transformed again into a degeneracy list. The implementation in Kenzo of this composition operation is central in the system as a whole. For example, the composition of the degeneracy lists $[3, 1]$ and $[5, 3, 0]$ is $\eta_3 \eta_1 \eta_5 \eta_3 \eta_0$, and applying repeatedly the equation $\eta_i \eta_j = \eta_{j+1} \eta_i$, when $i \le j$, we successively obtain $\eta_3 \eta_6 \eta_1 \eta_3 \eta_0$, $\eta_3 \eta_6 \eta_4 \eta_1 \eta_0$, $\eta_7 \eta_3 \eta_4 \eta_1 \eta_0$ and finally $\eta_7 \eta_5 \eta_3 \eta_1 \eta_0$, that is, the degeneracy list $[7, 5, 3, 1, 0]$.

The strategy Sergeraert devised was to interpret a degeneracy list $[i_0, \ldots, i_n]$ as a binary representation of an integer. He stores the degeneracies as integers (with the corresponding memory saving) and implements the composition of degeneracy lists by using very efficient Common Lisp primitives dealing with binary numbers (like logxor, ash, and so on). This is one of the reasons why Kenzo improves dramatically the performance of its predecessor EAT. Nevertheless, this efficient composition operator called dgop*dgop in Kenzo has a more obscure semantics than its corresponding in EAT, called cmp-ls-ls. This paper is devoted to describe the certification in ACL2 of the correctness of dgop*dgop, using cmp-ls-ls as a formal specification, and then proving additional properties like equation (1) of simplicial sets or associativity of dgop*dgop.

3   An Introduction to the ACL2 System

ACL2 ([10],[11]) stands for "A Computational Logic for an Applicative Common Lisp". Roughly speaking, ACL2 is a programming language, a logic and a theorem prover. Thus, the system constitutes an environment in which algorithms can be defined and executed, and their properties can be formally specified and proved with the assistance of a mechanical theorem prover.

As a programming language, it is an extension of an applicative subset of Common Lisp[1] [16]. The logic considers every function defined in the programming language as a first-order function in the mathematical sense. For that reason, the programming language is restricted to the applicative subset of Common Lisp. This means, for example, that there are no side-effects, no global variables, no destructive updates and no higher-order features. Even with these restrictions, there is a close connection between ACL2 and Common Lisp: ACL2 primitives that are also Common Lisp primitives behave exactly in the same way, and this means that, in general, ACL2 programs can be executed in any compliant Common Lisp.

The ACL2 logic is a first-order logic, in which formulas are written in prefix notation; they are quantifier–free and the variables in it are implicitly universally quantified. The logic includes axioms for propositional logic (with connectives implies, and,...), equality (equal) and those describing the behavior of a subset of primitive Common Lisp functions. Rules of inference include those for propositional logic, equality and instantiation of variables. The logic also provides a principle of *proof by induction* that allows to prove a conjecture splitting it into cases and inductively assuming some instances of the conjecture that are smaller with respect to some well–founded measure.

An interesting feature of ACL2 is that the same language is used to define programs and to specify properties of those programs. Every time a function is defined with defun, in addition to define a program, it is also introduced as an axiom in the logic (whenever it is proved to terminate for every input). Theorems and lemmas are stated in ACL2 by the defthm command, and this command also starts a proof attempt in the ACL2 theorem prover.

The main proof techniques used by ACL2 in a proof attempt are simplification and induction. The theorem prover is automatic in the sense that once defthm is invoked, the user can no longer interact with the system. However, in a deeper sense the system is interactive: very often non-trivial proofs are not found by the system in a first attempt and then it is needed to guide the prover by adding lemmas, suggested by a preconceived hand proof or by inspection of failed proofs. These lemmas are then used as rewrite rules in subsequent proof attempts. This kind of interaction with the system is called "The Method" by its authors.

4   From Kenzo and EAT to ACL2

Before giving the ACL2 definition of the composition of degeneracy lists (and the statements of the theorems we have proved), let us present the Kenzo code for

---

[1] In this paper, we will assume familiarity with Common Lisp.

that operation. As we have said before, Kenzo deals with degeneracy lists using a smart encoding. Basically, every degeneracy list can be seen as the natural number whose binary notation represents the characteristic function of the set of elements of the list. Let us explain this with an example: the degeneracy list $[5, 3, 0]$ can equivalently be seen as the binary list $[1, 0, 0, 1, 0, 1]$ in which 1 is in position i if the number i is in the degeneracy list, 0 otherwise. This list, seen as a binary number in the reverse order, is the natural number 41. Thus, Kenzo encodes the above degeneracy list as 41.

Let us now explain how Kenzo implements composition of degeneracy lists. This is better understood if we think first in the binary representation. Let us consider the composition of the degeneracy lists $[3, 1]$ and $[5, 3, 0]$. Applying repeatedly the equation $\eta_i \eta_j = \eta_{j+1} \eta_i$, when $i \leq j$, we obtain $[7, 5, 3, 1, 0]$. Using binary notation, this means that the composition of $[0, 1, 0, 1]$ and $[1, 0, 0, 1, 0, 1]$ is $[1, 1, 0, 1, 0, 1, 0, 1]$. In general (although it is not obvious), composition between two degeneracy lists in binary notation can be described as sequentially replacing the 0's in the first list by the successive elements of the second list, until one of the lists is exhausted; and then completing the result with the remaining elements of the other list.

As we have said before, Kenzo does not directly use the binary notation: it uses the natural number that this binary notation represents. Common Lisp logical operations on numbers, like logxor and ash, are used to reflect the corresponding manipulations on binary lists. The following is *the real* Common Lisp code of Kenzo for composition of degeneracy lists[2]:

```
(defun dgop*dgop (dgop1 dgop2)
  (declare (type fixnum dgop1 dgop2))
  (let ((dgop 0) (bmark 0))
    (declare (fixnum dgop bmark))
    (loop (when (zerop dgop1)
            (return-from dgop*dgop (logxor dgop (ash dgop2 bmark))))
          (when (zerop dgop2)
            (return-from dgop*dgop (logxor dgop (ash dgop1 bmark))))
          (cond ((evenp dgop1)
                 (when (oddp dgop2) (incf dgop (2-exp bmark)))
                 (setf dgop2 (ash dgop2 -1)))
                (t (incf dgop (2-exp bmark))))
          (setf dgop1 (ash dgop1 -1))
          (incf bmark))))
```

This definition receives as input two fixnum natural numbers dgop1 and dgop2 (encoding two degeneracy lists) and executes a loop that uses two local variables dgop and bmark storing respectively the (partially computed) result, and the number of elements of dgop already scanned. When one of the degeneracy lists is exhausted, it stops and returns the concatenation of dgop and the remaining elements of the other list. Otherwise, it updates the two local variables (according to the values of the first elements of dgop1 and dgop2) and executes again the body of the loop, removing the first element of dgop1, and eventually the first element of dgop2.

---

[2] In the following, to distinguish ACL2 code from general Common Lisp code, we will use italics for the latter.

Since the function `dgop*dgop` deals with natural numbers, we emphasize again that logical operators are used to treat them as binary lists. For example, computing (`logxor dgop (ash dgop2 bmark)`) is equivalent to "concatenate" `dgop` and `dgop2` (since `bmark` is the length of `dgop`). Or, for example, (`ash dgop1 -1`) is equivalent to remove "the first element" of `dgop1`. These logical operators on fixnum numbers are usually computed in Common Lisp very efficiently, and this is one of the reasons why Kenzo performs much better than EAT. On the negative side, the formal verification of `dgop*dgop` seems a hard task. In the rest of this section, we present a definition of `dgop*dgop` in ACL2 (trying to keep as close as possible to its original Common Lisp definition) and we state the theorem we want to prove in order to increase our confidence in the way Kenzo deals with degeneracy lists.

4.1 Definition of `dgop*dgop` in ACL2

Since the ACL2 programming language is a subset of Common Lisp, the definition of `dgop*dgop` in ACL2, based on the above Common Lisp code, is quite direct. Nevertheless, due to the applicative nature of ACL2, there are some things that have to be defined in a different (but equivalent) way. In particular, the only way to iterate in ACL2 is by means of recursion. Thus, we use an auxiliary recursive definition implementing the internal loop, trying to be as faithful as possible to the original version. Also, since destructive updates are not allowed in ACL2, we consider the local variables `dgop` and `bmark` as extra input parameters. Finally, since ACL2 functions have to be total, we have to define a result just in case the inputs were not of the intended type ((`type fixnum dgop1 dgop2`)). Taking all these considerations into account, the following is the ACL2 definition of the loop[3]:

```
(defun dgop*dgop-loop (dgop1 dgop2 dgop bmark)
  (if (and (natp dgop1) (natp dgop2))
      (cond ((zerop dgop1) (logxor dgop (ash dgop2 bmark)))
            ((zerop dgop2) (logxor dgop (ash dgop1 bmark)))
            ((evenp dgop1)
             (dgop*dgop-loop (ash dgop1 -1) (ash dgop2 -1)
                             (if (oddp dgop2)
                                 (+ dgop (ash 1 bmark))
                                 dgop)
                             (+ bmark 1)))
            (t (dgop*dgop-loop (ash dgop1 -1) dgop2
                               (+ dgop (ash 1 bmark)) (+ bmark 1)))))
      0))
```

Finally, the ACL2 definition of `dgop*dgop` is a call to the above auxiliary function, with suitable initial zero values for `dgop` and `bmark`:

```
(defun dgop*dgop (dgop1 dgop2)
  (dgop*dgop-loop dgop1 dgop2 0 0))
```

We claim that the ACL2 version is faithful with the original Kenzo definition, since we have tried to keep it as similar as possible. As we have said, the fact that

---

[3] (`2-exp n`) returns 2^n, the same as (`ash 1 n`); we will comment more on this in the conclusions.

ACL2 is a subset of Common Lisp makes this translation almost direct. Anyway, we strengthened our claim by an intensive testing. Since both definitions can be executed on any compliant Common Lisp, it was very easy to (successfully) test that they return the same result for all pairs of inputs n and m, with $n, m \leq 10000$.

4.2   Stating the Correctness Property of dgop*dgop

We now describe how we state the main theorem about the correctness of the above ACL2 definition. It is clear that we would like to prove that the function computes, using the natural number encoding, the composition of two degeneracy lists. Degeneracy lists have been defined in Section 2 as strictly decreasing lists of natural numbers.

Therefore, the first thing we have to define in ACL2 is the composition of degeneracy lists, represented as strictly decreasing lists. That will be our "specification" of the intended behavior of any implementation of composition of degeneracy lists. Note that, in principle, the computation carried out by dgop*dgop has nothing to do with the definition given in section 2. While the original definition is based on successive applications of degeneracy operators onto a degeneracy list, the function dgop*dgop makes some kind of "merge" between the binary representation of degeneracy lists. As we have said before, the EAT system (the Kenzo predecessor) used strictly decreasing lists of natural numbers to represent degeneracy lists. Thus, it seems a good idea to prove the equivalence (modulo the change of representation) of the Kenzo function with the corresponding EAT function.

In EAT, the composition of degeneracy lists is defined as an iterative application of the equation $\eta_i \eta_j = \eta_{j+1} \eta_i$, when $i \leq j$. The following is the real code for the EAT definition of composition. Note that the auxiliary function cmp-s-ls implements the application of a degeneracy operator to a degeneracy list; this function is iteratively used by the main function cmp-ls-ls to define composition:

```
(defun cmp-s-ls (s ls)
  (declare (type fixnum+ s) (type list ls))
  (do ((p ls (cdr p))
       (rsl (list ) (cons (1+ (car p)) rsl)))
      ((endp p) (nreverse (cons s rsl)))
    (declare (type list p rsl))
    (when (> s (car p)) (return (nreconc (cons s rsl) p))))))

(defun cmp-ls-ls (ls1 ls2)
  (declare (type list ls1 ls2))
  (do ((p (reverse ls1) (cdr p))
       (rsl ls2 (cmp-s-ls (car p) rsl)))
      ((endp p) rsl)
    (declare (type list p rsl)))))
```

We have defined ACL2 versions of these functions, trying to keep as faithful as possible with the original code. Analogously to the previous subsection, a do loop has to be replaced by auxiliary recursive functions. These are our ACL2 definitions for composition of degeneracy lists:

```
(defun cmp-s-ls-do (s p rsl)
  (cond ((endp p) (reverse (cons s rsl)))
        ((> s (car p)) (nreconc (cons s rsl) p))
        (t (cmp-s-ls-do s (cdr p) (cons (1+ (car p)) rsl)))))
(defun cmp-s-ls (s ls)
  (cmp-s-ls-do s ls nil))

(defun cmp-ls-ls-do (p rsl)
  (cond ((endp p) rsl)
        (t (cmp-ls-ls-do (cdr p) (cmp-s-ls (car p) rsl)))))
(defun cmp-ls-ls (ls1 ls2)
  (cmp-ls-ls-do (reverse ls1) ls2))
```

Again, the translation from the real Common Lisp code of EAT to the ACL2 version is quite straightforward. But in order to strengthen even more our confidence in this "model", we did intensive testing, checking that they compute the same results for 100000 inputs randomly generated.

We now have to define functions relating the encoding used by Kenzo and the representation of degeneracy list used by EAT. First, the function `dgop-ext-int` transforms a degeneracy list represented as a strictly decreasing list of natural numbers (checked by the function `dgl-p`) to its corresponding representation as a natural number. Note the use of logical arithmetic operators:

```
(defun dgop-ext-int (ext-dgop)
  (if (dgl-p ext-dgop)
      (if (endp ext-dgop)
          0
          (logxor (ash 1 (car ext-dgop))
                  (dgop-ext-int (cdr ext-dgop))))
      0))
```

We also define the function `dgop-int-ext`, its inverse. For that, we use an auxiliary recursive definition that simulates a `do` loop, with the input variables `rslt` and `bmark`, that work as extra parameters for storing respectively the result (partially) computed and the number of binary digits analyzed. The main function simply calls this auxiliary definition with suitable initial values for the extra parameters. This is our ACL2 definition:

```
(defun dgop-int-ext-do (dgop rslt bmark)
  (if (natp dgop)
      (if (zerop dgop)
          rslt
          (if (oddp dgop)
              (dgop-int-ext-do (ash dgop -1) (cons bmark rslt) (1+ bmark))
              (dgop-int-ext-do (ash dgop -1) rslt (1+ bmark))))
      nil))

(defun dgop-int-ext (dgop)
  (if (natp dgop)
      (dgop-int-ext-acc dgop nil 0)
      nil))
```

It should be emphasized that these definitions are defined trying to be as close as possible to the corresponding Kenzo definitions of these operations (although due to the lack of space we do not include here this part of the Kenzo code).

We have now defined all the functions that we need for stating the correctness property of `dgop*dgop`. This property expresses that for every pair of degeneracy lists represented as strictly decreasing lists of natural numbers, the result of computing `dgop*dgop` on their corresponding encoding as natural numbers is equal to the encoding of the result of the composition carried out by the EAT system. The following is the corresponding ACL2 theorem stating that property:

```
(defthm dgop*dgop-cmp-ls-ls
  (implies (and (natp dgn1) (natp dgn2))
           (equal (dgop*dgop dgn1 dgn2)
                  (dgop-ext-int (cmp-ls-ls (dgop-int-ext dgn1)
                                (dgop-int-ext dgn2)))))))
```

In the next section, we will explain how we carried out a mechanical proof of this theorem in ACL2.

5   The Proof: Transforming the Domain

The ACL2 proof of the above result is not simple, mainly for two reasons. Firstly, the functions `dgop*dgop` (Kenzo) and `cmp-ls-ls` (EAT) deal with different representations of degeneracy lists. Secondly, the Kenzo function implements an algorithm which is not intuitive and quite different from the algorithm of the EAT version, which is closely related to the mathematical definition. A suitable strategy to attack the proof is to try to solve the above two questions separately. Thus, it seems natural to consider an intermediate representation of degeneracy lists based on the binary lists described at the beginning of Section 4.

Our plan will be to define a function `dgb*dgb` implementing composition of degeneracy lists represented as binary lists, following the same algorithm than `dgop*dgop`, except for the use of this intermediate representation. This will allow us to prove the equivalence of `dgop*dgop` and `dgb*dgb` dealing only with the encoding aspects. After that, we will prove the equivalence of `dgb*dgb` and `cmp-ls-ls`, focusing only on the algorithmic aspects of the Kenzo definition. Schematically, if $\mathcal{D}_g^L$ denotes the set of strictly decreasing lists of natural numbers, $\mathcal{D}_g^B$ the set of binary lists and $\mathcal{D}_g^N$ the set of natural numbers, we will prove the commutativity of the following diagram (in which, for the sake of clarity, we have omitted the names for the encoding and decoding functions between the different representations):

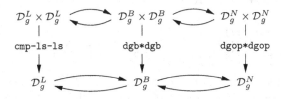

The rest of this section will be devoted to explain our proof. We will describe separately the properties concerning each of the three representations (or domains) considered, and finally we will show how we compose all these results to achieve the desired theorem.

5.1 The Domain $\mathcal{D}_g^L$

The tail-recursive definitions of functions `cmp-s-ls` and `cmp-ls-ls` that we
adopted (since we wanted to keep as close as possible to the EAT version) are not
the best option for reasoning in ACL2. Therefore, we proved that these functions
verify the following simple recursive schemata, which are much more suitable for
the induction heuristics of the ACL2 prover:

```
(defthm cmp-s-ls-recursive
  (equal (cmp-s-ls s ls)
         (cond ((endp ls) (list s))
               ((> s (car ls)) (cons s ls))
               (t (cons (1+ (car ls)) (cmp-s-ls s (cdr ls)))))))

(defthm cmp-ls-ls-recursive
  (equal (cmp-ls-ls ls1 ls2)
         (cond ((endp ls1) ls2)
               (t (cmp-s-ls (car ls1) (cmp-ls-ls (cdr ls1) ls2))))))
```

If we use these alternative recursive schemata, instead of the original versions,
it turns out that some properties of `cmp-ls-ls` (for example, its associativity)
can be proved very easily in ACL2.

5.2 The Domain $\mathcal{D}_g^B$

The following is the definition of the functions `dgl->dgb` and `dgb->dgl` implementing the change of representation between the domains $\mathcal{D}_g^L$ and $\mathcal{D}_g^B$:

```
(defun dgb-pos (n)
  (cond ((zp n) '(1))
        (t (cons 0 (dgb-pos (- n 1))))))

(defun dgb-app (dgb1 dgb2)
  (cond ((endp dgb1) dgb2)
        ((endp dgb2) dgb1)
        (t (cons (car dgb1) (dgb-app (cdr dgb1) (cdr dgb2))))))

(defun dgl->dgb (dgl)
  (cond ((endp dgl) nil)
        (t (dgb-app (dgl->dgb (cdr dgl)) (dgb-pos (car dgl))))))

(defun 1+ls (lst)
  (cond ((endp lst) nil)
        (t (cons (1+ (car lst)) (1+ls (cdr lst))))))

(defun dgb->dgl (dgb)
  (cond ((endp dgb) nil)
        ((eql (car dgb) 0) (1+ls (dgb->dgl (cdr dgb))))
        (t (append (1+ls (dgb->dgl (cdr dgb))) (list 0)))))
```

These functions are bijections between $\mathcal{D}_g^L$ and $\mathcal{D}_g^B$, and therefore they are
indeed a change of representation between two different encodings. The follo-
wing theorems proved in ACL2 establish that fact (where `dgl-p` and `dgb-p` are
respectively functions checking membership to $\mathcal{D}_g^L$ and $\mathcal{D}_g^B$):

```
(defthm dgb->dgl-dgl->dgb
  (implies (dgl-p dgl)
           (equal (dgb->dgl (dgl->dgb dgl)) dgl)))

(defthm dgl->dgb-dgb->dgl
  (implies (dgb-p dgb)
           (equal (dgl->dgb (dgb->dgl dgb)) dgb)))
```

Now we define the composition of degeneracy lists in $\mathcal{D}_g^B$, following the same algorithmic procedure than in the Kenzo version. Recall from section 4 that the result of composing two binary lists $d_{g\,1}^B \circ d_{g\,2}^B$, is obtained by sequentially replacing the 0's in the first list by the successive elements of the second list, until one of the two lists is exhausted; and then completing the result with the remaining elements of the other list. This is precisely what the recursive function dgb*dgb does:

```
(defun dgb*dgb (dgb1 dgb2)
  (if (and (dgb-p dgb1) (dgb-p dgb2))
      (cond ((endp dgb1) dgb2)
            ((endp dgb2) dgb1)
            ((eql (car dgb1) 0)
             (cons (car dgb2) (dgb*dgb (cdr dgb1) (cdr dgb2))))
            (t (cons 1 (dgb*dgb (cdr dgb1) dgb2))))
    nil))
```

As expected, the following theorem can be proved, establishing the equivalence of the function dgb*dgb in $\mathcal{D}_g^B$ and the function cmp-ls-ls, based on the EAT version:

```
(defthm dgb*dgb-cmp-ls-ls
  (implies (and (dgb-p dgb1) (dgb-p dgb2))
           (equal (dgb*dgb dgb1 dgb2)
                  (dgl->dgb (cmp-ls-ls (dgb->dgl dgb1)
                                       (dgb->dgl dgb2)))))))
```

5.3   The Domain $\mathcal{D}_g^N$

The following is the definition of the functions dgb->dgn and dgn->dgb implementing the change of representation between the domains $\mathcal{D}_g^B$ and $\mathcal{D}_g^N$. Recall that this is simply done by considering the elements of $\mathcal{D}_g^B$ as the reverse of the binary notation of a natural number:

```
(defun dgb->dgn (dgb)
  (cond ((endp dgb) 0)
        (t (+ (car dgb) (ash (dgb->dgn (cdr dgb)) 1)))))

(defun dgn->dgb (dgn)
  (cond ((zp dgn) nil)
        (t (cons (if (evenp dgn) 0 1) (dgn->dgb (ash dgn -1))))))
```

As in the previous subsection, it can be proved that these functions define a change of representation between two different encodings. That is, they are bijections between $\mathcal{D}_g^B$ and $\mathcal{D}_g^N$, as established by the following theorems:

```
(defthm dgb->dgn-dgn->dgb
  (implies (natp dgn)
           (equal (dgb->dgn (dgn->dgb dgn)) dgn)))

(defthm dgn->dgb-dgb->dgn
  (implies (dgb-p dgb)
           (equal (dgn->dgb (dgb->dgn dgb)) dgb)))
```

The following theorem proves the equivalence between the functions dgb*dgb and dgop*dgop (modulo the change of representation):

```
(defthm dgop*dgop-dgb*dgb
  (implies (and (natp dgn1) (natp dgn2))
     (equal (dgop*dgop dgn1 dgn2)
         (dgb->dgn (dgb*dgb (dgn->dgb dgn1)
                  (dgn->dgb dgn2))))))
```

5.4   Correctness of Kenzo Degeneracy Lists Composition

Having proved the equivalences of both dgop*dgop and cmp-ls-ls with the intermediate function dgb*dgb, the final step is to prove that the functions dgop-int-ext and dgop-ext-int are equivalent to the composition of the corresponding transformations between the domains $\mathcal{D}_g^L$, $\mathcal{D}_g^B$ and $\mathcal{D}_g^N$. That is:

```
(defthm dgop-int-ext-dgb->dgl-dgn->dgb
  (implies (natp dgn)
          (equal (dgop-int-ext dgn) (dgb->dgl (dgn->dgb dgn)))))
```

```
(defthm dgop-ext-int-dgb->dgn-dgl->dgb
  (implies (dgl-p dgl)
          (equal (dgop-ext-int dgl) (dgb->dgn (dgl->dgb dgl)))))
```

Now we only have to glue together the different pieces, using these last properties and the equivalences of the previous subsections, and finally obtaining the main correctness property we wanted to prove:

```
(defthm dgop*dgop-cmp-ls-ls
  (implies (and (natp dgn1) (natp dgn2))
          (equal (dgop*dgop dgn1 dgn2)
              (dgop-ext-int (cmp-ls-ls (dgop-int-ext dgn1)
                  (dgop-int-ext dgn2))))))
```

That is, we have established the correctness of the function dgop*dgop (based on the Kenzo code) with respect to the specification defined by the function cmp-ls-ls (based on the EAT code).

6   Translating Properties from $\mathcal{D}_g^L$ to $\mathcal{D}_g^N$

Once we have proved the main correctness theorem, it is easy to prove a property about dgop*dgop by first proving the property about cmp-ls-ls (which is usually much simpler) and then translating it to dgop*dgop, by means of the above theorem. Let us illustrate this with an example.

One of the properties assumed as an axiom in the definition of simplicial set is the following equation between degeneracy operators: $\eta_i\eta_j = \eta_{j+1}\eta_i, \forall i \leq j$. That is, for every pair of natural numbers $i \leq j$ and every degeneracy list d_g, we have $\eta_i(\eta_j(d_g)) = \eta_{j+1}(\eta_i(d_g))$. With respect to the composition of degeneracy lists, the property is stated as follows:

$$\forall i, j \in \mathbb{N}, \forall d_g \in \mathcal{D}_g : i \leq j \rightarrow [i] \circ ([j] \circ d_g) = [j+1] \circ ([i] \circ d_g)$$

This property should be true for any implementation of the composition operation. In particular, that is the case for the function cmp-ls-ls, as shown in the theorem below. ACL2 can prove this property immediately, using the simpler recursive schemata presented in subsection 5.1:

```
(defthm cmp-ls-ls-property
  (implies (<= i j)
           (equal (cmp-ls-ls (list i) (cmp-ls-ls (list j) dg))
                  (cmp-ls-ls (list (+ 1 j)) (cmp-ls-ls (list i) dg)))))
```

Now, this allows us to prove in a quite straightforward manner the corresponding version of this theorem for dgop*dgop. It is an easy consequence of the above theorem, the theorem dgop*dgop-cmp-ls-ls of the previous section, and the relations between the functions dgop-int-ext and dgop-ext-int and the transformations between the domains $\mathcal{D}_g^L$, $\mathcal{D}_g^B$ and $\mathcal{D}_g^N$. This results in the following:

```
(defthm dgop*dgop-property
  (implies (and (natp dgop) (natp i) (natp j) (<= i j))
           (equal (dgop*dgop (dgop-ext-int (list (+ 1 j)))
                             (dgop*dgop (dgop-ext-int (list i)) dgop))
                  (dgop*dgop (dgop-ext-int (list i))
                            (dgop*dgop (dgop-ext-int (list j)) dgop)))))
```

In a similar way, we have also proved that the function dgop*dgop is associative, from the same property for cmp-ls-ls, whose proof is very simple.

7   Conclusions and Further Work

In this paper we have described an ACL2 proof of the correctness of a first-order fragment of the Kenzo system. Concretely, the Kenzo programs dealing with degeneracy lists have been certified. Although the verified fragment is short in number of lines, it is important for efficiency reasons in Kenzo.

As for the proof effort, we followed "The Method" described in [10] and outlined in Section 3. We recall that although every proof attempt of the system is fully automatic, the system can be seen as interactive, since the appropriate lemmas has to be previously proved in order to obtain the proof. Thus, and following "The Method", when a proof attempt of a result failed, we inspected its output to discover which lemmas were needed to lead the prover to a successful proof. Most of the resulting proofs were carried out by induction and simplification. In most cases, the heuristics of the prover were able to automatically find a suitable induction scheme. Only in a few cases, we needed to supply an explicit induction scheme. All our interaction with the prover resulted in a collection of 27 definitions and 112 theorems. It is interesting to point out that we had no preconceived proof in mind, and that all we did was to follow the suggestions from the failed proof attempts. We urge the interested reader to consult the complete development in [13].

From a practical point of view, a library of results about the logical arithmetic operands was very useful. This library contains results previously proved by other ACL2 users and comes with the ACL2 distribution. Thus, we think this is a good example of reusability.

It is also worth pointing out the methodology devised to formally verify a system written in Common Lisp. Obviously, we are not directly verifying the actual code, due to limitations of the ACL2 programming language. But since ACL2 is a subset of Common Lisp, a "model" very closely related to the original code can be defined. And since ACL2 functions can be executed in any compliant Common Lisp, we could do intensive testing to strengthen even more the assumption that our model is faithful. Another remarkable point is our use of a previous version of Kenzo, called EAT, as a main component of the specification of the intended properties. This also increases the trust in the correctness of the methods appearing in both EAT and Kenzo.

In this paper, we have not dealt with efficiency issues. In fact, there are two technical details in the verified ACL2 function which make it less efficient than its Kenzo counterpart. The first one is that the ACL2 function `dgop*dgop-loop` has an explicit test in its body, checking that its first two arguments are natural numbers. This is needed to ensure termination of the function on *all possible inputs*, as required by the principle of definition of the ACL2 logic ([10]). That explicit condition has a negative impact on the efficiency of the ACL2 algorithm, since it is checked in *every* recursive call. The other technical detail that affects efficiency has to do with how 2^n is computed by Kenzo: at initialization, a lookup table is built, with the powers of two until the biggest fixnum; after that, every time a power of two is needed, the function `(2-exp n)` used by Kenzo simply retrieves the value from position `n` of the table. In contrast, our ACL2 function computes `(ash 1 n)`, which is an equivalent, but less efficient method. Although in a first stage we have not dealt with this issues, both technical details can be solved in ACL2, using the `defexec` and `stobj` features, respectively (see the users manual in [11] for details). We plan to introduce these improvements in our ACL2 code and formally verify them, in order to obtain a certified algorithm, comparable in efficiency with the Kenzo algorithm.

The work presented here is a first approach on using ACL2 with the purpose of certifying fragments of an already implemented system as Kenzo (that is, we do not want to reimplement the system, but to certify the existing code). This case study shows the benefits of the fact that both systems (Kenzo and ACL2) deal with the same programming language. Nevertheless, further research has to be done to test how ACL2 will behave with two important issues not addressed here. First, the mathematical theory underlying most Kenzo computations (algebraic topology) is more complex than the needed by this example. Second, Kenzo intensively uses higher order programming, not allowed in ACL2.

Thus, our future work will follow two lines of research: first, we intend to formalize in ACL2 some results of algebraic topology, which will allow us to tackle more difficult algorithms, such as the one extracted from the Eilenberg-Zilber theorem, where the combinatorial explosion of simplicial degeneracy *shuffles* appears [14]. Or for example to verify the Kenzo builders (to construct spheres, Moore spaces, projective spaces, . . .) which are used as primitives in the *reKenzo* graphical user interface [9]. Other line of research will be to study how we can model the higher-order features used by the Common Lisp Kenzo code, in a first-order Common Lisp ACL2 code; after that, we will be able to compare this approach with the alternative of using higher-order theorem provers like Coq, PVS or HOL.

Acknowledgements

In memoriam of Mirian Andrés, our colleague and, much more important, our friend.

References

1. Andrés, M., Lambán, L., Rubio, J.: Executing in Common Lisp, Proving in ACL2. In: Kauers, M., Kerber, M., Miner, R., Windsteiger, W. (eds.) MKM/ CALCULEMUS 2007. LNCS, vol. 4573, pp. 1–12. Springer, Heidelberg (2007)
2. Andrés, M., Lambán, L., Rubio, J., Ruiz-Reina, J.L.: Formalizing Simplicial Topology in ACL2. In: ACL2 Workshop 2007, University of Austin, pp. 34–39 (2007)
3. Aransay, J., Ballarin, C., Rubio, J.: A Mechanized Proof of the Basic Perturbation Lemma. Journal of Automated Reasoning 40, 271–292 (2008)
4. Aransay, J., Ballarin, C., Rubio, J.: Extracting Computer Algebra Programs from Statements. In: Moreno Díaz, R., Pichler, F., Quesada Arencibia, A. (eds.) EUROCAST 2005. LNCS, vol. 3643, pp. 159–168. Springer, Heidelberg (2005)
5. Coquand, T., Spiwack, A.: Towards Constructive Homological Algebra in Type Theory. In: Kauers, M., Kerber, M., Miner, R., Windsteiger, W. (eds.) MKM/ CALCULEMUS 2007. LNCS, vol. 4573, pp. 40–54. Springer, Heidelberg (2007)
6. Domínguez, C.: Formalizing in Coq Hidden Algebras to Specify Symbolic Computation Systems. In: Autexier, S., Campbell, J., Rubio, J., Sorge, V., Suzuki, M., Wiedijk, F. (eds.) AISC 2008, Calculemus 2008, and MKM 2008. LNCS, vol. 5144, pp. 270–284. Springer, Heidelberg (2008)
7. Domínguez, C., Lambán, L., Rubio, J.: Object Oriented Institutions to Specify Symbolic Computation Systems. Rairo - Theoretical Informatics and Applications 41, 191–214 (2007)
8. Dousson, X., Rubio, J., Sergeraert, F., Siret, Y.: The Kenzo Program, Institut Fourier (1999), http://www-fourier.ujf-grenoble.fr/~sergerar/Kenzo/
9. Heras, J., Pascual, V., Rubio, J.: Mediated Access to Symbolic Computation Systems. In: Autexier, S., Campbell, J., Rubio, J., Sorge, V., Suzuki, M., Wiedijk, F. (eds.) AISC 2008, Calculemus 2008, and MKM 2008. LNCS, vol. 5144, pp. 446–461. Springer, Heidelberg (2008)
10. Kaufmann, M., Manolios, P., Moore, J.S.: Computer-Aided Reasoning: An Approach. Kluwer Academic Publishers, Dordrecht (2000)
11. Kaufmann, M., Moore, J.S.: ACL2 Home Page, http://www.cs.utexas.edu/users/moore/acl2
12. Lambán, L., Pascual, V., Rubio, J.: An Object-Oriented Interpretation of the EAT System. Applicable Algebra in Engineering, Communication and Computing 14, 187–215 (2003)
13. Martín-Mateos, F.J., Ruiz-Reina, J.L., Rubio, J.: ACL2 verification of simplicial degeneracy programs in the Kenzo system, http://www.cs.us.es/~fmartin/acl2/kenzo
14. May, J.P.: Simplicial Objects in Algebraic Topology. Van Nostrand (1967)
15. Rubio, J., Sergeraert, F., Siret, Y.: EAT: Symbolic Software for Effective Homology Computation, Institut Fourier (1997), ftp://ftp-fourier.ujf-grenoble.fr/pub/EAT
16. Steele Jr., G.L.: Common Lisp The Language, 2nd edn. Digital Press (1990)

Combined Decision Techniques for the Existential Theory of the Reals*

Grant Olney Passmore and Paul B. Jackson

LFCS, University of Edinburgh
{s0793114,pbj}@inf.ed.ac.uk

Abstract. Methods for deciding quantifier-free non-linear arithmetical conjectures over $\mathbb{R}$ are crucial in the formal verification of many real-world systems and in formalised mathematics. While non-linear (rational function) arithmetic over $\mathbb{R}$ is decidable, it is fundamentally infeasible: any general decision method for this problem is worst-case exponential in the dimension (number of variables) of the formula being analysed. This is unfortunate, as many practical applications of real algebraic decision methods require reasoning about high-dimensional conjectures. Despite their inherent infeasibility, a number of different decision methods have been developed, most of which have "sweet spots" – e.g., types of problems for which they perform much better than they do in general. Such "sweet spots" can in many cases be heuristically combined to solve problems that are out of reach of the individual decision methods when used in isolation. RAHD ("Real Algebra in High Dimensions") is a theorem prover that works to combine a collection of real algebraic decision methods in ways that exploit their respective "sweet-spots." We discuss high-level mathematical and design aspects of RAHD and illustrate its use on a number of examples.

1   Introduction

RAHD ("Real Algebra in High Dimensions") is a tool for proving high-dimensional (many variable) quantifier-free non-linear theorems in the language of ordered fields over real closed fields (RCF)[1]. While the elementary theory of

---

* The authors would like to thank Bruno Dutertre, Sam Owre, John Rushby, N. Shankar, Hassen Saïdi, and Ashish Tiwari of SRI International for their ever helpful support and guidance for this project, including a visiting fellowship for the first author under which this work was originated. This fellowship was supported by NASA Cooperative Agreement NNX08AC59A and by NSF SGER Grant No. CNS-0823086.

[1]   A real closed field is a structure elementarily equivalent to the real number line with respect to a language of quantified boolean combinations of real polynomial equations and inequalities. This language is often referred to as the *language of ordered rings*.

the real numbers in this language[2] is decidable, it is fundamentally infeasible: any general decision method must take time exponential in the dimension of the formula being analysed. This is unfortunate, as many important applications of decision methods over RCF require reasoning about high-dimensional conjectures. To combat this difficulty, we focus not on the general decision problem, but instead upon deciding certain classes of sentences that arise in practice. We exploit the fact that most RCF decision methods have "sweet spots," e.g. types of problems for which they perform much better than they do in general, and such "sweet spots" can be heuristically combined to solve problems that are out of reach of the individual decision methods when used in isolation.

For the examples in this article, we focus especially upon the combination of a "sweet-spot" in the cylindrical algebraic decomposition procedure (for topologically open constraints), Gröbner basis calculations, Sturm chains, simple Positivstellensatz witnesses, and dimensional reduction techniques stemming from sound approximations to computations induced by real radical ideals.

1.1 Background

Since Tarski [17] established that the full elementary theory of RCF admits quantifier elimination (QE) by giving a QE procedure of non-elementary complexity, perhaps the most important practical[3] break-through for theorem proving over RCF has been the cylindrical algebraic decomposition (CAD) algorithm devised by Collins in the 1970s [3]. Collins' community of students have been prolific in their theoretical and practical improvements to CAD over the last twenty-five years, culminating in Brown's actively supported QEPCAD-B [1] system. In addition to QEPCAD-B, versions of CAD have also been implemented in Mathematica, REDLOG/Reduce, and Maple, and a perusal of the literature shows CAD implementations finding vigorous use in many sciences, both of applied and theoretical character. In addition to CAD, a number of other RCF QE procedures have been developed and implemented in working tools since the 1980s, including Weispfenning's method of virtual term substitution [21] (as implemented in Reduce/Redlog), and the Harrison-McLaughlin proof

---

[2] Classical work on RCF decision problems usually takes place over the language of *ordered rings,* not the language of *ordered fields,* as partial functions such as real division complicate the model theory. RAHD supports the language of ordered fields by pre-processing away division in literals in terms of equivalent multiplicative constraints.

[3] This is not to say that Collins's break-through was only of a practical nature: The geometrical insight contained within the CAD procedure has led to huge advances in the topological and model-theoretic understanding of ordered structures admitting quantifier elimination (e.g. o-minimality theory and tame topology). In these cases, the properties of RCF exploited by the CAD procedure have been generalized into the notion of "cellularly decomposable structures" and now bear rich mathematical fruits. [20]

producing version of the Cohen-Hörmander method (in the HOL Light proof assistant) [15]. The version of Weispfenning's method available in Reduce/Redlog (implemented and enhanced by Dolzmann and Sturm [7]) performs especially well on many difficult high-dimensional problems (see Section 3).

While work on improved RCF QE methods is of lasting importance, for many practical applications, full elementary QE is overkill. For these domains (such as program analysis [19], hardware verification [11], hybrid systems [18], and even ongoing large-scale projects in formalised mathematics [10]), simply *deciding the satisfiability* of boolean combinations of polynomial equations and inequalities over the real numbers is often sufficient. This problem is equivalent to QE for the purely ∃ (dually purely ∀) sentential fragment of the elementary theory of RCF, in which all formulas considered are sentences consisting only of a single block of non-alternating quantifiers. As will be discussed in Section 1.2, this fragment of the elementary theory of RCF admits an exponential speed-up over general RCF QE, though this fact has unfortunately not led to algorithms for this fragment that are in practice superior to the known general QE ones.

That said, the fundamental observation driving our current research is the following: While all RCF decision methods are constrained by the known complexity lower-bounds, most decision methods have types of problems for which they perform much better than their worst-case time complexity analysis would suggest. We refer to these more feasible fragments of a decision method's input domain as "sweet spots" of the decision method under investigation. We work in RAHD to orchestrate the heuristic combination of a number of decision methods for different RCF fragments by attempting to automatically massage difficult problem instances into equisatisfiable sequences of simpler problems that fit within known "sweet spots" of the decision methods RAHD provides. We will describe RAHD in more detail in Section 2.1.

1.2   Existential Decisions over Real Closed Fields

Let us make the fundamental decision problem in which we are interested precise.

Question 1 (Fundamental Decision Problem). *Let* $t_1(\boldsymbol{x}), \ldots, t_k(\boldsymbol{x}) \in \mathbb{R}(\boldsymbol{x})$

where $\mathbb{R}(\boldsymbol{x}) = \{\frac{p}{q} \mid p, q \in \mathbb{R}[x_1, \ldots, x_n], q \neq 0\}$. *Let* φ *be a quantifier-free boolean combination of atoms of the form* $(t_i \odot 0)$ *with* $\odot \in \{<, \leq, =, \geq, >\}$ $(1 \leq i \leq k)$. *Is* φ *satisfiable over* $\mathbb{R}^n$? *That is, does*

$$\langle \mathbb{R}, +, -, *, <, 0, 1 \rangle \models \exists x_1 \ldots x_n(\varphi)?$$

Though this decision problem has long been known to have a positive solution, available general purpose decision methods, such as the aforementioned CAD [3] or the Cohen-Hörmander procedure [2], have a very high worst-case time complexity. For instance, when given an n-dimensional existential RCF conjecture such as $\exists \boldsymbol{x} \varphi(\boldsymbol{x})$, the computing time for the CAD algorithm is dominated by a function doubly exponential in n. For the full theory of RCF (e.g., with arbitrary quantification), such doubly-exponential lower-bounds are known to be tight for quantifier elimination: Due to Davenport-Heinz [4], there are known families of n-dimensional RCF formulas of length $O(n)$ whose only quantifier-free equivalences must contain polynomials of degree $2^{2^{\Omega(n)}}$ and of length $2^{2^{\Omega(n)}}$.

For the question of decidability over the purely existential fragment, a number of more efficient algorithms have been proposed, including those of Grigor'ev-Vorobjov [9] and Renegar [16]. Both of these decision methods deliver a theoretical exponential speed-up over CAD for the existential fragment of RCF, requiring time dominated by a function only singly exponential in the underlying dimension. Despite their apparent complexity-theoretic advantages, neither procedure appears to have been implemented. Analysis by Hong [12] suggests that even though the procedures of Grigor'ev-Vorobjov and Renegar are theoretically advantageous over CAD (e.g., for sufficiently large inputs), for practically sized examples, CAD remains superior. This is due to infeasibly large constant factors lurking behind the asymptotic analyses of these singly exponential procedures.

2   The **RAHD** System

In this section, we begin by touching upon the mathematics of some of RAHD's different proof techniques. Then, we turn to the question of how these techniques can be fruitfully combined to solve problems beyond their reaches when they are used in isolation. This leads us to a discussion of design and control aspects of RAHD's automatic heuristic proof procedure, the so-called "waterfall."

2.1   RCF Decision-Theoretic "Sweet Spots"

We begin by describing some decision-theoretic "sweet spots" that are combined and exploited in RAHD.

Full and Fragmentary Open CAD (CAD-MD). In 1993, McCallum showed how CAD could be heavily modified and made much more efficient if the semialgebraic set defined by the formula being analysed was known to be an open set in the Euclidean topology on $\mathbb{R}^n$ [14]. This can be guaranteed if the formula under analysis consists only of strict inequality relations. The basic idea is that in these cases, rational sample points can be selected from the cells in the CAD, avoiding the need for costly irrational algebraic number computations that are in general required during normal CAD operation.

This more feasible fragment of CAD is used heavily in RAHD, and as will be discussed in Section 2.3, RAHD's design is centered around breaking difficult problems into simpler ones that are in a precise sense closer to being able to make use of this "sweet spot." When RAHD encounters a problem that it can see falls into this fragment, Brown's QEPCAD-B is used in a special mode[4] designed for making decisions about the emptiness of open sets.

Gröbner bases. Though one usually associates Gröbner basis calculations with decisions over algebraically closed fields, much use can be made out of them for existential RCF decisions. The reason is two-fold: First, if it can be shown that a collection of equational constraints in a formula being analysed is unsatisfiable over $\mathbb{C}^n$, then this collection is of course unsatisfiable over $\mathbb{R}^n$. Even if one is not so lucky, all term reductions induced by the equational constraints when interpreted over $\mathbb{C}^n$ are still valid over $\mathbb{R}^n$, and so one can make use of Gröbner basis calculations to simplify polynomials in the midst of RCF decisions. In practice, these reductions are often a tremendous boon, leading to simplified terms that are then more amenable to subsequently invoked decision methods. As will be seen in Section 2.2, RAHD also exploits Gröbner basis calculations for a number of other techniques centered around reducing a problem's dimension, and these techniques can derive new equations which then further enhance the reduction power of the Gröbner bases for the problem. RAHD uses a caching mechanism for sharing already computed Gröbner bases and term reductions among these different system components. RAHD has its own implementation of algorithms for computing with ideals that tend to work well for non-linear problems in six or less variables. When a non-linear problem is handed to RAHD in greater than six variables, RAHD will attempt to use the CoCoA computer algebra system for its ideal computations if it is available.

Simple Positivstellensatz certificates. Recent work exploiting Stengle's Weak Positivstellensatz, an RCF analogue of Hilbert's Nullstellensatz for algebraically closed fields, has led to a number of computational advances for the fundamental decision problem, and a simplified form of this result is currently used in RAHD.

Theorem 1 (Stengle's Weak Positivstellensatz). *Given* $p_i(\boldsymbol{x}), q_i(\boldsymbol{x}), s_i(\boldsymbol{x})$ $\in \mathbb{R}[\boldsymbol{x}] = \mathbb{R}[x_1, \ldots, x_n]$, *the conjunctive constraint system*

$$\varphi = \begin{pmatrix} p_1(\boldsymbol{x}) = 0 \wedge \ldots \wedge p_k(\boldsymbol{x}) = 0 \\ \wedge \; q_1(\boldsymbol{x}) \geq 0 \wedge \ldots \wedge \; q_l(\boldsymbol{x}) \geq 0 \\ \wedge \; s_1(\boldsymbol{x}) > 0 \wedge \ldots \wedge \; s_m(\boldsymbol{x}) > 0 \end{pmatrix}$$

is unsatisfiable over $\mathbb{R}^n$ *iff*

$$\exists \mathcal{P}(\boldsymbol{x}) \in Ideal(p_1, \ldots, p_k)$$
$$\exists \mathcal{Q}(\boldsymbol{x}) \in Cone(q_1, \ldots, q_l)$$
$$\exists \mathcal{R}(\boldsymbol{x}) \in Mon(s_1, \ldots, s_m)$$
$$s.t. \;\; \mathcal{P} + \mathcal{Q} + \mathcal{R}^2 = -1$$

---

[4] To use this mode, one replaces all existential quantifiers in the constraint in question with special "exists infinitely many" quantifiers which are equivalent over open sets and cause many irrational algebraic number calculations to be avoided.

where

$$Ideal(p_1,\ldots,p_k) = \left\{ \sum_{i=1}^{k} p_i q_i \mid q_i \in \mathbb{R}[\boldsymbol{x}] \right\},$$

$$Cone(q_1,\ldots,q_l) = \left\{ r + \sum_{i=1}^{v} t_i u_i \mid r, t_i \in \sum(\mathbb{R}[\boldsymbol{x}])^2, \ u_i \in Mon(q_1,\ldots,q_v) \mid v \in \mathbb{N} \right\},$$

$$Mon(s_1,\ldots,s_m) = \left\{ \prod_{i=1}^{m} (s_i)^j \mid j \in \mathbb{N} \right\},$$

$$\sum(\mathbb{R}[\boldsymbol{x}])^2 = \left\{ \sum_{i=1}^{v} (p_i)^2 \mid p_i \in \mathbb{R}[\boldsymbol{x}] \mid v \in \mathbb{N} \right\}.$$

Given such an unsatisfiable φ, the equation $\mathcal{P} + \mathcal{Q} + \mathcal{R}^2 = -1$ is called a *Positivstellensatz (Psatz) certificate* for φ's unsatisfiability. It is the finding of such certificates that has seen impressive modern advances: Building upon the work of Choi, Lam, Powers, Reznick, and Wörmann, Parrilo developed in his 2000 dissertation a feasible method for finding Psatz certificates, by translating the search for them into a convex optimization (semidefinite programming) problem that is in principle amenable to polynomial-time interior point methods. The complexity-theoretic difficulty lies in the fact that each polynomial-time solvable optimization problem only searches for a Psatz certificate up to a set multivariate total degree, and the known bounds on such degrees are at least triply exponential in φ. Still, many difficult problems are routinely solved by Psatz methods. In our experience, however, it is rare[5] to find a φ feasibly solvable by Psatz methods and not by CAD, and so we currently use only a very simplified restriction of Psatz methods in RAHD that does no convex optimization but is nevertheless useful for many practical problems.

The family of simple Positivstellensatz witnesses RAHD considers are those which contain constraints of the following form:

$$(p = 0) \ \text{s.t.} \ RC(p) > 0 \wedge p \in \left\{ \sum_{j=1}^{k} m^2 \mid m = \prod_{j=1}^{n} c_j x_j^{\alpha(j)} \mid c_j \in \mathbb{Q}, \alpha(j), k \in \mathbb{N} \right\}$$

where $p \in \mathbb{Q}[\boldsymbol{x}]$ and $RC(p)$ is the degree-zero rational coefficient of p. RAHD also looks for related Psatz certificates when p is constrainted via a strict inequality. Such certificates can be found simply by examining the degree parity of the monomials arising in p after polynomial canonicalization, which is a polynomial-time process all terms in RAHD undergo before Gröbner basis calculations.

---

[5] For those interested in foundational theorem proving, however, Psatz methods do have a clear advantage over CAD: The fact that Psatz certificates are simple algebraic identities guarantees that if found, they can be verified and translated into foundational proof objects easily. Harrison has made use of this fact for his REAL_SOS tactic in HOL Light [11].

Sturm chains for univariate constraints. Sturm's theorem prescribes a method for counting the number of real roots of a univariate polynomial p in a half-open interval through the analysis of the number of sign changes (SC) in the so-called *Sturm chain* induced by p and the interval in question. Despite the fact that Sturm chains have well-known pathological numerical properties, we have found them to be useful in RAHD on many practical problems. The family of formulas amenable to Sturm chain analysis in RAHD are those which contain constraints of the form $[(p > 0) \ \wedge \ (x > q_1) \ \wedge \ (x < q_2)]$ s.t. $p(\frac{q_2-q_1}{2}) \leq 0$, $\#SC(p, (q_1, q_2)) = 0$, and $q_1 < q_2$ with $p \in \mathbb{Q}[x]$.

2.2 Dimensional Reduction Techniques

As the time complexities of many algebraic decision methods applicable to RCF formulas are at least exponential in dimension, methods for reducing the dimension of a formula under analysis (e.g. the elimination of variables) are crucial to the feasible solubility of high-dimensional problems. This process is a central component of RAHD's automatic waterfall procedure, which spawns lower-dimensional equisatisfiable subgoals and calls itself recursively upon them when the dimension of a constraint under analysis has been reduced. Because of this subgoaling process, dimensional reductions in RAHD are allowed to eliminate a variable in terms of *any finite number* of lower-dimensional values, with each such value inducing a separate subgoal to be checked for satisfiability. For example, the transformation

$$\left(x^4 - 16 = 0 \ \wedge \ \mathcal{P}(x, y, z)\right) \ \rightarrow \ \left(\mathcal{P}(2, y, z) \ \vee \ \mathcal{P}(-2, y, z)\right)$$

is a valid[6] dimensional reduction for an RCF predicate $\mathcal{P}$ from $\mathbb{R}^3$ to $\mathbb{R}^2$.

Approximating real radical ideals. Over $\mathbb{C}$, the correspondence between ideals and varieties is elucidated by Hilbert's Strong Nullstellensatz.

Theorem 2 (Hilbert's Strong Nullstellensatz)

$$\mathcal{I}_{\mathbb{C}[\boldsymbol{x}]}(\mathcal{V}_{\mathbb{C}}(\mathcal{I}_{\mathbb{C}[\boldsymbol{x}]}(p_1, \dots, p_k))) = \sqrt{\mathcal{I}_{\mathbb{C}[\boldsymbol{x}]}(p_1, \dots, p_k)}$$
$$= \left\{p \in \mathbb{C}[\boldsymbol{x}] \mid \exists i \in \mathbb{N} \ s.t. \ p^i \in \mathcal{I}_{\mathbb{C}[\boldsymbol{x}]}(p_1, \dots, p_k)\right\}.$$

That is, given $p_i, q \in \mathbb{C}[\boldsymbol{x}]$ the decision problem for universal Horn formulas over $\mathbb{C}$ can be reduced to an ideal membership check for radical ideals as follows:

---

[6] It is interesting to note that while this reduction is valid over $\mathbb{R}$, it is not valid over $\mathbb{C}$: Consider the non-trivial quartic roots of unity $e^{\pm \frac{\pi i}{2}}$. Thus, this reduction would not be computed from any Gröbner basis for $\sqrt{\mathcal{I}(x^4 - 16)}$. This motivates the need for computations over *real* radical ideals, so that such reductions can be deduced.

$$\langle \mathbb{C}, +, -, *, 0, 1 \rangle \models \left(\bigwedge_{i=1}^{k} p_i = 0 \right) \implies q = 0$$

$$\Longleftrightarrow$$

$$q \in \sqrt{\mathcal{I}_{\mathbb{C}[\boldsymbol{x}]}(p_1, \ldots, p_k)}$$

which can then be effectively solved using Buchberger's algorithm. Modulo ideal membership checking, the important step here is the construction, from a set of generators for an ideal $\mathfrak{I}$ over $\mathbb{C}[\boldsymbol{x}]$, to a set of generators for the *radical ideal* containing $\mathfrak{I}$. This is a classically studied problem in algebraic geometry and most modern computer algebra systems provide efficient algorithms for complex ideal radicalization [13].

Over $\mathbb{R}$, however, things are not so simple. The algebraic structure analogous to a radical ideal for real algebraic varieties, the so-called *real radical ideal*, has to take into account the order structure of $\mathbb{R}$ by incorporating polynomial summands that are sums of squares. That is, letting $\mathfrak{I} = \mathcal{I}_{\mathbb{R}[\boldsymbol{x}]}(p_1, \ldots, p_k)$,

$$\mathcal{I}_{\mathbb{R}[\boldsymbol{x}]}(\mathcal{V}_{\mathbb{R}}(\mathfrak{I})) = \sqrt[\mathbb{R}]{\mathfrak{I}} = \left\{ p \in \mathbb{R}[\boldsymbol{x}] \mid p^{2i} + s \in \mathfrak{I} \mid s \in \sum (\mathbb{R}[\boldsymbol{x}])^2, \ i \in \mathbb{N} \right\}.$$

Known methods for transforming an ideal into its real radical are computationally infeasible for non-trivial problems, so we seek a method in **RAHD** that *approximates* real radicalization to obtain some practically useful membership decisions in an efficient way. The following is an example inference rule that captures some of this desired behavior:

$$\frac{(p_1 = 0), \ldots, (p_k = 0) \in \mathcal{C} \quad \exists m \in \mathbb{N} \text{ s.t. } (x^{2m} - q) \in \mathcal{I}(p_1, \ldots, p_k) \quad \sqrt[2m]{q} \in \mathbb{Q}}{\mathcal{C} \models (x = \sqrt[2m]{q} \ \vee \ x = -\sqrt[2m]{q})}$$

where $\mathcal{C}$ is a conjunctive RCF constraint. Note that the q above need not be guessed, as if the antecedent holds, one can obtain q by reducing x^{2m} modulo $GB(p_1, \ldots, p_k)$. This reduction process can be done incrementally for heuristically selected terms in a formula, with m ranging from 1 to some degree bound computed as a function of the generators of $GB(p_1, \ldots, p_k)$. Indeed, many important dimensional reductions in **RAHD** are accomplished in this way.

Reverse Rabinoswitch encodings. Over $\mathbb{R}$, the following equivalences hold:

$$pq = 0 \iff (p = 0 \ \vee \ q = 0) \tag{1}$$

$$\sum_{i=1}^{k} p_i^2 = 0 \iff \bigwedge_{i=1}^{k} p_i = 0. \tag{2}$$

Equivalence (1) can be used to reduce a constraint φ to an equisatisfiable disjunction $\varphi_1 \vee \varphi_2$ s.t. (i) $dim(\varphi_i) < dim(\varphi)$ in the special case that pq is a monomial, or (ii) each φ_i has additional polynomial vanishing assumptions (and hence Gröbner bases with enhanced reduction power) in the general case. Equivalence (2) can be used to reduce a constraint φ to an equisatisfiable constraint

φ' which contains new, simpler equations which enrich the equational structure of φ and increase the reduction power of the Gröbner bases it induces. It is interesting to note that the integral domain property (1) is valid over $\mathbb{C}$, while (2) is valid only over $\mathbb{R}$. As in the restricted version of the Psatz used by RAHD, instances of (2) are recognized only for sums of squares of monomials.

Gröbner bases induced solving for variables and interval techniques. RAHD contains a number of additional techniques that use Gröbner bases and term orderings to solve for variables by orienting equations, constructing elimination ideals, and dividing equations through by polynomials that RAHD can prove, via spawned subgoals, to be non-vanishing under constraint hypotheses. RAHD also has a number of simple interval arithmetic reasoning mechanisms that work together with Gröbner basis reductions. This is because interval-based inconsistencies in formulas are often easier to recognize when the polynomials appearing in formula inequalities have been canonicalized through Gröbner basis reductions.

2.3   Proof Strategy

Preprocessing of goals into goal-sets. RAHD sessions begin with the installation of a *goal*. In RAHD, a goal can be any quantifier-free formula in the language of ordered fields. All variables in RAHD goals are implicitly existentially quantified. We consider that a goal $\mathcal{G}$ in n variables has been "proved" if it has been shown to be *unsatisfiable* over $\mathbb{R}^n$. Equivalently, the semialgebraic set defined by $\mathcal{G}$ (e.g., the set $S = \{x \in \mathbb{R}^n \mid \langle \mathbb{R}, +, -, *, <, 0, 1 \rangle \models \mathcal{G}(x)\}$) is empty. Mathematically, the theorem proved is the universal dual of the existential goal being analysed. If in the process of trying to prove unsatisfiability of a goal RAHD instead finds it to be satisfiable, then this is reported as a *counter-example* to the conjecture installed as the current goal.

Once a goal is installed, its satisfiability cannot be decided until it has been pre-processed into a *goal-set*, $\mathcal{GS}$. A goal-set for a goal is an equivalent formula in disjunctive normal form (DNF). We consider the DNF formula as set of *cases*, each case being one of the conjunctions of literals. This pre-processing a goal into a goal-set also involves some normalisation transformations. For example, all occurrences of division are eliminated, each non-strict inequality is transformed into an equivalent disjunction consisting of an equation and a strict inequality, and each disequality is transformed into an equivalent disjunction of two strict inequalities. The result is that every literal in a goal-set is either an equality or a strict inequality, and every arithmetic expression is a polynomial. The next paragraph explains why we do this pre-processing.

Recall two important "sweet spots" mentioned previously: First, CAD can be made much more efficient if the semialgebraic set defined by the formula being analysed is known to be an open set. This openness can be guaranteed if the relation symbols in the formula being analysed are only strict inequalities. Second, terms appearing in formulas that contain equational constraints can be simplified by injecting those terms into the quotient ring induced by the ideal generated by the equational constraints. This can be done feasibly using Gröbner basis calculations. By decomposing non-strict inequalities into their requisite strict inequality and equational cases, we get closer to being able to exploit both "sweet spots": one of the two cases of our formula is now closer to being topologically open and thus suitable for CAD-MD, while the other case now has a richer equational structure inducing a potentially larger ideal, which can be exploited by Gröbner basis calculations resulting in more substantial term reductions[7]. Moreover, as members of the goal-set are purely conjunctive, we can exploit the following property (given $c \in \mathcal{GS}$):

$$\frac{\{l_1, \ldots, l_k\} \subseteq c \ \wedge \ \{\boldsymbol{x} \in \mathbb{R}^n \mid \langle \mathbb{R}, +, -, *, <, 0, 1 \rangle \models \bigwedge_{i=1}^{k} l_i(\boldsymbol{x})\} = \emptyset}{\{\boldsymbol{x} \in \mathbb{R}^n \mid \langle \mathbb{R}, +, -, *, <, 0, 1 \rangle \models c(\boldsymbol{x})\} = \emptyset}.$$

That is, to prove the unsatisfiability of some case $c \in \mathcal{GS}$, it suffices to prove the unsatisfiability of *any* subset of c. Considering this fact in conjunction with the aforementioned "sweet spots," we see now another way this strict-inequational/equational splitting can help us use the more feasible open fragment of CAD, even on the equational branch of a split non-strict inequality. We illustrate this by an example.

Example 1. Let $\varphi = ((p_1 \leq 0) \wedge \psi)$ s.t. (WLOG) ψ consists only of conjoined strict inequalities and equations. Let φ be split into $\varphi_1 = ((p_1 < 0) \wedge \psi)$ and $\varphi_2 = ((p_1 = 0) \wedge \psi)$, which will both be checked for satisfiability. Observe that the ideal generated by the equations in φ_2, $\mathcal{I}(\varphi_2)$, is a (possibly non-strict) superset of the corresponding ideals of φ and φ_1. Now, fix a term ordering and reduce all polynomials appearing in the strict inequalities in φ_2 with respect to $GB(\mathcal{I}(\varphi_2))$ to obtain an equisatisfiable formula φ_2'. Observe that the strict inequalities in φ_2' have now been potentially enriched with information contained in the equations of φ_2. We can now use the above observation on unsatisfiable subsets of conjoined constraints and examine the satisfiability of only the strict-inequational fragment of φ_2'. As this fragment is open, we may now use CAD-MD to decide its satisfiability, and an answer of "UNSAT" would imply the unsatisfiability of the equational branch of φ, φ_2.

Case manipulation functions. Each of the techniques discussed in Sec. 2.1 and Sec. 2.2 is embodied in one or more *case manipulation functions* (*CMF*s). A CMF operates on a single case, a conjunction of equalities and strict inequalities. A CMF first checks the structure of the case to see if it is of a form on which it can make progress. For example, the CMF for applying Full Open CAD first checks that all literals are strict inequalities. A CMF can have several outcomes.

– It can determine that the case is satisfiable. In this event, the initial goal is immediately also known to be satisfiable and RAHD terminates.

---

[7] Observe that super-ideals correspond to sub-varieties, and thus increasing an ideal takes one closer to the empty variety, which is the geometric object corresponding to an unsatisfiable formula.

- It can determine that the case is unsatisfiable
- It can return the case it was applied to unchanged, if its initial structure check fails, for example.
- It can make progress and return a simplified case, consisting again of a conjunction of equalities and strict inequalities, that is equisatisfiable with the case the CMF is applied to.
- It can return some Boolean formula equisatisfiable with the case the CMF is applied to. In general this formula might contain disjunctions, non-strict inequalities and division operations.

Sequencing proof steps. The set of cases in a goal-set can be considered the fringe of a partial proof tree of the initial goal. RAHD makes progress by applying CMFs to cases in the fringe. Our strategy for applying CMFs is simple but nevertheless effective.

We arrange our CMFs in a *master sequence* with the idea that we work on a case by applying each CMF from the master sequence in turn, so long as the CMFs return either identical cases or single simplified cases. This application of CMFs in sequence to a case extends the branch of the proof tree corresponding to that case. When the application of a CMF finds the case it is applied to unsatisfiable, the branch of the proof tree is closed or completed. In the event a general Boolean formula is output by a CMF, we apply the preprocessing step to generate new normalised cases to work on further. Usually there is more than one of these cases, so this introduces branching into the proof tree. When working on these new cases, we start again with the CMF at the beginning of the CMF sequence.

RAHD terminates either when some CMF finds the case it applied to satisfiable, in which case we have a counter-example to the original goal, or all proof branches are completed, in which case the original goal is proven.

We currently ensure that — at the level of CMFs — RAHD cannot diverge by requiring that, if a CMF outputs a more general boolean formula and restarts processing of the master sequence, it must reduce dimension: its output formula must contain fewer variables than the case it was applied to.

We refer sometimes to this master sequence of CMFs as the RAHD *waterfall* by analogy with the organisation of proof strategies in the NQTHM and ACL2 theorem provers.

Ordering of case manipulation functions. We describe here some of the main elements of our ordering of CMFs in the master sequence. The first principle of this ordering is that the more expensive CMFs should be postponed until later in the sequence, giving the less expensive CMFs priority in their attempts at closing and reducing cases. For instance, one should check if the univariate fragment of a case is unsatisfiable with the Sturm sequence analysis CMF before one tries any Open CAD CMF upon the case. The next principle is that given two CMFs of roughly equivalent expense, they should be ordered such that the result of one has the chance to inform and improve the result of the other, if possible. For example, the sums of squares based PD CMF, which will record the fact that a polynomial is positive definite if its canonicalization is a sum of squares of monomials plus a positive rational constant, should be run before the aforementioned Sturm sequence inequality CMF, as this has the potential to make more case inconsistencies explicit for the Sturm sequence CMF. The last principle we will mention is that a given CMF, A, should be included in the sequence in more than one place if the CMFs executed subsequent to A's last execution have a good chance of making the cases that remain more amenable to A. In this way, the lightest interval analysis CMF appears in the beginning of the master sequence, then again after Gröbner basis reductions have taken place as this may make inequality inconsistencies more explicit, and then again after reductions have taken place through the real radical ideal approximation CMF for the same reason.

At a very high level, the master sequence goes as follows: light arith. simp. and interval analysis $\rightarrow$ sos/positivstellensatz search $\rightarrow$ sturm chain analysis $\rightarrow$ full Open CAD $\rightarrow$ ideal triviality checking $\rightarrow$ term canonicalization $\rightarrow$ GB based rewriting $\rightarrow$ light arith. simp. and interval analysis $\rightarrow$ sos/positivstellensatz search $\rightarrow$ fragmentary Open CAD (as described in Example 1) $\rightarrow$ real radical ideal approximation $\rightarrow$ light arith. simp. and interval analysis $\rightarrow$ reverse rabinoswitch encodings $\rightarrow$ general CAD.

3   Experimental Results

Table 1 shows RAHD's performance on twenty-four example problems[8] and compares this performance to that of QEPCAD-B and two quantifier elimination procedures available in Reduce/Redlog: Rlqe, which is an enhanced implementation by Dolzmann and Sturm of Weispfenning's virtual term substitution (VTS) [21] , and Rlcad, which is an implementation by Seidl, Dolzmann and Sturm of Collins-Hong's partial CAD [6]. One interesting feature of this Rlqe procedure is that it performs VTS as long as it can (e.g., as long as the degree restrictions required for the method are not violated in a way irreparable by Redlog's simplification and degree-reduction mechanisms), and then sends the resulting formula to the Rlcad procedure if VTS alone was not sufficient. This approach is referred to as *fallback quantifier elimination*. Experiments were performed on an Intel Xeon Quad Core 2.8GHz machine with 4GB physical memory.

Table 2 presents a listing of seven of the twenty-four problems considered in Table 1.

The results of these experiments can be broadly summarized as follows:

- RAHD is able to solve a number of high-dimension, high-degree problems that QEPCAD-B, Redlog/Rlqe, and Redlog/Rlcad are not.
- Redlog/Rlqe is able to solve a number of high-dimension, high-degree problems that QEPCAD-B and Redlog/Rlcad are not.

---

[8] A copy of these problems may be obtained from `http://homepages.inf.ed.ac.uk/s0793114/calculemus09/`

Table 1. A comparison of RAHD, QEPCAD-B, Redlog/Rlqe and Redlog/Rlcad

	dim	deg	div	$|\mathcal{GS}|$	$|\mathcal{PT}|$	#p	#m	simp arith	simp GB	sos	strm	rad ideal	open CAD	gen CAD	RAHD	QB	RQ	RC
P0	5	4	N	1024	1024	42	55	1717	0	23	60	0	128	0	16.36	409*	36.6	-
P1	6	4	N	3072	3072	48	60	5371	3	99	156	0	378	0	74.09	-*	-	-
P2	5	4	N	768	768	40	61	1419	0	0	96	0	99	0	10.54	-*	-	-
P3	5	4	N	768	768	40	61	2187	0	0	96	0	99	0	10.87	-*	-	-
P4	5	4	N	768	768	40	61	1448	0	8	88	0	99	0	9.84	-*	-	-
P5	14	2	N	4	4	64	176	4	4	0	0	0	0	0	.89	-*	427	-
P6	11	5	N	8	14	24	31	124	20	6	0	6	0	2	28.23	-*	<.01	<.01
P7	8	2	N	1	1	8	18	1	0	1	0	0	0	0	<.01	.08	<.01	<.01
P8	7	32	N	64	94	30	34	352	152	0	0	30	34	0	182.98	9.72	<.01	-
P9	7	16	N	128	158	34	38	672	264	0	0	30	66	0	26.9	.29	.01	18.5
P10	7	12	N	32768	32795	66	165	78051	998	344	4	54	155	0	62.4	-*	-	-
P11	6	2	Y	32	32	28	28	14	16	0	0	0	31	1	2.99	.01	.01	.05
P12	5	3	N	16	16	22	23	86	4	0	0	0	2	4	.85	.02	.01	.07
P13	4	10	N	256	256	34	63	503	0	14	6	0	22	0	4.4	-*	<.01	<.01
P14	2	2	N	256	259	32	54	889	57	0	30	7	5	10	2.45	.01	-	-
P15	4	3	Y	8	8	14	15	8	0	0	0	0	0	8	1.32	.01	.06	.26
P16	4	2	N	128	132	28	44	662	128	17	0	1	39	9	6.11	.02	<.01	<.01
P17	4	2	N	4	6	14	19	34	3	3	0	0	0	1	.4	.28	.02	.61
P18	4	2	N	16	16	18	30	55	16	7	0	5	1	3	1.03	.01	.28	-
P19	3	6	Y	256	256	30	310	1248	0	0	0	0	256	0	24.69	.02	.01	.39
P20	3	4	N	16	16	18	21	77	18	0	0	1	3	5	1.19	.01	<.01	.23
P21	3	2	N	64	64	26	31	179	0	0	12	0	7	0	.6	.02	.04	.27
P22	2	4	N	2	2	8	10	3	1	0	0	0	1	0	.09	.01	<.01	.01
P23	2	2	Y	8	8	12	12	16	0	0	0	0	0	0	<.01	.01	<.01	<.01

Explanation of columns:

High-level problem features: [**dim**] dimension, [**deg**] maximal total multivariate degree of polynomials, [**div**] problem contains division operator.

Properties of RAHD*'s internal translation of the problem:* [$|\mathcal{GS}|$] # of cases in the generated goal-set, [$|\mathcal{PT}|$] # of leaves in constructed proof tree, [#p] # of polynomials, [#m] # of monomials.

Number of reduction or refutation steps made by RAHD *CMFs:* [**simp arith**] light weight arithmetical simplifiers and interval analysis, [**simp GB**] Gröbner bases based rewriting and canonicalization, [**sos**] sums of squares / real nullstellensatz / positivstellensatz witness extraction, [**strm**] sturm chain sign change analysis, [**rad ideal**] dimensional reduction by radical ideal approximations, [**open CAD**] open CAD or open fragmentary CAD (using QEPCAD-B and $\exists_\infty$), [**gen CAD**] general CAD (using QEPCAD-B and $\exists$).

Timing: (in seconds) [**RAHD**] RAHD , [**QB**] QEPCAD-B, [**RQ**] Redlog/Rlqe (fallback QE), [**RC**] Redlog/Rlcad (p-CAD)

A mark of (-) in any of the timing columns means the system listed was unable to solve the problem in 600 seconds. A mark of (*) in the **QB** column means that QEPCAD-B's default resource settings were raised in order to avoid reaching resource limits. For problems involving division, the Redlog translation flag RLNZDEN was used both for Rlqe and Rlcad runs as well as for generating the multiplicative translations of the problems for QEPCAD-B.

Table 2. Seven of the twenty-four problems considered in Table 1

P0 $a^2 + ab - 2ac + a + 21b^4 - 84b^3c + 126b^2c^2 - 84bc^3 + 21c^4 + c^2 + d^2 - 2de + d$
$+ e^2 < 0 \ \wedge \ e - 1 \leq 0 \ \wedge \ e \geq 0 \ \wedge \ d - 1 \leq 0 \ \wedge \ d \geq 0 \ \wedge \ c - 1 \leq 0 \ \wedge \ c \geq 0$
$\wedge \ b - 1 \leq 0 \ \wedge \ b \geq 0 \ \wedge \ a - 1 \leq 0 \ \wedge \ a \geq 0$

P1 $a^2b + a^2 - 2ac^2 + 3b^2 - 6bc + c^4 + 3c^2 + d^2 - 2de + d + e^2 + f + 1 < 0$
$\wedge \ (f - 2 = 0 \ \vee \ f - 1 = 0 \ \vee \ f = 0) \ \wedge \ e - 1 \leq 0 \ \wedge \ e \geq 0 \ \wedge \ d - 1 \leq 0$
$\wedge \ d \geq 0 \ \wedge \ c - 1 \leq 0 \ \wedge \ c \geq 0 \ \wedge \ b - 1 \leq 0 \ \wedge \ b \geq 0 \ \wedge \ a - 1 \leq 0 \ \wedge \ a \geq 0$

P5 $(y_6 \neq 0 \ \vee \ x_6 \neq 0) \ \wedge \ x_6^2 - 2x_6x_7 + x_7^2 + y_6^2 - 2y_6y_7 + y_7^2 - 4 > 0 \ \wedge \ x_5^2 - 2x_5x_7$
$+ x_7^2 + y_5^2 - 2y_5y_7 + y_7^2 - 4 = 0 \ \wedge \ x_5^2 - 2x_5x_6 + x_6^2 + y_5^2 - 2y_5y_6 + y_6^2 - 4 = 0$
$\wedge \ x_4^2 - 2x_4x_7 + x_7^2 + y_4^2 - 2y_4y_7 + y_7^2 - 4 = 0 \ \wedge \ x_4^2 - 2x_4x_6 + x_6^2 + y_4^2 - 2y_4y_6$
$+ y_6^2 - 4 = 0 \ \wedge \ x_4^2 - 2x_4x_5 + x_5^2 + y_4^2 - 2y_4y_5 + y_5^2 - 4 = 0 \ \wedge \ x_3^2 - 2x_3x_7 + x_7^2$
$+ y_3^2 - 2y_3y_7 + y_7^2 - 4 = 0 \ \wedge \ x_3^2 - 2x_3x_6 + x_6^2 + y_3^2 - 2y_3y_6 + y_6^2 - 4 = 0 \ \wedge \ x_3^2$
$- 2x_3x_5 + x_5^2 + y_3^2 - 2y_3y_5 + y_5^2 - 4 = 0 \ \wedge \ x_3^2 - 2x_3x_4 + x_4^2 + y_3^2 - 2y_3y_4 + y_4^2$
$- 4 = 0 \ \wedge \ x_2^2 - 2x_2x_7 + x_7^2 + y_2^2 - 2y_2y_7 + y_7^2 - 4 = 0 \ \wedge \ x_2^2 - 2x_2x_6 + x_6^2 + y_2^2$
$- 2y_2y_6 + y_6^2 - 4 = 0 \ \wedge \ x_2^2 - 2x_2x_5 + x_5^2 + y_2^2 - 2y_2y_5 + y_5^2 - 4 = 0 \ \wedge \ x_2^2 - 2x_2x_4$
$+ x_4^2 + y_2^2 - 2y_2y_4 + y_4^2 - 4 = 0 \ \wedge \ x_2^2 - 2x_2x_3 + x_3^2 + y_2^2 - 2y_2y_3 + y_3^2 - 4 = 0 \ \wedge$
$x_1^2 - 2x_1x_7 + x_7^2 + y_1^2 - 2y_1y_7 + y_7^2 - 4 = 0 \ \wedge \ x_1^2 - 2x_1x_6 + x_6^2 + y_1^2 - 2y_1y_6 + y_6^2$
$- 4 = 0 \ \wedge \ x_1^2 - 2x_1x_5 + x_5^2 + y_1^2 - 2y_1y_5 + y_5^2 - 4 = 0 \ \wedge \ x_1^2 - 2x_1x_4 + x_4^2 + y_1^2$
$- 2y_1y_4 + y_4^2 - 4 = 0 \ \wedge \ x_1^2 - 2x_1x_3 + x_3^2 + y_1^2 - 2y_1y_3 + y_3^2 - 4 = 0 \ \wedge \ x_1^2 - 2x_1x_2$
$+ x_2^2 + y_1^2 - 2y_1y_2 + y_2^2 - 4 = 0 \ \wedge \ x_7^2 + y_7^2 - 4 = 0 \ \wedge \ x_6^2 + y_6^2 - 4 = 0 \ \wedge \ x_5^2 + y_5^2$
$- 4 = 0 \ \wedge \ x_4^2 + y_4^2 - 4 = 0 \ \wedge \ x_3^2 + y_3^2 - 4 = 0 \ \wedge \ x_2^2 + y_2^2 - 4 = 0 \ \wedge \ x_1^2 + y_1^2 - 4$
$= 0$

P6 $45dxy - g + 45xy = 0 \ \wedge \ g - g_1 - g_2 - 82 > 0 \ \wedge \ w + 1 < 0 \ \wedge \ -x + y \geq 0 \ \wedge \ x$
$- 1 \geq 0 \ \wedge \ a = 0 \ \wedge \ -a + wz = 0 \ \wedge \ x^3y^2 - z = 0 \ \wedge \ -3g_1^2g_2 + 12g_1x_3x_7 - xy$
$- 11x \geq 0$

P10 $(-a + fg + 11f + g^2 + 13g + 22 = 0 \ \vee \ a^4b^2cd^3f^2 + 2a^4b^2cd^3fg + a^4b^2cd^3g^2 -$
$a^4b^2cdf^2 - 2a^4b^2cdfg - a^4b^2cdg^2 - a^4b^2d^2f^2 - 2a^4b^2d^2fg - a^4b^2d^2g^2 + a^4b^2f^2 +$
$2a^4b^2fg + a^4b^2g^2 - 2a^3b^3c^2d^2f^2 - 4a^3b^3c^2d^2fg - 2a^3b^3c^2d^2g^2 + 2a^3b^3c^2f^2 +$
$4a^3b^3c^2fg + 2a^3b^3c^2g^2 + 2a^3b^3d^2f^2 + 4a^3b^3d^2fg + 2a^3b^3d^2g^2 - 2a^3b^3f^2 - 4a^3b^3fg$
$- 2a^3b^3g^2 + a^2b^4c^3df^2 + 2a^2b^4c^3dfg + a^2b^4c^3dg^2 - a^2b^4c^2f^2 - 2a^2b^4c^2fg - a^2b^4c^2g^2$
$- a^2b^4cdf^2 - 2a^2b^4cdfg - a^2b^4cdg^2 + a^2b^4f^2 + 2a^2b^4fg + a^2b^4g^2 - 2a^2b^2c^3df^2 -$
$4a^2b^2c^3dfg - 2a^2b^2c^3dg^2 + 4a^2b^2c^2d^2f^2 + 8a^2b^2c^2d^2fg + 4a^2b^2c^2d^2g^2 - 2a^2b^2c^2f^2$
$- 4a^2b^2c^2fg - 2a^2b^2c^2g^2 - 2a^2b^2cd^3f^2 - 4a^2b^2cd^3fg - 2a^2b^2cd^3g^2 + 4a^2b^2cdf^2 +$
$8a^2b^2cdfg + 4a^2b^2cdg^2 - 2a^2b^2d^2f^2 - 4a^2b^2d^2fg - 2a^2b^2d^2g^2 + a^2c^3d^3f^2 +$
$2a^2c^3d^3fg + a^2c^3d^3g^2 - a^2c^2d^2f^2 - 2a^2c^2d^2fg - a^2c^2d^2g^2 - a^2cd^3f^2 - 2a^2cd^3fg$
$- a^2cd^3g^2 + a^2d^2f^2 + 2a^2d^2fg + a^2d^2g^2 - 2abc^3d^3f^2 - 4abc^3d^3fg - 2abc^3d^3g^2 +$
$2abc^3df^2 + 4abc^3dfg + 2abc^3dg^2 + 2abcd^3f^2 + 4abcd^3fg + 2abcd^3g^2 - 2abcdf^2 -$
$4abcdfg - 2abcdg^2 + b^2c^3d^3f^2 + 2b^2c^3d^3fg + b^2c^3d^3g^2 - b^2c^3df^2 - 2b^2c^3dfg -$
$b^2c^3dg^2 - b^2c^2d^2f^2 - 2b^2c^2d^2fg - b^2c^2d^2g^2 + b^2c^2f^2 + 2b^2c^2fg + b^2c^2g^2 < 0) \ \wedge$
$f^2g + 2g^5 - g = 0 \ \wedge \ f^4 - 1 = 0 \ \wedge \ e^3f^3 + g - 2 = 0 \ \wedge \ g - 1 \leq 0 \ \wedge \ f - 1 \leq 0$
$\wedge \ e - 1 \leq 0 \ \wedge \ d - 1 \leq 0 \ \wedge \ c - 1 \leq 0 \ \wedge \ b - 1 \leq 0 \ \wedge \ a - 1 \leq 0 \ \wedge \ g \geq 0 \ \wedge$
$f \geq 0 \ \wedge \ e \geq 0 \ \wedge \ d \geq 0 \ \wedge \ c \geq 0 \ \wedge \ b \geq 0 \ \wedge \ a \geq 0$

P16 $c^2 + cd - d^2 + 1 \leq 0 \ \wedge \ 2a + b - 1 \geq 0 \ \wedge \ a^2 + ab - b^2 - 1 \geq 0 \ \wedge \ d - 1 \geq 0 \ \wedge$
$c \geq 0 \ \wedge \ b \geq 0 \ \wedge \ ad + bc + bd \leq 0$

P19 $x \neq 1 \ \wedge \ y \neq 1 \ \wedge \ z \neq 1 \ \wedge \ x^2/(x - 1)^2 + y^2/(y - 1)^2 + z^2/(z - 1)^2 < 1 \ \wedge \ xyz = 1$

- Redlog/Rlqe is able to solve a number of problems significantly faster than RAHD, Redlog/Rlcad, and QEPCAD-B.
- For the problems QEPCAD-B is able to solve, using QEPCAD-B directly tends to be much faster than using RAHD's waterfall.

4 Future Work

We see many ways RAHD can be improved. First, based upon the fact that QEPCAD-B outperforms the RAHD waterfall on many low-dimension, low-degree problems, we should develop heuristics that use structural features of a problem to evaluate *a priori* its suitability for a direct handling by QEPCAD-B, causing RAHD in those cases to bypass both its inequality splitting pre-processing and all other CMFs in the waterfall.

Second, as the Redlog/Rlqe procedure solves a number of problems much faster than all of the others, it seems fruitful to investigate heuristics for incorporating Redlog/Rlqe into the RAHD waterfall.

Finally, in terms of RAHD's inequality splitting and translation of resulting problems into a (DNF) goal-set, the potential exponential blow-up this causes will become prohibitive for problems with large boolean structure. There are many more sophisticated techniques we will need to employ if we wish for RAHD to be applicable to these types of problems. The infeasibility of normal form conversions has motivated huge algorithmic advances in the SAT and SMT communities [8], and it would be very interesting to build a new version of RAHD that uses DPLL-like [5] case-analysis mechanisms instead of an explicit DNF conversion.

5 Conclusion

In closing, we have shown that a thoughtfully orchestrated heterogeneous combination of decision methods for fragments of the existential theory of the reals can be made to solve problems previously beyond the reach of automatic methods. In particular, we have shown one way that ideal-theoretic computations and restricted variants of CAD for topologically open predicates can be fruitfully combined. It is interesting that while this combination involves an exponential blow-up in its reliance on a DNF normalisation, for many problems the increase in complexity caused by this blow-up is overshadowed by the decrease in complexity of the CAD computations this process induces.

References

1. Brown, C.W.: QEPCAD B: a program for computing with semi-algebraic sets using CADs. SIGSAM Bull. 37(4), 97–108 (2003)
2. Cohen, P.J.: Decision procedures for real and p-adic fields. Comm. Pure and Applied Mathematics XXII(2), 131–151 (1969)

3. Collins, G.E.: Quantifier elimination for the elementary theory of real closed fields by cylindrical algebraic decomposition. In: Brakhage, H. (ed.) GI-Fachtagung 1975. LNCS, vol. 33, pp. 134–183. Springer, Heidelberg (1975)
4. Davenport, J.H., Heintz, J.: Real quantifier elimination is doubly exponential. Journal of Symbolic Computing 5(1-2), 29–35 (1988)
5. Davis, M., Putnam, H.: A computing procedure for quantification theory. J. ACM 7(3), 201–215 (1960)
6. Dolzmann, A., Seidl, A., Sturm, T.: Efficient projection orders for CAD. In: ISSAC 2004: Proceedings of the 2004 international symposium on Symbolic and algebraic computation, pp. 111–118. ACM Press, New York (2004)
7. Dolzmann, A., Sturm, T.: Redlog: computer algebra meets computer logic. SIGSAM Bull. 31(2), 2–9 (1997)
8. Dutertre, B., de Moura, L.: The YICES SMT solver (2006), http://yices.csl.sri.com/tool-paper.pdf
9. Yu Grigor'ev, D., Vorobjov Jr., N.N.: Solving systems of polynomial inequalities in subexponential time. Journal of Symbolic Computation 5(1,2), 37–64 (1988)
10. Hales, T.C.: Formalizing the proof of the Kepler Conjecture. In: Theorem Proving in Higher Order Logics (TPHOLs) (2004)
11. Harrison, J.: Verifying nonlinear real formulas via sums of squares. In: Schneider, K., Brandt, J. (eds.) TPHOLs 2007. LNCS, vol. 4732, pp. 102–118. Springer, Heidelberg (2007)
12. Hong, H.: Comparison of several decision algorithms for the existential theory of the reals. Technical report, RISC Linz (1991)
13. Laplagne, S.: An algorithm for the computation of the radical of an ideal. In: International Symposium on Symbolic and Algebraic Computation (2006)
14. McCallum, S.: Solving polynomial strict inequalities using cylindrical algebraic decomposition. The Computer Journal 36(5) (1993)
15. McLaughlin, S., Harrison, J.V.: A proof-producing decision procedure for real arithmetic. In: Nieuwenhuis, R. (ed.) CADE 2005. LNCS, vol. 3632, pp. 295–314. Springer, Heidelberg (2005)
16. Renegar, J.: On the computational complexity and geometry of the first-order theory of the reals (Part I). Technical Report 853, Cornell University (1989)
17. Tarski, A.: A decision method for elementary algebra and geometry. Technical report, Rand Corporation (1948)
18. Tiwari, A.: HybridSAL: Modeling and abstracting hybrid systems. Technical report, SRI International (2003)
19. Tiwari, A.: An algebraic approach for the unsatisfiability of nonlinear constraints. In: Ong, L. (ed.) CSL 2005. LNCS, vol. 3634, pp. 248–262. Springer, Heidelberg (2005)
20. van den Dries, L.: Tame topology and o-minimal structures. London Mathematical Society (1998)
21. Weispfenning, V.: Quantifier elimination for real algebra – the quadratic case and beyond. Applicable Algebra in Engineering Communication and Computing 8(2), 85–101 (1997)

Reasoning with Generic Cases
in the Arithmetic of Abstract Matrices

Alan P. Sexton[1], Volker Sorge[1], and Stephen M. Watt[2]

[1] School of Computer Science, University of Birmingham
`www.cs.bham.ac.uk/~aps|~vxs`
[2] Department of Computer Science, University of Western Ontario
`www.csd.uwo.ca/~watt`

Abstract. In previous work we have developed procedures to analyse, compute with and reason about abstract matrices, that is, matrices represented with symbolic dimensions and with a mixture of terms and ellipsis symbols to describe their structure. A central component in this are the so-called "support functions", which enable the representation of abstract matrices in closed forms. A key issue in making reasoning about such structures effective is controlling the complexity of the internal term structure of the closed form, which, in turn, hinges critically on the design of the support functions used.

Our earlier support functions were simple, easy to work with and sufficient to capture arithmetic of general partitioned matrices fully. They explicitly represent each potential homogeneous region, usually a triangle or a rectangle, of an abstract matrix with a single term. However, adding or multiplying a sequence of matrices can result in exponentially many different cases of possible regions that have to be represented, and the existence of many of these is mutually exclusive. As this representation can become unwieldy in certain situations, we experiment with a different type of support function that allows us to represent only one of the possible cases explicitly, and have all other cases captured by the representation implicitly.

In this paper we discuss this new support function and develop the full abstract matrix addition algorithm for this representation. We show that we indeed obtain much more concise and intuitive closed forms, retaining the properties necessary for reasoning with abstract matrices and being able to recover the human readable region structure from the combination of abstract matrices under addition. This representation reduces the time and space complexity of performing K abstract matrix additions from $O(N^{dK})$ to $O(K^d N^d)$, for d the number of boundary directions $(1 \leqslant d \leqslant 4)$ and N the maximum number of boundaries in any direction in the argument matrices.

1 Introduction

Through the contributions of many authors over the past few decades, computer algebra systems have become very effective at computing with values from a variety of mathematical domains. For example, polynomial, linear and differential

J. Carette et al. (Eds.): Calculemus/MKM 2009, LNAI 5625, pp. 138–153, 2009.

algebra can be performed effectively with coefficients being arbitrary precision rational numbers, elements of finite fields, algebraic numbers, and so on. Computer algebra systems have had more difficulty working, not with a particular values from one of these domains, but rather with objects representing sets of values of particular forms. Simple examples would be to factor $x^{2n} - 1$ or compute the determinant of the diagonal matrix $\text{diag}(1, 4, \ldots, n^2)$ without specifying n. We call this working with "symbolic" or "abstract" values. In previous work we have given algorithms for a number of problems on such symbolic polynomials [8,9,10] and abstract matrices [5,6,7]. The common feature of this work is that it allows computations that simultaneously treat a number of cases.

In this article we concentrate on the problem of expressing abstract matrix arithmetic in a manner that allows more effective reasoning about the cases represented. As before, we consider the problem of arithmetic on structured symbolic matrices. These matrices will be made up of various regions, where a region is a closed polygonal area that can be homogeneously evaluated by a single term. Non-zero regions must be convex. The line segments that compose the region boundaries define lines that we call boundary lines. We allow the size of the matrices and the locations of the boundaries between the regions to be given symbolically, for example, an $(h+k) \times (n+m)$ matrix made up of $h \times n, h \times m, k \times n$ and $k \times m$ blocks and defines a horizontal and a vertical boundary line (aside from the outer boundaries of the matrix). Our goal is to support automated reasoning and precisely stated algebraic algorithms on abstract matrices of this sort. We will refer throughout to special cases of block matrices, where the regions are rectangular submatrices, and banded matrices, where non-zero entries are confined to a diagonal band, comprising the main diagonal and zero or more diagonals on either side. We treat elsewhere the separate problem of obtaining the closed form expressions for each region from forms that express skipped entries with ellipses.

It would be a natural choice to represent these abstract matrices as having elements that are piecewise defined functions of the indices. However, matrix arithmetic quickly leads to an intractable situation. A chain of N arithmetic operations on matrices with K cases in their piecewise elements requires the analysis of K^N cases, each requiring simplification of boolean expressions to determine feasibility. Useful progress has been made in the scalar case for this problem [1], but we find that with matrices another approach is more fruitful.

Instead of representing the cases as logical conditions, we encode them algebraically with "support functions", in a manner we shall shortly make more precise. We are then able to handle multiple logical cases simultaneously via algebraic operations. We encounter one problem, however: when doing arithmetic without knowing how the region boundaries in one matrix relate to the region boundaries in the other, the generic case becomes overly complicated, with certain terms becoming mutually exclusive. For example, when adding an $(h_1 + k_1) \times (n_1 + m_1)$ to an $(h_2 + k_2) \times (n_2 + m_2)$ block matrix, the form of the result must satisfy all possible relative orderings $h_1 \lessgtr h_2, n_1 \lessgtr n_2$. Our earlier choice of support functions can cope with this issue, and, indeed, we have previously presented full abstract matrix addition and multiplication algorithms in this context, but at a cost of a high complexity in the number of terms in the result as a function of the number of terms in the abstract matrix operands. In this paper we explore using another class of support functions that allow the generic case to be written more more concisely (with asymptotically fewer terms) and with consequent advantages for space and time efficiency, and we present the solution for the addition case.

2   Motivation

Before developing the full *support function* we are aiming for, we motivate its need considering a simpler, interval based support function. While it is slightly different to the support function we have employed to define a full abstract matrix addition and multiplication algorithms in [7], it exhibits similar problems.

Definition 1. *Let $x, y, z \in \mathbb{N}$ then we define the support function $\hat{\xi}(x, y, z)$ as*

$$\hat{\xi}(x, y, z) = \begin{cases} 1 & \text{if } y \leqslant x < z \\ 0 & \text{otherwise} \end{cases}$$

Thus $\hat{\xi}$ acts as a *characteristic function* for a particular interval. We shall use $\hat{\xi}_{x,y,z}$ as a more compact notation for $\hat{\xi}(x, y, z)$.

Now consider the following two abstract vectors

$$u = [a_1, \ldots, a_{h-1}, b_h \ldots b_{n-1}] \qquad v = [c_1, \ldots, c_{k-1}, d_k \ldots d_{n-1}] \qquad (1)$$

Using our $\hat{\xi}$ functions, the elements of these vectors, u_j, v_j, for $j \in \{1..n - 1\}$ can be written:

$$u_j = \hat{\xi}_{j,1,h}\, a_j + \hat{\xi}_{j,h,n}\, b_j \qquad v_j = \hat{\xi}_{j,1,k}\, c_j + \hat{\xi}_{j,k,n}\, d_j \qquad (2)$$

Observe that j is here used as the horizontal index. Throughout this paper we will reserve i, j as *vertical and horizontal index variables*.

For simplicity, we will drop the indices on the a, b, c and d terms. If we add these two vectors, we get

$$(u + v)_j = \hat{\xi}_{j,1,h}\, a + \hat{\xi}_{j,h,n}\, b + \hat{\xi}_{j,1,k}\, c + \hat{\xi}_{j,k,n}\, d \qquad (3)$$

However, the four terms in these expressions define overlapping regions in the resultant vector, so we cannot easily visualise the shape of the regions in the result. To fix this, we need to identify the set of disjoint regions in the result, and calculate the term that is valid with each of these regions. We will do this in a slightly long-winded way in order to motivate the algorithms to follow. To find these disjoint regions we construct the characteristic functions of all intersections of regions from the original vectors. This can be done by multiplying the $\hat{\xi}$ components of each term from u with that of each term from v. Since the two regions from u are disjoint, as are the two regions from v, we do not have to worry about intersections between the a and b regions or between c and d regions and we are left with 4 regions, each of which we can describe with $\hat{\xi}$ functions:

$a \cap c$ region: $\hat{\xi}_{j,1,h}\,\hat{\xi}_{j,1,k} = \hat{\xi}_{j,1,\min(h,k)}$ $b \cap c$ region: $\hat{\xi}_{j,h,n}\,\hat{\xi}_{j,1,k} = \hat{\xi}_{j,h,k}$

$a \cap d$ region: $\hat{\xi}_{j,1,h}\,\hat{\xi}_{j,k,n} = \hat{\xi}_{j,k,h}$ $b \cap d$ region: $\hat{\xi}_{j,h,n}\,\hat{\xi}_{j,k,n} = \hat{\xi}_{j,\max(h,k),n}$

Having identified the disjoint regions, we need to find the terms that will be valid within each region. We can do this by multiplying each region characteristic function by the full term for the sum of u and v and simplifying. The characteristic functions will ensure that any component that does not belong with the requisite region will simplify to zero. We show the working in detail for the $a \cap c$ region:

$$\hat{\xi}_{j,1,\min(h,k)}\,(u+v)_j = \hat{\xi}_{j,1,\min(h,k)}\left(\hat{\xi}_{j,1,h}\,a + \hat{\xi}_{j,h,n}\,b + \hat{\xi}_{j,1,k}\,c + \hat{\xi}_{j,k,n}\,d\right)$$
$$= \hat{\xi}_{j,1,\min(h,k)}\,\hat{\xi}_{j,1,h}\,a + \hat{\xi}_{j,1,\min(h,k)}\,\hat{\xi}_{j,h,n}\,b +$$
$$\hat{\xi}_{j,1,\min(h,k)}\,\hat{\xi}_{j,1,k}\,c + \hat{\xi}_{j,1,\min(h,k)}\,\hat{\xi}_{j,k,n}\,d$$
$$= \hat{\xi}_{j,1,\min(h,k)}\,a + \hat{\xi}_{j,1,\min(h,k)}\,c$$
$$= \hat{\xi}_{j,1,\min(h,k)}\,(a+c)$$

Working out the terms for the other regions and adding, we get:

$$(u+v)_j = \hat{\xi}_{j,1,\min(h,k)}(a+c) + \hat{\xi}_{j,k,h}(a+d) + \hat{\xi}_{j,h,k}(b+c) + \hat{\xi}_{j,\max(h,k),n}(b+d) \quad (4)$$

Thus our expression contains four terms, apparently describing four disjoint regions. However, the two middle terms are mutually exclusive. In any concrete case, either $h < k$, $h = k$ or $h < k$. Depending on which case holds, one or both of the middle terms will reduce to 0. Thus we have a representation that, while correct, explicitly represents every region that can occur in any case obtainable by varying the ordering of the matrix parameters (in this case, h and k).

We can generalise this analysis. If, instead of two vectors with two regions each, we were to add K matrices with N regions each with all the boundaries parallel, we would arrive at a result with N^K terms describing potential regions. This is worse than just leaving the original sum with KN terms symbolic. Note that for any particular choice of region boundaries there will be at most $K(N-1)$ regions in the resulting sum. We next describe a choice of support functions that allow the resulting generic expression to have exactly this many terms. If we were to add K block matrices with $N \times M$ regions, using the first choice of support functions would give a result with $(N \times M)^K$ describing potential regions.

3   Definitions

At this point we can introduce the full ξ functions. The idea behind them is that, for the most part, they act like the $\hat{\xi}$ interval characteristic functions, but that they let us commit to a single case, i.e., a single ordering of the unknowns in the matrix expression, and, if that ordering is wrong, the negative part of the ξ function will compensate so that the correct result is obtained. In particular, the result will contain one term for each region that occurs in that single case, and hence shows directly the structure of the result for that case. However, if we wish to investigate a different case, we can apply a reordering of the matrix parameters, normalise the expression with respect to this new reordering, and then have our expression directly represent this new case.

Definition 2. *Let $x, y, z \in \mathbb{N}$ then we define the support function $\xi(x, y, z)$ as*

$$\xi(x, y, z) = \begin{cases} 1 & \text{if } y \leqslant x < z \\ -1 & \text{if } z \leqslant x < y \\ 0 & \text{otherwise} \end{cases}$$

We will again use $\xi_{x,y,z}$ as compact notation. The following properties follow immediately from the definition:

$$\xi_{u,x,x} = 0 \tag{5}$$

$$\xi_{u,x,y} = -\xi_{u,y,x} \tag{6}$$

$$\xi_{u,x,y} + \xi_{u,y,z} = \xi_{u,x,z} \tag{7}$$

In this paper, we will restrict our use of ξ support functions to a particular class thereof; namely that of parallel ξ support functions:

Definition 3. *A parallel ξ support function is a $\xi_{X,Y,Z}$ function where X is an integer expression restricted to one of the forms i, j, $i+j$ or $i-j$ for respective matrix row and column index variables i and j, and where Y and Z are integer expressions that do not contain any matrix index variables.*

We observe that ξ itself is not a characteristic function, as it can lead to negative values for the interval it represents. We therefore define a characteristic function based on ξ using its absolute value.

Definition 4. *Let $x, y, z \in \mathbb{N}$. The characteristic function of $\xi_{x,y,z}$ is $|\xi_{x,y,z}|$.*

In general, a $\xi_{x,y,z}$ function defines both an interval, $\hat{\xi}_{x,\min(y,z),\max(y,z)}$ or $|\xi_{x,y,z}|$, and a contribution factor, namely $+1$ or -1, depending on the strict ordering of y and z. We follow a similar procedure to that above, but start by already committing to an ordering, let us say $1 < h < k < n$. In fact, we are not changing the meaning of the expression, merely normalising it with respect to a particular ordering of the parameters. Now we get 3 regions as the fourth simplifies to 0:

$a \cap c$ region: $|\xi_{j,1,h}| \, |\xi_{j,1,k}| = |\xi_{j,1,h}|$ $b \cap c$ region: $|\xi_{j,h,n}| \, |\xi_{j,1,k}| = |\xi_{j,h,k}|$
$a \cap d$ region: $|\xi_{j,1,h}| \, |\xi_{j,k,n}| = 0$ $b \cap d$ region: $|\xi_{j,h,n}| \, |\xi_{j,k,n}| = |\xi_{j,k,n}|$

Now multiply these in to the full ξ version of the resultant vector. We again show the working in detail for the $a \cap c$ region, bearing in mind our choice of ordering of $1 < h < k < n$:

$$
\begin{aligned}
|\xi_{j,1,h}| \, (u+v)_j &= |\xi_{j,1,h}| \, (\xi_{j,1,h} \, a + \xi_{j,h,n} \, b + \xi_{j,1,k} \, c + \xi_{j,k,n} \, d) \\
&= |\xi_{j,1,h}| \, \xi_{j,1,h} \, a + |\xi_{j,1,h}| \, \xi_{j,h,n} \, b + |\xi_{j,1,h}| \, \xi_{j,1,k} \, c + |\xi_{j,1,h}| \, \xi_{j,k,n} \, d \\
&= \xi_{j,1,h} \, a + \xi_{j,1,h} \, c \\
&= \xi_{j,1,h} \, (a+c)
\end{aligned}
$$

Doing the same for the other two regions and adding the three terms we get a final result of:

$$
(u+v)_j = \xi_{j,1,h} \, (a+c) + \xi_{j,h,k} \, (b+c) + \xi_{j,k,n} \, (b+d) \tag{8}
$$

Now suppose our choice of ordering was incorrect and, in fact, $k < h$. We can show that Equation (8) is still correct by rearranging it with judicious applications of (6) and (7):

$$
\begin{aligned}
(u+v)_j &= \xi_{j,1,h} \, (a+c) + \xi_{j,h,k} \, (b+c) + \xi_{j,k,n} \, (b+d) \\
&= (\xi_{j,1,k} + \xi_{j,k,h}) \, (a+c) - \xi_{j,k,h} \, (b+c) + (\xi_{j,k,h} + \xi_{j,h,n}) \, (b+d) \\
&= \xi_{j,1,k} \, (a+c) + \xi_{j,k,h} \, ((a+c) - (b+c) + (b+d)) + \xi_{j,h,n} \, (b+d) \\
&= \xi_{j,1,k} \, (a+c) + \xi_{j,k,h} \, (a+d) + \xi_{j,h,n} \, (b+d)
\end{aligned}
$$

This final result is precisely the form that we would get if we had started with the assumption that $k < h$.

With this choice of support functions, our previous example of adding K matrices with N regions each and all boundaries parallel would give a result with $K(N-1)+1$ regions, corresponding to any arbitrary choice of relative order of the region boundaries among the arguments. Each region would have a sum of K elements, one from each matrix. So the total size of the generic element expression would be $K(K(N-1)+1)$. If each of the matrices had L elements, the cost to compute a specialisation of this abstract sum would be $O(LK^2N)$. This compares to a cost of $O(LK)$ if the specialisation is known in advance. If the specialisation were not known in advance, and the sum were not reorganised to represent a generic case, then the cost to compute a specialisation would be LKN. So the cost to present the generic form of the result is a factor of K. If we were to add K block matrices, each with $N \times M$ regions, using the ξ support functions would give $(K(N-1)+1)(K(M-1)+1)$ regions each with K terms.

4   Describing Regions

We need some tools to describe the spatial extent of regions and their complements. For brevity, we will henceforth refer to the characteristic function of a

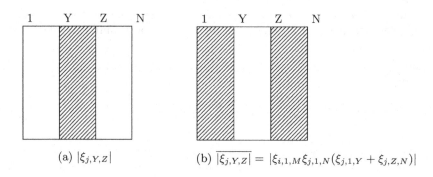

(a) $|\xi_{j,Y,Z}|$ (b) $\overline{|\xi_{j,Y,Z}|} = |\xi_{i,1,M}\xi_{j,1,N}(\xi_{j,1,Y} + \xi_{j,Z,N})|$

Fig. 1. ξ complement example

region or of the complement of a region simply as the region or the complement of the region.

We need the following equations to be satisfied by the complement $\overline{A}$ of a region A: (i) $\overline{A}\,A = 0$ (ii) $\overline{\overline{A}} = A$.

The complement of the region described by a single ξ expression will be the absolute value of the sum of the two ξ expressions that describe the spatial extent of the entire matrix minus the region described by the ξ (c.f. Fig. 1):

Definition 5. *The characteristic function of the complement $\overline{|\xi_{X,Y,Z}|}$ of the region described by a parallel ξ support function with respect to an abstract matrix of dimension $(M-1) \times (N-1)$ is defined to be:*

$$\overline{\xi_{X,Y,Z}} = \begin{cases} |\xi_{i,1,M}\,\xi_{j,1,N}\,(\xi_{i,1,Y} + \xi_{i,Z,M})| & \text{if } X \text{ has the form } i \\ |\xi_{i,1,M}\,\xi_{j,1,N}\,(\xi_{j,1,Y} + \xi_{j,Z,N})| & \text{if } X \text{ has the form } j \\ |\xi_{i,1,M}\,\xi_{j,1,N}\,(\xi_{i+j,1,Y} + \xi_{i+j,Z,M+N-1})| & \text{if } X \text{ has the form } i+j \\ |\xi_{i,1,M}\,\xi_{j,1,N}\,(\xi_{i-j,2-N,Y} + \xi_{i-j,Z,M-2})| & \text{if } X \text{ has the form } i-j \end{cases} \tag{9}$$

We can now define the complement of a region described by the product of ξ support functions as follows:

Definition 6. *Let $\Xi = \xi_1\xi_2\ldots\xi_n$, where each ξ_i is a parallel ξ support function of the form ξ_{X_i,Y_i,Z_i}. The complement of $|\Xi|$, $\overline{|\Xi|}$, is defined to be:*

$$\overline{|\Xi|} = \overline{|\xi_1|} + \left|\xi_1\overline{|\xi_2|}\right| + \left|\xi_1\xi_2\overline{|\xi_3|}\right| + \cdots + \left|\xi_1\ldots\xi_{n-1}\overline{|\xi_n|}\right| \tag{10}$$

Further expressions for complements can be found using elementary set theory: e.g., for sets A, B, $\overline{A \cup B} = \overline{A} \cap \overline{B}$, hence $\overline{|\xi_1| + |\xi_2|} = \left(\overline{|\xi_1|}\right)\left(\overline{|\xi_2|}\right)$, etc.

5   Abstract Matrix Addition

We now present a worked example considering the following abstract matrices:

$$A := \begin{bmatrix} a \dots a & \\ \vdots & \ddots & \\ a & & \\ & & \mathbf{0} \end{bmatrix} \qquad B := \begin{bmatrix} b \dots b & c \dots c \\ \vdots & \vdots & \vdots \\ \vdots & \vdots & \vdots \\ b \dots b & c \dots c \end{bmatrix}$$

$$A = \xi_{i,1,p}\,\xi_{j,1,p}\,\xi_{i+j,2,p+1}\; a \qquad B = \xi_{i,1,n}\,\xi_{j,1,q}\; b + \xi_{i,1,n}\,\xi_{j,q,n}\; c$$

Here both matrices are of dimension $n-1 \times n-1$, the a triangle ellipses are of length $p-1$, i.e., columns and rows $1\dots p-1$, the b rectangle fills columns $1\dots q-1$ and all rows and the c rectangle fills columns $q \dots n-1$ and all rows. All the -1 terms in these dimensions are just to make the ξ function subscripts simpler.

When we evaluate $A + B$, we can get a number of different cases. For any non-trivial matrix we always have that $1 < n$. However, we can consider p to be anywhere in the range $1..2n-1$ and q in the range $1..n$. In the extreme cases, $p = 1$, $p = 2n - 1$, $q = 1$ or $q = n$, one or more of the regions disappear. For the purposes of our example, we will restrict ourselves to two cases: $1 < p < q < n$ and $1 < q < p < n$, shown in Fig 2.

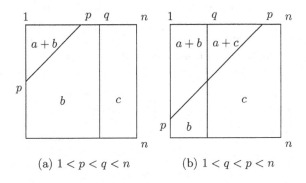

(a) $1 < p < q < n$ (b) $1 < q < p < n$

Fig. 2. Example: two of the possible orderings of $1, p, q, n$

Writing out the terms for the sum of the matrices, we get

$$(A + B)_{i,j} = \xi_{i,1,p}\,\xi_{j,1,p}\,\xi_{i+j,2,p+1}\; a + \xi_{i,1,n}\,\xi_{j,1,q}\; b + \xi_{i,1,n}\,\xi_{j,q,n}\; c \qquad (11)$$

This has three terms, but the terms describe potentially overlapping regions. To be able to extract the shape of the resulting regions, as shown in Fig. 2, we need to choose an ordering and normalise the expression with respect to that ordering. To normalise to this order, we start by identifying every potential region as the intersection of every region from A with every region from B.

Note that A has a 0 region, which is not explicitly represented in the expression for A. To get the characteristic function for this region, we need to find the characteristic function of the complement of the a region. Since the a region is $|\xi_{i,1,p}\,\xi_{j,1,p}\,\xi_{i+j,2,p+1}|$, we can calculate its complement as

$$\overline{|\xi_{i,1,p}\,\xi_{j,1,p}\,\xi_{i+j,2,p+1}|} = \overline{|\xi_{i,1,p}|} + \left|\xi_{i,1,p}\,\overline{|\xi_{j,1,p}|}\right| + \left|\xi_{i,1,p}\,\xi_{j,1,p}\,\overline{|\xi_{i+j,2,p+1}|}\right|$$

$$= |\xi_{i,1,n}\,\xi_{j,1,n}\,(\xi_{i,1,1} + \xi_{i,p,n})| + \left|\xi_{i,1,p}\,\overline{|\xi_{j,1,p}|}\right| + \left|\xi_{i,1,p}\,\xi_{j,1,p}\,\overline{|\xi_{i+j,2,p+1}|}\right|$$

$$= |\xi_{i,p,n}\,\xi_{j,1,n}| + \left|\xi_{i,1,p}\,\overline{|\xi_{j,1,p}|}\right| + \left|\xi_{i,1,p}\,\xi_{j,1,p}\,\overline{|\xi_{i+j,2,p+1}|}\right|$$

$$= |\xi_{i,p,n}\,\xi_{j,1,n}| + |\xi_{i,1,p}\,|\xi_{j,1,n}\,\xi_{j,1,n}\,(\xi_{j,1,1} + \xi_{j,p,n})|| + \left|\xi_{i,1,p}\,\xi_{j,1,p}\,\overline{|\xi_{i+j,2,p+1}|}\right|$$

$$= |\xi_{i,p,n}\,\xi_{j,1,n}| + |\xi_{i,1,p}\,\xi_{j,p,n}| + \left|\xi_{i,1,p}\,\xi_{j,1,p}\,\overline{|\xi_{i+j,2,p+1}|}\right|$$

$$= |\xi_{i,p,n}\,\xi_{j,1,n}| + |\xi_{i,1,p}\,\xi_{j,p,n}| + |\xi_{i,1,p}\,\xi_{j,1,p}\,|\xi_{i,1,n}\,\xi_{j,1,n}(\xi_{i+j,2,2} + \xi_{i+j,p+1,2n-1})||$$

$$= |\xi_{i,p,n}\,\xi_{j,1,n}| + |\xi_{i,1,p}\,\xi_{j,p,n}| + |\xi_{i,1,p}\,\xi_{j,1,p}\,\xi_{i+j,p+1,2n-1}| \tag{12}$$

$$= |\xi_{i,1,n}\,\xi_{j,1,n}\,\xi_{i+j,p+1,2n-1}| \tag{13}$$

The step from (12) to (13) is valid, but obscure. Consider the middle summand: $\xi_{i,1,p}\,\xi_{j,p,n}$. We can multiply this by $\xi_{i+j,p+1,2n-1}$ without changing its value for any i, j in the range of the matrix, as it already implies that $1 + p \leqslant i + j < \min(p + n, 2n - 1) \leqslant 2n - 1$. If we do that multiplication, we can then combine it with the final summand to replace both with $\xi_{i,1,p}\,\xi_{j,1,n}\,\xi_{i+j,p+1,2n-1}$. We can similarly fold in the first summand to get the required result.

Note that (12) shows that the complement of the a is constructed it as a sum (union) of three disjoint regions (c.f. Fig. 3). The step to (13) is simply reconstructing the single region from the three partial ones.

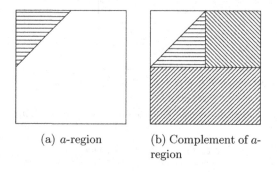

(a) a-region   (b) Complement of a-region

Fig. 3. a region and its complement

Now that we can produce a expression for every region, including 0 regions, in the component matrices, we can choose an order and normalise to it. We choose $1 < q < p < n$. We can construct the disjoint region descriptions of the result by intersecting each of the two regions of A with each of the two regions of B and simplifying under the assumption that $1 < q < p < n$:

$a \cap b$ **region:** $\left|\xi_{i,1,p}\,\xi_{j,1,p}\,\xi_{i+j,2,p+1}\,\xi_{i,1,n}\,\xi_{j,1,q}\right| = \left|\xi_{i,1,p}\,\xi_{j,1,q}\,\xi_{i+j,2,p+1}\right|$

$a \cap c$ **region:** $\left|\xi_{i,1,p}\,\xi_{j,1,p}\,\xi_{i+j,2,p+1}\,\xi_{i,1,n}\,\xi_{j,q,n}\right| = \left|\xi_{i,1,p}\,\xi_{j,q,p}\,\xi_{i+j,q+1,p+1}\right|$

$\bar{a} \cap b$ **region:** $\left|\xi_{i,1,n}\,\xi_{j,1,n}\,\xi_{i+j,p+1,2n-1}\,\xi_{i,1,n}\,\xi_{j,1,q}\right| = \left|\xi_{i,1,n}\,\xi_{j,1,q}\,\xi_{i+j,p+1,n+q}\right|$

$\bar{a} \cap c$ **region:** $\left|\xi_{i,1,n}\,\xi_{j,1,n}\,\xi_{i+j,p+1,2n-1}\,\xi_{i,1,n}\,\xi_{j,q,n}\right| = \left|\xi_{i,1,n}\,\xi_{j,q,n}\,\xi_{i+j,p+1,2n-1}\right|$

Thus we have now identified that in the case that $1 < q < p < n$ there are four disjoint regions. We still have to obtain the full term for each region, but that is easily done by multiplying each region expression by the full original abstract expression to get the full projection of the abstract matrix expression onto each disjoint region. We show the detailed derivation for the $a \cap b$ region:

$$\left|\xi_{i,1,p}\,\xi_{j,1,q}\,\xi_{i+j,2,p+1}\right|\,\left(\xi_{i,1,p}\,\xi_{j,1,p}\,\xi_{i+j,2,p+1}\,a + \xi_{i,1,n}\,\xi_{j,1,q}\,b + \xi_{i,1,n}\,\xi_{j,q,n}\,c\right)$$

$$= \xi_{i,1,p}\,\xi_{j,1,q}\,\xi_{i+j,2,p+1}\,a + \xi_{i,1,p}\,\xi_{j,1,q}\,\xi_{i+j,2,p+1}\,b + 0 \tag{14}$$

$$= \xi_{i,1,p}\,\xi_{j,1,q}\,\xi_{i+j,2,p+1}\,(a+b) \tag{15}$$

The final result is then obtained by adding the terms obtained.

$$\begin{aligned}
&\xi_{i,1,p}\,\xi_{j,1,q}\,\xi_{i+j,2,p+1}\,(a+b)+\\
&\xi_{i,1,p-q}\,\xi_{j,q,p}\,\xi_{i+j,q+1,p+1}\,(a+c)+\\
&\xi_{i,1,n}\,\xi_{j,1,q}\,\xi_{i+j,p+1,n+q-1}\,b+\\
&\xi_{i,1,n}\,\xi_{j,q,n}\,\xi_{i+j,p+1,2n-1}\,c
\end{aligned} \tag{16}$$

Equation (16) clearly corresponds to Fig. (2b). As a final check, we reconsider (16) in the alternative case of $1 < p < q < n$. We try reversing the order of p and q and depend upon the reversing properties of the ξ support functions to retrieve the situation.

$$\begin{aligned}
\text{Eq. (16)} = {}&\xi_{i,1,p}\,\xi_{j,1,q}\,\xi_{i+j,2,p+1}\,(a+b)+\\
&\xi_{i,1,n}\,\xi_{j,1,q}\,\xi_{i+j,p+1,n+q-1}\,b + \xi_{i,1,n}\,\xi_{j,q,n}\,\xi_{i+j,p+1,2n-1}\,c \tag{17}\\
= {}&\xi_{i,1,p}\,(\xi_{j,1,p} + \xi_{j,p,q})\,\xi_{i+j,2,p+1}\,(a+b)+\\
&\xi_{i,1,n}\,\xi_{j,1,q}\,\xi_{i+j,p+1,n+q-1}\,b + \xi_{i,1,n}\,\xi_{j,q,n}\,\xi_{i+j,p+1,2n-1}\,c \tag{18}
\end{aligned}$$

$$\begin{aligned}
= {}&\xi_{i,1,p}\,\xi_{j,1,p}\,\xi_{i+j,2,p+1}\,(a+b) + \xi_{i,1,p}\,\xi_{j,p,q}\,\xi_{i+j,2,p+1}\,(a+b)+\\
&\xi_{i,1,n}\,\xi_{j,1,q}\,\xi_{i+j,p+1,n+q-1}\,b + \xi_{i,1,n}\,\xi_{j,q,n}\,\xi_{i+j,p+1,2n-1}\,c \tag{19}\\
= {}&\xi_{i,1,p}\,\xi_{j,1,p}\,\xi_{i+j,2,p+1}\,(a+b)+\\
&\xi_{i,1,n}\,\xi_{j,1,q}\,\xi_{i+j,p+1,n+q-1}\,b + \xi_{i,1,n}\,\xi_{j,q,n}\,\xi_{i+j,p+1,2n-1}\,c \tag{20}\\
= {}&\xi_{i,1,p}\,\xi_{j,1,p}\,\xi_{i+j,2,p+1}\,(a+b)+\\
&\xi_{i,1,n}\,\xi_{j,1,q}\,\xi_{i+j,p+1,n+q-1}\,b + \xi_{i,1,n}\,\xi_{j,q,n}\,c \tag{21}
\end{aligned}$$

The steps involved in this derivation can be explained as follows:

(16) $\longrightarrow$ **(17):** The coefficient of $(a+c)$ is 0 within the matrix if $p < q$ because $p - q$ is negative and $\xi_{i,1,p-q}$ is only non-zero outside the matrix bounds.

(17) $\longrightarrow$ **(18):** use of (7)

(**18**) $\longrightarrow$ (**19**): distributing sum

(**19**) $\longrightarrow$ (**20**): The coefficient of the second $(a + b)$ term is 0 because $\xi_{i,1,p}\,\xi_{j,p,q}$ and $p < q$ implies that $i + j$ is in the range $p + 1 \ldots p + q$, but $\xi_{i+j,2,p+1}$ is always zero in this range.

(**20**) $\longrightarrow$ (**21**): $\xi_{i,1,n}\,\xi_{j,q,n}$ already limits $i + j$ to the range $q + 1 \ldots 2n - 1$, and $p < q$, so the $\xi_{i+j,p+1,2n-1}$ is redundant and can be dropped.

Equation (21) clearly corresponds to Fig. (2a) and so we have been able to represent the matrix using one term for each region of only one case, but still capture, with this expression, the information required for the alternative case.

However, it is important to note that this transformation can not be achieved in all possible cases, only certain representations contain implicitly all possible shapes of the abstract matrix. For example, if we had chosen the ordering $1 < p < q < n$ initially we could have not reached the matrix shape for the ordering $1 < q < p < n$. The desired implicit representation needs to have a certain maximality property with respect to the reachability of all cases for the abstract matrix. We will define this notion more precisely in the next section.

6   Matrix Addition Algorithm

The primary advantage of our choice of ξ support functions is that they allow the entire set of possible relative orderings to be represented immediately by one generic order choice for certain important classes of matrices. If all boundaries are vertical and horizontal or all boundaries are diagonal or all boundaries are anti-diagonal, then the form of the generic sum can be written down by picking an arbitrary relative order of the boundaries of each kind and directly writing the resulting generic term. In particular, this situation includes the classes of block matrices and banded matrices as special cases. The class of block matrices is closed under addition (adding block matrices gives a block matrix, possibly with smaller blocks) as is the class of banded matrices (with all boundaries along diagonals with $i - j = constant$). It is not necessary in these cases to attempt to merge regions or use region complements. We need the following definitions before giving the algorithm for this matrix addition:

Definition 7. *A* generic placement *of a boundary in an abstract matrix is one that intersects all boundary lines to which it is not parallel in the abstract matrix and does not cause an intersection of three boundaries at one point.*

Definition 8. *An abstract matrix is in a* generic configuration *if all its boundary lines have generic placement.*

Algorithm 1 (Special Matrix Addition)
INPUT:

> Two abstract matrices A and B for which one of the following holds
> 1. both are abstract block matrices
> 2. both are abstract banded matrices with diagonal region boundaries
> 3. both are abstract banded matrices with anti-diagonal region boundaries.

OUTPUT:

An abstract matrix of the same form as the two inputs, with an expression for a generic entry.

METHOD:

1. *Form the sets* $v(A)$, $h(A)$, $d(A)$, $a(A)$ *containing the expressions for* $j = expr$, $i = expr$, $i - j = expr$ $i + j = expr$ *defining the region boundaries in* A. *Likewise for* B. *If* $d(A)$ *or* $d(B)$ *are non empty, then the other sets must be empty. If* $a(A)$ *or* $a(B)$ *are non empty, then the other sets must be empty. If any of* $v(A)$, $v(B)$, $h(A)$, $h(B)$ *are non empty then the diagonal and anti-diagonal sets will be empty.*

2. *Impose an arbitrary order on each of these sets (if one has not been imposed already).*

3. *Form the sets* $v(C) = v(A) \cup v(B)$, $h(C) = h(A) \cup h(B)$, $d(C) = d(A) \cup d(B)$, *and* $a(C) = a(A) \cup a(B)$

   Impose an arbitrary order on the sets $v(C)$, $h(C)$, $d(C)$ *and* $a(C)$ *that*

   (a) *is consistent with the orders on* $v(X)$, $h(X)$, $d(X)$ *and* $a(X)$ *for* $X = A$ *and* $X = B$

   (b) *places the resulting abstract matrix in generic configuration.*

   Let $v(C)_k, h(C)_k, d(C)_k$ *and* $a(C)_k$ *denote the k-th element of the corresponding set in the imposed order.*

   Let $v(C)_0 = 1$, $v(C)_{\#v(C)+1} = \#\text{cols} + 1$

   Let $h(C)_0 = 1$, $v(C)_{\#h(C)+1} = \#\text{rows} + 1$

   Let $d(C)_0 = 1 - \#\text{cols}$, $d(C)_{\#d(C)+1} = \#\text{rows}$

   Let $a(C)_0 = 2$, $a(C)_{\#d(C)+1} = \#\text{rows} + \#\text{cols} + 1$

4. *Form the regions, depending on input cases 1 to 3 respectively,*

   1: $\xi_{i,h(C)_h,h(C)_{h+1}} \xi_{j,v(C)_k,v(C)_{k+1}}$ *for* $h = 0..\#h(C), k = 0..\#v(C)$,

   2: $\xi_{i-j,d(C)_h,d(C)_{h+1}}$ *for* $h = 0..\#d(C)$,

   3: $\xi_{i-j,a(C)_h,a(C)_{h+1}}$ *for* $h = 0..\#a(C)$

5. *Construct the expression. Multiply the* ξ *term for each region by the corresponding sum of a term from* A *and a term from* B.

Theorem 1. *Algorithm 1 gives an expression that correctly evaluates all elements of* $A+B$ *under any permitted reordering of the boundaries.*

We omit the proof for theorem 1 as it is a lengthy and tedious case analysis. Instead we now consider the general case.

In order to generalise algorithm 1, we observe that to represent the maximal number of possible regions in a matrix, we have to choose boundary placements such that at most two boundary lines intersect at a single point. However, the generic configuration of an abstract matrix then possibly has intersection points of two boundary lines outside the boundaries of the matrix. But it is easy to see that we can always embed a given abstract matrix as a rectangular submatrix into a larger abstract matrix that contains the intersection points, such that at most two boundary lines intersect, as summarised in the following two results:

Lemma 1. *Every abstract matrix may be written as a submatrix of an abstract matrix* in generic configuration, *possibly after a shift of indices.*

Theorem 2. *Given two abstract matrices A, B in generic configuration. Then let $(\mathbf{v}(A), \mathbf{h}(A), \mathbf{d}(A), \mathbf{a}(A))$ and $(\mathbf{v}(B), \mathbf{h}(B), \mathbf{d}(B), \mathbf{a}(B))$ be the corresponding vertical, horizontal, diagonal, and anti-diagonal boundary sets as defined in algorithm 1. It is then possible to impose orders on the unions of the boundary sets $(\mathbf{v}(A) \cup \mathbf{v}(B), \mathbf{h}(A) \cup \mathbf{h}(B), \mathbf{d}(A) \cup \mathbf{d}(B), \mathbf{a}(A) \cup \mathbf{a}(B))$ such that the resulting boundaries are in generic placement, possibly in a submatrix of a larger matrix.*

Again we omit the proofs and now present a method for the addition of general abstract matrices based on the results. Although we have not discovered any counterexamples, we have not as yet proven its correctness in all cases.

Algorithm 2 (General Matrix Addition)
INPUT:
>   Two abstract matrices A and B in generic configuration.

OUTPUT:
>   An abstract matrix in generic configuration and an expression for the generic entry.

METHOD:

1. *Form the sets $\mathbf{v}(A)$, $\mathbf{h}(A)$, $\mathbf{d}(A)$, and $\mathbf{a}(A)$ containing the expressions for $j = expr$, $i = expr$, $i - j = expr$ $i + j = expr$ defining the region boundaries in A. Likewise for B.*

2. *Impose an arbitrary order on each of these sets (if one has not yet been imposed).*

3. *Form the sets $\mathbf{v}(C) = \mathbf{v}(A) \cup \mathbf{v}(B)$, $\mathbf{h}(C) = \mathbf{h}(A) \cup \mathbf{h}(B)$, $\mathbf{d}(C) = \mathbf{d}(A) \cup \mathbf{d}(B)$, and $\mathbf{a}(C) = \mathbf{a}(A) \cup \mathbf{a}(B)$*

   Impose an arbitrary order on the sets $\mathbf{v}(C)$, $\mathbf{h}(C)$, $\mathbf{d}(C)$ and $\mathbf{a}(C)$ that
   (a) is consistent with the orders on $\mathbf{v}(X)$, $\mathbf{h}(X)$, $\mathbf{d}(X)$ and $\mathbf{a}(X)$ for $X = A$ and $X = B$

   (b) places the resulting abstract matrix in generic configuration.
   It may be necessary to embed in a larger matrix to do this.
   Let $\mathbf{v}(C)_k, \mathbf{h}(C)_k, \mathbf{d}(C)_k$ and $\mathbf{a}(C)_k$ denote the k-th element of the corresponding set in the imposed order.
   Let $\mathbf{v}(C)_0 = 1$, $\mathbf{v}(C)_{\#\mathbf{v}(C)+1} = \#\text{cols} + 1$
   Let $\mathbf{h}(C)_0 = 1$, $\mathbf{v}(C)_{\#\mathbf{h}(C)+1} = \#\text{rows} + 1$
   Let $\mathbf{d}(C)_0 = 1 - \#\text{cols}$, $\mathbf{d}(C)_{\#\mathbf{d}(C)+1} = \#\text{rows}$
   Let $\mathbf{a}(C)_0 = 2$, $\mathbf{a}(C)_{\#\mathbf{d}(C)+1} = \#\text{rows} + \#\text{cols} + 1$

4. *Form the regions*
   $\xi_{i,\mathbf{h}(C)_h,\mathbf{h}(C)_{h+1}} \, \xi_{j,\mathbf{v}(C)_k,\mathbf{v}(C)_{k+1}} \, \xi_{i-j,\mathbf{d}(C)_l,\mathbf{d}(C)_{l+1}} \, \xi_{i-j,\mathbf{a}(C)_m,\mathbf{a}(C)_{m+1}}$
   for $h = 0..\#\mathbf{h}(C), k = 0..\#\mathbf{v}(C), l = 0..\#\mathbf{d}(C), m = 0..\#\mathbf{a}(C)$
   Note: if any of the sets are empty, the corresponding ξ factors are 1.

5. *Construct the expression.*
   Multiply the ξ term for each region by the corresponding sum of a term from A and a term from B.

7   Block Matrix Example

As a further example of our matrix addition algorithm we consider the problem of adding two 2×2 block matrices $A_{n-1 \times m-1} + B_{n-1 \times m-1}$:

$$
\begin{bmatrix} a \ldots a\ b \ldots b \\ \vdots \quad \vdots\ \vdots \quad \vdots \\ a \ldots a\ b \ldots b \\ c \ldots c\ d \ldots d \\ \vdots \quad \vdots\ \vdots \quad \vdots \\ c \ldots c\ d \ldots d \end{bmatrix}
+
\begin{bmatrix} a' \ldots a'\ b' \ldots b' \\ \vdots \quad \vdots\ \vdots \quad \vdots \\ a' \ldots a'\ b' \ldots b' \\ c' \ldots c'\ d' \ldots d' \\ \vdots \quad \vdots\ \vdots \quad \vdots \\ c' \ldots c'\ d' \ldots d' \end{bmatrix}
=
\begin{bmatrix} a_{k \times l} & b_{k \times m-l} \\ c_{n-k \times l} & d_{n-k \times m-l} \end{bmatrix}
+
\begin{bmatrix} a'_{k' \times l'} & b'_{k' \times m-l'} \\ c'_{n-k' \times l'} & d'_{n-k' \times m-l'} \end{bmatrix}
$$

We get 4 different shapes for the combination of the blocks in the sum matrix:

$$
(a) \begin{bmatrix} -\,-\!\!\mid\!\!-\,- \end{bmatrix} \quad
(b) \begin{bmatrix} -\,-\!\!\mid\!\!-\,- \\ -\,-\!\!\mid\!\!-\,- \end{bmatrix} \quad
(c) \begin{bmatrix} -\,-\!\!\mid\!\!-\,-\!\!\mid\!\!-\,- \end{bmatrix} \quad
(d) \begin{bmatrix} -\,-\!\!\mid\!\!-\,-\!\!\mid\!\!-\,- \\ -\,-\!\!\mid\!\!-\,-\!\!\mid\!\!-\,- \end{bmatrix}
$$

which leads to nine different cases, depending on the order of the index variables $k, l, , k', l'$:

$$
\begin{aligned}
(a):&\quad k = k', l = l' \\
(b):&\quad k < k', l = l' \text{ or } k > k', l = l' \\
(c):&\quad k = k', l < l' \text{ or } k = k', l > l' \\
(d):&\quad k < k', l < l' \text{ or } k > k', l < l' \text{ or } k < k', l > l' \text{ or } k > k', l > l'
\end{aligned}
$$

Observe that we assume that the index variables are strictly within the boundaries, i.e., none of the blocks in the original matrices vanish.

After constructing all possible disjoint regions we get the following sum of region terms:

$$
\begin{aligned}
&\xi_{i,1,k}\xi_{j,1,l}\xi_{i,1,k'}\xi_{j,1,l'}(a+a') \;+\; \xi_{i,1,k}\xi_{j,1,l}\xi_{i,1,k'}\xi_{j,l',m}(a+b') \\
+\;&\xi_{i,1,k}\xi_{j,1,l}\xi_{i,k',n}\xi_{j,1,l'}(a+c') \;+\; \xi_{i,1,k}\xi_{j,1,l}\xi_{i,k',n}\xi_{j,l',m}(a+d') \\
+\;&\xi_{i,1,k}\xi_{j,l,m}\xi_{i,1,k'}\xi_{j,1,l'}(b+a') \;+\; \xi_{i,1,k}\xi_{j,l,m}\xi_{i,1,k'}\xi_{j,l',m}(b+b') \\
+\;&\xi_{i,1,k}\xi_{j,l,m}\xi_{i,k',n}\xi_{j,1,l'}(b+c') \;+\; \xi_{i,1,k}\xi_{j,l,m}\xi_{i,k',n}\xi_{j,l',m}(b+d') \\
+\;&\xi_{i,k,n}\xi_{j,1,l}\xi_{i,1,k'}\xi_{j,1,l'}(c+a') \;+\; \xi_{i,k,n}\xi_{j,1,l}\xi_{i,1,k'}\xi_{j,l',m}(c+b') \\
+\;&\xi_{i,k,n}\xi_{j,1,l}\xi_{i,k',n}\xi_{j,1,l'}(c+c') \;+\; \xi_{i,k,n}\xi_{j,1,l}\xi_{i,k',n}\xi_{j,l',m}(c+d') \\
+\;&\xi_{i,k,n}\xi_{j,l,m}\xi_{i,1,k'}\xi_{j,1,l'}(d+a') \;+\; \xi_{i,k,n}\xi_{j,l,m}\xi_{i,1,k'}\xi_{j,l',m}(d+b') \\
+\;&\xi_{i,k,n}\xi_{j,l,m}\xi_{i,k',n}\xi_{j,1,l'}(d+c') \;+\; \xi_{i,k,n}\xi_{j,l,m}\xi_{i,k',n}\xi_{j,l',m}(d+d')
\end{aligned}
$$

Our algorithm now selects one maximal expression, which is one of the 3×3 matrices under case (d), e.g., the first case with order $(k < k', l < l')$. This corresponds to the left matrix below. Observe, that the choice of variable orderings is independent for the vertical and horizontal parameters in this particular case.

$$1 < k < k' < n;\ 1 < l < l' < m \qquad\qquad 1 < k' < k < n;\ 1 < l' < l < m$$

$$\begin{bmatrix} a + a''_{\,|}b + a'_{\,|}b + b' \\ \hline c + a'_{\,|}d + a'_{\,|}d + b' \\ \hline c + c'_{\,|}d + c'_{\,|}d + d' \end{bmatrix} \qquad\qquad \begin{bmatrix} a + a''_{\,|}a + b'_{\,|}b + b' \\ \hline a + c'_{\,|}a + d'_{\,|}b + d' \\ \hline c + c'_{\,|}c + d'_{\,|}d + d' \end{bmatrix}$$

We now observe what happens if we rewrite the expression for the left matrix above into the right matrix above by changing the variable ordering. Here the right hand side to the last case in (d), respectively. Firstly, after fixing the order for the chosen maximal representation we get the following reduced sum of region expressions.

$$\xi_{i,1,k}\xi_{j,1,l}(a + a') \; + \; \xi_{i,1,k}\xi_{j,l,l'}(b + a') \; + \; \xi_{i,1,k}\xi_{j,l',m}(b + b')$$
$$+ \; \xi_{i,k,k'}\xi_{j,1,l}(c + a') \; + \; \xi_{i,k',n}\xi_{j,1,l}(c + c') \; + \; \xi_{i,k,k'}\xi_{j,l,l'}(d + a')$$
$$+ \; \xi_{i,k,k'}\xi_{j,l',m}(d + b') \; + \; \xi_{i,k',n}\xi_{j,l,l'}(d + c') \; + \; \xi_{i,k',n}\xi_{j,l',m}(d + d')$$

After changing the ordering of variables the ξ terms are rewritten using the algorithm of the previous section, which will introduce sums of ξ terms as well as negative regions:

$$(\xi_{i,1,k'}\xi_{j,1,l'} + \xi_{i,1,k'}\xi_{j,l',l} + \xi_{i,k',k}\xi_{j,1,l'} + \xi_{i,k',k}\xi_{j,l',l})(a + a')$$
$$+ \qquad\qquad\qquad (-\xi_{i,1,k'}\xi_{j,l',l} - \xi_{i,k',k}\xi_{j,l',l})(b + a')$$
$$+ \; (\xi_{i,1,k'}\xi_{j,l',l} + \xi_{i,1,k'}\xi_{j,l,m} + \xi_{i,k',k}\xi_{j,l',l} + \xi_{i,k',k}\xi_{j,l,m})(b + b')$$
$$+ \qquad\qquad\qquad (-\xi_{i,k',k}\xi_{j,1,l'} - \xi_{i,k',k}\xi_{j,l',l})(c + a')$$
$$+ \; (\xi_{i,k',k}\xi_{j,1,l'} + \xi_{i,k,n}\xi_{j,1,l'} + \xi_{i,k',k}\xi_{j,l',l} + \xi_{i,k,n}\xi_{j,l',l})(c + c')$$
$$+ \qquad\qquad\qquad\qquad\qquad (\xi_{i,k',k}\xi_{j,l',l})(d + a')$$
$$+ \qquad\qquad\qquad (-\xi_{i,k',k}\xi_{j,l',l} - \xi_{i,k',k}\xi_{j,l,m})(d + b')$$
$$+ \qquad\qquad\qquad (-\xi_{i,k',k}\xi_{j,l',l} - \xi_{i,k,n}\xi_{j,l',l})(d + c')$$
$$+ \; (\xi_{i,k,n}\xi_{j,l',l} + \xi_{i,k',k}\xi_{j,l,m} + \xi_{i,k',k}\xi_{j,l',l} + \xi_{i,k,n}\xi_{j,l,m})(d + d')$$

Finally subsequent simplification of the above sum yields a new expression that corresponds indeed to the desired result matrix. Moreover, the result is again a maximal representation and thus would allow for further transformation into any one of the 8 other cases.

$$\xi_{i,1,k'}\xi_{j,1,l'}\,(a + a') + \xi_{i,1,k'}\xi_{j,l',l}\,(a + b') + \xi_{i,k',k}\xi_{j,1,l'}\,(a + c')$$
$$+ \; \xi_{i,k',k}\xi_{j,l',l}\,(a + d') + \xi_{i,1,k'}\xi_{j,l,m}\,(b + b') + \xi_{i,k',k}\xi_{j,l,m}\,(b + d')$$
$$+ \; \xi_{i,k,n}\xi_{j,1,l'}\,(c + c') + \xi_{i,k,n}\xi_{j,l',l}\,(c + d') + \xi_{i,k,n}\xi_{j,l,m}\,(d + d')$$

8   Conclusion

We have presented an approach to abstract matrix addition, that enables reasoning on arithmetic closure properties for classes of structured matrices. The approach addresses the shortcomings of our previous approach as presented in [7] that uses half-plane constraints as support functions and that can suffer combinatorial problems in that all regions of all resulting matrix cases have to be represented explicitly. The new approach instead works with a more compact representation that implicitly contains all possible cases and therefore scales up better during a sequence of computations.

The work can use our previously developed parsing procedure for abstract matrices that determines their meaning and represents them in terms of the homogeneous regions they contain, thereby making them available as templates for concrete matrices [4]. Our work is related to previous work by Fateman in Macsyma [2], in which indefinite matrices can be subjected to some basic algebraic manipulations. While his matrices are indefinite in size, their elements are fixed to one particular functional expression and cannot be of arbitrary composition. The work is also similar in spirit to earlier work by Watt [9,10], which presented algorithms for GCD, factorisation and functional decomposition of polynomials with terms of symbolic degree and to work by Knauers and Schneider [3] on indefinite symbolic summation using unspecified sequences of summands.

While in [7] we have presented a full system for abstract matrix arithmetic, i.e., addition, multiplication, and their combination, our approach presented in this paper so far only extends to addition. Since the primary property of ξ relies on cancellation of region elements when orders of parameters are reversed, a comparable approach for multiplication would need to involve division by the region elements. A first approach of lifting ξ expressions to abstract matrix multiplication has been presented in [6], which had the drawback that abstract matrix representations for addition and multiplication were incompatible. This is a problem to be addressed further in the future.

References

1. Carette, J.: A canonical form for some piecewise defined functions. In: Proc. ISSAC 2007, pp. 77–84. ACM Press, New York (2007)
2. Fateman, R.: Manipulation of matrices symbolically (2003), http://http.cs.berkeley.edu/~fateman/papers/symmat2.pdf
3. Kauers, M., Schneider, C.: Application of unspecified sequences in symbolic summation. In: Proceedings of ISSAC 2006, pp. 177–183 (2006)
4. Sexton, A.P., Sorge, V.: Abstract matrices in symbolic computation. In: Proceedings of ISSAC 2006, pp. 318–325. ACM Press, New York (2006)
5. Sexton, A.P., Sorge, V., Watt, S.M.: Abstract matrix arithmetic. In: International Symposium on Symbolic and Algebraic Computation (ISSAC 2008) (2008) (submitted)
6. Sexton, A.P., Sorge, V., Watt, S.M.: Abstract matrix arithmetic. In: Proceedings of SYNASC 2008. IEEE Computer Society Press, Los Alamitos (2009)
7. Sexton, A.P., Sorge, V., Watt, S.M.: Computing with abstract matrix structures. In: ISSAC 2009 (submitted, 2009)
8. Watt, S.M.: Making computer algebra more symbolic. In: Transgressive Computing, pp. 43–49 (2006)
9. Watt, S.M.: Two families of algorithms for symbolic polynomials. In: Computer Algebra 2006: Advances in Symb. Algorithms, pp. 193–210. World Scientific, Singapore (2006)
10. Watt, S.M.: Functional decomposition of symbolic polynomials. In: Proc. of ICCSA, pp. 353–362. IEEE Computer Society Press, Los Alamitos (2008)

Invariant Properties of Third-Order Non-hyperbolic Linear Partial Differential Operators

Ekaterina Shemyakova

Research Institute for Symbolic Computation (RISC),
J.Kepler University,
Altenbergerstr. 69, A-4040 Linz, Austria
kath@risc.uni-linz.ac.at,
http://www.risc.uni-linz.ac.at

Abstract. A test in terms of invariants for the existence of a factorization of a bivariate, non-hyperbolic third-order Linear Partial Differential Operator (LPDO) which has a given factorization of its principal symbol is found. The invariants that are used are with respect to known gauge transformations, which is together with constructive factorization itself are essentially involved in modern exact integration algorithms. The invariants, and even a generating system of those were found in previous paper using Moving Frames methods.

In order to find the expressions in terms of invariants that guarantee the existence of a factorization of a certain type, we show that the operation of taking the formal adjoint can be also defined in terms of invariants, that is for equivalence classes of LPDOs, and explicit formulae defining this operation in the space invariants are obtained. The operation of formal adjoint is highly interesting for factorization of LPDOs for if the initial operator has a factorization, its adjoint has also one, and they are related. (informally, the factorization types are symmetric).

1 Introduction

The factorization of Linear Partial Differential Operators (LPDOs) is an essential part of recent algorithms for the exact solution for Linear Partial Differential Equations (LPDEs). Examples of such algorithms include numerous generalizations and modifications of the 18th-century Laplace Transformations Method [1,2,3,4,5,6,7,8], the Loewy decomposition method [9,10,11], and others.

Thus a constructive factorization algorithm for a general LPDO is in heavy need for it will be a part of the exact integration algorithms of LPDEs implemented in a computer algebra system and then perhaps "trusted" by a proof assistant. In the present paper we suggest a test, or in other words a set of constrains on invariants of the given LPDO, which if satisfied, guarantee existence of a factorization of this or another type. We use the invariants, or more exactly, the generating systems of invariants obtained in [12]. These invariants

J. Carette et al. (Eds.): Calculemus/MKM 2009, LNAI 5625, pp. 154–169, 2009.
© Springer-Verlag Berlin Heidelberg 2009

are invariants under the gauge transformations of the operator L, $L \to L^g = g(x, y)^{-1} \circ L \circ g(x, y)$. The transformations are widely used in the mentioned integration algorithms.

In the present paper we use generating systems of invariants found in [12]. In addition we give a constructive prove that the generating systems of invariants obtained there indeed generate all other possible differential invariants, while the moving frames paper [12] based its results on the theoretical foundation of the Moving Frames theory [13,14,15,16]. Given two LPDOs L_1, L_2 that have exactly the same values of the corresponding invariants of the system of invariants of [12], we show how to construct $g = g(x, y)$ such that $L = L^g = g(x, y)^{-1} \circ L \circ g(x, y)$.

Or many properties appearing in connection with the factorization of an LPDO are invariant. Given a generating set of (differential) invariants under such transformations, these invariant properties can be expressed in terms of the generating invariants only. Factorization itself is invariant under the gauge transformations. Indeed, if for some LPDO L, $L = L_1 \circ L_2$, then $L^g = L_1^g \circ L_2^g$. The highest order terms in the factors of such factorizations are invariant also.

The classical Laplace-Darboux-Transformations Method [17] for hyperbolic operators of the second order,

$$L = D_x \circ D_y + aD_x + bD_y + c \; , \; a = a(x, y) \; , \; b = b(x, y) \; , \; c = c(x, y) \; , \quad (1)$$

is an example of the use of invariants (with respect to the gauge transformations) to describe certain invariant properties. If L has a factorization, the equation corresponding to L, $L(z) = 0$ can be solved in quadratures. If L has no factorization, only two incomplete factorizations of (1) are possible:

$$L = (D_x + b) \circ (D_y + a) + h \; , \quad (2)$$
$$= (D_y + a) \circ (D_x + b) + k \; , \quad (3)$$

where $h = c - a_x - ab$ and $k = c - b_y - ab$. These functions are invariants of (1) with respect to gauge transformations, and are called the *Laplace invariants*. Also it can be proved that two Laplace invariants form together a minimal generating set of invariants of (1) with respect to the gauge transformations. Then one can easily define the property of the existence of a factorization (of certain factorization type) invariantly. The definition of factorization type is given in Sec. 2. We have that type $(X)(Y)$, *viz.* (2), requires $h = 0$, and type $(Y)(X)$, *viz.* (3), requires $k = 0$.

In Laplace's case, the generating invariants h and k could be simply obtained from the incomplete factorizations, and moreover it was easy to derive the invariant definition of the property of the existence of a factorization (of certain factorization type). For *hyperbolic* operators of higher order, the situation is drastically different: the "remainder" of an incomplete factorization is not invariant in the generic case; the invariant conditions of the existence of a factorization are not trivial. The third-order case was first addressed by [18], who found a complete generating set for third-order bivariate hyperbolic operators, and later [19] found, for the same operators, invariant necessary and sufficient conditions for factorizations extending given factorizations of the operator's principal symbol.

In the present paper we study the problem of (constructive) factorization of *non-hyperbolic* LPDOs, which is very rarely a subject of investigations. The main difficulty lies in the non-uniqueness of factorizations of such LPDOs. This is in contrast to other situations. For example, given two irreducible factorizations of a Linear *Ordinary* Differential Operator, they have the same number of factors and the factors are pairwise "similar" in some transposed order [20]; for a hyperbolic LPDO, the factorization is determined uniquely by a factorization of the operator's (principal) symbol [9]. This is contrasted by the case of non-hyperbolic, non-ordinary LPDOs, for which there is an interesting example given by Landau [21]: the operator $L = D_x^3 + x D_x^2 D_y + 2 D_x^2 + (2x + 2) D_x D_y + D_x + (2 + x) D_y$ has two factorizations into different numbers of irreducible factors:

$$L = Q \circ Q \circ P = R \circ Q \, ,$$

for the operators $P = D_x + x D_y$, $Q = D_x + 1$, $R = D_{xx} + x D_{xy} + D_x + (2+x) D_y$. Parametric factorizations can appear, and their structure for LPDOs of orders $2, 3, 4$ has been described in [22]. No invariants of non-hyperbolic LPDOs of orders higher than two were been known before the recent work [12], where methods of regularized moving frames have been applied to describe a way to obtain generating sets of invariants for hyperbolic and non-hyperbolic LPDOs of arbitrary orders and also for pairs of operators.

The present paper uses the results above to describe the property of the existence of a factorization of all possible factorization types for the equivalence classes of third-order bivariate non-hyperbolic LPDOs, or in other words in terms of invariants. Since the coefficients of LPDOs cannot be expressed in terms of invariants (since the coefficients are not invariants), then the constraints on the invariants that guarantee the existence of a factorization of some factorization type cannot be obtained trivially.

Also as it was already noticed in [9], the problem of factorizations of even hyperbolic operators can be solved algebraically up to solution of a Ricatti equation. Here some of the results have similar requirements.

Also in the scope of the paper we investigate the classical operation of taking the formal adjoint of an operator, define it on the equivalence classes of the considered LPDOs, and obtain explicit formulae in the space of invariants. Some instances of the latter result allow us to reduce the number of case considerations when finding an invariant definition of the property of the existence of a factorization.

The paper is organized as follows. In Section 2 preliminaries facts and definitions are given. In Sections 3 and 4, operators with the (principal) symbols $X^2 Y$ and X^3 are considered, respectively. Alternative constructive proofs (to ones from [12]) of the theorems that certain sets of invariants form a complete generating set of invariants for such operators are given. The properties of the existence of a factorization of some main factorization types are described in the space of invariants. In Section 5, the operation of taking of the formal adjoint is considered. Such an operation can be defined on the equivalence classes of operators, and explicit formulae for all the cases of non-hyperbolic third-order operators are given. Based on these formulae and the invariant definitions of the properties of the existence of a factorization of the main factorization types obtained in the Sections 3 and 4, we compute the invariant definitions for the rest factorization types.

2   Preliminary Facts

Consider a field K with commuting derivations ∂_x, ∂_y acting on it. Consider the ring of linear differential operators $K[D] = K[D_x, D_y]$, where D_x, D_y correspond to the derivations ∂_x, ∂_y, respectively. Any operator $L \in K[D]$ is of the form $L = \sum_{i+j=0}^{d} a_{ij} D_x^i D_y^j$, where $a_{ij} \in K$. In $K[D]$ the variables D_x, D_y commute with each other but not with elements of K. For $f \in K$ we have the relation $D_i f = f D_i + \partial_i(f)$ for $i = \{x, y\}$, and, therefore, there is non-commutativity, which starts with the second-highest order terms. Thus, the highest order terms of LPDOs in $K[D]$ commute, and the *(principal) symbol* Sym(L) of an operator, which is a polynomial representation of the highest-order terms within the operator is a basic concept of the theory. The commutativity of symbols implies that any factorization of an LPDO extends some factorization of the polynomial Sym(L). These factorizations have to be considered as non-commutative due to the non-commutativity of the considered LPDOs. Thus, if some factorization $L = F_1 \circ \cdots \circ F_k$, $F_i \in K[D]$ extends Sym$(L) = S_1 \cdot \cdots \cdot S_k$, we say that this factorization of L is of the *factorization type* $(S_1) \ldots (S_k)$. In general, given a factorization of the polynomial Sym(L), an extension to a factorization of L is not unique.

An operator $L \in K[D]$ is said to be *hyperbolic* if its symbol is completely factorable (all factors are of first order) and each factor has multiplicity one. Otherwise the operator is *non-hyperbolic*.

Let K^* denote the set of invertible elements in K. Then for $L \in K[D]$ and every $g \in K^*$ consider the gauge transformation $L \to g^{-1} \circ L \circ g$. Then an algebraic differential expression I in coefficients of L is *invariant* under the gauge transformations (we consider only such invaraints in the present paper) if it is unaltered by these transformations. Trivial examples of an invariant are coefficients of the symbol of the operator. A complete generating set of invariants is a basis in which all possible differential invariants can be expressed. Also the property of having a factorization (or a factorization extending a certain factorization of the symbol) is invariant. Indeed, let $L = F_1 \circ F_2 \circ \cdots \circ F_k$, for some operators $F_i \in K[D]$. Then for every $g \in K^*$ we have $g^{-1} \circ L \circ g = (g^{-1} \circ F_1 \circ g) \circ (g^{-1} \circ F_2 \circ g) \circ \cdots \circ (g^{-1} \circ F_k \circ g)$.

In the present paper we consider LPDOs that are of third-order and non-hyperbolic and their principal symbols has a factorization into first-order factors. For such operators there is a normalized form that the (principal) symbol becomes $X^2 Y$ or X^3.

3   Operators of Symbol X^2Y

Consider class of operators of the form

$$L = D_{xxy} + \sum_{i+j=0}^{2} a_{ij} D_x^i D_y^j , \quad a_{ij} = a_{ij}(x,y) . \tag{4}$$

Theorem 1. *[12] For LPDOs (4) the followings form a generating set of invariants.*

$$\left.\begin{array}{l} I_1 = a_{02} , \\ I_2 = a_{11y} - 2a_{20x} , \\ I_3 = a_{10} - a_{20}a_{11} - 2a_{20x} , \\ I_4 = a_{01} - \frac{1}{4}a_{11}^2 - 2a_{02}a_{20} - \frac{1}{2}a_{11x} , \\ I_5 = a_{00} - \frac{1}{2}a_{10}a_{11} - a_{01}a_{20} + \frac{1}{2}a_{20}a_{11}^2 - a_{02}a_{20y} + a_{02}a_{20}^2 - a_{20xx} . \end{array}\right\} \tag{5}$$

As we promise in the introductory section, we give (right below) a constructive proof of this theorem, while the work [12] provides a non-constructive one using the properties of the regularized moving frames.

Proof. One can verify by straightforward computations that the five functions from the statement of the theorem are invariants. Then if

$$g^{-1}Lg = L_1 , \tag{6}$$

holds for two operators L, L_1 of the form (4), then they belong to the same equivalence class, and the corresponding invariants are the same.

Now prove the opposite direction. Suppose we have two operators L, L_1 of the form (4) (let the coefficients of L be denoted by a_{ij} and those of L_1 by b_{ij}), and the values of the invariants are same correspondingly. We construct such g that (6) holds. It is convenient to consider g in the form $g = \exp(f)$, $f = f(x,y) \in K$. The equality of the invariants I_1 of the operators implies $b_{02} = a_{02}$. Then using this substitutions for the equalities of the invariants I_3, I_4, I_5, we have $b_{10} = -a_{10} + a_{20}a_{11} - b_{20}b_{11} + 2(a_{20} - b_{20})_x$, $b_{01} = -a_{01} + 1/4(a_{11}^2 - b_{11}^2) + 2a_{02}(a_{20} - b_{20}) + 1/2(a_{11} - b_{11})_x$, $b_{00} = a_{00} + a_{02}(b_{20} - a_{20})^2 + a_{01}(b_{20} - a_{20}) + b_{11}(b_{20} - a_{20})_x + a_{02}(b_{20} - a_{20})_y + (b_{20} - a_{20})_{xx} + 1/4b_{20}(b_{11}^2 - a_{11}^2) + 1/2(a_{10} - a_{11}a_{20})(b_{11} - a_{11}) + 1/2b_{20}(b_{11} - a_{11})_x$. Using these for (6), the equality of the second order terms on the both sides of (6) implies

$$f_x = (b_{11} - a_{11})/2 , \quad f_y = b_{20} - a_{20} . \tag{7}$$

As the equality of the invariants I_2 of the operators L and L_1 implies

$$(b_{11} - a_{11})_y = 2(b_{20} - a_{20})_x , \tag{8}$$

we have that the equalities (7) define f, and therefore g, uniquely up to a multiple.

Now we verify that such f connects operators L and L_1. Using the expressions for f_x and f_y, we have that the equality (6) holds if and only if $0 = 2b_{20x} - 2a_{20x} - b_{11y} + a_{11y}$, $0 = b_{20xx} - 1/2b_{11y}b_{11} + 1/2a_{11y}b_{11} - a_{20xx} - 1/2b_{11xx} + 1/2a_{11xy} + b_{11}b_{20x} - b_{11}a_{20x}$. The first equality holds as an immediate application of (8). The right hand side of the second one equals $C_x/2 + Cb_{11}/2$, where C is the difference of left and right hand sides of (8), and, therefore, the second equality holds.

Necessary and sufficient conditions of existence of a factorization of certain factorization type can be easily written out in terms of the coefficients of the given LPDO. As, in general, the coefficients are not invariant, and therefore, cannot be expressed in terms of invariants only, which means that we cannot have invariant conditions gratis (as it is in the Laplace case).

Proposition 1 (factorizations of the types $(Y)(X^2)$ and $(Y)(X)(X)$). *Consider an equivalence class of (4) defined by the values I_1, I_2, I_3, I_4, I_5 of invariants (5). The operators of the class has a factorization of the factorization type $(Y)(X^2)$ if and only if*

$$I_5 = -I_{1yy} + \frac{1}{2}I_{2x} + I_{4y} \quad \& \quad I_2 = I_3 . \tag{9}$$

Such factorization is unique when exists. For the existence of a factorization of the type $(Y)(X)(X)$ in addition to the two conditions above it is necessary that

$$I_1 = 0$$

holds for the class.

Proof. First we express all the coefficients of L but a_{20} and a_{11} in terms of these invariants:

$$\left.\begin{array}{l} a_{02} = I_1 , \quad a_{10} = I_3 + a_{20}a_{11} + 2a_{20x} , \\ a_{01} = I_4 + 2a_{20}I_1 + \frac{1}{4}(a_{11}^2 + 2a_{11x}) , \\ a_{00} = I_5 + (a_{20}^2 + a_{20y})I_1 + \frac{a_{11}}{2}I_3 + a_{20}I_4 + \\ \quad a_{11}a_{20x} + a_{20xx} + (a_{20}a_{11}^2 + 2a_{20}a_{11x})/4 . \end{array}\right\} \tag{10}$$

Then a factorization $F_{(Y)(X^2)} = (D_y + r) \circ (D_{xx} + aD_x + bD_y + c)$, where all the coefficients are functions of x and y, takes place if and only if $L - F_{(Y)(X^2)} = 0$. The latter occurs if and only if $a = a_{11}$, $r = a_{20}$, $b = I_1$, $c = I_4 + a_{20}I_1 - I_{1y} + (a_{11}^2 + 2a_{11x})/4$ (and therefore, all the coefficients of the factorization have been determined), and the following two conditions hold: $0 = I_5 + I_3a_{11}/2 - I_{4y} + I_{1yy} - (a_{11}a_{11y} + a_{11xy} - 2a_{11}a_{20x} - 2a_{20xx})/2$, $0 = I_3 + 2a_{20x} - a_{11y}$. Employing the expression for I_2 from (5), one can prove that this system is equivalent to the system (9). The second factor of $F_{(Y)(X^2)}$ may be factorable only if $b = 0$, that is $I_1 = 0$ hold.

Proposition 2 (factorizations of the types $(X)(XY)$, $(X)(X)(Y)$, and $(X)(Y)(X)$). *Consider an equivalence class of (4) defined by the values of the* *invariants I_1, I_2, I_3, I_4, I_5. Operators of the class are factored with the factorization type $(X)(XY)$ if and only if*

$$
\left[
\begin{array}{llll}
A: & I_3 \neq 0 & \& \quad I_1 = 0 & \& \quad 0 = I_4 + I_6^2 + I_{6x}, \quad \text{or} \\
B: & I_3 = I_1 = I_5 = 0 & \& \quad \text{Ricatti equation} \quad 0 = I_4 + u^2 + u_x \quad \text{can be solved};
\end{array}
\right.
$$

with the factorization type $(X)(X)(Y)$ if and only if the set of conditions B holds, and with the factorization type $(X)(Y)(X)$ if and only if

$$
\left[
\begin{array}{lll}
A & \& & I_2/2 - I_3 - I_{6y} = 0, \quad \text{or} \\
B & \& & I_2 \neq 0 \quad \& \quad I_2 - I_{7y} = I_2/2, \quad \text{or} \\
B & \& & I_2 = 0 \quad \& \quad I_{4y} = 0,
\end{array}
\right.
$$

where denote invariants $I_6 = (I_5 - I_{3x})/I_3$, $I_7 = (I_{4y} + I_{2x}/2)/I_2$.

Proof. Consider when a factorization $F_{(X)(XY)} = (D_x + r) \circ (D_{xy} + a D_x + b D_y + c)$ exist for for an operator L of the equivalence class.

Let $I_3 \neq 0$ for the given equivalence class, then the latter equality holds if and only if $I_1 = 0$, $a = a_{20}$, $b = a_{11} - r$, $c = I_3 + a_{20} a_{11} + a_{20x} - r a_{20}$, $r = (2I_5 + a_{11}I_3 - 2I_{3x})/(2I_3)$, and $0 = I_4 + (I_5^2 - 3I_5 I_{3x} + 2I_{3x}^2)/I_3^2 + (I_{5x} - I_{3xx})/I_3$. The latter can be rewritten in the form $0 = I_4 + (I_6)^2 + (I_6)_x$, where we denote $I_6 = (I_5 - I_{3x})/I_3$

The second factor of the factorization $F_{(X)(XY)}$ is a second order hyperbolic LPDO in the normalized form, and therefore, it can be decomposed if and only if one of the Laplace invariants h and k is zero. Substituting obtained above expressions, we compute $h = -I_3$, $k = -I_3 - a_{20x} + a_{11y} - (2I_{5y} + a_{11y}I_3 + a_{11}I_{3y} - 2I_{3xy})/(2I_3) + 1/2(2I_5 + a_{11}I_3 - 2I_{3x})/I_3^2 I_{3y}$. As $I_3 \neq 0$, we have $h \neq 0$, and, therefore, the operators of the given equivalence class can never have a factorization of the factorization type $(X)(X)(Y)$, provided $I_3 \neq 0$ holds for this class. The expression for k can be simplified employing (5) as $k = I_2/2 - I_3 - I_{5y}/I_3 + I_{3xy}/I_3 + I_{3y}I_5/I_3^2 - I_{3y}I_{3x}/I_3^2$, which can be rewritten as $k = I_2/2 - I_3 - (I_6)_y$. Thus, the operators of an equivalence class for which $I_3 \neq 0$ holds have a factorization of the factorization type $(X)(Y)(X)$ if and only if $I_2/2 - I_3 - I_{6y} = 0$.

Let now $I_3 = 0$. Then the factorization $F_{(X)(XY)}$ of L takes place if only if $I_1 = 0$, $a = a_{20}$, $b = a_{11} - r$, $c = a_{20}a_{11} + a_{20x} - r a_{20}$, and the following two equalities hold: $I_5 = 0$, and $0 = I_4 + a_{11}^2/4 - a_{11x}/2 - r a_{11} + r^2 + r_x$. The latter can be rewritten in the form

$$
0 = I_4 + u^2 + u_x, \quad u = r - \frac{a_{11}}{2}, \tag{11}
$$

and this is a defining equation for r.

Compute the Laplace invariants for the second factor of the factorization $F_{(X)(XY)}$ substituting the obtained above expressions, we have: $h = 0$, $k = -r_y + \frac{a_{11y}}{2} + I_2/2 = -u_y + I_2/2$. Thus, for the considered equivalence classes, the second factor of $F_{(X)(XY)}$ has a factorization of the factorization type $(X)(Y)$ always, and that of the factorization type $(Y)(X)$ if and only if $u_y = I_2/2$. Consideration of factorizations of the second factor of $F_{(X)(XY)}$ makes sense only if there is a factorization of the operators of the class of the factorization type $(X)(XY)$, the condition (11) should be satisfied. Differentiating the both sides of this condition with respect to y, we have $0 = I_{4y} + 2uu_y + (u_y)_x = I_{4y} + uI_2 + I_{2x}/2$, and, therefore, $u = -(I_{4y} + I_{2x}/2)/I_2$, provided $I_2 \neq 0$. Thus, the condition of factorization of the type $(Y)(X)$ in this case is $-((I_{4y} + I_{2x}/2)/I_2)_y = I_2/2$.

Suppose now $I_2 = 0$, then the second factor of $F_{(X)(XY)}$ has a factorization of the factorization type $(Y)(X)$ if and only if $k = -u_y = 0$, that is $u = u(x)$. In this case condition (11) implies that I_4 does not depends on y, that is $I_{4y} = 0$. Then the solution of such equations is of the form $u = u_0(x) + F_1(y)$, where $u_0(x)$ is solution of the equation as an ordinary differential equation, and F_1 is an arbitrary function. Thus for every $I_4 = I_4(x)$ we have $u = u_0$ which satisfy condition $u = u(x)$.

4   LPDOs of Symbol X^3

Consider class of operators of the form

$$L = D_x^3 + \sum_{i+j=0}^{2} a_{ij}(x,y) D_x^i D_y^j . \tag{12}$$

Further below we denote $a_{ij} = a_{ij}(x,y)$. The class admits gauge transformations, and a complete generating sets of invariants has been recently obtained [12]:

Theorem 2. *Consider class of operators (12). Then $I^{a_{02}} = a_{02}$ and $I^{a_{11}} = a_{11}$ are invariants.*

1. *For the equivalence classes of (12) having the property $I^{a_{02}} \neq 0$ five invariants*

$$\left.\begin{array}{l}
I^{a_{11}} = a_{11} , \quad I^{a_{02}} = a_{02} , \\
I^{a_{10}} = a_{10} - \frac{1}{2} a_{01} a_{11}/a_{02} - \frac{1}{3} a_{20}^2 + \frac{1}{6} a_{11}^2 a_{20}/a_{02} - a_{20x} , \\
I_x^{a_{01}} = a_{01x} + a_{02x}(\frac{1}{3} a_{11} a_{20} - a_{01})/a_{02} - \frac{1}{3}(a_{11} a_{20})_x - \frac{2}{3} a_{02} a_{20y} , \\
I^{a_{00}} = a_{00} + a_{11} a_{20}(a_{01} - \frac{1}{6} a_{11} a_{20})/(6a_{02}) - \frac{1}{3} a_{10} a_{20} - \frac{1}{4a_{02}} a_{01}^2 \\
\quad + \frac{2}{27} a_{20}^3 - \frac{a_{01y}}{2} + \frac{a_{11y} a_{20} - a_{11} a_{20y}}{6} - \frac{a_{20xx}}{3} \\
\quad + a_{02y}(\frac{1}{2} a_{01} - \frac{1}{6} a_{11} a_{20})/a_{02} .
\end{array}\right\} \tag{13}$$

   form together a complete generating set of invariants.
2. *For the equivalence classes of (12) possessing the both properties $I^{a_{02}} = 0$ and $I^{a_{11}} \neq 0$ four invariants*

$$\left.\begin{array}{l}
I^{a_{11}} = a_{11} , \\
I_x^{a_{10}} = a_{10x} - \frac{2}{3} a_{20} a_{20x} + a_{11x}(\frac{1}{3} a_{20}^2 - a_{10} + a_{20x})/a_{11} - \\
\quad \frac{1}{3} a_{11} a_{20y} - a_{20xx} , \\
I^{a_{01}} = a_{01} - \frac{1}{3} a_{11} a_{20} , \\
I^{a_{00}} = a_{00} + a_{01}(a_{20x} - a_{10})/a_{11} - \frac{1}{27} a_{20}^3 - \\
\quad \frac{1}{3}(a_{11} a_{20y} + a_{20} a_{20x} + a_{20xx} - a_{01} a_{20}^2/a_{11}) .
\end{array}\right\} \tag{14}$$

   form together a complete generating set of invariants.

3. *For the equivalence classes of (12) possessing the both properties $I^{a_{02}} = 0$ and $I^{a_{11}} = 0$ three invariants*

$$\left.\begin{aligned}
I^{a_{10}} &= a_{10} - \tfrac{1}{3}a_{20}^2 - a_{20x} , \quad I^{a_{01}} = a_{01} , \\
I_x^{a_{00}} &= a_{00x} - \tfrac{1}{3}(a_{10}a_{20} + a_{20xx})_x - \tfrac{1}{3}a_{01}a_{20y} + \tfrac{2}{9}a_{20}^2 a_{20x} \\
&\quad + \tfrac{1}{3}\,a_{01x}(a_{10}a_{20} + a_{20xx} - \tfrac{2}{9}\,a_{20}^3 - 3a_{00})/a_{01} .
\end{aligned}\right\} \quad (15)$$

form together a complete generating set of invariants.

Again we give a constructive proof of the theorem, while a non-constructive one, as well as the formulae for the invariants have been obtained in [12].

Proof. It is easy to verify that the functions from the statement of the theorem are invariants. Then if (6) holds for two operators L, L_1 of the form (12), then they belong to the same equivalence class, and the corresponding invariants are the same.

Now prove the opposite direction. Suppose we have two operators L, L_1 of the form (12) (let the coefficients of L be denoted by a_{ij} and those of L_1 by b_{ij}), and the values of the invariants are same correspondingly. We construct such g that (6) holds. It is convenient to consider g in the form $g = \exp(f)$, $f = f(x, y) \in K$.

1. The equality of the invariants $I^{a_{11}}, I^{a_{02}}$ of the operators implies $b_{02} = a_{02}$, $b_{11} = a_{11}$. Then using this substitutions for the equalities of the invariants $I^{a_{10}}, I^{a_{00}}$, we (by means of straightforward computation) have expressions for b_{10} and b_{00} in terms of b_{20}, b_{01} and the coefficients a_{ij}. When we apply all the obtained substitutions for (6), the equality of the terms at D_{xx} and D_y on the both sides of (6) implies

$$f_x = (b_{20} - a_{20})/3 , \quad f_y = (b_{01} - a_{01} - a_{11}(b_{20} - a_{20})/3)/(2a_{02}) . \quad (16)$$

The equality of the invariants I_4 of the operators L and L_1 guarantees that if we differentiate the first equation with respect to y, the second with respect to x, and then subtract, then we will have a true equality. Thus, f is defined uniquely up to a multiple.

Such f connects operators L and L_1, which one can (in this case straightforwardly) verify by the substitution of the expressions (16) for f_x and f_y into (6).

2. Analogously, one can prove that some function f is defined uniquely up to a multiple by the equalities $f_x = (b_{20} - a_{20})/3$, $f_y = (b_{10} - a_{10} + (a_{20}^2 - b_{20}^2)/3 - (b_{20} - a_{20})_x)/a_{11}$, and such f connects L and L_1.

3. Analogously, one can prove that some function f is defined uniquely up to a multiple by the equalities $f_x = (b_{20} - a_{20})/3$, $f_y = (3b_{00} - 3a_{00} - a_{10}(b_{20} - a_{20}) - (b_{20} - a_{20})_{xx} - b_{20}(b_{20} - a_{20})_x + a_{20}^2 b_{20}/3 - 2a_{20}^3/9 - b_{20}^3/9)/(3a_{01})$, and such f connects L and L_1.

Proposition 3 (factorizations of the type $(X)(X^2)$). *Consider the equivalence classes of (12) given by the values of the invariants from Theorem 2.*
1. Operators of the classes having the property $I^{a_{02}} \neq 0$ have no factorizations of the factorization type $(X)(X^2)$.

2. *Operators of a class having the both properties $I^{a_{02}} = 0$ and $I^{a_{11}} \neq 0$, and are given by the values I_1, I_2, I_3, I_4 of the invariants $I^{a_{11}}, I^{a_{01}}, I^{a_{10}}_x, I^{a_{00}}$, correspondingly, have a factorization of the factorization type $(X)(X^2)$ if and only if*

$$(I_{1xxx} - I_{2xx})I_1^2 + (6I_2 - 11I_{1x})I_{1x}I_2 + (5I_{2x} - 6I_{1xx})I_1I_{1x} - 3I_2I_{2x}I_1 +$$
$$4I_2I_{1xx}I_1 + (I_4 - I_3)I_1^3 - I_2^3 + 6I_{1x}^3 = 0 . \tag{17}$$

3. *If the operators of a class having the both properties $I^{a_{02}} = 0$ and $I^{a_{11}} = 0$, and is given by the values of the invariants $I^{a_{10}}, I^{a_{01}}, I^{a_{00}}_x$ can be factored with the factorization type $(X)(X^2)$ then*

$$I^{a_{01}} = 0$$

holds for this class. That is only equivalence classes of ordinary differential operators may have such factorizations.

Proof. 1. Consider an operator (12), and (formally) its factorization $F_{(X)(X^2)} = (D_x + r) \circ (D_{xx} + aD_x + bD_y + c)$, where all the coefficients are functions of x and y. The equality $L - F_{(X)(X^2)} = 0$ implies $I^{a_{02}} = 0$.

2. Let $I^{a_{02}} = 0$ and $I^{a_{11}} \neq 0$. Then, by the Theorem 2, the values of four invariants (14) uniquely define such equivalence class. We start with expressing as many as possible coefficients (12) in terms of these basic invariants and the remained coefficients. Thus, we have, for example,

$$\left.\begin{array}{l} a_{11} = I_1 , \quad a_{01} = I_2 + \frac{1}{3}a_{20}I_1 , \\ a_{00} = I_4 + (a_{10} - \frac{1}{3}a_{20}^2 - a_{20x} - \frac{2}{27}a_{20}^3)I_2/I_1 + \frac{1}{3}a_{10}a_{20} \\ \quad + \frac{1}{3}a_{20y}I_1 + \frac{1}{3}a_{20xx} , \end{array}\right\} \tag{18}$$

provided $I_1 \neq 0$. Employing these expressions, we have that $L - F_{(X)(X^2)} = 0$ holds if and only if $a = a_{20} - r$, $b = I_1$, $c = a_{10} - ra_{20} + r^2 - a_{20x} + r_x$, $r = \frac{1}{3}(3I_2 + a_{20}I_1 - 3I_{1x})/I_1$ (that is we determined all the coefficients of the factorization) and some equality in terms of I_1, I_2, I_3, I_4, a_{20}, a_{10} and their derivatives holds. One can simplify the latter constraint using the one for the invariant I_3, which is the only uninvolved yet invariant, and have (17).

If for the operators of an equivalence class of (12) possessing the property $I_1 = 0$ factorizations of the form $F_{(X)(X^2)}$ exist, then $I^{a_{01}} = 0$, that is all the operators of the class are ordinary differential operators.

Remark 1. Factorizations of the type $(X)(X)(X)$ can exist only for operators of the equivalence classes having the properties $I^{a_{02}} = 0$ and $I^{a_{11}} = 0$, and $I^{a_{01}} = 0$.

5   Formal Adjoint

In this section we consider the operation of taking the formal adjoint of an LPDO, and define such operation on the equivalence classes of third-order bivariate non-hyperbolic LPDO. At the end of the section we apply this knowledge to complete the cases' consideration in the finding of invariant condition of the property of the existence of a factorization of certain factorization type.

For an operator $L = \sum_{|J| \leq d} a_J D^J$, where $a_J \in K$, $J \in \mathbf{N}^n$ and $|J|$ is the sum of the components of J, the *formal adjoint* is defined as

$$L^\dagger(f) = \sum_{|J| \leq d} (-1)^{|J|} D^J(a_J f) , \quad \forall f \in K .$$

The formal adjoint possesses the following useful factorization-theoretic properties:

$$(L^\dagger)^\dagger = L , \quad (L_1 \circ L_2)^\dagger = L_2^\dagger \circ L_1^\dagger , \quad \mathrm{Sym}_L = (-1)^{\mathrm{ord}(L)} \mathrm{Sym}_{L^\dagger} .$$

The property of having a factorization is invariant under the operation of taking the formal adjoint, while the property of having a factorization of certain factorization type is not invariant, and an operator L has a factorization of some factorization type $(S_1)(S_2)$ (where $\mathrm{Sym}_L = S_1 S_2$) if and only if $L^\dagger$ has that of factorization type $(S_2)(S_1)$.

Lemma 1. *The operation of taking the formal adjoint can be defined on the equivalence classes of LPDOs.*

Proof. Show that operation of taking the formal adjoint and the gauge transformations of LPDOs commute. For every $g \in K^*$, and $f = g^{-1}$ we have $(g^{-1} \circ L \circ g)^\dagger = g^\dagger \circ L^\dagger \circ (g^{-1})^\dagger = g \circ L^\dagger \circ g^{-1} = f^{-1} \circ L^\dagger \circ f$. This finishes the proof. Indeed, every operator L_2 that belongs to the same equivalence class as L has the form $L_2 = g^{-1} \circ L \circ g$ for some $g = g(x, y)$. The commutativity proved above implies that $L_2^\dagger = f^{-1} \circ L^\dagger \circ f$, and, therefore, $L_2^\dagger$ belong to the same equivalence class as $L^\dagger$. Thus under the operation of taking the formal adjoint an equivalence class transforms into another equivalence class.

Example 1 (LPDOs of order 2). For operators of the form $L = D_{xy} + a D_x + b D_y + c$ there is a complete generating set of invariants that consists of first-order invariants: $h = c - a_x - ab$ and $k = c - b_y - ab$. For the formal adjoint $L^\dagger = D_{xy} - a D_x - b D_x + c - a_x - b_y$ they are $h^\dagger = c - b_y - ab$ and $k^\dagger = c - a_x - ab$, and so $h^\dagger = k$, $k^\dagger = h$.

For LPDOs of the odd orders the symbol changes the sign under the operation of taking the formal adjoint. On the other hand, formulae of invariants are usually given for the symbol in a normalized form. Thus, when considering the common adjoint for the operator of odd orders, the normalized form of symbols is not preserved. On the other hand, invariants of the operator L and $-L$ are not the same (or the same up to a sign) in general case. The following lemma shows how to treat this problem.

Lemma 2. *Let a function I of the coefficients a_J of an operator L of some family of operators with the same symbol Sym be an invariant. Then given $p \in K$, $p \neq 0$, the result of the substitution a_J/p for a_J for every J in I is an invariant of the family of operators $\{pL\}$.*

Proof. Since $p \neq 0$ we can multiply the operator pL by p^{-1} on the left, and get some new operator $L_1 = \mathrm{Sym}_L + \sum_{|J|<d} \frac{a_J}{p} D^J$. The invariants of the operator L and L_1 are the same, and the latter are known.

Proposition 4 (formal adjoint for classes with $\mathrm{Sym} = X^2 Y$). *Consider the equivalence classes of (4) given by the values of the invariants I_1, I_2, I_3, I_4, I_5 (5). Then the operation of taking of the formal adjoint is defined by the following formulae*

$$
\left.
\begin{aligned}
I_1^\dagger &= -I_1 \ , \\
I_2^\dagger &= -I_2 \ , \\
I_3^\dagger &= I_3 - I_2 \ , \\
I_4^\dagger &= I_4 - 2I_{1y} \ , \\
I_5^\dagger &= -I_5 + I_{3x} + I_{4y} - I_{1yy} - \tfrac{1}{2} I_{2x} \ .
\end{aligned}
\right\}
\tag{19}
$$

Proof. Consider an operator L (4) of some equivalence class. and use the obtained above expressions (10) to substitute some of the coefficients of L by the expressions of the five invariants and the remained coefficients. Then compute the formal adjoint: $L^\dagger = -D_{xxy} + a_{20} D_{xx} + a_{11} D_{xy} + I_1 D_{yy} - (I_3 + a_{20} a_{11} - a_{11y}) D_x + (-I_4 - \frac{1}{4} a_{11}^2 - 2 a_{20} I_1 + \frac{1}{2} a_{11x} + 2 I_{1y}) D_y + I_5 + a_{20} I_4 + \frac{1}{2} a_{11} I_3 + (-a_{20y} + a_{20}^2) I_1 - I_{3x} - \frac{1}{2} a_{20} a_{11x} + \frac{1}{4} a_{20} a_{11}^2 - I_{4y} - \frac{1}{2} a_{11} a_{11y} - 2 a_{20} I_{1y} + \frac{1}{2} a_{11xy} + I_{1yy}$. By lemma 2 and Theorem 1 compute the invariants of a complete generating set of invariants. Then it is easy to guess expressions in terms of basic invariants of L: $I_1^\dagger = -I_1$, $I_2^\dagger = -a_{11y} + 2a_{20x} = -I_2$, $I_3^\dagger = I_3 - a_{11y} + 2a_{20x} = I_3 - I_2$, $I_4^\dagger = I_4 - 2I_{1y}$, $I_5^\dagger = -I_5 + I_{3x} + I_{4y} - \frac{1}{2} a_{11xy} - I_{1yy} + a_{20xx} = -I_5 + I_{3x} + I_{4y} - I_{1yy} - \frac{1}{2} I_{2x}$.

Corollary 1 (factorizations of the types $(X^2)(Y)$ and $(X)(X)(Y)$). *Consider an equivalence class of (4) given by the values of the invariants I_1, I_2, I_3, I_4, I_5. Operators of the class has a factorization of the factorization type $(X^2)(Y)$ if and only if*

$$ I_5 = I_3 = 0 $$

Such factorization is unique when exists. If in addition

$$ I_1 = 0 \ , $$

then we have necessary conditions of the existence of the factorization type $(X)(X)(Y)$ for the operators of the class.

Proof. Operators of the given equivalence class are factored with the factorization type $(X^2)(Y)$ if and only if their formal adjoints are factored with the factorization type $(Y)(X^2)$. The equivalence class of their formal adjoints is given by the values $I_1^\dagger, \ldots, I_5^\dagger$ of (19) computed using the given values of $I_1, \ldots, I_5$. By the proposition 1, the operators of this class are factored with the factorization type $(Y)(X^2)$ if and only if $I_5^\dagger = -I_{1yy}^\dagger + \frac{1}{2} I_{2x}^\dagger + I_{4y}^\dagger$ & $I_2^\dagger = I_3^\dagger$, which after the substitutions have the forms $0 = -I_5 + I_{3x}$, $I_3 = 0$, and further $I_5 = I_3 = 0$. Analogously we get the last statement of the corollary.

Corollary 2 (factorizations of the types $(XY)(X)$ and $(Y)(X)(X)$). *Consider an equivalence class of (4) given by the values I_1, I_2, I_3, I_4, I_5 of the invariants (5). Operators of the class are factored with the factorization type $(XY)(X)$ if and only if*

$$\left[\begin{array}{llll} A: & I_3 \neq I_2 & \& \quad I_1 = 0 & \& \quad 0 = I_4 + (I_8)^2 + I_{8x} , \quad \text{or} \\ B: & I_3 = I_2 & \& \quad I_1 = 0 & \& \quad I_9 = 0 , \end{array} \right.$$

with the factorization type $(X)(X)(Y)$ if and only if the set of conditions B holds, where $I_8 = I_9/(I_3 - I_2)$ and $I_9 = -I_5 + I_{4y} + \frac{1}{2}I_{2x}$.

Proof. The proof is analogous to the proof of the corollary (1)

Lemma 3. *Consider the equivalence classes of (12), which are uniquely defined by the values of the invariants $I^{a_{02}}$ and $I^{a_{11}}$ and some other invariants given in the Theorem 2. The property of $I^{a_{02}}$ and $I^{a_{11}}$ to be zero is invariant under the operation of the taking of the formal adjoint.*

Proof. The formal adjoint of an operator L (12) has the form $L^\dagger = -D_{xxx} + a_{20}D_{xx} + a_{11}D_{xy} + a_{02}D_{yy} + T_1$, where T_1 denotes the terms of orders lower than two. Then the statement implies from the fact that $I^{a_{02}} = a_{02}$ and $I_{a_{11}} = a_{11}$.

Proposition 5 (formal adjoint for classes with $\text{Sym} = X^3$). *Consider the equivalence classes of (12), which is defined by the values of the invariants $I^{a_{02}}$ and $I^{a_{11}}$, and the rest invariants as they are given in the Theorem 2.*

1. For an equivalence class that possesses the property $I^{a_{02}} \neq 0$, and the corresponding values I_1, I_2, I_3, I_4, I_5 of the basic invariants $I^{a_{02}}, I^{a_{11}}, I^{a_{10}}, I^{a_{01}}_x, I^{a_{00}}$, the following formulae define the operation of taking of the formal adjoint:

$$\left. \begin{array}{l} I_1^\dagger = -I_1 , \\ I_2^\dagger = -I_2 , \\ I_3^\dagger = I_3 - I_{2y} + I_2(\frac{1}{2}I_{2x} + I_{1y})/I_1 , \\ I_4^\dagger = I_4 - I_{2xx} - 2I_{1xy} + I_{1x}(I_{2x} + 2I_{1y})/I_1 , \\ I_5^\dagger = -I_5 + \frac{1}{2}I_2I_4/I_1 + I_{3x} - \frac{1}{2}I_{2xy} + I_{2x}(\frac{1}{4}I_{2x} + \frac{1}{2}I_{1y})/I_1 . \end{array} \right\}$$

2. For an equivalence class that possesses the both properties $I^{a_{02}} = 0$ and $I^{a_{11}} \neq 0$, and the corresponding values I_1, I_2, I_3, I_4 of the basic invariants $I^{a_{11}}, I^{a_{01}}, I^{a_{10}}_x, I^{a_{00}}$, the following formulae define the operation of taking of the formal adjoint:

$$\left. \begin{array}{l} I_1^\dagger = -I_1 , \\ I_2^\dagger = I_2 - I_{1x} , \\ I_3^\dagger = I_3 - I_{1xy} + I_{1x}I_{1y}/I_1 , \\ I_4^\dagger = -I_4 + I_3 + I_{2y} - I_{1xy} + I_{1y}(I_{1x} - I_2)/I_1 . \end{array} \right\} \qquad (20)$$

3. For an equivalence class that possesses the both properties $I^{a_{02}} = 0$ and $I^{a_{11}} = 0$, and the corresponding values I_1, I_2, I_3 of the basic invariants $I^{a_{01}}, I^{a_{10}}, I^{a_{00}}_x$, the following formulae define the operation of taking of the formal adjoint:

$$I_1^\dagger = I_1 \;,$$
$$I_2^\dagger = I_2 \;,$$
$$\left.\begin{array}{l} \\ \\ I_3^\dagger = -I_3 - I_{1x}(I_{2x} - I_{1y})/I_1 + I_{2xx} + I_{1xy} \;. \end{array}\right\}$$

Proof. Consider the first case. Expressing as many as possible coefficients of an operator L of the class in terms of these basic invariants and the remained coefficients, we can have $a_{02} = I_1$, $a_{11} = I_2$, $a_{10} = I_3 + \frac{1}{3}a_{20}^2 + \frac{1}{2}a_{01}I_2/I_1 - \frac{1}{6}a_{20}I_2^2/I_1 + a_{20x}$, $a_{00} = I_5 + \frac{1}{3}a_{20}I_3 + \frac{1}{27}a_{20}^3 + (\frac{1}{4}a_{01}^2 - \frac{1}{36}a_{20}^2I_2^2 + \frac{1}{61}a_{20}I_{1y}I_2 - \frac{1}{21}a_{01}I_{1y})/I_1 + \frac{1}{3}a_{20}a_{20x} + \frac{1}{6}I_2a_{20y} + \frac{1}{2}a_{01y} - \frac{1}{6}I_{2y}a_{20} + \frac{1}{3}a_{20xx}$. We substitute these expressions for the coefficient of L and compute the formal adjoint $L^\dagger$: $L^\dagger = -D_{xxx} + a_{20}D_{xx} + I_2D_{xy} + I_1D_{yy} + (-a_{01} + I_{2x} + 2I_{1y})D_y + (a_{20x} - I_3 - \frac{1}{3}a_{20}^2 + \frac{1}{6}I_2(I_2a_{20} - 3a_{01})/I_1 + I_{2y})D_x + c_{00}^\dagger$, where the free coefficient $c_{00}^\dagger$ is $I_5 + (\frac{1}{18}I_2^2(6a_{20x} - a_{20}^2) + \frac{1}{3}I_2a_{20}(I_{2x} + 2I_{1y}) - (I_2a_{01})_x - I_{1y}a_{01} + \frac{1}{2}a_{01}^2)/(2I_1) + \frac{1}{6}I_{1x}I_2(3a_{01} - I_2a_{20})/I_1^2 + \frac{1}{3}a_{20xx} - I_{3x} - \frac{1}{3}a_{20}a_{20x} + \frac{1}{3}a_{20}I_3 + \frac{1}{27}a_{20}^3 + \frac{1}{6}I_2a_{20y} - \frac{1}{2}a_{01y} - \frac{1}{6}I_{2y}a_{20} + I_{2xy} + I_{1yy}$. Applying lemma 2 to the case 1. of Theorem 2 we can find the formulae for invariants of a complete generating set of invariants for such operators. Computing the invariants of $L^\dagger$, and then having some easy guessing, we can obtain expressions of the invariants in the terms of exclusively $I_1, \ldots, I_5$.

Consider the second case of the statement of the theorem. Substituting for the coefficients of an operator L of such equivalence class expressions (18), compute $L^\dagger$: $L^\dagger = -D_{xxx} + a_{20}D_{xx} + I_1D_{xy} + (2a_{20x} - a_{10} + I_{1y})D_x - (I_2 + \frac{1}{3}a_{20}I_1 - I_{1x})D_y + I_4 - I_2(\frac{1}{3}a_{20}^2 - a_{10} + a_{20x})/I_1 - I_{2y} - \frac{1}{3}a_{20}I_{1y} + I_{1xy} + \frac{4}{3}a_{20xx} - a_{10x} + \frac{1}{3}a_{10}a_{20} - \frac{2}{27}a_{20}^3$. Applying lemma 2 to the case 2. of Theorem 2 we can find the formulae for invariants of a complete generating set of invariants for such operators. Then computing the invariants of $L^\dagger$, and then having some easy guessing, we can obtain expressions of the invariants in the terms of exclusively $I_1, \ldots, I_4$.

Consider the third case of the statement of the theorem. Expressing maximal number of coefficients in terms of the remained ones and the basic invariants, we can get $a_{01} = I_1$, $a_{10} = I_2 + \frac{1}{3}a_{20}^2 + a_{20x}$. Using these expressions, we compute $L^\dagger$: $L^\dagger = -D_{xxx} + a_{20}D_{xx}(a_{20x} - I_2 - \frac{1}{3}a_{20}^2)D_x - I_1D_y - I_{2x} - \frac{2}{3}a_{20}a_{20x} + a_{00} - I_{1y}$. Applying lemma 2 to the case 3. of Theorem 2 we can find the formulae for invariants of a complete generating set of invariants for such operators. Then computing the invariants of $L^\dagger$, and involving some easy guessing, we can obtain expressions of the invariants in the terms of exclusively I_1, I_2, I_3.

Corollary 3 (factorizations of the type $(X^2)(X)$). *Consider the equivalence classes of (12), which are defined by the values of the invariants $I^{a_{02}}$ and $I^{a_{11}}$, and the rest invariants as they are given in the Theorem 2.*
1. Operators of the classes having the property $I^{a_{02}} \neq 0$ has no factorizations of the factorization type $(X^2)(X)$.
2. Operators of a class having the both properties $I^{a_{02}} = 0$ and $I^{a_{11}} \neq 0$, and is given by the values I_1, I_2, I_3, I_4 of the invariants $I^{a_{11}}, I^{a_{01}}, I_x^{a_{10}}, I^{a_{00}}$, correspondingly, have a factorization of the factorization type $(X^2)(X)$ if and only if

$$-3I_1I_2I_{2x} - I_1^2I_2I_{1y} + I_2^3 + 3I_2^2I_{1x} + 2I_2I_{1x}^2 - I_1^3I_4 + I_1^3I_{2y}$$
$$+I_1^2I_{2xx} - I_1I_2I_{1xx} - 2I_1I_{1x}I_{2x} = 0 \; . \tag{21}$$

3. *If the operators of a class having the both properties* $I^{a_{02}} = 0$ *and* $I^{a_{11}} = 0$, *and is given by the values of the invariants* $I^{a_{10}}, I^{a_{01}}, I_x^{a_{00}}$ *can be factored with the factorization type* $(X^2)(X)$ *then*

$$I^{a_{01}} = 0$$

holds for this class.

Proof. The proof is analogous to the proof of the corollary (1)

6   Conclusions

In the present paper we found a test, in other words some constraints on invariants of a bivariate, non-hyperbolic third-order Linear Partial Differential Operator to guarantee its factorization for every factorization type. The test can be implemented (in fact has been implemented) in a computer algebra system and be a part of an exact integration algorithm for LPDEs and then perhaps "trusted" by a proof assistant. The invariants that are used are with respect to known gauge transformations, which is together with constructive factorization itself are heavily involved in modern exact integration algorithms. Therefore, they are a natural basis for such a test. The expression for invariants in terms of the coefficients an LPDO are found explicitly before in [12] and we used them. However, as one can consult with [12], it is easy to recompute them at any time using computer algebra systems. As a by-product of the investigations we determined how invariants of an equivalence class of an LPDO is changed under the operation of taking the formal adjoint. In fact explicit formulae have been found. This subresult is interesting by itself for the formulae are attractive. Also it can be used in some factorization connected algorithms due to the factorization property of adjoint operators.

Acknowledgements

The author was supported by the Austrian Science Fund (FWF) under project DIFFOP, Nr. P20336-N18.

References

1. Tsarev, S., Shemyakova, E.: Differential transformations of parabolic second-order operators in the plane (submitted, 2008), http://arxiv.org/abs/0811.1492
2. Tsarev, S.: Generalized laplace transformations and integration of hyperbolic systems of linear partial differential equations. In: ISSAC 2005: Proceedings of the 2005 international symposium on Symbolic and algebraic computation, pp. 325–331. ACM Press, New York (2005)

3. Tsarev, S.: Factorization of linear partial differential operators and darboux' method for integrating nonlinear partial differential equations. Theo. Math. Phys. 122, 121–133 (2000)
4. Anderson, I., Juras, M.: Generalized Laplace invariants and the method of Darboux. Duke J. Math. 89, 351–375 (1997)
5. Anderson, I., Kamran, N.: The variational bicomplex for hyperbolic second-order scalar partial differential equations in the plane. Duke J. Math. 87, 265–319 (1997)
6. Athorne, C.: A $z \times r$ toda system. Phys. Lett. A. 206, 162–166 (1995)
7. Zhiber, A.V.: Integrals, solutions and existence of the laplace transformations for a linear hyperbolic system of equations. Math. Notes 74(6), 848–857 (2003)
8. Startsev, S.: Cascade method of laplace integration for linear hyperbolic systems of equations. Mathematical Notes 83 (2008)
9. Grigoriev, D., Schwarz, F.: Factoring and solving linear partial differential equations. Computing 73(2), 179–197 (2004)
10. Grigoriev, D., Schwarz, F.: Generalized loewy-decomposition of d-modules. In: IS-SAC 2005: Proceedings of the 2005 international symposium on Symbolic and algebraic computation, pp. 163–170. ACM, New York (2005)
11. Grigoriev, D., Schwarz, F.: Loewy decomposition of third-order linear pde's in the plane. In: ISSAC 2008: Proceedings of the 2005 international symposium on Symbolic and algebraic computation, pp. 277–286. ACM, New York (2008)
12. Shemyakova, E., Mansfield, E.: Moving frames for laplace invariants. In: Proceedings of ISSAC 2008 (The International Symposium on Symbolic and Algebraic Computation), pp. 295–302 (2008)
13. Fels, M., Olver, P.J.: Moving coframes. I. A practical algorithm. Acta Appl. Math. 51(2), 161–213 (1998)
14. Fels, M., Olver, P.J.: Moving coframes. II. Regularization and theoretical foundations. Acta Appl. Math. 55(2), 127–208 (1999)
15. Olver, P., Pohjanpelto, J.: Pseudo-groups, moving frames, and differential invariants. In: Eastwood, M., Miller, W.J. (eds.) Symmetries and Overdetermined Systems of Partial Differential Equations. IMA Volumes in Mathematics and Its Applications, vol. 144, pp. 127–149. Springer, New York (2007)
16. Olver, P., Pohjanpelto, J.: Moving frames for Lie pseudo-groups. Canadian J. Math (to appear) (2007)
17. Darboux, G.: Leçons sur la théorie générale des surfaces et les applications géométriques du calcul infinitésimal, vol. 2. Gauthier-Villars (1889)
18. Shemyakova, E., Winkler, F.: A full system of invariants for third-order linear partial differential operators in general form. In: Ganzha, V.G., Mayr, E.W., Vorozhtsov, E.V. (eds.) CASC 2007. LNCS, vol. 4770, pp. 360–369. Springer, Heidelberg (2007)
19. Shemyakova, E., Winkler, F.: On the invariant properties of hyperbolic bivariate third-order linear partial differential operators. In: Kapur, D. (ed.) ASCM 2007. LNCS (LNAI), vol. 5081. Springer, Heidelberg (2008)
20. Loewy, A.: Ueber reduzible lineare homogene differentialgleichungen. Math. Annalen 56, 549–584 (1903)
21. Blumberg, H.: Über algebraische Eigenschaften von linearen homogenen Differentialausdrücken. PhD thesis, Göttingen (1912)
22. Shemyakova, E.: The Parametric Factorizations of Second-, Third- and Fourth-Order Linear Partial Differential Operators on the Plane. Mathematics in Computer Science 1(2), 225–237 (2007)

A Groupoid of Isomorphic Data Transformations

Paul Tarau

Department of Computer Science and Engineering
University of North Texas
tarau@cs.unt.edu

Abstract. As a variation on the known theme of Gödel numberings, isomorphisms defining data type transformations in a strongly typed functional language are organized as a finite groupoid using a higher order combinator language that unifies popular data types as diverse as natural numbers, finite sequences, digraphs, hypergraphs and finite permutations with more exotic ones like hereditarily finite functions, sets and permutations.[1]

Keywords: computational mathematics in Haskell, data type transformations, ranking/unranking,Gödel numberings, higher order combinators, hylomorphisms.

1 Introduction

Analogical/metaphorical thinking routinely shifts entities and operations from a field to another hoping to uncover similarities in representation or use [1]. Compilers convert programs from human centered to machine centered representations - sometime reversibly. Complexity classes are defined through compilation with limited resources (time or space) to similar problems [2]. Mathematical theories often borrow proof patterns and reasoning techniques across close and sometime not so close fields. A relatively small number of universal data types are used as basic building blocks in programming languages and their runtime interpreters, corresponding to a few well tested mathematical abstractions like sets, functions, graphs, groups, categories etc.

In their simplest form, isomorphisms between data types show up as *encodings* to some canonical representation, for instance natural numbers. Such encodings can be traced back to Gödel numberings [3,4] associated to formulae, but a wide diversity of common computer operations, ranging from data compression and serialization to wireless data transmissions and cryptographic codes are indirectly related.

---

[1] A (very) long version of this paper is available at http://arXiv.org/abs/0808.2953. Like this paper, it is organized as a literate Haskell program while also including Haskell sources as a separate file.

J. Carette et al. (Eds.): Calculemus/MKM 2009, LNAI 5625, pp. 170–185, 2009.

We will show in this paper how such isomorphisms can be organized naturally as a finite groupoid i.e. a category [5] where every morphism is an isomorphism, with objects provided by the data types and morphisms provided by their transformations.

One can see these encodings as a first step towards a "theory of everything" meant to provide a uniform view of the basic building blocks of various computational artifacts. We hope this can help refactoring the enormous ontology exhibited by computer science and engineering fields that have resulted over a relatively short period of evolution in unnecessarily steep learning curves limiting communication and synergy between fields.

The paper is organized as follows: section 2 describes our data transformation framework, section 3 introduces isomorphisms between finite sequences, sets and natural numbers and section 4 shows their lifting to hereditarily finite structures. Ranking/unranking of permutations and hereditarily finite permutations as well as Lehmer codes and factoradics are covered in section 5. Section 6 describes encodings for digraphs and hypergraphs. Sections 7 and 8 discuss related work and conclusions.

2   An Embedded Data Transformation Language

We will start by designing an embedded transformation language as a set of operations on a groupoid of isomorphisms. We will then extended it with a set of higher order combinators mediating the composition of the encodings and the transfer of operations between data types.

2.1   The Groupoid of Isomorphisms

We implement an isomorphism between two objects X and Y as a Haskell data type encapsulating a bijection f and its inverse g. We will call the *from* function the first component and the *to* function the second component defining the isomorphism.

$$X \xrightleftharpoons[g = f^{-1}]{f = g^{-1}} Y$$

```
data Iso a b = Iso (a→b) (b→a)

from (Iso f _) = f
to (Iso _ g) = g

compose :: Iso a b → Iso b c → Iso a c
compose (Iso f g) (Iso f' g') = Iso (f' . f) (g . g')
itself = Iso id id
invert (Iso f g) = Iso g f
```

Assuming that for any pair of type Iso a b, $f \circ g = id_b$ and $g \circ f = id_a$, we have:

Proposition 1. *The data type* Iso *defines a groupoid, i.e. for all cases when it is defined, the* compose *operation is associative, itself acts as an identity element and* invert *computes the inverse of an isomorphism.*

We can see the combinators from, to, compose, itself, invert as part of an *embedded data transformation language*. In the general case, as composition is only a partial function (i.e. $f : A \rightarrow B$ can be composed with $g : B' \rightarrow C$ only if $B = B'$), the resulting finite groupoid can be seen as a disjoint union of connected categories corresponding to the *weakly connected components* of the underlying graph.

Choosing a Root in a connected groupoid Within each connected groupoid, to avoid defining $n(n-1)/2$ isomorphisms between n objects, we can choose a *Root* object to/from which we will actually implement isomorphisms. Then we can extend our embedded combinator language using the groupoid structure of the isomorphisms to connect *any* two objects through isomorphisms to/from the *Root*.

2.2   The Gödel groupoid

Let us, from now on, focus on the connected groupoid of isomorphisms mapping various data types to *natural numbers (Nat)*. It makes sense to call it the *Gödel groupoid* as traditionally such mappings have been investigated in his work on arithmetization of formulae in the proofs of his incompleteness theorems.

Within each connected groupoid, choosing a *Root* object is somewhat arbitrary, but it makes sense to pick a representation that is relatively easily convertible to various others, efficiently implementable and, last but not least, scalable to accommodate large objects up to the runtime system's actual memory limits.

With this in mind, instead of the obvious choice *Nat*, let us chose as our *Root* object for the Gödel groupoid as the set of *Finite Sequences of Natural Numbers*. They can also be seen as as finite functions from an initial segment of *Nat*, say $[0..n]$, to *Nat*, or as words on an alphabet with an infinite supply of symbols. We will represent them as lists i.e. their Haskell type is $[Nat]$. As we will show in subsection 3.1 such sequences will be mapped one to one to *Nat* while accommodating large objects more efficiently.

```
type Nat = Integer
type Root = [Nat]
```

We can now define an *Encoder* as an isomorphism connecting an object to *Root*

```
type Encoder a = Iso a Root
```

together with the combinators *with* and *as* providing an *embedded transformation language* for routing isomorphisms through two *Encoders*.

```
with :: Encoder a→Encoder b→Iso a b
with this that = compose this (invert that)

as :: Encoder a → Encoder b → b → a
as that this thing = to (with that this) thing
```

The combinator **with** turns two Encoders into an arbitrary isomorphism, i.e. acts as a connection hub between their domains. The combinator **as** adds a more convenient syntax such that converters between A and B.

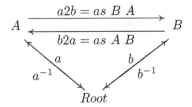

We will provide use cases for these combinators as we populate our groupoid of isomorphisms. Given that $[Nat]$ has been chosen as the root, we will define our finite function data type *fun* simply as the identity isomorphism on sequences in $[Nat]$.

```
fun :: Encoder [Nat]
fun = itself
```

3 Extending the Groupoid of Isomorphisms

We will now populate our groupoid of isomorphisms with combinators based on a few primitive encoders.

3.1 A Ranking/Unranking Algorithm for Finite Sequences

A *ranking/unranking* function defined on a data type is a bijection to/from the set of natural numbers (denoted *Nat* through the paper). We start with an unusually simple but (at our best knowledge) novel ranking/unranking algorithm for finite sequences of arbitrary unbounded size natural numbers. Given the definitions

```
cons :: Nat→Nat→Nat
cons x y  = (2^x)*(2*y+1)

hd :: Nat→Nat
hd n | n>0 = if odd n then 0 else 1+hd  (n 'div' 2)

tl :: Nat→Nat
tl n = n 'div' 2^((hd n)+1)

nat2fun :: Nat→[Nat]
nat2fun 0 = []
nat2fun n = hd n : nat2fun (tl n)

fun2nat :: [Nat]→Nat
fun2nat [] = 0
fun2nat (x:xs) = cons x (fun2nat xs)
```

Proposition 2. `fun2nat` *is a bijection from finite sequences of natural numbers to natural numbers and* `nat2fun` *is its inverse.*

This follows from the fact that `cons` and the pair (`hd`, `tl`) define a bijection between $Nat - \{0\}$ and $Nat \times Nat$ and that the value of `fun2nat` is uniquely determined by the number of applications of `tl` and the sequence of values returned by `hd`.

```
*ISO> hd 2008
3
*ISO> tl 2008
125
*ISO> cons 3 125
2008
```

We can define the `Encoder`

```
nat :: Encoder Nat
nat = Iso nat2fun fun2nat
```

working as follows

```
*ISO> as fun nat 2008
[3,0,1,0,0,0,0]
*ISO> as nat fun [3,0,1,0,0,0,0]
2008
```

Note also that this isomorphism preserves "list processing" operations i.e. if one defines:

```
app 0 ys = ys
app xs ys = cons (hd xs) (app (tl xs) ys)
```

then the isomorphism commutes with operations like list concatenation:

Proposition 3. *(as fun nat n) ++ (as fun nat m) ≡ as fun nat (app n m)*
  as nat fun (ns ++ ms) ≡ app (as nat fun ns) (as nat fun ms)

Given the definitions:

```
unpair z = (hd (z+1),tl (z+1))
pair (x,y) = (cons x y)-1
```

shifting by 1 turns `hd` and `tl` in total functions on Nat such that $unpair\ 0 = (0,0)$ i.e.

Proposition 4. $unpair : Nat \rightarrow Nat \times Nat$ *is a bijection and* $pair = unpair^{-1}$.

Note that unlike `hd` and `tl`, `unpair` is defined for all natural numbers:

```
*ISO> map unpair [0..7]
[(0,0),(1,0),(0,1),(2,0),(0,2),(1,1),(0,3),(3,0)]
```

As the cognoscenti might notice, this turns out to be in fact a classic *pairing/unpairing function* that has been used, by Pepis, Kalmar and Robinson in some fundamental work on recursion theory, decidability and Hilbert's Tenth Problem in [6,7,8] and `hd,tl,cons,0` define on *Nat* an algebraic structure isomorphic to the one introduced by `CAR,CDR,CONS,NIL` in John McCarthy's classic LISP paper [9].

With the isomorphism defined by `pair` and `unpair` we obtain the `Encoder`:

```
type Nat2 = (Nat,Nat)
nat2 :: Encoder Nat2
nat2 = compose (Iso pair unpair) nat
```

working as follows:

```
*ISO> as nat2 nat 123
(2,15)
*ISO> as nat nat2 (2,15)
123
```

3.2   An Isomorphism to Finite Sets of Natural Numbers

We can *rank* a set represented as a list of distinct natural numbers by mapping it into a single natural number, and, reversibly, by observing that it can be seen as the list of exponents of 2 in the number's base 2 representation. We obtain the `Encoder`:

```
set :: Encoder [Nat]
set = compose (Iso set2nat nat2set) nat

nat2set n | n≥0 = nat2exps n 0 where
  nat2exps 0 _ = []
  nat2exps n x = if (even n) then xs else (x:xs) where
    xs=nat2exps (n 'div' 2) (x+1)

set2nat ns = sum (map (2^) ns)
```

Note that in this case sets are sharing with sequences the underlying list representation `[Nat]`. The injection between sets represented by ordered sequences of distinct numbers and arbitrary sequences requires implementing a predicate `is_set` (see [10]) to enforce such constraint on each set argument. To keep our code simpler, we will assume in this paper that such constraints implicitly hold when required.

4   Generic Unranking and Ranking Hylomorphisms

The *ranking problem* for a family of combinatorial objects is finding a unique natural number associated to it, called its *rank*. The inverse *unranking problem* consists of generating a unique combinatorial object associated to each natural number.

4.1   Pure Hereditarily Finite Data Types

The unranking operation is seen here as an instance of a generic *anamorphism* (an *unfold* operation), while the ranking operation is seen as an instance of the corresponding catamorphism (a *fold* operation) [11]. Together they form a mixed transformation called *hylomorphism*. We will use such hylomorphisms to *lift* isomorphisms between lists and natural numbers to isomorphisms between a derived tree data type and natural numbers.

The data type representing such *hereditarily finite* structures will be a generic multi-way tree with a single leaf type [].

```
data T = H Ts deriving (Eq,Ord,Read,Show)
type Ts = [T]
```

The two sides of our hylomorphism rank and unrank are parameterized by two transformations f and g forming an isomorphism Iso f g:

```
unrank :: (a → [a]) → a → T
unranks :: (a → [a]) → [a] → Ts

unrank f n = H (unranks f (f n))
unranks f ns = map (unrank f) ns

rank :: ([b] → b) → T → b
ranks :: ([b] → b) → Ts → [b]

rank g (H ts) = g (ranks g ts)
ranks g ts = map (rank g) ts
```

Both combinators can be seen as a form of "structured recursion" that propagates a simpler operation guided by the structure of the data type. We can now combine an anamorphism+catamorphism pair into an isomorphism hylo defined with rank and unrank on the corresponding hereditarily finite data types:

```
hylo :: Iso b [b] → Iso T b
hylo (Iso f g) = Iso (rank g) (unrank f)

hylos :: Iso b [b] → Iso Ts [b]
hylos (Iso f g) = Iso (ranks g) (unranks f)
```

As its most general type shows, hylo lifts an isomorphism from b to [b] to an isomorphism between trees of type T and b. In our case b is Nat but the mechanism is more general - for instance it would also work if b is instantiated to Church numerals or bitstrings instead of Nat.

Hereditarily Finite Sets. Hereditarily Finite Sets [12] will be represented as an Encoder for the tree type T:

```
hfs :: Encoder T
hfs = compose (hylo (Iso nat2set set2nat)) nat
```

Otherwise, hylomorphism induced isomorphisms work as usual with our embedded transformation language:

```
*ISO> as hfs nat 2008
H [H [H [],H [H []]],H [H [H [H []]]],H [H [H []],H [H [H []]]],
   H [H [],H [H []], H [H [H []]]],H [H [H [],H [H []]]],H [H [],
   H [H [],H [H []]]],H [H [H []],H [H [],H [H []]]]]
```

One can notice that we have just derived as a "free algorithm" Ackermann's encoding [13] from Hereditarily Finite Sets to Natural Numbers:

$$f(x) = \texttt{if } x = \{\} \texttt{ then } 0 \texttt{ else } \sum_{a \in x} 2^{f(a)}$$

together with its inverse

```
ackermann = as nat hfs
inverse_ackermann = as hfs nat
```

Hereditarily Finite Functions. The same tree data type can host a hylomorphism derived from finite functions instead of finite sets:

```
hff :: Encoder T
hff = compose (hylo nat) nat
```

The hff Encoder can be seen as a "free algorithm", providing data compression/succinct representation for Hereditarily Finite Sets. Note, for instance, the significantly smaller tree size in:

```
*ISO> as hff nat 2008
H [H [H [],H []],H [],H [H []],H [],H [],H [],H []]
```

that can be also expressed as $((()())()(())()()()())$ using a parenthesis language [10].

One can represent the action of a hylomorphism unfolding a natural number into a hereditarily finite set as a directed graph with outgoing edges induced by

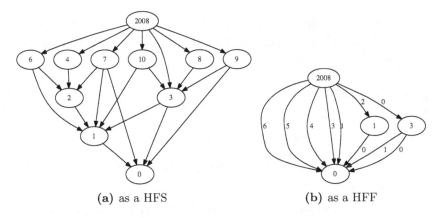

(a) as a HFS   (b) as a HFF

Fig. 1. Hereditarily finite representations of 2008

by applying the `inverse_ackermann` function as shown in **Fig. 1 (a)**. Similarly, a hereditarily finite function is expressed as a directed ordered multi-graph as shown in **Fig. 1 (b)**. Note that in this case the mapping `as fun nat` generates a sequence where the order of the edges matters. This order is indicated by integers starting from 0 labeling the edges.

5   Permutations and Hereditarily Finite Permutations

We have seen that finite sets and their derivatives represent information in an *order* independent way, focusing exclusively on information *content*. We will now look at data representations that focus exclusively on *order* in a *content* independent way - finite permutations and their hereditarily finite derivatives. To obtain an encoding for finite permutations we will first review a ranking/unranking mechanism for permutations that involves an unconventional numeric representation, *factoradics*.

5.1   The Factoradic Numeral System

The factoradic numeral system [14] replaces digits multiplied by a power of a base n with digits that multiply successive values of the factorial of n. In the increasing order variant `fr` the first digit d_0 is 0, the second is $d_1 \in \{0,1\}$ and the n-th is $d_n \in [0..n]$. For instance, $42 = 0*0! + 0*1! + 0*2! + 3*3! + 1*4!$. The left-to-right, decreasing order variant `fl` is obtained by reversing the digits of `fr`.

```
fr 42
  [0,0,0,3,1]
rf [0,0,0,3,1]
  42
fl 42
  [1,3,0,0,0]
lf [1,3,0,0,0]
  42
```

The function `fr` generating the factoradics of n, right to left, handles the special case of 0 and calls a local function `f` which recurses and divides with increasing values of n while collecting digits with `mod`:

```
fr 0 = [0]
fr n = f 1 n where
    f _ 0 = []
    f j k = (k 'mod' j) :
            (f (j+1) (k 'div' j))
```

The function `fl`, with digits left to right is obtained as follows:

```
fl = reverse . fr
```

The function `lf` (inverse of `fl`) converts back to decimals by summing up results while computing the factorial progressively:

```
rf ns = sum (zipWith (*) ns factorials) where
  factorials=scanl (*) 1 [1..]
```

Finally, `lf`, the inverse of `fl` is obtained as:

```
lf = rf . reverse
```

5.2   Ranking and Unranking Permutations of Given Size with Lehmer Codes and Factoradics

The Lehmer code of a permutation f of size n is defined as the sequence $l(f) = (l_1(f)\ldots l_i(f)\ldots l_n(f))$ where $l_i(f)$ is the number of elements of the set $\{j > i|f(j) < f(i)\}$ [15].

Proposition 5. *The Lehmer code of a permutation determines the permutation uniquely.*

The function `perm2nth` computes a `rank` for a permutation `ps` of `size>0`. It starts by first computing its Lehmer code `ls` with `perm2lehmer`. Then it associates a unique natural number `n` to `ls`, by converting it with the function `lf` from factoradics to decimals. Note that the Lehmer code `ls` is used as the list of digits in the factoradic representation.

```
perm2nth ps = (1,lf ls) where
  ls=perm2lehmer ps
  l=genericLength ls

perm2lehmer [] = []
perm2lehmer (i:is) = l:(perm2lehmer is) where
  l=genericLength [j|j←is,j<i]
```

The function `nat2perm` provides the matching *unranking* operation associating a permutation `ps` to a given `size>0` and a natural number `n`. It generates the n-th permutation of a given size.

```
nth2perm (size,n) =
  apply_lehmer2perm (zs++xs) [0..size-1] where
    xs=fl n
    l=genericLength xs
    k=size-1
    zs=genericReplicate k 0
```

The following function extracts a permutation from a "digit" list in factoradic representation.

```
apply_lehmer2perm [] [] = []
apply_lehmer2perm (n:ns) ps@(x:xs) =
   y : (apply_lehmer2perm ns ys) where
   (y,ys) = pick n ps

pick i xs = (x,ys++zs) where
  (ys,(x:zs)) = genericSplitAt i xs
```

Note also that `apply_lehmer2perm` is used this time to reconstruct the permutation `ps` from its Lehmer code, which in turn is computed from the permutation's factoradic representation.

One can try out this bijective mapping as follows:

```
*ISO> nth2perm (5,42)
  [1,4,0,2,3]
*ISO> perm2nth [1,4,0,2,3]
  (5,42)
*ISO> nth2perm (8,2008)
  [0,3,6,5,4,7,1,2]
*ISO> perm2nth [0,3,6,5,4,7,1,2]
  (8,2008)
```

5.3   A Bijective Mapping from Permutations to Natural Numbers

One more step is needed to to extend the mapping between permutations of a given length to a bijective mapping from/to *Nat*: we will have to "shift towards infinity" the starting point of each new block of permutations in *Nat* as permutations of larger and larger sizes are enumerated.

First, we need to know by how much - so we compute the sum of all factorials up to $n!$.

```
sf n = rf (genericReplicate n 1)
```

This is done by noticing that the factoradic representation of [0,1,1,..] does just that.

To know by how much we have to shift our mapping, we want to decompose n into the distance to the last sum of factorials smaller than n, n_m and the index in the sum, k.

```
to_sf n = (k,n-m) where
  k=pred (head [x|x←[0..],sf x>n])
  m=sf k
```

Unranking of an arbitrary permutation is now easy - the index k determines the size of the permutation and n-m determines the rank. Together they select the right permutation with `nth2perm`.

```
nat2perm 0 = []
nat2perm n = nth2perm (to_sf n)
```

Ranking of a permutation is even easier: we first compute its size and its rank, then we shift the rank by the sum of all factorials up to its size, enumerating the ranks previously assigned.

```
perm2nat ps = (sf l)+k where
  (l,k) = perm2nth ps
```

It works as follows:

```
*ISO> nat2perm 2008
  [0,2,3,1,4]
*ISO> perm2nat [0,2,3,1,4]
  42
*ISO> nat2perm 2008
  [1,4,3,2,0,5,6]
*ISO> perm2nat [1,4,3,2,0,5,6]
  2008
```

We can now define the Encoder as:

```
perm :: Encoder [Nat]
perm = compose (Iso perm2nat nat2perm) nat
```

The Encoder works as follows:

```
*ISO> as perm nat 2008
[1,4,3,2,0,5,6]
*ISO> as nat perm it
2008
*ISO> as perm nat 1234567890
[1,6,11,2,0,3,10,7,8,5,9,4,12]
*ISO> as nat perm it
1234567890
```

5.4 Hereditarily Finite Permutations

By using the generic unrank and rank functions defined in section 4 we can extend the isomorphism defined by nat2perm and perm2nat to encodings of Hereditarily Finite Permutations (HFP).

```
nat2hfp = unrank nat2perm
hfp2nat = rank perm2nat
```

The encoding works as follows:

```
*ISO> nat2hfp 42
H [H [],H [H [],H [H []]],H [H [H []],H []],
   H [H []],H [H [],H [H []],H [H [],H [H []]]]]
*ISO> hfp2nat it
  42
```

We can now define the Encoder as:

```
hfp :: Encoder T
hfp = compose (Iso hfp2nat nat2hfp) nat
```

The Encoder works as follows:

```
*ISO> as hfp nat 42
H [H [],H [H [],H [H []]],H [H [H []],H []],
   H [H []],H [H [],H [H []],H [H [],H [H []]]]]
*ISO> as nat hfp it
42
```

6   Encoding Directed Graphs and Hypergraphs

We will now show that more complex data types like digraphs and hypergraphs
have extremely simple encoders.

6.1   Encoding Directed Graphs

We can find a bijection from directed graphs (with no isolated vertices, corresponding to their view as binary relations), to finite sets by fusing their list of
ordered pair representation into finite sets with a pairing function:

```
digraph2set ps = map pair ps
set2digraph ns = map unpair ns
```

The resulting Encoder is:

```
digraph :: Encoder [Nat2]
digraph = compose (Iso digraph2set set2digraph) set
```

working as follows:

```
*ISO> as digraph' nat 2009
[(0,0),(2,0),(0,2),(0,3),(3,0),(0,4),(1,2),(0,5)]
*ISO> as nat digraph it
2009
```

Fig. 2 shows the digraph associated to 2009.

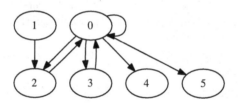

Fig. 2. 2009 as a digraph

6.2   Encoding Hypergraphs

A hypergraph (also called *set system*) is a pair $H = (X, E)$ where X is a set
and E is a set of non-empty subsets of X. We can derive a bijective encoding of
hypergraphs, represented as sets of sets:

```
set2hypergraph = map nat2set
hypergraph2set = map set2nat
```

The resulting Encoder is:

```
hypergraph :: Encoder [[Nat]]
hypergraph = compose (Iso hypergraph2set set2hypergraph) set
```

working as follows

```
*ISO> as hypergraph nat 2008
[[0,1],[2],[1,2],[0,1,2],[3],[0,3],[1,3]]
*ISO> as nat hypergraph it
2008
```

Discussion. For digraphs understood as subsets of $N \times N$, i.e. provided that a canonical mapping of vertices to an initial segment of N is assumed, our representations are clearly bijections. Once the mapping is fixed, a digraph is seen as a list of edges, each mapped to a distinct natural number using a pairing function. As all edges are distinct, the resulting list represents a set - which then is mapped to a unique natural number. Note also that digraphs can be disconnected and isolated vertices can be represented simply as vertices not occurring in the list of edges, assuming that the canonical mapping to vertices is such that the last vertex is connected to at least one other vertex. Under these assumptions, "digraphs" consisting only of isolated vertices collapse to the same encoding as the empty digraph. To avoid this problem, one can pair the representation of a digraph with a number indicating how many such vertices are considered part of the digraph after the last connected vertex occurring in an edge.

Similar reasoning applies to hypergraphs which are represented as lists of (distinct) hyperedges - i.e. sets of natural numbers that are then mapped to a unique natural number.

7   Related Work

The closest reference on encapsulating bijections as a Haskell data type is [16] and Conal Elliott's composable bijections module [17], where, in a more complex setting, Arrows [18] are used as the underlying abstractions. While our Iso data type is similar to the *Bij* data type in [17] and BiArrow concept of [16], the techniques for using such isomorphisms as building blocks of an embedded composition language centered around encodings as Natural Numbers are new.

Ranking functions can be traced back to Gödel numberings [3,4] associated to formulae. Together with their inverse *unranking* functions they are also used in combinatorial generation algorithms [19,20,21]. However the generic view of such transformations as hylomorphisms obtained compositionally from simpler isomorphisms, as described in this paper, is new. Note also that Gödel numberings are typically injective but not *onto* applications, and can only be turned into bijections by exhaustive enumeration of their range. By contrast our ranking/unranking functions are designed to be "genuinely" bijective, usually with computational effort linear in the size of the data types.

Pairing functions have been used in work on decision problems as early as [8]. A typical use in the foundations of mathematics is [22]. An extensive study of various pairing functions and their computational properties is presented in [23].

Natural Number encodings of Hereditarily Finite Sets have triggered the interest of researchers in fields ranging from Axiomatic Set Theory and Foundations of Logic to Complexity Theory and Combinatorics [12,24,25]. Contrary to the well known hereditarily finite sets, the concepts of hereditarily finite functions and permutations as well as their encodings, are likely to be new, given that our sustained search efforts have not lead so far to anything similar.

8   Conclusion

We have described encodings for various data types in a uniform framework as data type isomorphisms with a groupoid structure. The framework has been extended with hylomorphisms providing generic mechanisms for encoding hereditarily finite sets, functions and permutations. In addition, by using pairing/unpairing functions we have also derived unusually simple encodings for graphs, digraphs and hypergraphs.

While we have focused on the Gödel groupoid providing isomorphisms to/from natural numbers and finite sequences of natural numbers, similar techniques can be used to organize bijective transformations in fields ranging from compilation and complexity theory to data compression and cryptography.

We refer to [10] for implementations of a number of other encoders, covering data types as diverse as functional binary numbers, BDDs, multigraphs, parenthesis languages, multisets, primes, Gauss integers, as well as applications ranging from succinct encodings and generation of random instances of complex data types to experiments in number theory, boolean logic and circuit minimization.

Acknowledgements

The author thanks the anonymous referees of CALCULEMUS 2009 for their helpful suggestions and comments.

References

1. Lakoff, G., Johnson, M.: Metaphors We Live By. University of Chicago Press, Chicago (1980)
2. Cook, S.: Theories for complexity classes and their propositional translations. In: Complexity of computations and proofs, pp. 1–36 (2004)
3. Gödel, K.: Über formal unentscheidbare Sätze der Principia Mathematica und verwandter Systeme I. Monatshefte für Mathematik und Physik 38, 173–198 (1931)
4. Hartmanis, J., Baker, T.P.: On simple goedel numberings and translations. In: Loeckx, J. (ed.) ICALP 1974. LNCS, vol. 14, pp. 301–316. Springer, Heidelberg (1974)
5. Mac Lane, S.: Categories for the Working Mathematician. Springer, New York (1998)
6. Pepis, J.: Ein verfahren der mathematischen logik. The Journal of Symbolic Logic 3(2), 61–76 (1938)
7. Kalmar, L.: On the reduction of the decision problem. first paper. ackermann prefix, a single binary predicate. The Journal of Symbolic Logic 4(1), 1–9 (1939)

8. Robinson, J.: General recursive functions. Proceedings of the American Mathematical Society 1(6), 703–718 (1950)
9. McCarthy, J.: Recursive functions of symbolic expressions and their computation by machine, part i. Commun. ACM 3(4), 184–195 (1960)
10. Tarau, P.: Declarative Combinatorics: Isomorphisms, Hylomorphisms and Hereditarily Finite Data Types in Haskell, 104 pages (January 2009) (unpublished draft), http://arXiv.org/abs/0808.2953
11. Meijer, E., Hutton, G.: Bananas in Space: Extending Fold and Unfold to Exponential Types. In: FPCA, pp. 324–333 (1995)
12. Takahashi, M.O.: A Foundation of Finite Mathematics. Publ. Res. Inst. Math. Sci. 12(3), 577–708 (1976)
13. Ackermann, W.F.: Die Widerspruchsfreiheit der allgemeinen Mengenlhere. Mathematische Annalen (114), 305–315 (1937)
14. Knuth, D.E.: The art of computer programming, 3rd edn. seminumerical algorithms, vol. 2. Addison-Wesley/ Longman Publishing Co., Inc., Boston (1997)
15. Mantaci, R., Rakotondrajao, F.: A permutations representation that knows what "eulerian" means. Discrete Mathematics & Theoretical Computer Science 4(2), 101–108 (2001)
16. Alimarine, A., Smetsers, S., van Weelden, A., van Eekelen, M., Plasmeijer, R.: There and back again: arrows for invertible programming. In: Haskell 2005: Proceedings of the 2005 ACM SIGPLAN workshop on Haskell, pp. 86–97. ACM Press, New York (2005)
17. Conal Elliott: Data.Bijections Haskell Module, http://haskell.org/haskellwiki/TypeCompose
18. Hughes, J.: Generalizing Monads to Arrows Science of Computer Programming 37, 67–111 (2000)
19. Martinez, C., Molinero, X.: Generic algorithms for the generation of combinatorial objects. In: Rovan, B., Vojtáš, P. (eds.) MFCS 2003. LNCS, vol. 2747, pp. 572–581. Springer, Heidelberg (2003)
20. Knuth, D.: The Art of Computer Programming, vol. 4, draft (2006), http://www-cs-faculty.stanford.edu/~knuth/taocp.html
21. Myrvold, W., Ruskey, F.: Ranking and unranking permutations in linear time. Information Processing Letters 79, 281–284 (2001)
22. Cégielski, P., Richard, D.: Decidability of the theory of the natural integers with the cantor pairing function and the successor. Theor. Comput. Sci. 257(1-2), 51–77 (2001)
23. Rosenberg, A.L.: Efficient pairing functions - and why you should care. International Journal of Foundations of Computer Science 14(1), 3–17 (2003)
24. Kaye, R., Wong, T.L.: On Interpretations of Arithmetic and Set Theory. Notre Dame J. Formal Logic Volume 48(4), 497–510 (2007)
25. Kirby, L.: Addition and multiplication of sets. Math. Log. Q. 53(1), 52–65 (2007)

Algorithms for the Functional Decomposition of Laurent Polynomials

Stephen M. Watt

University of Western Ontario, London, Ontario, CANADA N6A 5B7
http://www.csd.uwo.ca/~watt

Abstract. Recent work has detailed the conditions under which univariate Laurent polynomials have functional decompositions. This paper presents algorithms to compute such univariate Laurent polynomial decompositions efficiently and gives their multivariate generalization.

One application of functional decomposition of Laurent polynomials is the functional decomposition of so-called "symbolic polynomials." These are polynomial-like objects whose exponents are themselves integer-valued polynomials rather than integers. The algebraic independence of $X, X^n, X^{n^2/2}$, *etc*, and some elementary results on integer-valued polynomials allow problems with symbolic polynomials to be reduced to problems with multivariate Laurent polynomials. Hence we are interested in the functional decomposition of these objects.

1 Introduction

Determining whether a univariate polynomial may be written as the functional composition of others of lower degree is a question that has been studied for almost a century. Ritt [1] considered the case of polynomials with complex coefficients and showed the decomposition factors and their degrees are unique up to certain transformations. Engstrom [2] and Levi [3] generalized Ritt's results, showing they hold for arbitrary fields of characteristic zero.

Polynomial decompositions can be useful because they reveal the structure of a problem. This may allow certain problems to be solved explicitly that otherwise could not be. Decomposable polynomials of a given degree form a low-dimensional subspace of the space of all polynomials of that degree. A polynomial that is the composition of two others of degrees r and s has degree rs, but instead of requiring $rs + 1$ coefficients to describe, it can be specified by the $r + s$ independent coefficients of its composition factors.

Algorithms by Barton and Zippel [4] and more recently by Kozen and Landau [5] have been incorporated in many computer algebra systems. Generalizations have been studied for functional decomposition of rational functions [6], algebraic functions [7] and multivariate polynomials [8]. More recent work by Zieve [9] has shown the conditions under which univariate Laurent polynomials may be decomposed, and gives results analogous to those of Ritt.

J. Carette et al. (Eds.): Calculemus/MKM 2009, LNAI 5625, pp. 186–200, 2009.

Separately, we have been interested in the problem of reasoning about and performing algebraic operations on families of polynomials parameterized their exponents [10,11,12]. This work explores algorithms that work in the generic case, and can be specialized uniformly by evaluating the exponent parameters. Other work considers case-based structure [13,14,15].

As defined more precisely in [10,11], the so-called "symbolic polynomials" resemble ordinary polynomials with exponents that are integer-valued polynomials. For example, $X^{(n^2-n)/2} - X^{2nm}Y^m - 4$ would belong to a particular ring of symbolic polynomials. Taking the integer-valued polynomials as an abelian group gives the symbolic polynomials an obvious group ring structure. Using the fact that integer-valued polynomials have integer coefficients when written in a binomial basis (in this example, $\binom{n}{i}\binom{m}{j}$ for $i, j \geq 0$) and on the algebraic independence of the polynomial variables raised to different monomial powers (in this example $X, X^n, X^{\binom{n}{2}}, X^{nm}, Y^m$), it is possible to reduce many problems on symbolic polynomial to problems on multivariate Laurent polynomials.

Most recently, the problem of functional decomposition of symbolic polynomials has been studied, and reduced to the functional decomposition of multivariate Laurent polynomials [16]. In this article we now explore the algorithmic aspects of finding such functional decompositions of Laurent polynomials. We present two algorithms for univariate Laurent polynomial decomposition: one that reduces the problem to polynomial decomposition and one that solves the problem directly. We also present their multivariate generalization.

The paper is organized as follows: First, Section 2 gives some initial definitions and notations. Then Section 3 presents the decomposition problem for univariate Laurent polynomials. We note that the important case, from an algorithms point of view, is when a Laurent polynomial f decomposes as $f = g \circ h$ with g a polynomial and h a Laurent polynomial. The main body of the paper is devoted to showing how to compute such decompositions. The first method uses the leading and trailing coefficients of f to find the leading and trailing coefficients of h. Section 4 gives the required mathematical justification for the method and Section 5 gives the algorithm. This method has the advantage that it may be implemented using an existing polynomial decomposition library, but it has the disadvantage that it may require trying multiple candidate values for h. The second method avoids this problem and determines the coefficients of h from a single triangular system involving the leading coefficients of f. Section 6 gives the mathematical background for the method and Section 7 presents the algorithm. The multivariate generalization of these methods is discussed in Section 8 and Section 9 concludes the paper.

2   Preliminaries

We begin by establishing certain notation and conventions we use throughout.

Notation 1 (Univariate Laurent polynomials). *For a ring R R, we denote by $R[(X)]$ the ring of Laurent polynomials $R[X, X^{-1}]/\langle XX^{-1} - 1\rangle$.*

Notation 2 (Multivariate Laurent polynomials). *For a ring R R, we denote by $R[(X_1, \ldots, X_n)]$ the ring of multivariate Laurent polynomials*

$$R[X_1, \ldots, X_n, X_1^{-1}, \ldots, X_n^{-1}]/\langle X_1 X_1^{-1} - 1, \ldots, X_n X_n^{-1} - 1 \rangle.$$

Notation 3 (Coefficient). *Given $f \in R[(X)]$, we denote the coefficient of X^k in f as $[X^k]f$ or f_k.*

Notation 4 (Multiplication time). *We denote by $M(m, n)$ the time to multiply polynomials of degrees m and n. If $m = n$, we write $M(n)$.*

Definition 1 (Degree of univariate Laurent polynomial)
Let $h \in R[(X)]$ be a Laurent polynomial with a pole of order t at 0 and of order s at ∞. Then the degree of h is $\deg h = \langle -t, s \rangle$.

Definition 2 (Degree of multivariate Laurent polynomial)
Let $h \in R[(X_1, \ldots, X_v)]$ be a Laurent polynomial with poles in X_i of order t_i at 0 and of order s_i at ∞. Then the degree of h is $\deg h = \langle (-t_1, \ldots, -t_v), (s_1, \ldots, s_v) \rangle$.

Definition 3 (Total degree of multivariate Laurent polynomial)
Let $h = \sum_i c_i X_1^{e_{1i}} \cdots X_n^{e_{ni}} \in R[(X_1, \ldots, X_n)]$ and $w \in \mathbb{Z}_{>0}^n$. Then the total degree of h with weight vector w is $\mathrm{tdeg}_w \, h = \max_i \sum_{j=1}^n e_{ji} w_j$. If no weight vector is specified, then $w = (1, \ldots, 1)$ is assumed.

Convention 1 (Empty sequence). *The sequence $h_a, \ldots, h_b$ is empty if $b < a$.*

3   Univariate Decomposition

We phrase the functional decomposition problem for univariate Laurent polynomials over a field K as follows:

Problem 1 (Univariate Laurent polynomial decomposition).
Given $f \in K[(X)]$, K a field, and $r \geq 2 \in \mathbb{Z}$, do there exist $g \in K[X]$ of degree r and $h \in K[(X)]$ such that $f = g \circ h$? If so, find such g and h.

We justify below why we consider $g \in K[X]$ and $h \in K[(X)]$ as opposed to $g, h \in K(X)$ or $g, h \in K[(X)]$.

  For the discussion in later sections we fix the following: We let $\deg f = \langle -rt, rs \rangle$. Supposing g and h exist, we let

$$g = \sum_{i=0}^r g_i X^i, \quad h = \sum_{i=-t}^s h_i X^i \quad f = \sum_{i=-rt}^{rs} f_i X^i. \tag{1}$$

We place certain conditions on r, s and t to concentrate on the problem of interest. We assume $t > 0$ since otherwise $f, g, h \in K[X]$ and we have the usual polynomial decomposition. We require the inverse of r in K. In the following, we let $\ell = s + t$ and $N = \ell r$. Then h has $\ell + 1$ coefficients, g has $r + 1$ and f has $N + 1$.

  We now discuss our restriction that $g \in K[X], h \in K[(X)]$. This relates to the ways in which a Laurent polynomial may decompose. The following result of Zieve [9] describes the situation when $K = \mathbb{C}$.

Lemma 1 (Zieve). *For $f \in \mathbb{C}(X)\backslash\mathbb{C}$, the fields between $\mathbb{C}(X)$ and $\mathbb{C}(f)$ are precisely the fields $\mathbb{C}(h)$, where $g, h \in \mathbb{C}(X)$ satisfy $f = g \circ h$; moreover, for $h, H \in \mathbb{C}(X)$, we have $\mathbb{C}(h) = \mathbb{C}(H)$ if and only if there is a degree-one $\mu \in \mathbb{C}(X)$ such that $h = \mu \circ H$. If f is a Laurent polynomial (respectively, polynomial) and $f = g \circ h$ with $g, h \in \mathbb{C}(X)$, then there is a degree-one $\mu \in \mathbb{C}(X)$ such that both $g \circ \mu$ and $\mu^{-1} \circ h$ are Laurent polynomials (respectively, polynomials).*

With this result we may show the following:

Lemma 2. *For $f = g \circ h \in \mathbb{C}[(X)], g, h \in \mathbb{C}(X)$, there is a degree-one $\mu \in \mathbb{C}(X)$ such that both $g \circ \mu \in \mathbb{C}[(X)]$ and $\mu^{-1} \circ h \in \mathbb{C}[(X)]$ and either (i) $g \circ \mu \in \mathbb{C}[X]$ or (ii) $\mu^{-1} \circ h = X^s$ for some $s \in \mathbb{Z}$, or both.*

Proof. Let $\hat{g} = g \circ \mu, \hat{h} = \mu^{-1} \circ h \in \mathbb{C}[(X)]$ which exist by Lemma 1. Suppose that $\hat{h}$ is not a monomial and $\hat{g} \notin \mathbb{C}[X]$. We then have $\hat{g} = X^{-n}G, G \in \mathbb{C}[X], n > 0 \in \mathbb{Z}$. Because $\hat{h}$ is not a monomial, it will have a finite non-zero root. This will be a pole of f due to the X^{-n} factor of $\hat{g}$. This contradicts the fact that f can have poles only at zero and infinity. □

The case where h is a monomial may be handled trivially, so we restrict our attention to the situation where $g \in K[X]$.

4   Facts about Univariate Laurent Polynomials

We now present some elementary facts about Laurent polynomials that are required to justify our first algorithm.

Engstrom [2] observed that for polynomial composition the leading coefficients of f and $g_r h^r$ agree and, if $h(0) = 0$, give a triangular system for the coefficients of h. The polynomial decomposition algorithm of Kozen and Landau [5] is based on this fact. We develop generalizations of these ideas for Laurent polynomials. We begin by showing that both the leading s terms and trailing t terms of f and $g_r h^r$ agree.

Lemma 3. *The coefficients of X^i in f and $g_r h^r$ agree for $i > rs - s$ and for $i < -rt + t$.*

Proof. Let $f = g_r h^r + F$ for $F = \sum_{i=0}^{r-1} g_i h^i$. The degree of F is $\langle -t(r-1), s(r-1)\rangle$ so F has vanishing support for $X^i, i > rs - s$ and $i < -rt + t$. □

Next we show that the leading and trailing terms of f depend, respectively, only on the positive and negative degree terms of h.

Lemma 4. *Let $h_+ = \sum_{i=1}^s h_i X^i$ and $h_- = \sum_{i=1}^t h_{-i} X^{-i}$ so $h = h_+ + h_0 + h_-$. Then the coefficients of X^i in f and $g_r h_+^r$ agree for $i > rs - s$. Likewise, the coefficients of X^i in f and $g_r h_-^r$ agree for $i < -rt + t$.*

Proof. Let $f = g_r h^r + F$. The only f_i with $i < -rt + t$ or $i > rs - s$ arise from $g_r h^r = g_r \sum_{r_+ + r_0 + r_- = r} \binom{r}{r_+ \ r_0 \ r_-} h_+^{r_+} h_0^{r_0} h_-^{r_-}$. If both r_+ and $r_0 + r_-$ are non-zero, then $1 \leq \deg(h_+^{r_+}) \leq (r-1)s$ and $-(r-1)t \leq \deg(h_0^{r_0} h_-^{r_-}) \leq 0$ so $-rt+t+1 \leq \deg(h_+^{r_+} h_0^{r_0} h_-^{r_-}) \leq rs-s$. Only when $r_+ = r$ can the degree exceed $rs-s$. Therefore $[X^i]f = [X^i]g_r h_+^r$ when $i > rs-s$. Similarly, $[X^i]f = [X^i]g_r h_-^r$ when $i < rt - t$. □

The following is the observation of Engstrom, where we leave h_s unrestricted in order to make certain statements easier later.

Lemma 5. *The coefficients of h_+ are determined, up to a choice of h_s, by the triangular system*

$$\left. \begin{array}{ll} g_r &= f_{rs}/h_s^r \\ h_{s-i} &= f_{rs-i}/(rg_r h_s^{r-1}) + P_{s-i}(h_s, \ldots, h_{s-i+1}, g_r), \, 1 \leq i \leq s - 1 \end{array} \right\} \quad (2)$$

where P_{s-i} is a polynomial function of $i+1$ variables.

Proof. Lemma 4 and multinomial expansion of $g_r h_+^r$. □

A similar result holds for the trailing terms:

Lemma 6. *The coefficients of h_- are determined, up to a choice of h_{-t}, by the triangular system*

$$\left. \begin{array}{ll} g_r &= f_{-rt}/h_{-t}^r \\ h_{-t+i} &= f_{-rt+i}/(rg_r h_{-t}^{r-1}) + P_{-t+i}(h_{-t}, \ldots, h_{-t+i-1}, g_r), \, 1 \leq i \leq t - 1 \end{array} \right\} \quad (3)$$

where P_{-t+i} is a polynomial function of $i+1$ variables.

Proof. As for Lemma 5. □

We will also require the following simple fact.

Lemma 7. *Given $k \neq 0 \in K$, there exist $\hat{g} \in K[X]$, $\hat{h} \in K[(X)]$, such that $f = \hat{g} \circ \hat{h}$, $\hat{h}_s = k$, $\hat{h}_0 = 0$, $\deg g = \deg \hat{g}$, $\deg h = \deg \hat{h}$.*

Proof. Take $\hat{g} = g \circ (\frac{X}{a} - \frac{b}{a})$ and $\hat{h} = (aX + b) \circ h$ where $a = k/h_s$ and $b = -h_0/h_s$. Then $\hat{h}_s = k, \hat{h}_0 = 0$ as desired, and $\hat{g} \circ \hat{h} = g \circ h$ by the associativity of $\circ$. □

5 The Two-Ended Algorithm

5.1 Finding h

It is possible to find the decomposition of Laurent polynomials using the ideas presented in Section 4. Given $f \in K[(X)]$ of degree $\langle -rt, rs \rangle$ we may find a candidate inner composition factor h_{cand} of degree $\langle -t, s \rangle$ by independently finding the positive degree terms, $h_{\text{cand}+}$, and negative degree terms, $h_{\text{cand}-}$. By Lemma 7, the constant term, $h_{\text{cand}0}$, can be set to zero. Once h_{cand} is chosen, the outer composition factor g, if it exists, may be found easily by a number of methods.

There is one point that requires particular attention, however. While it is possible to specify an arbitrary leading coefficient or trailing coefficient for h_{cand}, they may not be chosen independently. Lemmas 5, 6 and 7 show that we are free to choose h_0 and we can find all the other coefficients of h if we know h_s and h_{-t}. We choose $h_s = 1$ and set $h_0 = 0$. Then requiring g_r to be the same in the systems for both leading and trailing coefficients gives

$$h_{-t}^r = f_{-rt}/f_{rs}.\tag{4}$$

Depending on the field K, there may be up to r possible values for h_{-t} satisfying this equation. These do not normally all lead to decompositions of f.

Example 1. Let

$$f = X^4 + 4X^3 + 4X^2 + 6X + 3 - 20X^{-1} + 9X^{-2} - 30X^{-3} + 25X^{-4}.$$

We set $r = 2$, $h_2 = 1$ and find $h_+ = X^2 + 2X$. The possibilities for h_{-2} are then $\pm\sqrt{25}$. Choosing $h_{-2} = -5$ gives $f = (X^2 + 1) \circ (X^2 + 2X + 3X^{-1} - 5X^{-2})$. Choosing $h_{-2} = +5$ gives $h_{\text{trial}} = X^2 + 2X - 3X^{-1} + 5X^{-2}$. Composing with generic g and equating coefficients with f gives an inconsistent system. There is therefore no g such that $f = g \circ h$ with $h_2 = 1$ and $h_{-2} = 5$. □

It is possible to try each of the r possible choices for h_{-t} until one leads to a decomposition. This is the main idea of our "two-ended" algorithm. We shall explain this in more detail shortly. We first present a few pre-requisites.

The first component is an algorithm to find a candidate h_+, given f the degree and the desired leading coefficient for h. This is used twice in the two-ended algorithm — once to find h_+ from f and once to find $\hat{h}_- = h_-/h_{-t}$ from $f(1/X)$.

Algorithm 1 (Positive Degree Terms of h)

INPUT:
  $f \in K[(X)]$ *of degree* $\langle -rt, rs \rangle$ *and* $r \geq 2 \in \mathbb{Z}$.
OUTPUT:
  A monic polynomial $h_+ \in K[X]$, *such that if there exist* $g \in K[X]$,
  $h \in K[(X)]$, $\deg g = r$, $f = g \circ h$, *then a choice of* h *has* $\sum_{i=1}^{s} h_i X^i = h_+$.
  Note, it may be that there do not exist g, h *of the required degrees such*
  that $f = g \circ h$.
METHOD:
1. *Let* $p := X^s$.
2. *For k from 1 to $s - 1$,*
   (a) *Let* $c := \frac{1}{r}[X^{rs-k}](f/f_{rs} - p^r)$.
   (b) *Let* $p := p + cX^{s-k}$.
3. *Return* $h_+ = p$.

Theorem 1. *Algorithm 1 solves the polynomial system (2).*

Proof. Let $c_{(k)}$ and $p_{(k)}$ be the values of c and p after k iterations of the loop. We have $f_{rs-k} = [X^{rs-k}]g_r h_+^r$ by Lemma 4 so step 2a computes

$$c_{(k)} = \frac{1}{r}[X^{rs-k}]\left(\left(p_{(k-1)} + h_{s-k}X^{s-k} + O(X^{s-k-1})\right)^r - p_{(k-1)}^r\right)$$

Induction on k shows $p_{(k)} = \sum_{i=0}^{k} h_{s-i} X^{s-i}$ so $p_{(s-1)} = h_+$. The system (2) is triangular and introduces each variable linearly so the solution is unique.   □

5.2   Finding g

In the case of polynomials, Kozen and Landau find g by solving the triangular linear system $\mathbf{A} \cdot \mathbf{g} = \mathbf{f}$ with entries $\mathbf{A}_{ij} = [X^{is}] h^j$, $\mathbf{g}_i = g_i$, $\mathbf{f}_i = f_{is}$, $0 \le i, j \le r$. They observe that the coefficients $\mathbf{A}_{ij}$ can be saved during the construction of h and that $h_{(k)}^r$ may be computed using values from previous iterations.

For Laurent polynomials, finding g by solving a linear system would require the coefficients $\mathbf{A}_{ij} = [X^{is}](h_+ + h_{-t} \hat{h}_-)^j$ for a choice of h_{-t}. These are not immediately available as the two applications of Algorithm 1 produce $[X^{is}] h_+^j$ and $[X^{is}] \hat{h}_-^j$. We may nevertheless compute the matrix $\mathbf{A}$, given h_+, $\hat{h}_-$ and h_{-t}, but the advantage of using saved values from the construction of h is lost. Moreover, we generally need to construct this matrix for several choices of h_{-t}. While it is possible to do this, depending on the field, it may be more convenient to find g by interpolation.

Finding g by linear system solving

Suppose we have h_+, $\hat{h}_-$, h_{-t} and f and wish to find g by solving a linear system.

1. Find the $(r+1)^2$ coefficients $\mathbf{A}_{ij} = [X^{is}](h_+ + h_{-t} \hat{h}_-)^j$ for $0 \le i, j \le r$. Computing $\mathbf{A}_{ij}$ can be done in time $\sum_{i=1}^{r} M(\ell, i\ell)$. This can be done in time $O(r^2 \ell^2) = O(M(r\ell))$ with classical polynomial multiplication or time $O(r^2 \ell \log(r\ell)) = O(r M(r\ell))$ with fast arithmetic.
2. Solve the triangular system $\mathbf{A} \cdot \mathbf{g} = \mathbf{f}$, which can be done in time $O(r^2)$.

If up to r such systems must be solved, with $\mathbf{A}_{ij}$ being computed afresh each time, then time $O(r^3 \ell^2)$ is required for classical arithmetic or $O(r^3 \ell \log(r\ell))$ for fast arithmetic.

Finding g by interpolation

Suppose we have h_+, $\hat{h}_-$, h_{-t} and f and wish to find g by interpolation

1. Evaluate h_+, $\hat{h}_-$ and f at points, $\alpha_1, \ldots, \alpha_q \in K$, until $r+1$ distinct values are found for $h_+ + h_{-t} \hat{h}_-$. This requires $2q(r+1)\ell$ operations.
2. Interpolate the points $\{(h_+(\alpha_j) + h_{-t} \hat{h}_-(\alpha_j), f(\alpha_j)) \mid 1 \le j \le q\}$ to obtain g. This requires $O(r^2) = O(M(r))$ operations with classical arithmetic or $O(r \log^2 r) = O(\log r M(r))$ operations for fast arithmetic.

If multiple such interpolations must be performed, the values of $h_+, \hat{h}_-, f$ need not be recomputed. Only the q sums $h_+ + h_{-t} \hat{h}_-$ need be recomputed, requiring $2q$ operations. If up to r interpolations are required, the total time is then $O(qr\ell + r^3)$ for classical arithmetic or $O(qr\ell + r^2 \log^2 r)$ for fast arithmetic. If the field is large enough, $q = r + 1$ with high probability. Thus we have expected time $O(r^2\ell + r^3)$ with classical arithmetic and expected time $O(r^2\ell + r^2\log^2 r)$ with fast arithmetic. In the worst case, because there may be up to ℓ values of X such that $h = \alpha$, it is theoretically possible to require as many as $q = (r+1)\ell$ evaluations of h. The worst case is thus $O(r^2\ell^2 + r^3)$ for classical arithmetic or $O(r^2\ell^2 + r^2\log^2 r)$ for fast arithmetic.

Comparison

The complexity of finding the outer composition factor g by linear system solving and by interpolation is summarized in the table below. The first two columns give the time complexity if only one candidate for h is tried and the second pair of columns give the time complexity if $O(r)$ possibilities for h_{-t} must be tried.

	1 Linear Sys.	1 Interp.	r Linear Sys.	r Interp.
Expected Classical	$O(r^2\ell^2)$	$O(r^2\ell)$	$O(r^3\ell^2)$	$O(r^2\ell + r^3)$
Expected Fast	$O(r^2\ell\log(r\ell))$	$O(r^2\ell)$	$O(r^3\ell\log(r\ell))$	$O(r^2\ell + r^2\log^2 r)$
Worst Case Class.	$O(r^2\ell^2)$	$O(r^2\ell^2)$	$O(r^3\ell^2)$	$O(r^2\ell^2 + r^3)$
Worst Case Fast	$O(r^2\ell\log(r\ell))$	$O(r^2\ell^2)$	$O(r^3\ell\log(r\ell))$	$O(r^2\ell^2 + r^2\log^2 r)$

Provided the field has sufficiently many elements, the only situation where solving a linear system is superior to interpolation is when *all* of the following conditions hold:

1. the worst case number of evaluations is required (unlikely),
2. $O(r)$ candidates for h must be tried,
3. fast arithmetic is used, and
4. $O(r\log(r\ell)) < O(\ell)$, e.g. when searching for g of fixed low degree.

Under normal circumstances, therefore, interpolation should be used. This may be done as described in Algorithm 2.

Algorithm 2 (Interpolation of g)

INPUT:

    $f \in K[(X)]$ *with* $\deg f = \langle -rt, rs \rangle$,
    $h_+ \in K[X]$ *with* h_+ *monic,* $\deg h_- = s$,
    $\hat{h}_- \in K[(X)]$ *with* $\hat{h}_-(X^{-1}) \in K[X]$, $\hat{h}_-(X^{-1})$ *monic,* $\deg \hat{h}_-(X^{-1}) = t$
    T *a finite set of values* $\{\tau_i \in K\}$.

OUTPUT:

    If there exit $g \in K[X]$ *and* $\tau \in T$, *such that* $f = g \circ (h_+ + \tau\hat{h}_-)$, *then returns* g *and* τ. *Otherwise returns FAIL.*

METHOD:

1. *Choose* $r + 1$ *values* $\alpha_j \in K$, *and compute* $F_j = f(\alpha_j)$, $H_{+j} = h_+(\alpha_j)$, $H_{-j} = \hat{h}_-(\alpha_j)$, $j = 1, \ldots, r + 1$.

2. *For each value $\tau_i \in T$,*
   (a) *Compute the values $H_j = H_{+j} + \tau_i H_{-j}, j = 1, \ldots, r+1$*
   (b) *While the values H_j are not all distinct, say $H_{j_1} = H_{j_2}$, choose a new for α_{j_1} and recompute $F_{j_1}, H_{+j_1}, H_{-j_1}, H_{j_1}$.*
   (c) *Form g by interpolating the points $(H_j, F_j), j = 1, \ldots, r+1$.*
   (d) *Test whether $f = g \circ (h_+ + \tau_i \hat{h}_-)$. If so, return g and τ_i.*
3. *Return FAIL.*

5.3 Two-Ended Univariate Laurent Polynomial Decomposition

The above results may be combined to give an algorithm for the decomposition of univariate Laurent polynomials. The leading coefficients for h_+ and the trailing coefficients for a multiple of h_- are found, and the possible values of h_{-t} are tried to put them together.

Algorithm 3 (Two-Ended Univariate Laurent Polynomial Decomposition)

INPUT:
  $f \in K[(X)]$ *of degree* $\langle -rt, rs \rangle$ *and* $r \geq 2 \in \mathbb{Z}$.
OUTPUT:
  If there exist $g \in K[X], h \in K[(X)]$ such that $\deg g = r$, $f = g \circ h$, returns a choice of g and h. Otherwise, returns FAIL.
METHOD:
  1. *Apply Algorithm 1 to $f(X)$ and r to compute monic $h_+(X) \in K[X]$.*
  2. *Apply Algorithm 1 to $f(\frac{1}{X})$ and r to compute monic $\hat{h}_-(\frac{1}{X}) \in K[X]$.*
  3. *Compute the set $T = \{\tau \in K \mid \tau^r = f_{-rt}/f_{rs}\}$.*
  4. *Apply Algorithm 2 to $f(X), h_+(X), \hat{h}_-(X)$ and T to find g and τ. If Algorithm 2 returns FAIL, return FAIL.*
  5. *Let $h_{-t} = \tau$ and return g and $h_+ + h_{-t}\hat{h}_-$.*

Although this method requires up to r attempts to find the inner composition factor h, it is easy to implement in a setting where polynomial decomposition is already provided. Also, in some important cases the trailing coefficient equation has only a few solutions, and possibly only one. For example, when $K = \mathbb{R}$ there are one or two alternatives for h_{-t} according as r is odd or even.

If implementing Laurent polynomial decomposition *ab initio*, it is possible to find a candidate for h by examining only the leading coefficients of f and without having to try alternatives. For this we need a few more properties of Laurent polynomials.

6 Further Facts about Univariate Laurent Polynomials

For the second algorithm for Laurent polynomial decomposition it is useful to consider more leading and trailing coefficients than contemplated by Lemma 3. The following obviously generalizes to $i > (r - k)s$ and $i < -(r - k)t$, but we need only $k = 2$.

Lemma 8. *The coefficients of X^i in f and $g_r h^r + g_{r-1} h^{r-1}$ agree for $i > (r-2)s$ and for $i < -(r-2)t$.*

Proof. As for Lemma 3.

The leading coefficients are related as follows:

Lemma 9. *Let $T = \min(t, s-1)$. The coefficients of g, h and the leading $s+T+1$ coefficients of f are related by a system of polynomial equations of the form*

$$
\begin{aligned}
f_{rs} &= g_r h_s^r \\
f_{rs-i} &= r g_r h_s^{r-1} h_{s-i} & + P_{s-i}(h_s, \ldots, h_{s-i+1}, g_r), & \quad 1 \le i \le s-1 \\
f_{rs-i} &= r g_r h_s^{r-1} h_{s-i} + g_{r-1} h_s^{r-1} + P_{s-i}(h_s, \ldots, h_{s-i+1}, g_r) & & \quad i = s \\
f_{rs-i} &= r g_r h_s^{r-1} h_{s-i} & + P_{s-i}(h_s, \ldots, h_{s-i+1}, g_r, g_{r-1}), & \quad s+1 \le i \le s+T.
\end{aligned}
$$

Proof. Lemma 8 and multinomial expansion of $g_r h^r + g_{r-1} h^{r-1}$. □

The key observation that allows a one-ended algorithm is that the triangular system (2) can be extended, *as a triangular system*, if h_0 is restricted to be 0. We see this as follows: From Lemma 8 we know $f_{rs-s} = [X^{rs-s}](g_r h^r + g_{r-1} h^{r-1})$. A degree counting argument shows that this coefficient can depend only on $h_i, i \ge 0$, g_r and g_{r-1}. Higher degree coefficients of f give all of these but h_0 and g_{r-1} by (2). Then restricting $h_0 = 0$ determines g_{r-1}. We then have a triangular system that introduces each of the coefficients of h and g_{r-1} linearly.

Lemma 10. *If $f \in K[(X)]$ and $r \ge 2 \in \mathbb{Z}$ invertible in K, such that $f = g \circ h$ for some $g \in K[X]$ of degree r and $h \in K[(X)]$ of degree $\langle -t, s \rangle$, then g_r, g_{r-1} and all coefficients of h, save possibly h_{-t}, can be determined by a triangular system of the form:*

$$
\left.
\begin{aligned}
g_r &= Q_s & (f_{rs}) & \\
h_{s-i} &= Q_{s-i}(f_{rs-i}, h_{s-1}, \ldots, h_{s-i+1}, g_r^{-1}, g_r) & & \quad 1 \le i \le s-1 \\
g_{r-1} &= Q_0 & (f_{rs-s}, h_{s-1}, \ldots, h_1, \quad g_r^{-1}, g_r) & \\
h_{s-i} &= Q_{s-i}(f_{rs-i}, h_{s-1}, \ldots, h_1, h_{-1}, \ldots, h_{s-i+1}, g_r^{-1}, g_r, g_{r-1}) & & \\
& & & \quad s+1 \le i \le s+T
\end{aligned}
\right\}
$$
$$(5)$$

where $T = \min(t, s-1)$ and each Q_{s-i} is a polynomial function of $i+1$ variables. The coefficient h_{-t} is also determined if $t < s$.

Proof. As allowed by Lemma 7, we set $h_s = 1$, $h_0 = 0$ and specialize the system of Lemma 9. □

The above results are sufficient for our purposes when $t < s$, but the following will be necessary when $t = s$.

Lemma 11. *If $f \in K[(X)]$ is of degree $\langle -rs, rs \rangle$, and $f = g \circ (h_s X^s + h_{-s} X^{-s})$, then*

$$f_{is} = \sum_{n=0}^{\lfloor \frac{r-i}{2} \rfloor} \binom{2n+i}{n+i} g_{2n+i} h_s^{n+i} h_{-s}^n \qquad\qquad 0 \leq i \leq r, \qquad (6)$$

$$h_{-s}^i f_{is} = h_s^i f_{-is} \qquad\qquad -r \leq i \leq r, \qquad (7)$$

$$f_j = 0 \qquad\qquad j \neq is, \ -r \leq i \leq r. \qquad (8)$$

Proof. Use induction on r, noting $\sum_{n=\lfloor \frac{r-1-i}{2} \rfloor + 1}^{\lfloor \frac{r-i}{2} \rfloor}$ is empty if $r - i$ is odd and otherwise gives one term with $n = (r - i)/2$. $\qquad\square$

7   The One-Ended Algorithm

We now show how to decompose a Laurent polynomial by solving a triangular system derived from its leading coefficients. In the following we assume $0 < t \leq s$. This does not exclude any Laurent polynomials: If $t = 0$, the problem reduces to ordinary polynomial decomposition. If $t > s$, the algorithm can be applied to $f(\frac{1}{X})$. Under these assumptions, we are able to determine all the coefficients of h, except possibly h_{-t}, from the leading $2s$ coefficients of f. The coefficient h_{-t} is also found if $t < s$. The following algorithm computes h, possibly minus its trailing term.

Algorithm 4 (Determining $h - \eta$)

INPUT:
> $f \in K[(X)]$ and $r \geq 2 \in \mathbb{Z}$, with $\deg f = \langle -rt, rs \rangle, s \geq t$.

OUTPUT:
> *If there exist $g \in K[X]$, $\deg g = r$ and $h \in K[(X)]$ such that $f = g \circ h$, returns a choice of $h - \eta$, where $\eta = h_{-s} X^{-s}$. (Note $\eta = 0$ if $s > t$.)*

METHOD:

1. Let $p := X^s$.

2. For k from 1 to $s - 1$,
   (a) Let $c := \frac{1}{r}[X^{rs-k}](f/f_{rs} - p^r)$.
   (b) Let $p := p + cX^{s-k}$.

3. Let $g1 := [X^{rs-s}](f/f_{rs} - p^r)$.

4. For k from $s + 1$ to $s + \min(s - 1, t)$,
   (a) Let $c := \frac{1}{r}[X^{rs-k}](f/f_{rs} - p^{r-1}(p + g1))$.
   (b) Let $p := p + cX^{s-k}$.

5. Return $h = p$.

Theorem 2. *Algorithm 4 solves the polynomial system (5).*

Proof. We take $h_s = 1, h_0 = 0$. Step 2 gives the values for $h_{s-1}, \ldots, h_1$ by Theorem 1. A similar argument shows that Step 3 computes $g1 = g_{r-1}$ and that Step 4a computes $c_{(k)} = [X^{rs-k}] (f/f_{rs} - (h^r + g_{r-1}h^{r-1}))$. By Lemma 8, these give the unique values for $h_{-1}, \ldots, h_{-T}$, $T = \min(s-1, t)$. $\square$

Algorithm 4 gives h if $s > t$, but if $s = t$ the coefficient h_{-t} is not found. Depending on the form of h, it is possible to find this remaining coefficient in one of two ways. If the $h - \eta$ computed by Algorithm 4 has more than one term, then we may compute decompositions of $f(X)$ and $f(\frac{1}{X})$ and use the ratio of a pair of corresponding interior coefficients to determine h_{-t}. Otherwise, a special method is used for $h = X^s + h_{-s}/X^s$. These two procedures are described below.

Algorithm 5 (Determining h_{-s} when $s = t$, $h \neq h_s X^s + h_0 + h_{-s} X^{-s}$)

INPUT:

$\quad$ $f \in K[(X)]$ *of degree* $\langle -rs, rs \rangle$, $h - h_{-s} X^{-s} \in K[(X)]$ *such that* $f = g \circ h$, $g \in K[X]$, $h \neq h_s X^s + h_0 + h_{-s} X^{-s}$.

OUTPUT:

$\quad$ *Returns* h_{-s}.

METHOD:

1. *Find the smallest* i, $s - 1 \leq i \leq -s + 1$, *such that* $h_i \neq 0$.
2. *Apply Algorithm 4 to compute* $\bar{h} - \bar{h}_{-s} X^{-s}$ *from* $f(\frac{1}{X})$ *and* r. *Algorithm 4 may be terminated early, as soon as* $\bar{h}_{-i}$ *is computed.*
3. *Return* $h_{-s} = h_{-i}/\bar{h}_i$.

Note that here $h_0 = \bar{h}_0 = 0$ and one $h_i \neq 0$ by the input requirements.

Algorithm 6 (Determining h_{-1} when $h = X + h_{-1} X^{-1}$)

INPUT:

$\quad$ $f \in K[(X)]$ *of degree* $\langle -r, r \rangle$ *such that* $f = g \circ h$ *for some* $g \in K[X]$ *and* $h = X + h_{-1} X^{-1}$.

OUTPUT:

$\quad$ *Returns* h_{-1}.

METHOD:

1. *Let* $m = \gcd_{i \in I}(i)$ *where* $I = \{i \mid i > 0, f_i \neq 0\}$.
2. *If* $m = 1$,
   (a) *Compute* $c_i = f_{-i}/f_i$, $i \in I$. *Note* $c_i = h_{-1}{}^i$, *by (7)*.
   (b) *Use the extended Euclidean algorithm to find* m_i, $\sum_{i \in I} m_i i = 1$.
   (c) *Return* $h_{-1} = a$ *where* $a = \prod_{i \in I} c_i^{m_i}$. *Note* $\prod_{i \in I} c_i^{m_i} = h_{-1}^{\sum_{i \in I} m_i i}$.
3. *If* $m > 1$,
   (a) *Recursively find* $G \circ H = \sum_{i=-r/m}^{r/m} f_{mi} X^i$, $\deg G = r/m$, $H = X + A/X$.

(b) Return $h_{-1} = a$ for any a such that $a^m = A$.

We now have all the ingredients of the one-ended algorithm for univariate Laurent polynomial decomposition. We require $s \geq t$ so that, with the restriction $h_0 = 0$ and $h_s = 1$, the first $2s$ coefficients of f give a triangular system for g_r, g_{r-1} and all the coefficients of h, except possibly h_{-s}. As stated earlier, if $s < t$ we apply the algorithm to $f(\frac{1}{X})$.

Algorithm 7 (One-Ended Univariate Laurent Polynomial Decomposition)

INPUT:

   $f \in K[(X)]$ of degree $\langle -rt, rs \rangle$, $s \geq t$ and $r \geq 2 \in \mathbb{Z}$.

OUTPUT:

   If there exist $g \in K[X], h \in K[(X)]$ such that $\deg g = r$, $f = g \circ h$, returns a choice of g and h. Otherwise, returns FAIL.

METHOD:

1. Apply Algorithm 4 to f and r to obtain $h - \eta$.
2. If $s > t$, then $\eta = 0$ and we have h.
3. If $s = t$, then
   (a) If $h - \eta$ is a monomial, then
      i. If any $f_j \neq 0$ for $s \nmid j$, return FAIL.
      ii. Form $F = \sum_{i=-r}^{} r f_{is} X^i$.
      iii. Apply Algorithm 6 to F to compute h_{-s}
   (b) If $h - \eta$ is not a monomial, then
      i. Apply Algorithm 5 to f and $h - \eta$ to compute h_{-s}.
   We now have a candidate for h.
4. Construct the corresponding g by interpolation or by solving the linear system $\mathbf{A} \cdot \mathbf{g} = \mathbf{f}$ where $\mathbf{A}_{ij} = [X^{is}]h^j$, $\mathbf{g}_i = g_i$, $\mathbf{f}_i = f_{is}$, $0 \leq i, j \leq r$.
   The coefficients $\mathbf{A}_{ij}$ computed by Algorithm 4 in Step 1 may be reused.
5. Test whether $f = g \circ h$. If so, return g and h. Otherwise return FAIL.

8   Multivariate Laurent Polynomial Decomposition

The functional decomposition of Laurent polynomials can be extended to the multivariate case. We consider the following problem:

Problem 2 (Multivariate Laurent polynomial decomposition)
Given $f \in K[(X_1, \ldots, X_v)]$, K a field, and $r \geq 2 \in \mathbb{Z}$, do there exist $g \in K[Y]$ of degree r and $h \in K[(X_1, \ldots, X_v)]$ such that $f = g \circ h$? If so, find such g and h.

We reduce this to univariate Laurent polynomial decomposition. The reduction is not entirely trivial because the univariate algorithm sets $h_0 = 0$ and the usual multivariate reduction techniques may require $h_0 \neq 0$.

   To discuss the problem we set the following notation. Let $f \in K[(X_1, \ldots, X_v)]$. We seek a decomposition $f = g \circ h$ with $g \in K[Y]$ with $\deg g = r$. We require that r have an inverse in K and let $\deg f = \langle(-rt_1, \ldots, -rt_v), (rs_1, \ldots, rs_v)\rangle$. We use the notation $p_{i_1 \ldots i_v} = [X_1^{i_1} \cdots X_v^{i_v}]\, p$ where convenient.

Our univariate decomposition methods are based on the degrees of monomials. We will therefore employ techniques that preserve monomial degree. The first problem is then to find a weight vector such that no term of f, other than the constant term, has weighted total degree 0. This gives the following problem.

Problem 3 (Finding a constant-isolating weight vector)
Given a finite set of vectors $\mathbf{v}^{(1)}, \ldots, \mathbf{v}^{(N)} \in \mathbb{Z}^n$, find a vector $\mathbf{w} \in \mathbb{Z}^n$ such that $\mathbf{v}^{(j)} \cdot \mathbf{w} \Leftrightarrow \mathbf{v}^{(j)} = 0$.

Finding such a weight vector is straightforward. Finding such a weight vector that, for efficiency, minimizes the weighted degree of f requires more attention.

Once such a weight vector is found, we may make substitutions $X_i \mapsto \alpha_i X_1^{w_i}$, $\alpha_i \in K, 2 \le i \le v$ to obtain a univariate problem. Because of the choice of w, setting $h_0 = 0$ in the univariate image omits only the constant term in the multivariate problem. Finding multiple images of h under different substitutions allows h to be constructed by dense or sparse interpolation. The outer composition factor g need be computed only once. As before, it is necessary to test whether the candidate h gives $f = g \circ h$ since not all of the coefficients of f were examined to construct the composition factors.

In practice, we have found it to be more convenient avoid interpolation and to construct a multivariate h candidate directly. This can be achieved by adapting Algorithm 4 to use polynomials of homogeneous weighted degree d wherever a monomial of degree d is used in the original algorithm.

9   Conclusions

Motivated by the desire to reason about symbolic polynomials, we have studied the problem of Laurent polynomial decomposition. We have presented two algorithms to find the functional decomposition, if one exists, of a Laurent polynomial f as $g \circ h$, where g is a polynomial of a specified degree. The "two-ended" method constructs h from the leading and trailing coefficients of f and can be implemented in terms of an existing polynomial decomposition library. The "one-ended" method is more efficient and constructs h from only the leading coefficients of f. Multivariate Laurent polynomial decomposition can be given in terms of either of these methods.

These methods may be used to give the complete decomposition of a Laurent polynomial into irreducible composition factors. Both of these methods are susceptible to the same techniques to improve asymptotic complexity as the polynomial decomposition method of Kozen and Landau. Test implementations have been made in the Maple computer algebra system.

References

1. Ritt, J.: Prime and composite polynomials. Trans. American Math. Society 23(1), 51–66 (1922)
2. Engstrom, H.T.: Polynomial substitutions. American Journal of Mathematics 63(2), 249–255 (1941)

3. Levi, H.: Composite polynomials with coefficients in an arbitrary field of characteristic zero. American Journal of Mathematics 64(1), 389–400 (1942)
4. Barton, D.R., Zippel, R.E.: A polynomial decomposition algorithm. In: Proc. 1976 ACM Symposium on Symbolic and Algebraic Computation, pp. 356–358. ACM Press, New York (1976)
5. Kozen, D., Landau, S.: Polynomial decomposition algorithms. J. Symbolic Computation 22, 445–456 (1989)
6. Zippel, R.E.: Rational function decomposition. In: Proc. ISSAC 2001, pp. 1–6. ACM Press, New York (1991)
7. Kozen, D., Landau, S., Zippel, R.: Decomposition of algebraic functions. J. Symbolic Computation 22(3), 235–246 (1996)
8. von zur Gathen, J., Gutierrez, J., Rubio, R.: Multivariate polynomial decomposition. Applied Algebra in Engineering, Communication and Computing 14, 11–31 (2003)
9. Zieve, M.E.: Decompositions of Laurent polynomials (2007) Preprint: arXiv.org:0710.1902v1
10. Watt, S.M.: Making computer algebra more symbolic. In: Proc. Transgressive Computing 2006: A conference in honor of Jean Della Dora, pp. 43–49 (2006)
11. Watt, S.M.: Two families of algorithms for symbolic polynomials. In: Kotsireas, I., Zima, E. (eds.) Computer Algebra 2006: Latest Advances in Symbolic Algorithms – Proceedings of the Waterloo Workshop, pp. 193–210. World Scientific, Singapore (2007)
12. Watt, S.M.: Symbolic polynomials with sparse exponents. In: Proc. Milestones in Computer Algebra 2008: A conference in honour of Keith Geddes' 60th birthday, Stonehaven Bay, Trinidad and Tobago, University of Western Ontario, pp. 91–97 (2007) ISBN 978-0-7714-2682-7
13. Weispfenning, V.: Gröbner bases for binomials with parametric exponents. Technical report, Universität Passau, Germany (2004)
14. Yokoyama, K.: On systems of algebraic equations with parametric exponents. In: Proc. ISSAC 2004, pp. 312–319. ACM Press, New York (2004)
15. Pan, W., Wang, D.: Uniform gröbner bases for ideals generated by polynomials with parametric exponents. In: Proc. ISSAC 2006, pp. 269–276. ACM Press, New York (2006)
16. Watt, S.: Functional decomposition of symbolic polynomials. In: Proc. International Conference on Computatioanl Sciences and its Applications (ICCSA 2008), pp. 353–362. IEEE Computer Society, Los Alamitos (2008)

A Linear Grammar Approach to Mathematical Formula Recognition from PDF

Josef B. Baker, Alan P. Sexton, and Volker Sorge

School of Computer Science, University of Birmingham
J.B.Baker|A.P.Sexton|V.Sorge@cs.bham.ac.uk
www.cs.bham.ac.uk/~jbb|aps|vxs

Abstract. Many approaches have been proposed over the years for the recognition of mathematical formulae from scanned documents. More recently a need has arisen to recognise formulae from PDF documents. Here we can avoid ambiguities introduced by traditional OCR approaches and instead extract perfect knowledge of the characters used in formulae directly from the document. This can be exploited by formula recognition techniques to achieve correct results and high performance.

In this paper we revisit an old grammatical approach to formula recognition, that of Anderson from 1968, and assess its applicability with respect to data extracted from PDF documents. We identify some problems of the original method when applied to common mathematical expressions and show how they can be overcome. The simplicity of the original method leads to a very efficient recognition technique that not only is very simple to implement but also yields results of high accuracy for the recognition of mathematical formulae from PDF documents.

1 Introduction

In this paper we consider the problem of extracting mathematical formulae from Adobe PDF files, analysing their content and generating LaTeX output that reliably reflects the presentation of the formulae in the document. Furthermore, it is our intention that the LaTeX that we produce should not be dissimilar to that which a human user who commonly uses LaTeX might produce. In particular, this means that

1. we reject the option of simply independently placing every character found in the document at its correct location using LaTeX's picture environment. This would produce results that are only a visually accurate reproduction of the original but that lose a human writer's intention in the source text.
2. the produced LaTeX is clean and simple, often cleaner and simpler than the author's original source, if that source was indeed in LaTeX.
3. the produced PDF may actually improve upon the original because of LaTeX features that may not have been included in the original, or indeed, because the original was not formatted with LaTeX.

J. Carette et al. (Eds.): Calculemus/MKM 2009, LNAI 5625, pp. 201–216, 2009.
© Springer-Verlag Berlin Heidelberg 2009

It is our hope that such a system would be of benefit to the sight-impaired, who are otherwise excluded from reading the mathematical content of normal PDF documents, as well as providing some first steps towards improved usability of scientific documents to scientists, engineers, teachers and students; namely via the ability to easily extract potentially complicated formulae from documents and enter them into software tools such as computer algebra systems, function graphing packages, program code generation tools or theorem provers.

There is a moderately large and growing body of work on mathematical formula recognition from optically scanned images of documents. However, there is also a large number of scientific papers and texts available in PDF and, to date, very little work on taking advantage of the PDF document format to improve the accuracy, reliability and speed of formula recognition. Indeed, the most sophisticated and widely available tool for mathematical formula recognition at this time, Suzuki's Infty system [15], currently processes PDF documents by rendering the pages to an image format and applying its image analysis on that image. However, we claim that there are considerable benefits that can be obtained, albeit after a certain investment of effort, by analysing the PDF contents directly, rather than just analysing its rendered image. For a certain wide range of PDF documents we have the following advantages:

1. PDF documents contain proper character names for each character, obviating the need for the naturally error-prone and complex task of identifying characters from their shapes.
2. PDF documents unambiguously identify the font names and families that the characters are from. This is a particular source of complexity in mathematical formula recognition from scanned images, as font differences can be subtle but much more significant in mathematical texts than in normal text.
3. Other font metrics are directly available from the PDF document that can be extremely difficult to robustly obtain from images. These include the baseline position, the font weight, the italic angle, the capital and x height.
4. Mapping to Unicode can be obtained via the Adobe Glyph List, which, in particular, would simplify translation to MathML.

Unfortunately, not all PDF files provide these advantages. Some PDF documents store their page content only as images, in which case no advantage can accrue to the PDF analyser. Also, different versions of the PDF format require different algorithms for analysing them. Finally, PDF supports different font types. Type 1 and true type fonts are embedded in the PDF document with the meta information available as described above. Type 3 fonts, however, contain only rendered versions of the characters and the meta-information in not usually obtainable. Our research prototype currently works only with PDF versions 1.3 and 1.4 [1] using type 1 fonts.

By default TeX and LaTeX produce files suitable for our analysis, but other document processing systems (e.g., Troff) do so as well. Of course, if the source PDF has been originally produced from LaTeX, one could argue that it might be preferable to immediately work with the LaTeX source rather than the PDF, thus avoiding the entire recognition problem. A counterargument to this is that, firstly, most documents are only available as PDF files without the corresponding sources, even if generated from LaTeX. Secondly, analysing a LaTeX document with possibly multiple, nested layers of author defined macros might turn out to be more difficult and potentially less precise than working with the rendered result in form of a PDF document. This is especially the case when authors indulge in constructing symbols by overlaying multiple characters with explicit positioning — we have found this to be, unfortunately, relatively common in papers in Logic and Computer Science, even though correct symbols are available in the appropriate fonts.

Previous work in this area includes work by Yang and Fateman [16], who worked with mathematics contained in postscript files. By using font information contained within the file and heuristics based on changing fonts, sizes and using certain symbols, they were able to detect mathematics, which could then be recognised and parsed. Yuan and Liu [17] and Anjwierden [4] have both analysed the contents of PDF files in order to extract content and structure, however neither considered recognition of mathematics. Blostein and Grbavec [6] and Chan and Yeung [7] have written general reviews on mathematical formula recognition.

Our process to recognise mathematical formulae from PDF documents begins with identifying a clip region to analyse and extracting the information about the glyphs in the clip region from the PDF file, c.f. Section 2. We employ a two phase approach to parsing the formula itself, described in Section 3. The first phase is based on Anderson's original linearizer [3], adapted and extended to overcome some of its limitations, to turn a two dimensional mathematical formula into a linear representation, followed by a standard Yacc-style LALR parser to analyse the resulting expression into an abstract syntax tree. In Section 4 we present our LaTeX driver that walks this tree to generate the LaTeX output. We summarise our experiments in Section 5.1, discuss the issues resulting from this work (Section 5 and present our conclusions and future work in Section 6.

2   Extracting Information from PDF Files

We have previously presented the problems of, and our solutions to, extracting precise information about the characters from the PDF file [5], but summarise our approach here. PDF documents are normally presented in a compressed format. We currently use the open source Java program, Multivalent [11] to decompress them. At this point we can extract the PDF's bounding box (PDFBB) information about the characters as well as their font and Unicode metadata. Unfortunately, the PDFBB data obtained is a gross overestimate of each character's true size and only a rough guide to its position. This information is good enough for the analysis of normal text but inadequate for the fine distinctions required for two dimensional mathematical formula recognition. In particular, the PDFBBs for characters overlap significantly, even if the underlying characters are fully disjoint. In order to obtain the true bounding boxes, we render the PDF page to a tiff image and identify the true glyph bounding boxes (GBBs) from the image. Then we need to register the GBBs with the PDFBBs from the PDF file to produce the final symbol structures, which contain the character information together with a true, minimal bounding box.

The overlap in PDFBBs is great enough that, even in simple cases, the true character bounding boxes will intersect with a number of different PDF bounding boxes, making identification of the correct registration difficult. To overcome this problem we uniformly shrink the PDF bounding boxes by calculating the standard PDFBBs for the characters using the standard algorithm, but on the basis of a font size that is ten times smaller than the true one. This ensures that baseline information is preserved but also that the PDFBBs no longer overlap in most cases. For many cases, checking for intersection between this reduced PDF bounding boxes and the true bounding boxes is sufficient to identify the correct registration between glyph and character. However there are a number of special cases that still need to be dealt with. Some glyphs are composed of multiple overlapping characters, e.g., extended brackets or parentheses. Some characters are composed of multiple separate glyphs, e.g., the equals sign. The true bounding box for some symbols will necessarily intersect the bounding boxes of some different characters, e.g., the true bounding box for a square root symbol will typically intersect that of all of the symbols in the expression under it.

We handle these cases using the following algorithm where a *syntactic unit* is a structure identifying the symbol and its true bounding box and is the analogue of a single character in a one dimensional parser. The resulting set of syntactic units forms the input for the next step in the process.

Algorithm 1 (GlyphMatch)

INPUT: *A set of glyph bounding boxes and a set of PDF characters*
OUTPUT: *The set of syntactic units with exact bounding boxes and metadata.*
METHOD:
1. *Extenders: The fence extenders have indicative names, so use the names and the fact that their reduced PDF bounding boxes intersect the glyph bounding box of the fence glyph to register, and consume, the connected set of characters with the fence glyph.*
2. *Roots: A root symbol is composed of a radical character and a horizontal line. The former is clearly identified in the PDF file but, because its glyph bounding box is large and may contain many other characters, including nested root symbols, some care is required. The reduced PDFBB for the radical is always contained within the GBB for the root symbol, although the appropriate GBB may not be the smallest GBB that encloses it. Iterate through the radical characters in the clip in topmost, leftmost order. For each such symbol, register with it, and consume, the largest enclosing GBB.*
3. *One-One: Now we can safely register and consume every single glyph with a single character where the GBB of the glyph intersects only the PDFBB of the character and vice versa.*
4. *One-Many: Any sets of characters whose PDFBBs intersect only the same single glyph are registered and consumed.*

5. *Many-Many: This usually occurs in cases such as the definite integral, where the integral and the limits do not touch, but the PDFBB of the limits intersect the GBB of the much larger integral character. For a group where more than one GBB intersects, identify a character whose PDFBB intersects only one of the GBBs, Register and consume that character with that GBB. If all characters have not yet been consumed, repeat from Step 3.*

3 2D Parser

3.1 Anderson's System

In Anderson's thesis [3], which describes a coordinate grammar approach, he presented two algorithms. The first is a backtracking algorithm and does not scale well with large mathematical expressions. The second was far more efficient and it is this upon what we have based our work. This approach produces a single string representing a 2-d mathematical expression using a recursive function called LINEARIZE. It takes as input a list of syntactic units, ordered by left-to-right and top-to-bottom bounds. Each symbol in the list is either output or used to partition the remainder of the list into sets that are recursively processed by LINEARIZE in a strict order and output with special characters which identify their spatial relationships, which we call a *linearised structure string*. This string can then be parsed by a normal one-dimensional grammar to produce a parse tree. Unfortunately, it was only designed to work with a relatively simple algebra, working on a subset of the rules for mathematics described in his thesis.

The grammar itself has many restrictions, and relies on very carefully typeset mathematics, e.g., upper and lower limits in symbols such as $\sum$ had to be bounded horizontally by the symbol itself. Hence limits which occur to the right of the symbol, common in inline mathematics, or which extend past the right or left horizontal extent of the $\sum$ symbol itself, would not be correctly recognised. It was limited in the number of operators it recognised and could not cope with multi-line expressions at all. Despite these limitations, it provides a base that can be extended and modified to deal with a far larger set of mathematics.

3.2 Linearizer for PDF Data

In this section we present our modified LINEARIZE algorithm, extending that of Anderson to manage a much larger range of mathematical expressions. We start by grouping some syntactic units into *terminal symbols*. In many cases, the terminal symbols are just syntactic units, but a set of syntactic units that together make up an integer, a floating point number or a mathematical keyword (e.g., sin, cos, log etc.) are grouped together to form single terminal symbols.

Algorithm 2 (Lex)

INPUT: *A set of syntactic units*
OUTPUT: *A set of terminal symbols*
METHOD:

1. *Find groups of syntactic units whose baseline is common and whose horizontal displacement is within a predefined grouping threshold*

2. *For each group, if their syntactic units match a regular expression pattern for an integer, a floating point number or a mathematical keyword, construct the corresponding grouped terminal symbol and add it to the output set*

3. *All remaining syntactic units are added to the output set*

Next the LINEARIZE algorithm transforms the set of terminal symbols produced by LEX to a linearised structure string for our one-dimensional LALR parser:

Algorithm 3 (Linearize)

INPUT: *A set of terminal symbols*
OUTPUT: *A linearised structure string for single or multiple line formulae*
METHOD:

1. *The set of terminal symbols is maximally partitioned by horizontal bars of a predefined width of unbroken white space and each group is sorted lexicographically by increasing leftmost and decreasing topmost boundary position.*

2. *If the partition contains more than one group (i.e., line), note the horizontal position of the first symbol of the second group, output the token* multiline, *"(", and call* LINEARIZE *recursively on each group, inserting an* alignat *token at the noted position in the first, and at the start of each remaining group, finally output a terminating ")"*

3. *Otherwise, call* LINEARIZEGROUP *on the single group*

The LINEARIZE algorithm uses a utility method, LINEARIZEGROUP, that processes the specific cases that can occur within a single group of tokens:

Algorithm 4 (LinearizeGroup)

INPUT: *A list of terminal symbols, in left-to right, top-to-bottom order for a single line formula*
OUTPUT: *A linearised structure string for the single line formula*
METHOD:

1. *Consume the elements of the input list in order, taking the following action depending on the value of the first element:*

   Symbol with limits, e.g., $\sum$ or $\int$: *If a symbol which often has limits associated with it is identified, then the remaining list is scanned and the symbols partitioned into 3 sets: upper, lower and others. The head symbol is then output with the appropriate limits.*

   Horizontal line: *This signals a division. The symbols forming the numerator and denominator are partitioned. Then* LINEARIZE *is run on each partition followed by the remainder of the list.*

Radical: *If a radical symbol occurs then all symbols occurring within its bounding box are collected — typically, the extreme leftmost tip of the radical is to the left of any index symbol of the root. Any symbols in the top left corner of this bounding box are identified as the index of the root and the rest are passed to* LINEARIZE *as the body of the root.*

Fence symbol: *Search for the closing fence.*

(a) *If one exists, and the symbols bounded by these fences can be split into multiple lines, it is treated as a matrix and processed line by line, identifying column boundaries by horizontal whitespace. Each cell is processed as a single group by* LINEARIZE

(b) *Otherwise, if no matching fence was found and all of the remaining symbols can be partitioned into more than one line, then it is treated as a case statement and each line in the case is processed by* LINEARIZE.

(c) *Otherwise, the fence is treated as a standard symbol and output.*

None of the above: *If none of the above cases apply then a lookahead check is made on the next terminal symbol in the input*

(a) *if the next symbol is directly above or below the current one (normally such a case indicates an accent, bar, underbrace, etc.), the current symbol and all subsequent symbols that are similarly covered by the same accent are collect into a group, passed to* LINEARIZEGROUP *to be output and an* UNDER *or* OVER *token is output followed by the symbol identified to be placed or over under the group.*

(b) *Otherwise, if the the baseline of the next symbol differs from that of the current by a predefined minimum and maximum threshold, and the horizontal positions differ by no more than a predefined threshold, the next symbol is assumed to define a superscript or subscript group and this group is identified, partitioned and processed by* LINEARIZE

(c) *Otherwise the current symbol is treated as a standard symbol and output*

Our modified LINEARIZE algorithm can now recognise everything listed in Anderson's grammar, along with A. case statements, which are discussed, but not included in his grammar, B. accents, underbraces, underlines and overlines, C. limits, whether they occur as sub/superscripts or above or below a symbol, D. more mathematical operators, such as det and lim, E. Formulae spanning multiple lines, including simple alignment.

4 Drivers

Once we have the extracted the available mathematical content in linearized form we can further process it to regain the intended mathematical structure for both syntactic and semantic analysis. Furthermore, parsing the linearized expressions into a parse tree can already expose problems in the recognised expression, such as formulae that have been composed without using standard command structures (see Section 5 for more details).

Currently we focus primarily on the faithful reconstruction of formulae for presentation purposes. We first generate parse trees that are used as an intermediate representation for subsequent translation into mathematical markup. Concretely we have implemented drivers for LATEX and MathML.

4.1   Syntax Trees

The parse trees we generate from the linearised expressions contains nodes of different types that reflect the different structures we have recognised during the linearization algorithm. We define the data structure **STree** of parse trees via its single components as follows:

Leaf Nodes: The following leaf nodes are of type **STree**.

Empty: is the empty node.
Alignat: is a marker node to mark alignment positions in multiline expressions.
Number(d)**:** where d is either an integer or a floating point number.
Name(n)**:** where n is a string composed of alphanumeric characters.

Inner Nodes: Let $s, s', s'', s_1, \ldots, s_n$ be structures of type **STree** that are not **Empty**, let t, t' be structure of type **STree** that are potentially **Empty**, let $l_1, \ldots, l_n$ be lists of **STree** structures and let n, n' be strings composed of alphanumeric characters. Then the following are of type **STree**:

Linear(s, s')**:** meaning that s is followed by s'.
Div(s, s')**:** s is divided by s'.
Functor(n, s)**:** n contains s.
Super(s, s')**:** s' is superscript of s.
Sub(s, s')**:** s' is subscript of s.
SuperSub(s, s', s'')**:** s' is superscript of s and s'' is subscript of s.
Limit(s, t', t'')**:** s is an expression with possibly empty limits t and t'.
Over(s, s')**:** s' is on top of s.
Under(s, s')**:** s' is underneath s.
Case($n, s_1, \ldots, s_m$)**:** where $s_i, i = 1, \ldots, m$ represent vertical lines and n represents a, possible empty, left fence.
Multiline($s_1, \ldots, s_m$)**:** where $s_i, i = 1, \ldots, m$ represent stacked expression lines.
Matrix($n, n', l_1, \ldots, l_m$)**:** where $l_i, i = 1, \ldots, m$ are rows in a matrix that is has left fence n and right fence n'.

4.2   LATEX Driver

The concrete syntax trees are particularly well suited to generate LATEX code, and its translation is straightforward. The tree is recursively descended and replaced with proper LATEX expressions. Leaf nodes are either translated into the empty string (**Empty**), a number (**Number**), or mapped using a lookup table (**Name**). This lookup either translates the given name into a corresponding LATEX command or leaves it unchanged if it can not find one. We constructed the lookup table by extracting the Adobe names from a special PDF file composed of 579 commonly used characters taken from a database of LaTeX symbols [13]. While this special file is currently constructed by hand, and is therefore incomplete, we plan for a more exhaustive, automatic mechanism in the future.

As for the inner nodes, the translation of **Super**, **Sub**, and **SuperSub** is straightforward. **Limit** nodes are translated in a similar manner to super-subscript nodes. **Linear** represents linear concatenation and **Div** is translated with the `\frac` command. A node of the form **Functor**(n, s) is translated by taking n as a prefix command for s. Thus if n represents the square root symbol, we generate `\sqrt{s}`. Expressions in **Under** nodes are vertically stacked.

Over nodes on the other hand are interpreted as accents. Here the translation algorithm has to explicitly handle the case of multi-accented characters: While in PDF accents are stacked bottom up, in LaTeX, multi-accent characters are constructed recursively from the inside out. For example, the character $\overset{..}{\vec{\omega}}$ has to be translated from the syntax tree **Over**(omega, **Over**(vector, dotaccent)) into the LaTeX command `\dot{\vec{\omega}}`.

Case nodes are translated into left aligned arrays with the single fence character to the left. **Matrix** nodes are likewise translated into arrays with their corresponding left and right fences. The column number of the array is determined by the maximal number of expressions given in a single row. Finally, **Multiline** nodes are translated into amsmath split environments, with each **Alignat** nodes translated into & symbols to handle the alignment.

4.3   MathML Driver

The MathML driver is similar to the LaTeX driver, but has some significant differences. **Empty** nodes are again translated to empty strings and numbers are marked up with the `<mn>` tag. **Name** nodes are again mapped using a lookup table as before, but we employ a translation table[1] that maps all of Adobe's 4281 PDF characters to their corresponding Unicode values. This has the advantage that we should not come across any character that is not mapped. On the other hand, mapping to Unicode values, rather than to actual characters or commands as in LaTeX, looses information that could be useful for a future, more detailed semantic analysis. The result of this mapping is uniformly put between `<mi>` tags, thus operators, normally marked up by `<mo>` tags, are currently not distinguished. This could be achieved with another lookup table. However, we believe this is best left to a proper semantic markup such as an OpenMath driver, as we can then exploit the semantic knowledge given in content dictionaries rather than employing a handcrafted lookup table.

We combine consecutive **Linear** nodes recursively to put them into a single `<mrow>` tag. **Div** nodes are translated into `<mfrac>` tags and **Sub**, **Super** and **Supersub** nodes are mapped to the MathML environments `<msub>`, `<msup>`, and `<msubsup>`, respectively. **Over** and **Under** nodes are translated to `<mover>` and `<munder>` tags, where we set the parameter `accent` to true for the former

---

[1] http://partners.adobe.com/public/developer/en/opentype/glyphlist.txt

and false for the latter. As opposed to the LaTeX driver, in MathML we have to explicitly sort out nested over and under expressions in order to put them into <munderover>. Similarly, **Limit** nodes are mapped to <munderover> environments rather than represented as sub- and superscripts.

In terms of **Functor** nodes we currently only handle root symbols, which are either mapped to <msqrt>, or to <mroot> if the expression is combined with an additional **Sup** node, where the latter is then taken as the index value. Again this analysis is not necessary in the LaTeX case as it is handled automatically by LaTeX's conventions.

Finally, **Case**, **Matrix**, and **Multiline** nodes are all handled by <mtable> environments. For the latter the alignment is achieved by using MathML's special alignment tags <maligngroup/>.

5   Discussion

We present our experimental setup to test the effectiveness our developed approach and discuss the obtained results as well as some of the general advantages and deficiencies of the current procedure.

5.1   Experiments

While we developed the PDF extraction and matching algorithms with bespoke, hand-crafted examples, for the design and debugging of our grammar we have used a document of LaTeX samples [12]. The document contains 22 expressions, covering a broad range of mathematical formulae of varying complexity. For our experiments we then chose parts of two electronic books from two complementary areas of Mathematics:

1. **Sternberg's "Semi-Riemannian Geometry and General Relativity"** [14]. We have extracted all the 79 displayed mathematical expressions on the first 22 pages of that book.
2. **Judson's "Abstract Algebra – Theory and Applications"** [8]. We have taken 49 mathematical expressions from the first 31 pages.

Note, that we had to choose books that are not only freely available, but also in the right format, that is, they needed to be in the right PDF format and have accessible content in the sense that it was created from LaTeX and not given as embedded images or encrypted. Note also that from Judson's book we have used a selection of expressions concentrating on complex and thus from our point of view interesting formulae, as many of the expressions on these pages are of similar structure or fairly trivial (e.g., simple sequences of elements or linear formulae) and we still do the clipping manually.

The evaluation of the results was carried out using the LaTeX output, as it is more easily comparable with the original expressions and therefore gives a better indication as to the faithfulness of the recognition.

$\displaystyle\int\sqrt{\sum_{i,j=1}^{n-1}Q_{ij}(y(t))\frac{dy^i}{dt}(t)\frac{dy^j}{dt}(t)}\ \ dt$	$\displaystyle\int\sqrt{\sum_{i,j=1}^{n-1}Q_{ij}\,(y\,(t))\frac{dy^i}{dt}\,(t)\,\frac{dy^j}{dt}\,(t)}\,dt$
$\displaystyle\gamma'(t)=\sum_{j=1}^{n-1}X_j(y(t))\frac{dy^j}{dt}(t)$	$\displaystyle\gamma'\,(t)=\sum_{j=1}^{n-1}X_j\,(y\,(t))\frac{dy^j}{dt}\,(t)$
$y(t)=(y^1(t),\ldots,y^{n-1}(t))$	$y\,(t)=(y^1\,(t),\ldots,y^{n-1}\,(t))$
$\displaystyle\lVert\gamma'(t)\rVert^2=\sum_{i,j=1}^{n-1}Q_{ij}(y(t))\frac{dy^i}{dt}(t)\frac{dy^j}{dt}(t)$	$\displaystyle\lVert\gamma'\,(t)\rVert^2=\sum_{i,j=1}^{n-1}Q_{ij}\,(y\,(t))\frac{dy^i}{dt}\,(t)\,\frac{dy^j}{dt}\,(t)$
$\displaystyle\int\lVert\gamma'(t)\rVert dt$	$\displaystyle\int\lVert\gamma'\,(t)\rVert dt$
$Q\ =\ \begin{pmatrix}E & F\\ F & G\end{pmatrix}$	$Q=\begin{pmatrix}E & F\\ F & G\end{pmatrix}$
$\begin{aligned}e\ &=\ N\cdot X_{uu}\\ &=\ \frac{1}{\sqrt{EG-F^2}}X_{uu}\cdot(X_u\times X_v)\\ &=\ \frac{1}{\sqrt{EG-F^2}}\det(X_{uu},X_u,X_v)\end{aligned}$	$\begin{aligned}e &= N\cdot X_{uu}\\ &= \frac{1}{\sqrt{EG-F^2}}X_{uu}\cdot(X_u\times X_v)\\ &= \frac{1}{\sqrt{EG-F^2}}\det(X_{uu},X_u,X_v)\end{aligned}$
$\det Q\ =\ EG-F^2$	$\det Q=EG-F^2$

Fig. 1. Formulae from [14]. Left column contains rendered images from the PDF, right column contains the formatted latex output of the generated results.

In Figure 1, we show the images of a sample of equations as clipped from rendered images of pages of this book together with the equations as extracted to LATEX and subsequently formatted. In Figure 2, we show the generated latex code for the first expression in Figure 1. We have tidied up the white space in this code for presentation purposes, but not modified any non white space characters.

From the 79 expressions of the first book, only 1 failed to be recognised when creating the parse tree. An additional 13 were rendered slightly differently to the original, but with no loss of semantic information. From the 49 expressions of book two, 2 could be recognised but produced incorrect LATEX and a further 5 had rendering differences with respect to font inconsistencies.

```
\[ \int ^{}_{} \sqrt{ \sum ^{ n - 1 }_{ i , j = 1 }
Q _{ i j } \left( y \left( t \right) \right)
\frac{ d y ^{ i }}{ d t } \left( t \right)
\frac{ d y ^{ j }}{ d t } \left( t \right) } d t \]
```

Fig. 2. Sample generated LᴬTEX code for first equation in Figure 1

$J \; := \; \left(\begin{smallmatrix} \frac{\partial u}{\partial u'} & \frac{\partial u}{\partial v'} \\ \frac{\partial v}{\partial u'} & \frac{\partial v}{\partial v'} \end{smallmatrix} \right)$	$J \colon = \left(\begin{smallmatrix} \frac{\partial u}{\partial_{\partial_v} u'} & \frac{\partial u}{\partial_{\partial_v} v'} \\ \partial u' & \partial v' \end{smallmatrix} \right)$
$L_{ij} = -(N, \dfrac{\partial^2 X}{\partial y_i \partial y_j})$	$L_{ij} = - \left(N, \dfrac{\partial^2 X}{\partial y_i \partial y_j} \right)$

Fig. 3. Some of the incorrectly recognised formulae; original rendered image on the left, formatted LᴬTEX output of the generated results on the right

A more detail analysis of the results for both books show:

Fences: Within the sample formulae were 186 pairs of fences, of which 182 were rendered correctly. The other 4 pairs were rendered larger than those in the sample formulae. However, even though they were a different size, it actually improved the readability of the mathematics. This is shown in the bottom formula of Fig. 3 where the parentheses now enclose the whole expression.

Horizontal Whitespace: Of 137 lines of formulae, 122 were spaced equivalently to the original samples. Of the 14 cases where spacing was different, 5 did not include appropriate spacing in between pairs of equations separated by commas, 8 had too much spacing between the : and = symbols, and 2 had too much spacing between a function denoted by a Greek letter and its bracketed argument. All formulae that spanned several lines were aligned correctly.

Matrices: All but two of the 19 matrices were identified and rendered correctly. One could not be translated into a syntax tree as the right bracket had a superscript that is not yet handled by our second phase grammar that parses the linearized expressions. The second incorrect matrix, given in the lower formula of Fig. 3, contained no whitespace between the two rows. Therefore the matrix was recognised as a bracketed expression, with the elements being recognised as superscripts and subscripts of each other. This case will often occur when text has been badly manipulated for formatting purposes.

Superscripts and Subscripts: Over 250 super and subscripts occurred, all of which were recognised correctly. Also no text was incorrectly identified as being a script. Two expressions could not be formatted in LᴬTEX as they contained accent characters in unexpected places, which caused problems with the generic LᴬTEX translation. See the next section for more details.

Font Problems: Except for 5 expressions all formulae rendered in the correct Math fonts. 2 of the formulae contained blackboard characters for number sets which rendered as normal Roman characters and a further 3 contained interspersed text, which was not recognised as such. For a more general discussion of this and other shortcomings of our current procedure see Sec. 5.3.

5.2   Advantages

We have already discussed the improvements of our algorithm over the original approach by Anderson previously. Some additional advantages are:

Super- and subscript detection: Since our algorithm for the detection of super- and subscripts is based on the characters' true baseline and not on their centre points on the vertical axis, we gain a reliable method to recognise sub- and superscript relations. Our experimental results confirm that the algorithm does indeed yield perfect results, even in the case where the author has used unusual ways of producing superscripts (e.g., by abusing an accent character). This is not only a clear improvement of the original, threshold based procedure of Anderson, but also over comparable approaches. For example, Aly, Uchida, and Suzuki present an elaborate approach for the detection of super and subscripts in images [2]. While it yields very good results, it is still based on statistical data and cannot compete with the advantages of having true baseline information.

Limits: As with super- and subscripts, we also obtain limits of operators like summations and integrals purely via baseline analysis, yielding perfect results.

Characters vs. Operators: A common problem for regular OCR systems is to distinguish alphabetical characters representing operators such as sums or products from their counterpart representing the actual character, for example, recognising the difference between $\sum_{i=0}^{n}$ from Σ^*. In PDF these symbols are usually flagged by character name such as "summationtext" or "summationdisplay" as opposed to "Sigma", which makes their distinction easy, yielding the their semantics automatically. But, in case the author has not adhered to the normal LaTeX conventions, a "Sigma" can still be given upper and lower limits as they will be caught as super and subscripts.

Enclosing symbols: These pose a traditional problem for OCR systems. An example is the square root symbol for which it is generally difficult either to determine their extension or to get to the enclosed characters in the first place. However, both pieces of information are straight forward to collect from PDF and our experiments yield perfect results on square roots so far.

5.3   Shortcomings

We have identified some shortcomings of our current procedure, both from the experimental results and from general considerations of the algorithm.

Matrices: Matrices are not identified if there is no whitespace between rows, instead, they are recognised as an expression enclosed by fences. This can lead to undesired formatting, in which elements are recognised as superscripts and subscripts of each other.

Character abuse and manual layout: Problems can occur if authors have used LaTeX commands contrary to their intended purposes. For example we have come across expressions of the form A' where we have recognised the prime character $'$ to be in fact an accent character `acute`. In other words the author has most probably used a command combination like `A\acute{}` to achieve the desired effect. Our grammar, however, views the character as a superscript rather than an accent, since the character is in the right top corner rather than above another character. As a consequence our mapping leads to a subsequent LaTeX error. On the other hand a direct translation of the recognised character into Unicode and translating into MathML as superscript would not yield a problem. These situations can occur if expressions have been manufactured by moving characters manually into place (e.g., by using explicit positioning commands) to achieve a desired presentational effect or if single characters have been created by overlapping several characters. Then the likelihood to recognise the corresponding character is higher using a conventional OCR engine than our technique.

Brackets and fences: In our current approach we simply translate bracket symbols into corresponding LaTeX commands and pre-attach a `\left` or `\right` modifier depending on the orientation of the bracket. Obviously this does not necessarily correspond to the actual form or size of brackets in the original presentation and it could also pair brackets that are not meant to be opening and closing to each other, in particular if the author has inserted some solitary brackets.

In case there is an imbalanced number of fences, we add the necessary `\right.` or `\left.` at the beginning or end of the expression, respectively, to avoid LaTeX errors. Obviously this form of error correction is prone to introduce presentation errors, as it is not evident which superfluous brackets have to be matched up and where.

Moreover, not all potential fence symbols can be identified in this way. In particular, neutral fence symbols (i.e., symbols for which the left and right version are identical) like bars but also customary fence symbols can not be handled this way. A simple heuristic could aim to identify all characters in the PDF with vertical extenders, excluding some specialist symbols like integral signs. However, since sometimes even characters of small vertical extension can contain extenders, this heuristic could not be failsafe. Moreover, one would still have to pair fences in order to recognise which is a left and which is a right fence, thus even if fences were always recognised, in case some fence occurs alone, as is often the case with single bars for example, it is not yet clear how to judge whether the symbol functions as a left or a right fence to some expression.

Matrix alignment: Matrices are aligned by putting them into bracketed arrays. The horizontal extension of the array is determined by the maximal number of expressions given in a single row. Since the length of each row can indeed vary, e.g., in case the author has omitted elements and left free space, the matrix will appear left aligned and some of the elements of the recognised matrix are not necessarily at a the position originally intended. This problem can be overcome by extending the purely grammatical approach and exploiting the actual spatial information on the elements in the matrix that can be obtained from the PDF. We plan to adapt Kanahori and Suzuki's approach to correctly align OCRed matrices [9,10] to work with the additional special character information.

Multiline formula alignment: We currently employ a simple method to align multiline formulae. This works well in most common cases of equational alignments. However, we anticipate that it will not necessarily yield good results for more customised alignments chosen by authors. A more advanced approach will have to take more detailed spatial information from the PDF into account.

Interspersed text: Regular text within mathematical expressions is currently not recognised as such and therefore not properly grouped. We intend to employ improved segmentation techniques that will identify large portions of text between mathematical expressions. Segmentation would, however, not work for small areas of text as can often be found, for example, in a definition by cases. Here a promising approach is to perform text grouping by recognising the font as non-mathematical.

Specialist fonts: In general we are not yet making use of the specific font information that we acquire during the PDF extraction phase. The grammatical recognition phase is purely based on the character information pertaining to size and relative special positions. In the future we intend to attach the font information to the recognised symbols and exploit it in the drivers by mapping it to the appropriate LaTeX or MathML fonts.

6   Conclusions

We have presented an approach at recognising mathematical content directly from PDF documents rather than going the route via traditional OCR. As a continuation of our previous work in which we revisit traditional heuristic formula recognition techniques and turn them into more analytical approaches in the light of perfect data, we have presented an adaptation and extension of Anderson's original linear grammar to process the PDF data. The result yields a faithful recognition of the formulae in a predominant number of cases in both LaTeX and MathML translation. Our experiments so far have shown that the approach is very effective, and although there are some current shortcomings, they are not of a type that appear to be insolvable within the linear grammar approach.

To address these shortcomings we want to exploit more of the information extracted from the PDF explicitly, in particular information on fonts and multiline alignments. This would also help to identify additional operator names, similar to sin, cos, etc., that are not mapped directly to LaTeX commands.

We subsequently want to run a case study with a much larger selection of books and expressions. Thereby the major drawback is still that the segmentation of the mathematical expressions has to be done manually. However, we want to combine our approach with an automatic segmentation algorithm for PDF documents, such as the one used in Infty [15].

References

1. Adobe Systems. PDF Reference fifth edition Adobe Portable Document Format Version 1.6 (2004)
2. Aly, W., Uchida, S., Suzuki, M.: Identifying subscripts, superscripts in mathematical documents. Mathematics in Computer Science (2008)
3. Anderson, R.H.: Syntax-Directed Recognition of Hand-Printed Two-dimensional Mathematics. PhD thesis, Harvard University, Cambridge, MA (January 1968)
4. Anjwierden, A.: Aidas: Incremental logical structure discovery in PDF documents. In: Proc. of ICDAR 2001, p. 374. IEEE Computer Society, Los Alamitos (2001)
5. Baker, J., Sexton, A.P., Sorge, V.: Extracting precise data on the mathematical content of PDF documents. In: Proc. of DML 2008. Masaryk University Press (2008)
6. Blostein, D., Grbavec, A.: Handbook on Optical Character Recognition and Document Image Analysis, Recognition of Mathematical Notation. World Scientific, Singapore (1996)
7. Chan, K., Yeung, D.: Mathematical expression recognition: a survey. International Journal on Document Analysis and Recognition (2000)
8. Judson, T.: Abstract algebra — theory and applications (February 2009), http://abstract.ups.edu/download.html
9. Kanahori, T., Suzuki, M.: A recognition method of matrices by using variable block pattern elements generating rectangular areas. In: Blostein, D., Kwon, Y.-B. (eds.) GREC 2001. LNCS, vol. 2390, pp. 320–329. Springer, Heidelberg (2002)
10. Kanahori, T., Suzuki, M.: Detection of matrices and segmentation of matrix elements in scanned images of scientific documents. In: ICDAR 2003, pp. 433–437 (2003)
11. Phelps, T.: Multivalent, http://multivalent.sourceforge.net/
12. Roberts, T.: LATEX mathematics examples (May 2004), http://www.sci.usq.edu.au/staff/aroberts/LaTeX/Src/maths.pdf
13. Sexton, A., Sorge, V.: Database-driven mathematical character recognition. In: Liu, W., Lladós, J. (eds.) GREC 2005. LNCS, vol. 3926, pp. 218–230. Springer, Heidelberg (2006)
14. Sternberg, S.: Semi-riemann geometry and general relativity (September 2003), http://www.math.harvard.edu/~shlomo/docs/semi_riemannian_geometry.pdf
15. Suzuki, M., Tamari, F., Fukuda, R., Uchida, S., Kanahori, T.: Infty — an integrated OCR system for mathematical documents. In: Proceedings of ACM Symposium on Document Engineering, pp. 95–104. ACM Press, New York (2003)
16. Yang, M., Fateman, R.: Extracting mathematical expressions from postscript documents. In: Proc. of ISSAC 2004, pp. 305–311. ACM Press, New York (2004)
17. Yuan, F., Liu, B.: A new method of information extraction from PDF files. In: Proc. of Machine Learning and Cybernetics, pp. 1738–1742. IEEE Computer Society, Los Alamitos (2005)

Formal Proof:
Reconciling Correctness and Understanding

Cristian S. Calude[1] and Christine Müller[1,2]

[1] Department of Computer Science, University of Auckland, New Zealand
[2] Department of Computer Science, Jacobs University, Bremen, Germany
www.cs.auckland.ac.nz/~cristian, kwarc.info/cmueller

A good proof is a proof that makes
us wiser. Manin [41, p. 209].

Abstract. Hilbert's concept of formal proof is an ideal of rigour for
mathematics which has important applications in mathematical logic,
but seems irrelevant for the practice of mathematics. The advent, in the
last twenty years, of proof assistants was followed by an impressive record
of deep mathematical theorems formally proved. Formal proof is practically achievable. With formal proof, correctness reaches a standard that
no pen-and-paper proof can match, but an essential component of mathematics — the insight and understanding — seems to be in short supply.
So, what makes a proof understandable? To answer this question we first
suggest a list of symptoms of understanding. We then propose a vision of
an environment in which users can write and check formal proofs as well
as query them with reference to the symptoms of understanding. In this
way, the environment reconciles the main features of proof: correctness
and understanding.

1 Introduction

From Pythagoras and Euclid to Hilbert and Bourbaki, mathematical proofs were
essentially based on axiomatic-deductive reasoning. This view was repeatedly
expressed by the most prominent mathematicians. For Bourbaki [11], *Depuis
les Grecs, qui dit Mathématique, dit démonstration*, and for Mac Lane [37], *If
a result has not yet been given valid proof, it isn't yet mathematics: we should
strive to make it such.*

A *formal proof* is written in a formal language consisting of certain strings
of symbols from a fixed alphabet. Formal proofs are precisely specified without
any ambiguity because all notions are explicitly defined, no steps (no matter
how small) are omitted, no appeal to any kind of intuition is made. They satisfy
Hilbert's criterion of mechanical testing:

> *The rules should be so clear, that if somebody gives you what they claim
> is a proof, there is a mechanical procedure that will check whether the
> proof is correct or not, whether it obeys the rules or not.*

J. Carette et al. (Eds.): Calculemus/MKM 2009, LNAI 5625, pp. 217–232, 2009.

By making sure that every step is correct, one can tell once and for all whether a proof is correct or not, i.e. whether a theorem has been proved.

Hilbert's concept of formal proof is an ideal of rigour for mathematics which has important applications in mathematical logic (computability theory and proof theory), but seems irrelevant for the practice of mathematics.

An *informal* (pen-on-paper) *proof* is a rigorous argument expressed in a mixture of natural language and formulae (for some mathematicians an equal mixture is the best proportion) that is intended to convince a knowledgeable mathematician of the truth of a statement, the theorem. Routine logical inferences are omitted. "Folklore" results are used without proof. Depending on the area, arguments may rely on intuition. Informal proofs are the standard of presentation of mathematics in textbooks, journals, classrooms, and conferences. They are the product of a social process.

In theory, each informal proof can be converted into a formal proof. However, this is rarely, almost never, done in practice[1]. Bourbaki, who came closer to formal proving than most mathematicians, still declared that *formalized mathematics cannot in practice be written down in full*, a goal that is an *absolutely unrealizable* program.

Gödel's Incompleteness Theorem [25] shows that in every formal system satisfying a modicum of natural assumptions certain statements are true but not provable. In this sense, the formal approach to mathematics is not universal, not everything can be formally proved. Still, no universal alternative is available. Although a formal proof cannot guarantee 100% correctness because, for example, one cannot prove the correctness of the formal prover itself (a well-known result in computability theory, [24]) the certainty achieved is close to "certain"[2].

The advent, in the last twenty years, of proof assistants was followed by an impressive record of deep mathematical theorems formally proved. The list includes Gödel Incompleteness Theorem (1986)[3], the Fundamental Theorem of Calculus (1996), the Fundamental Theorem of Algebra (2000), the Four Colour Theorem (2004), Jordan's Curve Theorem (2005), the Prime Number Theorem (2008), see [32]. The December 2008 issue of the *Notices of AMS* includes four papers on formal proof: three general overviews [32, 33, 66] and one study case, the formal proof of the Four-Colour Theorem [29]. *Hilbert's standard of proof is practicable, it's becoming reality.*

An automatic prover can be used not only to check the validity of a formalised proof of a known mathematical result (as in the list of famous theorems enumerated above), but also to interactively help to "prove" new theorems. The informal proof of the main result in [19] benefited substantially from the process

---

[1] Russell and Whitehead 2,500-page opus *Principia Mathematica* [65] is a famous exception: a fully formalised mathematical book. Russell believed that *no human being will ever read through it.*

[2] Not all agree. *Practically, I am "certain" that the HW+OS+ML/Compiler/Runtime + Isabelle implementation is* not *fully trustable*, [62].

[3] It's ironic to have this theorem — which limits the power of formal proving — as the first formally proved important theorem.

of formalisation in the interactive theorem prover Isabelle [48] of one of the key results in algorithmic information theory, the Kraft-Chaitin Theorem.

The correctness achieved by formal proofs cannot be matched by pen-and-paper proofs. However, an essential component of mathematics — the insight and understanding — seems to be in short supply. So, what makes a proof understandable? While correctness can be formally defined, understanding is subjective, so much more difficult to pin down. Our solution is to suggest a list of symptoms of understanding and, with reference to these symptoms, to propose a framework that reconciles correctness and understanding.

The paper is organised as follows. In Section 2 we discuss a list of symptoms of understanding. In Section 3 we present an envisioned environment that provides services regarding these symptoms. Section 4 concludes the paper with a summary of services supporting symptoms and describes future work.

2   Understanding Mathematical Proof

The gap between correctness and understanding seems to be widening (see [15, 18, 16]). Should one abandon the axiomatic-deductive model, should one sacrifice understanding for efficiency, should one try other avenues?

Although most mathematicians agree that understanding is paramount to mathematics there is little consensus regarding the understanding of understanding. Understanding in mathematics may mean many things, but, usually, mathematicians have no difficulties in recognising it. In contrast with correctness, understanding is subjective and probably cannot be rigorously defined.

Inspired by the analysis in [8, p. 9–10] we propose a list of *symptoms* for detecting the understanding of a proof. We use the term *symptom* in analogy with its medical meaning. The list is not exhaustive, not all symptoms are necessary to identify understanding, and some symptoms overlap. Not all symptoms are equally important and ranking seems almost impossible. Many mathematicians may argue that the first two symptoms are the most important ones. Symptoms are discussed and illustrated sometimes with reference to the following lemma, which is presented with three proofs, one informal and two formal ones. The formal proofs were generated with Isabelle. The complete proof script, written by N. Hay, appears in [34]; it is part of a more complex proof for the Kraft-Chaitin Theorem [19]. The formal proof in Isar [64] was written by M. Wenzel [62].

Lemma 1. *For all (binary) strings x, y, xy extends x.*

Proof. (Informal) The relation 'n extends v' (written $u \supset v$) is defined by the following two rules: a) for every string u, $u \supset u$, b) for every strings u,v, if $u \supset v$ then $ui \supset v$, for every $i \in \{0,1\}$.

Take two strings x, y. If y is the empty string then $xy = x \supset x$ by a). If $xy \supset x$ and $i \in \{0,1\}$, then $x(yi) = (xy)i \supset x$ by associativity of concatenation, hypothesis, and b).   □

Proof. (Formal: Isabelle proof script using recursive definitions)

```
fun extends :: "'A list ⇒ 'A list ⇒ bool"
where
"extends [] [] = True"
| "extends [] (y#ys) = False"
| "extends x [] = True"
| "extends (x#xs) (y#ys) = ((x=y) & (extends xs ys))"

lemma extends1 : "extends A [] "
  apply (induct A) apply(simp_all)
done
lemma extends2 : "extends (A@B) A "
  apply (induct A) apply(simp_all) apply(simp only: extends1)
done
```

Proof. (Formal: Isar proof using Natural Deduction rules)

```
theory Extends
imports Main
begin
definition extends :: "'a list ⇒ 'a list ⇒ bool"
  where "extends A B ⟷ (∃C. A = B @ C)"
lemma extendsI [intro]:
  assumes "A = B @ C" shows "extends A B"
  using assms unfolding extends_def by blast
lemma extendsE [elim]:
  assumes "extends A B" obtains C where "A = B @ C"
  using assms unfolding extends_def by blast
lemma extends0: "extends A A"
proof
  show "A = A @ []" by simp
qed

lemma extends1: "extends A []"
proof
  show "A = [] @ A" by simp
qed

lemma extends2: "extends (A @ B) A"
proof
  show "A @ B = A @ B" by simp
qed

lemma extends3:
  assumes "extends A' A"
  shows "extends (A' @ B) A"
proof -
  have "extends A' A" by fact
  then obtain C' where "A' = A @ C'" ..
  then have "A' @ B = A @ (C' @ B)" by simp
  then show ?thesis ..
qed
```

```
lemma "extends A A" by auto
lemma "extends A []" by auto
lemma "extends (A @ B) A" by auto
end
```

□

The following list describes symptoms of understanding a mathematical proof as the ability to perform various tasks. These symptoms have been motivated by the understanding of mathematics in general.

Symptom 1: *Fill in simple details of the proof, like explication of notation and definitions.* Understanding implies the ability to answer questions about concepts, their properties, and relations: What is the domain of variables x, y used in the informal proof? What happens when x has a value outside its domain, say $x = 102, y = 10$? For the first formal proof one can query: What is bool? What is the relation between True or False or 0 or 1? What is list, @, [], etc.? What is = and # and what properties do these two relations have?

The level of detail in definitions and concepts is different in the two proofs. For example, one can query the definitions and properties of =, #, or *lists* in the Isabelle formal proof, but hardly in the informal proof.

Symptom 2: *Justify other results implicitly used in the proof and inferences.* The property of associativity and the proof by induction are assumed to be known in the informal proof for Lemma 1 above.

Symptom 3: *Give presentations of the proof for different audiences having various degrees of expertise.* Users can be experts in the subject, experts in the area but not in the subject, professional mathematicians, graduate students, undergraduate students, non-mathematicians with interest in the subject, readers of a science magazine, etc. For example, for an expert the proof of Lemma 1 above is too detailed, in fact the lemma itself may be omitted. For a beginner, the detailed proof for the irrationality of $\sqrt{2}$ is suitable, see [30, p. 37]. *A mathematical theory is not to be considered complete until you have made it so clear that you can explain it to the first man whom you meet on the street* says Hilbert.

Symptom 4: *Cast the proof in different terms.* For example, by varying the proportion of natural language and formulae, by varying the level of detail, or by using the language of a different area of mathematics[4]. The irrationality of the golden ratio can be presented from various perspectives, geometrical, algebraic, [30, p. 41–45].

Symptom 5: *Motivate the proof.* Explain why certain notions/constructions are natural, necessary, in contrast with other potential candidates. For example, one may ask what is the natural representation of strings in Isabelle and what are the basic operations with strings [19].

---

[4] The word language does not only refer to the terminology and notation only, but to the whole "spirit" of an area of mathematics.

Symptom 6: *Indicate key or novel points in the argument.* The solution of Post's Problem [24, p.237] requires a new ingredient, the priority argument. The argument is highlighted several times in the proof of the Friedberg-Muchnik Theorem presented in [24, p.238].

Symptom 7: *Give natural examples and counter-examples for various notions used in the proof.* Follow the proof of Lemma 1 and justify why $xy \supset x$ for various groups of strings like $x = 10111, y = 10$ or $x = 102, y = 10$.

Symptom 8: *Indicate where certain hypotheses are needed.* The Banach-Tarski Paradox shows how to cut a solid sphere into pieces and then reassemble them without bending, stretching or distorting them to finally obtain two (or ten) solid spheres equal in volume with the original one — the analog of the "Ponzi-type" effect in economics. This is possible because by cutting the sphere into non-measurable pieces one loses information about the initial volume. Is this possible? Solovay [58] proved that if one doesn't use the Axiom of Choice one can construct a set theory in which the Banach-Tarski Paradox is impossible because every set of reals is measurable. However, in the standard set theory with the Axiom of Choice the answer is affirmative.

Symptom 9: *View the proof in a broader context, for example, as a generalisation or adaptation of another proof.* Many results in different areas of mathematics, from theoretical computer science to dynamics, can be seen as some kind of fixed-point construction [31] and, as a consequence, their proof can be phrased in this general type of argument.

Symptom 10: *Discuss interesting generalisations of the proof.* Category theory is one of the important tools for generalisations. Goguen [28] showed that the construction of the minimal Moore automaton can be lifted to a pair of adjunct functors between the category of Moore automata and the category of their behaviours, a more general/deep presentation of minimisation.

Symptom 11: *Discuss interesting modifications of hypotheses and their corresponding modifications of conclusions.* Solovay's result discussed above shows the existence of two set theories, one in which there are non-measurable sets of reals, and another one in which all sets of reals are measurable.

Symptom 12: *Explore alternative proofs.* One correct proof is enough to justify a theorem, but different proofs illustrate the same mathematical phenomenon from different angles. Pythagoras' Theorem has at least 367 essential different proofs [40], Pythagoras' proof, Euclid's proof, algebraic proof, various types of geometric proof, proof by re-arrangement, proof using differential equations, even a proof by an American President, James A. Garfield. There are four different proofs for the completeness of the predicate calculus, leading to four techniques to build models.

Symptom 13: *Discuss analogies between notions involved in the proof, between proofs, between theories, analogies between analogies.* This symptom was

discussed by many authors, see for example [51]. The notion of Hilbert space —
which evolved into a branch of mathematics [68] — appeared when David Hilbert
realised that some important mathematical proofs were structurally the same,
so at an appropriate level of generality they could be regarded as the same type
of argument. Algorithmic information theory shows that the quantity of information can be equally defined in complexity terms, without using Shannon's
probabilistic approach [20, 14].

Symptom 14: *Calculate a quantity used in the proof.* Chaitin's Omega number
is a well-defined mathematical real which is random, hence non-computable,
and, as a consequence, only finitely many digits of its binary expansion *may* be
calculated; in [17] exact values for the initial 40 bits of an Omega number have
been calculated offering a new understanding of its uncomputability.

Symptom 15: *Provide an explicit description of an object whose existence is
guaranteed by the theorem.* Chaitin's Omega numbers are computably enumerable and (algorithmically) random, but are there other such reals? The answer
is negative, every computably enumerable random real is the Omega number
of a prefix-free Turing machine [14], a more "concrete" description of a general
notion.

Symptom 16: *Provide a diagram or visual argument illustrating the proof.* The
diagram used in in [30, p. 49] for illustrating a short proof of Pythagoras' Theorem is very useful in understanding the proof.

Symptom 17: *Identify the main idea of the proof and use it in other contexts.*
For example, the standard proof of the irrationality of $\sqrt{2}$, a widely discussed
proof, can be easily adapted for infinitely many other reals, $\sqrt{3}$, $\sqrt{5}$, etc., but it
fails for π (why?).

Symptom 18: *Apply the theorem in different contexts.* Solovay's Theorem uses
"forcing" — a technique invented by Cohen [23] for proving consistency and
independence results in set theory — for a different type of problem. In fact,
the important results in mathematics re-appear in contexts different from the
original one. Group theory [55] sprang from number theory into the theory of algebraic equations, and from geometry, developed as an abstract subject, and has
many applications not only in mathematics, but also in physics and chemistry,
even in image processing and arts.

Symptom 19: *Recognise the constructive or non-constructive character of a
proof.* A constructive proof gives more insight than a non-constructive argument [12]. To illustrate this delicate point we consider the following

Theorem 1. *There exist two irrationals $x, y > 0$ such that x^y is
rational.*

Proof. The proof indicates how to 'construct' the reals x and y subject
to the conditions of Theorem 1. We distinguish two cases: a) the real
$\sqrt{2}^{\sqrt{2}}$ is irrational, b) the real $\sqrt{2}^{\sqrt{2}}$ is rational.

In case a) we choose $x = \sqrt{2}^{\sqrt{2}}, y = \sqrt{2}$; in case b) we choose $x = y = \sqrt{2}$. To verify that our choice is correct we proceed again by cases. In case a) x is irrational by hypothesis, $y = \sqrt{2}$ is well known to be irrational and $x^y = (\sqrt{2}^{\sqrt{2}})^{\sqrt{2}} = \sqrt{2}^2 = 2$. In case b) $x = y = \sqrt{2}$ are irrationals and $x^y = \sqrt{2}^{\sqrt{2}}$ is rational by hypothesis.   □

The proof depends on whether $\sqrt{2}^{\sqrt{2}}$ is irrational or not, and for the time being this is an open problem. This proof is not constructive as it doesn't return values for the reals x, y: the proof produces two pairs of reals, exactly one satisfying the requirements of the theorem. The proof above is less non-constructive than the following one [39]:

> *Proof.* Take the equation $x^y = 2$, and let x run through all irrational numbers greater than 1. This gives uncountably many corresponding values of y, which are all different (as x increases y decreases). The conclusion follows as it is not possible for all these values of y to be rational because there are only countably many rationals.   □

The second proof gives more information than the first one, as it shows that there are infinitely many pairs satisfying the conditions of the theorem.

Symptom 20: *Program (parts of) the proof in a programming language.* Formalisation of a proof requires full understanding; once formalised, the correctness of the proof can be verified and checked. In the process of formalisation, authors can debug their proof, e.g. they can fix syntactical mistakes, check whether they used the correct definition and description of all symbols, and whether all symbols have been used correctly and consistently. *To me, you understand something only if you can program it. (You, not someone else!) ... programming something forces you to understand it better, it forces you to really understand it, since you are explaining it* **to a machine** says Chaitin [21, p. xiiii].

The symptoms discussed above are just illustrative for the diversity of meaning of understanding of proof. Even by contrasting our extremely simple proofs for Lemma 1 one can see that the informal proof is more intuitive while the formal proofs are more rigorous. The first formal proof is more compact than the second one, which is closer in spirit to the informal proof. For deeper proofs this divide is sharper: see for example the informal proof of the Kraft-Chaitin Theorem in [14] and the Isabelle proof in [34]. It is worth observing that the Kraft-Chaitin Theorem has two "roles": one to be executed as an algorithm, the other to be analysed and validated. Previous formalisation efforts focused only on the first part [20]; the work in [19] was directed towards the second.

Informal proofs have many problems (correctness, for example), but also a glorious history of achievements. What are the problems with formal proofs? Some may argue that there is no problem whatsoever. Nelson [46] makes a strong point that syntax is all:

As to whether or not a string of formulas is a proof there is no dispute: one simply checks the rules of formation. This is the syntax of mathematics. Is that all there is to mathematics? Yes, and it is enough.

We believe that *understanding* of formal proofs is the main obstacle. Because of high complexity, most formal proofs cannot be checked by humans, so we can ask with Graham: *If no human being can ever hope to check a proof, is it really a proof?* Bluntly, can we understand formal proofs to the extent they can be used instead of pen-on-paper proofs in the practice of mathematics?

3   An Environment for Correctness and Understanding

Our answer to the last question is emphatically affirmative and is described in the form of an envisioned environment, called *active proof environment* (APE), in which users can write and check formal proofs as well as query them with reference to the symptoms of understanding. In particular, an APE supports the following services:

- discover or verify a formal proof,
- query for theorems and their proofs,
- formalise informal proofs for verification and checking,
- explore proofs at different levels of abstraction and in various forms of presentation,
- support queries corresponding to our symptoms of understanding,
- publish mathematical results at an appropriate level of detail and formality.

Various technologies supporting the above services already exist, but no unique system providing *all* services. An APE includes a *proof assistant* and an *intelligent interface*

Much research is done in the field of proof assistants, such as Mizar [44], Isabelle, Coq [10], or Ωmega [56], which include libraries of highly interlinked formal proofs and theorems [44, 1], the backbone of our environment.

Fig. 1 illustrates three alternative intelligent interfaces: proof assistant interfaces, web applications that are integrated with the proof assistant, and web applications that rely on a mathematical knowledge base that communicates with a proof assistant. Proof assistant interface implement some features of an APE's intelligent interface: Isabelle uses the *Proof General* Emacs Interface [2] and the *Intelligible semi-automated reasoning* language (Isar) [64] producing structured proof documents. Coq provides a TeXMacs [59] interface for authoring [3] and a MathML integration for publishing mathematics on the web. The mediator PlatΩ [6] connects TeXMacs with Ωmega and allows users to develop, publish, formalise, and query mathematical proofs.

However, the list of symptoms discussed in the previous section shows the necessity of enhancing existing interfaces to implement services for understanding. Web application implement additional features that can support some of

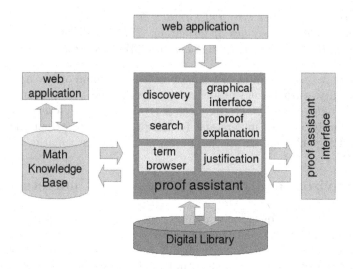

Fig. 1. An Active Proof Environment

our symptoms. Users of the `vdash` wiki [60] can collaboratively formalise mathematical proofs. Initial submissions can be verified or marked as sketches to be fully formalised later. Other mathematical web applications, such as the semantic wiki SWiM [38] and the document reader *panta rhei* [45], are not integrated with proof assistants but work with a central repository of mathematical knowledge represented in OpenMath [50] and OMDoc [35].

In the following we enumerate available services of proof assistants and propose new services, focusing on web-based technologies for OMDoc.

Service 1: *Query different levels of details for a proof.* (Symptom 1)
Authors of formal proofs write *proof scripts*, including instructions and definitions, for the proof assistant with just enough information to generate a formal proof for a given theorem. Users are usually presented with an extract of the fully formalised proof, but can explore the proof on different levels of abstractions (e.g. see the *proof plan* data structure PDS [7, 22] in Ωmega [4, 56] or *high-level proofs* in [13]), or request the system to print all steps, declarations, and definitions.

OMDoc representations can include the fully formalised proof. Folding and elisions of proofs allow one to hide and display different steps and to interactively adapt the level of detail. SWiM includes static cross-links between OMDoc and OpenMath representations of symbols and their definitions.

Service 2: *Query for justification of proving steps.* (Symptom 2)
Proof assistants can automatise the generation of justification for any proving step. OMDoc uses a markup of informal justification that allows users to link manually supplied justification with automatically generated inferences.

Service 3: *Write the proof in different levels of expertise.* (Symptom 3)
Isabelle/HOL generates formal representations of formal proofs for further computation and export the proof into human-readable formats. Ωmega provides a graphical map of the proof tree, a linearised presentation of the proof nodes with their formulae and justifications, a term browser, and a natural language presentation of the proof [26], [57, p. 370]. Furthermore, interactive natural language explanation and justifications of proofs can be generated. PlatΩ supports the presentations of proofs (generated in Ωmega) for different audiences and can adapt the level of detail, the proportion of formal and natural languages, or the mathematical notations [6, 54], as well as interactive stepwise explorations of mathematical proofs [6, 7].

Service 4: *Produce the proof in a specific natural language with different proportions of text and formulae.* (Symptoms 3, 4)
Isabelle is a generic framework for human-readable formal proof documents generated by Isar [63]. PlatΩ supports automatic translations from programming code to natural language [6, 54]. OMDoc includes a multilingual markup for proofs [35]. The conversion from OMDoc to XHTML is based on a collection of XSLT [36] stylesheets, which can be parametrised with the user's preferred language or level of formality.

Service 5: *Query the motivation for the proof.* (Symptom 5)
In OMDoc all fragments of a document are uniquely identified with an *Uniform Resource Identifier* [9] and are classified and interlinked according to mathematical categories [35]. This system allows to interlink fragments of a proof with complementary information, such as a motivation, and can answer corresponding queries. For example, SWiM represents these categories and relations as RDF triples [42] and uses SPARQL [52] to process the queries.

Service 6: *Query novel points.* (Symptom 6)
OMDoc can be extended by rhetorical markup for "novel points" or "obstacle" [27, 35]; XSLT stylesheets can be extended with appropriate visual markers.

Service 7: *Query examples and counter examples.* (Symptom 7)
The *Archive of Formal Proofs* (AFP) [1] is a collection of proof libraries, including examples, formally verified with Isabelle. Isabelle/HOL provides a counter-example search based on Quickcheck [47] and Refute [61]. OMDoc supports the annotation of proofs with examples and counter-examples. SWiM includes static cross-links to examples and dynamically embeds a list of examples into a page.

Service 8: *Query why certain hypotheses are needed and explore consequences if they are changed or omitted.* (Symptoms 8, 11)
Isabelle can be used to experiment and explore consequences of changing various hypotheses.

Service 9: *Produce other proofs that relate to the proof and apply theorems in different contexts.* (Symptoms 9, 10, 18)
Proof assistants organise their mathematical theorems and proofs into contexts

(or mathematical theories) and morphisms, which are used to transfer entities from one context to another. Isabelle theories are organised as a graph, so users can apply a theorem in a new context and explore related theorems and proofs.

OMDoc uses the markup of mathematical theories and theory morphisms [35, 53, 49]. Theories are interlinked via theory morphisms, supporting the reuse of previously defined concepts. These logical dependencies build up a theory graph from which one can infer relations between theorems, proofs, or examples.

Service 10: *Query for alternative proofs.* (Symptom 12)
The formal proofs for Lemma 1 use different techniques: the first one recursion, the second one natural deduction. Proof assistants can be used to discover or verify alternative proofs. In OMDoc requests such as "retrieve all proofs for lemma X" can be processed (Service 6).

Service 11: *Query for analogies.* (Symptom 13)
Analogies are very important in mathematics and more work, such as [43], has to be invested to deliver them in APE.

Service 12: *Query for calculations.* (Symptom 14)
Ωmega integrates external systems such as *computer algebra systems* (CAS) for symbolic computation; see [5] for an overview of further technologies.

Service 13: *Produce visual illustrations.* (Symptom 16)
Some proof assistant generate diagrams and other visual illustrations [67].

Service 14: *Query for the main idea.* (Symptom 17)
For OMDoc, [27] proposes the markup of "nucleus", a compact presentation of the proof from which the full argument can be easily reconstructed. For example, the nucleus of the second proof of Theorem 1 is "examine the cardinality of the solutions of the equation $x^y = 2$, when x runs through all irrational positive numbers greater than one".

Service 15: *Query whether a proof is constructive or not.* (Symptom 19)
For Coq, which uses constructive logic, this query is easy to answer. In general, partial answers can be obtained by identifying non-constructive rules of inference.

Service 16: *Program parts of the proof.* (Symptom 20)
This symptom is automatically satisfied by formal proofs.

4   Conclusion

The main features of mathematical proof are correctness and understanding. Correctness is easy to define, but there is little consensus regarding the understanding of understanding. To address this, we have proposed a list of symptoms for detecting the understanding of proofs. We have presented a vision of an environment that provides services addressing the symptoms of understanding, in which users can write, check, an query formal proofs. In such an environment, formal proofs are not only theoretical concepts. Because they guarantee a high level of certainty and provide understanding, formal proofs can become the standard of mathematical proof.

In the table below we summarise the technologies that support understanding of formal proofs, which should be integrated into a *unique* system. The proposal is preliminary and needs more extensive experimentation, implementation, and evaluation. Our analysis mainly refers to the proof assistants Isabelle and Ωmega, and the OMDoc projects. Our choice doesn't imply any value judgement on technologies.

The second author is developing an active document environment [45] in which users can produce, edit, query their documents, and use the following services:

- configurable layouts and output formats,
- presentations with varying level of detail, expertise, or formality, including multilingual presentations and consistent use of mathematical notations,
- enrichment of the documents with definitions, motivations, examples, or justifications and clarification of novel points and main ideas.

Symptom	Services	Symptom	Services
1	PDS/ Ωmega, High-level proofs, OMDoc	9	Isabelle, OMDoc
		10	Isabelle, OMDoc
2	proof assistants, OMDoc	11	Isabelle
3	Isabelle/HOL, PlatΩ/Ωmega	12	proof assistants, OMDoc (RDF/SparQL)
4	Isabelle/Isar, PlatΩ/Ωmega, OMDoc/ XSLT	13	[43]
5	OMDoc/ SWiM/ RDF/ SparQL	14	CAS
		16	[67]
6	extension of OMDoc	17	OMDoc
7	AFP, Isabelle/HOL, OMDoc/ SWiM	18	Isabelle, OMDoc
		19	Coq
8	Isabelle	20	formal proofs

Acknowledgements. We thank Greg Chaitin, Fulya Horozal, Michael Kohlhase, Christoph Lange, Charles Leedham-Green, Radu Nicolescu, Marc Wagner, Makarius Wenzel for enlightening discussions, suggestions and critical comments. We also thank the anonymous referees for useful suggestions. Our work was supported by the Centre for Discrete Mathematics and Theoretical Computer Science, Auckland, New Zealand. The second author has been partially funded by JEM-Thematic-Network ECP-038208.

References

[1] AFP. Archive of formal proofs (March 2009), http://afp.sourceforge.net
[2] Aspinall, D.: Proof general: A generic tool for proof development. In: Proceedings of the 6th International Conference on Tools and Algorithms for Construction and Analysis of Systems, pp. 38–42. Springer, Heidelberg (2000)

[3] Audebaud, P., Rideau, L.: TeXMacs as authoring tool for publication and dissemination of formal developments. In: Aspinall, D., Lueth, C. (eds.) User-Interface for Theorem Provers. Electronic Notes in Theoretical Computer Science, vol. 103, pp. 27–48 (2004)

[4] Autexier, S., Benzmüller, C., Dietrich, D., Meier, A., Wirth, C.-P.: A generic modular data structure for proof attempts alternating on ideas and granularity. In: Borwein, J.M., Farmer, W.M. (eds.) MKM 2006. LNCS (LNAI), vol. 4108, pp. 126–142. Springer, Heidelberg (2006)

[5] Autexier, S., Campbell, J., Rubio, J., Sorge, V., Suzuki, M., Wiedijk, F. (eds.): AISC 2008, Calculemus 2008, and MKM 2008. LNCS (LNAI), vol. 5144. Springer, Heidelberg (2008)

[6] Autexier, S., Fiedler, A., Neumann, T., Wagner, M.: Supporting user-defined notations when integrating scientific text-editors with proof assistance systems. In: Kauers, M., Kerber, M., Miner, R., Windsteiger, W. (eds.) MKM/CALCULEMUS 2007. LNCS (LNAI), vol. 4573, pp. 176–190. Springer, Heidelberg (2007)

[7] Autexier, S., Benzmüller, C., Dietrich, D., Wagner, M.: Organisation, transformation, and propagation of mathematical knowledge in Ωmega. Journal of Mathematics in Computer Science (accepted, 2009)

[8] Avigad, J.: Understanding proofs. In: Mancosu, P. (ed.) The Philosophy of Mathematical Practice, pp. 317–353. Oxford University Press, Oxford (2008)

[9] Berners-Lee, T., Fielding, R., Masinter, L.: Uniform Resource Identifier (URI): Generic Syntax. RFC 3986, Internet Engineering Task Force (2005)

[10] Bertot, Y., Castéran, P.: Interactive Theorem Proving and Program Development – Coq'Art: The Calculus of Inductive Constructions. Springer, Heidelberg (2004)

[11] Bourbaki, N.: Theory of Sets. Elements of Mathematics. Springer, Heidelberg (1968)

[12] Bridges, D., Richman, F.: Varieties of Constructive Mathematics. Cambridge University Press, Cambridge (1987)

[13] Bundy, A.: A science of reasoning. In: Lassez, J.L., Plotkin, G. (eds.) Computational Logic – Essays in Honor of A. Robinson, pp. 178–198. MIT Press, Cambridge (1989)

[14] Calude, C.S.: Information and Randomness: An Algorithmic Perspective, 2nd edn. Springer, Heidelberg (2002)

[15] Calude, C.S., Calude, E., Marcus, S.: Passages of proof. Bull. Eur. Assoc. Theor. Comput. Sci. EATCS 84, 167–188 (2004)

[16] Calude, C.S., Calude, E., Marcus, S.: Proving and programming. In: Calude, C. (ed.) Randomness and Complexity, From Leibniz to Chaitin, pp. 310–321. World Scientific, Singapore (2007)

[17] Calude, C.S., Dinneen, M.J.: Exact approximations of Omega numbers. Int. Journal of Bifurcation & Chaos 17(6), 1937–1954 (2007)

[18] Calude, C.S., Marcus, S.: Mathematical proofs at a crossroad? In: Karhumäki, J., Maurer, H., Păun, G., Rozenberg, G. (eds.) Theory Is Forever. LNCS, vol. 3113, pp. 15–28. Springer, Heidelberg (2004)

[19] Calude, C.S., Hay, N.J.: Every Computably Enumerable Random Real Is Provably Computably Enumerable Random. Research Report of CDMTCS-328 (July 2008)

[20] Chaitin, G.J.: The Limits of Mathematics. Springer, Heidelberg (1998)

[21] Chaitin, G.J.: Meta Math! Pantheon (2005)

[22] Cheikhrouhou, L., Sorge, V.: PDS – a three-dimensional data structure for proof plans. In: Proceedings of the International Conference on Artificial and Computational Intelligence for Decision, Control and Automation in Engineering and Industrial Applications (ACIDCA 2000), pp. 144–149 (2000)

[23] Cohen, P.J.: Set Theory and the Continuum Hypothesis. Addison-Wesley, Reading (1966)

[24] Cooper, S.B.: Computability Theory. Chapman &Hall/CRC (2004)

[25] Feferman, S., Dawson Jr., J., Kleene, S.C., Moore, G.H., Solovay, R.M., van Heijenoort, J., Velleman, D.J. (eds.): Kurt Gödel Collected Works. Oxford University Press, Oxford (1986)

[26] Fiedler, A.: User-adaptive Proof Explanation. PhD Thesis, Universität des Saarlandes, Saarbrücken, Germany (2001)

[27] Giceva, J.: Capturing Rhetorical Aspects in Mathematical Documents using OMDoc and SALT. Technical Report, Jacobs University, Germany (2008)

[28] Goguen, J.A.: Realization is universal. Theory of Computing Systems 6(4), 359–374 (1973)

[29] Gonthier, G.: Formal proof – The Four-Color Theorem. Notices of the AMS (11), 1382–1393 (2008)

[30] Gowers, T.: Mathematics. A Very Short Introduction. Oxford University Press, Oxford (2002)

[31] Granas, A., Dugundji, J.: Fixed Point Theory. Springer, Heidelberg (2003)

[32] Hales, T.C.: Formal proof. Notices of the AMS (11), 1370–1380 (2008)

[33] Harrison, J.: Formal proof – theory and practice. Notices of the AMS (11), 1395–1406 (2008)

[34] Hay, N.J.: Formal Proof for the Kraft-Chaitin Theorem (March 2009), http://www.cs.auckland.ac.nz/~nickjhay/KraftChaitin.thy

[35] Kohlhase, M.: OMDoc – An Open Markup Format for Mathematical Documents [version 1.2]. LNCS, vol. 4180. Springer, Heidelberg (2006)

[36] Kohlhase, M.: XSLT Stylesheet for converting OMDoc documents into XHTML (January 2008), http://kwarc.info/projects/xslt

[37] Lane, S.M.: Despite physicists, proof is essential in mathematics. Synthese 111(2), 147–154 (1997)

[38] Lange, C.: SWiM – a semantic wiki for mathematical knowledge management. In: Bechhofer, S., Hauswirth, M., Hoffmann, J., Koubarakis, M. (eds.) ESWC 2008. LNCS, vol. 5021, pp. 832–837. Springer, Heidelberg (2008)

[39] Leedham-Green, C.: Personal communication to C. Müller, March 7 (2009)

[40] Loomis, E.S.: The Pythagorean Proposition, 2nd edn. Oxford University Press, Oxford (1968)

[41] Manin, Y.I.: Mathematics as Metaphor. American Mathematical Society (2007)

[42] Manola, F., Miller, E.: RDF Primer. W3C Recommendation, World Wide Web Consortium (February 2004)

[43] Melis, E., Whittle, J.: Analogy in inductive theorem proving. Journal of Automated Reasoning 22(2), 117–147 (1999)

[44] Mizar (March 2009), http://web.cs.ualberta.ca/~piotr/Mizar

[45] Müller, C., Kohlhase, M.: Panta rhei. In: Hinneburg (ed.) LWA Conference Proceedings, pp. 318–323. Martin-Luther-University (2007)

[46] Nelson, E.: Syntax and Semantics (March 2009), http://www.math.princeton.edu/~nelson/papers/s.pdf

[47] Berghofer, S., Nipkow, T.: Random testing in Isabelle/HOL. In: Cuellar, J., Liu, Z. (eds.) Software Engineering and Formal Methods (SEFM 2004), pp. 230–239. IEEE Computer Society, Los Alamitos (2004)

[48] Nipkow, T., Paulson, L.C., Wenzel, M.T.: Isabelle/HOL. LNCS, vol. 2283. Springer, Heidelberg (2002)

[49] Normann, I.: Theory Morphisms. PhD Thesis, Jacobs University, Germany (2008)

[50] OpenMathHome (March 2007), http://www.openmath.org
[51] Pólya, G.: Mathematics and Plausible Reasoning Volume I: Induction and Analogy in Mathematics. Princeton University Press, Princeton (1969)
[52] Prud'hommeaux, E., Seaborne, A.: SPARQL Query Language for RDF. W3C Recommendation (March 2009),
     http://www.w3.org/TR/2008/REC-rdf-sparql-query-20080115/
[53] Rabe, F.: Representing Logics and Logic Translations. PhD Thesis, Jacobs University, Germany (2008)
[54] Schiller, M., Dietrich, D., Benzmüller, C.: Proof step analysis for proof tutoring – a learning approach to granularity. Teaching Mathematics and Computer Science (in press) (2009)
[55] Scott, W.R.: Group Theory. Dover, New York (1987)
[56] Siekmann, J., Benzmüller, C., Fiedler, A., Meier, A., Normann, I., Pollet, M.: Proof development in OMEGA: The irrationality of square root of 2. In: Kamareddine, F. (ed.) Thirty Five Years of Automating Mathematics, pp. 271–314. Kluwer, Dordrecht (2003)
[57] Siekmann, J., Benzmüller, C., Fiedler, A., Meier, A., Pollet, M.: Proof development with Omega: Sqrt(2) is irrational. In: Baaz, M., Voronkov, A. (eds.) LPAR 2002. LNCS (LNAI), vol. 2514, pp. 367–387. Springer, Heidelberg (2002)
[58] Solovay, R.M.: A model of set-theory in which every set of reals is Lebesgue measurable. Annals of Mathematics 38(3), 1–56 (1970)
[59] Gnu TEXMACS (March 2009), http://www.texM.s.org
[60] vdash: A Formal Math Wiki (March 2009), http://www.vdash.org
[61] Weber, T.: Bounded model generation for isabelle/hol. Electronic Notes in Theoretical Computer Science 125(3), 103–116 (2005)
[62] Wenzel, M.: Personal communication to C. Müller, March 7 (2009)
[63] Wenzel, M.: Isabelle/Isar – a generic framework for human-readable proof documents. In: Matuszewski, R., Zalewska, A. (eds.) From Insight to Proof: Festschrift in Honour of Andrzej Trybulec. Studies in Logic, Grammar and Rhetoric, vol. 10(23), pp. 277–298. University of Białystok (2007)
[64] Wenzel, M.T.: Isar – a generic interpretative approach to readable formal proof documents. In: Bertot, Y., Dowek, G., Hirschowitz, A., Paulin, C., Théry, L. (eds.) TPHOLs 1999. LNCS, vol. 1690, pp. 167–184. Springer, Heidelberg (1999)
[65] Whitehead, A.N., Russell, B.: Principia Mathematica, 2nd edn., vol. I. Cambridge University Press, Cambridge (1910)
[66] Wiedijk, F.: Formal proof – getting started. AMS Notices (11), 1408–1414 (2008)
[67] Winterstein, D., Bundy, A., Gurr, C., Jamnik, M.: An experimental comparison of diagrammatic and algebraic logics. In: Blackwell, A.F., Marriott, K., Shimojima, A. (eds.) Diagrams 2004. LNCS, vol. 2980, pp. 432–434. Springer, Heidelberg (2004)
[68] Young, N.: An Introduction to Hilbert Space. Cambridge University Press, Cambridge (1988)

A Review of Mathematical Knowledge Management*

Jacques Carette and William M. Farmer

Department of Computing and Software
McMaster University
Hamilton, Ontario, Canada
{carette,wmfarmer}@mcmaster.ca

Abstract. Mathematical Knowledge Management (MKM), as a field, has seen tremendous growth in the last few years. This period was one where many research threads were started and the field was defining itself. We believe that we are now in a position to use the MKM body of knowledge as a means to define what MKM is, what it worries about, etc. In this paper, we review the literature of MKM and gather various metadata from these papers. After offering some definitions surrounding MKM, we analyze the metadata we have gathered from these papers, in an effort to cast more light on the field of MKM and its evolution.

1 Introduction

In 2001 Bruno Buchberger and Olga Caprotti organized the *First International Workshop on Mathematical Knowledge Management* [10, 11] which was held September 24–26, 2001 at the Research Institute for Symbolic Computation (RISC) in Hagenberg, Austria. The MKM 2001 workshop, attended by 60 or so participants from 10 countries, launched the field of Mathematical Knowledge Management (MKM)[1] and was the first in a series of international [10,11,4,3,29, 7,27,6] and regional [31,33,34,35] conferences and workshops on MKM. Since its inception, the MKM community has struggled with questions like "What does it mean to manage mathematical knowledge?", "What should the field of MKM be?", "Should MKM have a wide focus?", if not, "What topics should MKM focus on?", "In what direction is MKM heading?", and "Is MKM making progress?". We agree with those who point out that this field is about "(MK)M" rather than "M(KM)".

In this paper we seek to answer these and similar questions by reviewing the literature of MKM, particularly the papers presented at the previous seven international MKM conferences (MKM 2001, 2003, 2004–8). By gathering and

---

* This research was supported by NSERC.
[1] We will use "MKM" exclusively to mean the field of Mathematical Knowledge Management that started with MKM 2001 and "mathematical knowledge management" to mean the activity of managing mathematical knowledge that started centuries before MKM 2001.

J. Carette et al. (Eds.): Calculemus/MKM 2009, LNAI 5625, pp. 233–246, 2009.

analyzing various metadata about the MKM papers of the past we would like to show where MKM is today and lay the groundwork for future work that can trace its evolution.

Our aim with this paper is both to survey the current state of MKM and to give future surveyors clear data (and hopefully a clear analysis) of the beginnings of MKM. We also want to offer a tested framework for classifying and analyzing future MKM research.

In the next section, we cover our understanding of MKM, and in section 3, we review the history of "mathematical knowledge management", as a survey of the context in which we understand the field. In section 4, we outline our data gathering and data analysis methodology. In the following section, we lay out the raw results we have obtained, and in section 6, we analyze them. We close with a conclusion.

2   What Is MKM?

In 2004 in the article [20], we described MKM as follows:

> MKM is a new interdisciplinary field of research in the intersection of mathematics, computer science, library science, and scientific publishing. The objective of MKM is to develop new and better ways of managing mathematical knowledge using sophisticated software tools. MKM is expected to serve mathematicians, scientists, and engineers who produce and use mathematical knowledge; educators and students who teach and learn mathematics; publishers who offer mathematical textbooks and disseminate new mathematical results; and librarians and mathematicians who catalog and organize mathematical knowledge.

Although mathematical knowledge possesses several characteristics that sharply distinguish it from other kinds of knowledge, MKM also has a nontrivial intersection with the field of general *knowledge management* [21].

MKM is indeed a new field of research, but mathematicians have been concerned with managing mathematical knowledge for hundreds, if not thousands, of years. A short history of mathematical knowledge management is given in the next section. However, mathematical knowledge management is now a much greater concern to mathematicians and other mathematics practitioners than it ever was before. There are several reasons for a new heightened interest in managing mathematical knowledge.

First, since World War II there has been an explosion in the mathematical knowledge produced by mathematicians. The evidence for this statement is abundant. One only has to examine the growth in mathematics articles, reviews, journals, conferences, etc.

Second, there has also been a parallel explosion in the mathematical knowledge produced by scientists and engineers as a by-product of their work. Perhaps the best example of this explosive growth is seen in software development. Computer scientists and software engineers produce millions of software artifacts—requirements specifications, design documents, pieces of computer code—that are essentially mathematical objects. The development and analysis of these artifacts generates an overwhelming amount of highly specific, but still quite valuable, mathematical knowledge.

Third, due to the rise in computer and communication systems, how mathematical knowledge is managed—that is, articulated, organized, disseminated, and accessed—is in the midst of a profound transformation. One example is that a large, and quickly growing, body of mathematical knowledge is now represented either axiomatically by logical theories or algorithmically by symbolic computation programs. Another example is the many new ways that mathematical knowledge is being disseminated, particularly involving the web.

The field of MKM was established to address the large and increasing need for effective mathematical knowledge management. In the eight years since MKM 2001, researchers have approached the task of managing mathematical knowledge from different points of view and have pursued different topics. It is our contention that the collection of these *views* and *topics* is a strong indication of what MKM is and where it is heading. Consequently, our review will focus on extracting from the MKM literature the dominant MKM views and topics.

3   History

While mathematical knowledge management has been named as a separate endeavor only recently, its history goes back much further at least to Euclid's great and extraordinarily influential *Elements*.

For the formalist, certainly one important milestone is Frege's *Begriffsschrift* [22], to whom we owe modern logic. In Hilbert's hands, this became his famous *Program*, while Russell and Whitehead produced the *Principia Mathematica* [42], to which we owe type theory. While Gödel's *incompleteness theorem* [23] certainly put an understandable damper on these developments, luckily many nevertheless persevered. Of course, one must mention the *Bourbaki* project as extolling the virtues of a formal library of mathematics.

But Bourbaki was hardly the first to try to design such a library. Leibnitz, frequently credited as having founded both library science and information theory [16], deserves first-mover credit here. The issues of managing large amounts of information (including substantial parts of mathematics) were already brough to the fore by Denis Diderot's *Encyclopédie ou dictionnaire raisonné des sciences, des arts et des métiers* [18].

Other aspects of mathematical knowledge management have a similarly extended history. Those interested in mathematical presentation would be well advised to read Cajori's monumental 1929 *A History of Mathematical Notations* [14]. For the ones more concerned with interactivity, watching Douglas Engelbart's 1968 *Mother of All Demos* [19] is humbling.

For those most interested in mechanizing mathematics, it is well worth revisiting the early pioneers like Turing and von Neumann (in particular [39]). Completely indispensable is a thorough reading of the *Automath* papers [17,36]— some recent MKM work just "rediscovers" some of de Bruijn's early insights.

Similarly, the *QED Manifesto* [9] has helped frame the discussion around formalized mathematics for a very long time (see [43] as an enlightening and readable example).

The more recent history of many parts of MKM have been covered elsewhere (although a unified treatment is still missing), and we will not repeat that here. However, we felt that it was important to remind our readers that mathematical knowledge management actually has a very long history, if one just knows where to look. This history is for us the proper context in which to evaluate the recent work explicitly labeled as *Mathematical Knowledge Management*.

4   Methodology

Before writing this paper, we first agreed on the methodology we should follow. First and foremost, although our results will inevitably be colored by some of our biases, we wanted our results to reflect the field itself. This meant that we have to carefully follow a bottom-up data gathering process where we would systematically review the MKM literature for metadata.

We decided that the refereed proceedings of the previous seven international MKM conferences should be considered the "primary sources". The refereeing process serves two purposes: insuring a minimal level of quality as well as asserting that the contributions are "on topic". While there are secondary sources of useful information on MKM, choosing amongst these would have required too much subjective judgment on our part. We will come back to this issue in a later section.

More specifically, this meant that we had to review all 143 papers contained in [11, 4, 3, 29, 7, 27, 6] (which also contain papers for co-located conferences but which are not counted here). A first pass was done to extract the main "topics" which were discussed in every paper, in the author's vernacular. Although at least one of us has looked through every page of every paper (more than once!), we relied heavily on the abstract to extract these "topics". We then formed groups of topics which seemed closely related: for example, some authors speak of libraries, while others of repositories. We came up with labels and descriptions for each of these.[2] At no point did we ever discuss whether any topic was important (or not), interesting (or not), relevant, etc. When abstracting from the specifics to get general topics, the only criterion was: *Is "mathematical knowledge" a crucial aspect?* In some cases, for example issues relating to distributed systems, we decided that the topic (as it appeared in the papers under review) was core computer science rather than containing specific MKM issues.

As we still ended up with a rather long list of topics, it was natural to try to *organize* the list somehow. At first, we naïvely attempted to create a hierarchy[3] out of these topics—and failed miserably. This is when we realized that we were oversimplifying the problem and, firmly inspired by the field of software

---

[2] Although we believe the extraction of important topics was objective, the grouping and labeling is inevitably somewhat more subjective.

[3] Especially naïve as both of us had read [38].

Document	Library	Formal	Digital	Interactive	Process

Fig. 1. The 6 views of MKM

architecture, we saw that these papers differed not only in their topics, but also in their *points of view*. The next section explains this in more detail. We then had to re-review each paper to extract the author's point-of-view, as this information could not be obtained from the list of topics. We again tried to shorten the list of topics, and although some topics seem to overlap, each seemed to be about a separate enough concern that we did not feel justified in narrowing the list further.

5   Results

This section presents the results of our investigation of the MKM literature. More specifically, we present the *points of view*, *topics* and quantitative data relating to these.

5.1   Views

In our investigation of the MKM literature we identified six major lenses through which researchers view MKM. These *views* are not incompatible; more than one view is often exhibited in the same research paper.

1. **Document.** Mathematical knowledge is traditionally communicated via mathematical documents. The *document* view of MKM sees the management of mathematical knowledge as largely happening inside documents, and managing these documents is a central concern. The documents, however, can have several forms. Some examples are articles in journals, hypertext documents on the web, and theory files produced using theorem provers. An example of a recent MKM 2008 paper written from the document view is "On Correctness of Mathematical Texts from a Logical and Practical Point of View" by K. Verchinine et al. [41]. It is concerned with formalized mathematical documents. Other examples are [37, 2].

2. **Library.** One major view of mathematics is that it is a huge body of mathematical facts. According to the *library* view of MKM, the main objective of MKM is to design and implement libraries, repositories, and archives in which a part of the body of mathematical facts is assembled, organized, and made accessible in various ways. How a mathematical library works is the primary concern; what is held in a library and how it is represented are secondary concerns. The MKM 2008 paper "Cross-Curriculum Search for Intergeo" by P. Libbrecht [30] takes a library view of MKM. It describes how a library of interactive geometry resources is organized so that it facilitates search. Other examples are [40, 44].

3. **Formal.** Mathematical knowledge is highly structured and interrelated. In the *formal* view of MKM, mathematical knowledge is managed according to how it is structured and interrelated. Deduction and computation are a very important part of this view since they are the principal means by which the structure of mathematical knowledge is created, discovered, and communicated. A formal view is taken in the MKM 2007 paper "Formal Representation of Mathematics in a Dependently Typed Set Theory" by F. F. Horozal and C. E. Brown [26]. It studies the relationship between an informal presentation of introductory real analysis and a formal presentation of it in the Scunak type theory. Other examples are [13, 15].

4. **Digital.** Like almost all other kinds of knowledge, there is a strong impetus to digitize mathematical knowledge so that it can be handled by computer and communication systems. The *digital* view of MKM considers the essence of managing mathematical knowledge to be managing digital objects that encode mathematical knowledge. The digital view, in particular, is concerned with how mathematical knowledge can be put on and accessed via the web. A. S. Youssef's MKM 2007 paper "Methods of Relevance Ranking and Hit-Content Generation in Math Search" [45] takes a digital view. It proposes techniques for searching digital mathematics libraries. Other examples are [1, 24].

5. **Interactive.** Mathematical knowledge is created, discovered, and communicated by human-to-human and human-to-tool interaction. The basis of the *interactive* view of MKM is that mathematical knowledge can only be properly managed within the context of this interaction. This view emphasizes the central role of mathematical knowledge in how mathematics is learned, produced, and applied. The MKM 2008 paper "Specifying Strategies for Exercises" by B. Heeren et al. [25] exhibits an interactive view. It investigates the specification of strategies for use in exercise-solving systems. Other examples are [32, 5].

6. **Process.** Another major view of mathematics is that it is a process in which mathematical models are created, explored, and interconnected. The *process* view of MKM focuses on how mathematical knowledge is produced. Managing mathematical knowledge is thus seen as managing the process that produces mathematical knowledge. This view includes a concern for the community of mathematicians, scientists, and engineers who produce mathematical knowledge. Process is the dominant view taken in A. Bundy's MKM 2008 paper " Automated Signature Evolution in Logical Theories" [12]. It argues that logical theories evolve over time and, as a consequence, their signatures need to be managed. Other examples are [8, 28].

5.2   Topics

A great many topics have been addressed in the MKM literature. From the topics our investigation has found, we have consolidated a list of 25 topics which the MKM community, through MKM literature, has concerned itself with. It is

Representation	Case-study	Extraction	Markup	Presentation
Mechanized	Interactivity	Search	Practice	Web
Translation	Organization	Library	Usability	Document
Education	Environment	Integrity	Process	Framework
Communication	Maintenance	Philosophy	Natural-language	Publishing

Fig. 2. The 25 topics of MKM

important to remember that these topics were chosen because some papers made the point that these topics were of special concern for "mathematical knowledge management".

1. **Representation.** Techniques and devices for representing mathematical knowledge including data structures, logics, formal theories, normalization, diagrams, etc.
2. **Case-study.** Work that focuses on a particular example of mathematical knowledge, most often as a requirements gathering and analysis exercise.
3. **Mechanized.** Systems, such as theorem provers and computer algebra systems, that provide mathematical services that mechanize certain aspects of the mathematics process.
4. **Markup.** Markup languages for expressing mathematics such as XML, MathML, OpenMath, and OMDoc.
5. **Presentation.** Techniques and devices for presenting mathematical knowledge (like notation and diagrams).
6. **Extraction.** Techniques for extracting or inferring mathematical knowledge (like AMS classification or internal but implicit cross-references) from mathematical documents and other sources.
7. **Search.** Searching and querying collections of mathematical knowledge as well as mathematical services.
8. **Practice.** Today's practice of mathematics by mathematicians, scientists, and engineers including issues like the mathematical vernacular, mathematics communities, and the role of context and convention.
9. **Process.** The process of creating, discovering, exploring, and applying mathematical knowledge.
10. **Translation.** The meaning-preserving translation of mathematical knowledge from one representation to another, including parsing techniques.
11. **Usability.** Techniques for making mathematical knowledge more usable.
12. **Web.** The fundamental use of the web to communicate mathematical knowledge and to support mathematics practice.
13. **Organization.** The organization of mathematical knowledge, including the use of ontologies and metadata.
14. **Natural-language.** Mathematical knowledge expressed via natural languages.
15. **Library.** Libraries, repositories, and archives of mathematical knowledge.

16. **Document.** Mathematical documents of all forms.
17. **Education.** MKM in, and for, mathematical education.
18. **Integrity.** The consistency, correctness, and certification of mathematical knowledge.
19. **Environment.** The development and use of software environments for managing mathematical knowledge.
20. **Maintenance.** The maintenance and version control of collections of mathematical knowledge.
21. **Philosophy.** The impact of the philosophy of mathematics on MKM.
22. **Communication.** The communication of mathematical knowledge between systems, particularly heterogeneous systems.
23. **Framework.** Frameworks for managing mathematical knowledge.
24. **Publishing.** Issues concerning the publication of mathematical knowledge.
25. **Interactivity.** Human-to-human and human-to-tool interaction involving mathematical knowledge.

5.3 Statistics

In Figure 3, we see the sorted distribution of weighted *views* for all papers. Each paper is assigned a total weight of 1, and this weight is divided evenly amongst all points of view espoused by the paper. Figure 4 is the similar histogram for topics. We also looked at the unweighted data, and for both views and topics, the ordering was essentially the same, i.e. the only changes were when views/topics already had statistically indistinguishable counts.

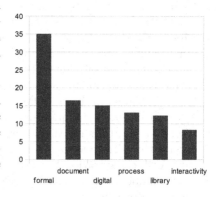

Fig. 3. Views

We have also broken down the data in these two figures by year, rescaling the results as percentages per year. We can extract some information from the view-per-year data (see Appendix A), but there is not enough data (143 papers in $25 \times 7 = 175$ bins) to extract meaningful results from a similar breakdown of the topics data. We were unable to find a meaningful clustering of the topics that might allow trends (if any) to become visible.

6 Analysis

What can we extract from this data? It is very clear that the community tends to favor a *formal* view of mathematics. While that is not totally unexpected, looking that the PROBLEM of MKM, it would probably be healthier if the points of view were more uniformly distributed. Statistically speaking, the *document* and *digital* views are tied for second, and *process* and *library* third, with *interactivity* getting the least attention. We believe that the large ratio (4 : 1) between *formal* and *interactivity* is mainly due to the current makeup of the community (many coming from formal backgrounds and otherwise working on highly mathematical problems) and the current state of the field (it is difficult to build a novel interactive system atop quicksand and convince formalists of its worth). In between, considering the amount of time and energy it takes to build a reasonable library of mathematics, it is probably unsurprising that this viewpoint has not received equal attention, especially since MKM has not attracted many system builders.

The distribution of topics clearly indicates that *representation* issues get the highest share of the community's attention (with the related issues surrounding *markup* joining in at number 4). More interesting is the second-place showing of *case-study*: we take this as a sign of a burgeoning field which takes the scientific method seriously and is doing some amount of requirements analysis before diving in with solutions.[4]

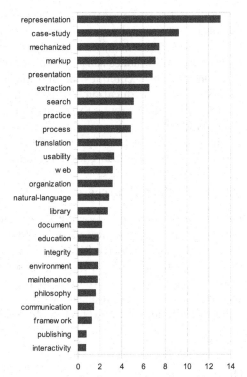

Fig. 4. Topics

We can also analyze the correlations between views (seen as depending on the topics) and vice-versa (raw data is shown in Appendix B). For the views, the most significant correlation (0.7) is between the *library* and *digital* views, which basically says that no one today is looking at large repositories of mathematics outside the digital domain. There is no correlation (0.0) between *digital* and *interactive*; this is potentially an artifact of how we chose to assign views, but not clearly so: the emphasis in the *digital* view is on mathematical knowledge being digital, while the *interactive* view emphasizes human interaction (most often on computers). It is reassuring that there are no negative correlations, which would have indicated a real flaw in our choices!

Analyzing the correlations between topics, there is a very strong pairwise correlation (> 0.87) between the 4 topics *representation, case-study, mechanized* and *usability*. In other words, regardless of point of view, these topics tend to appear together. This can also be interpreted to indicate that MKM has a strong affinity for the topics covered by the Calculemus conference, and would

---

[4] A lack of requirements analysis very often leads to interesting solutions to problems which did not need solving.

further justify the co-location of these conferences for $2007, 2008$ and 2009. At the other extreme, the pairs (*markup, education*), (*extraction, education*), and (*process, web*) are strongly negatively correlated ($-0.9, -0.95$ and -0.89 respectively). This also makes sense as, no matter how one looks at MKM topics, neither markup nor techniques for information extraction are (currently) relevant to MKM issues in education, nor is the advances in web technology as discussed in MKM papers (currently) relevant to the process of creating mathematics.[5]

Looking at the per-year data, only a couple of trends appear to be statistically significant: the *process* view is gaining some traction, while the *formal* view appears to be slowly losing its dominance.

6.1   Secondary Sources

Did we miss something important by ignoring some secondary sources? If we look at the topics and views covered in the different regional workshops and less formal conference proceedings [10,31,33,34,35], we see[6] that this is not the case. In fact, the topics and views of the talks at these other meetings seem to fall even more neatly into our categories than many papers in the MKM proceedings! What we do notice is a different emphasis, with the formal view being less prominent, but otherwise all views and essentially all topics are represented.

6.2   Discussion

One must remember that neither our "views" nor our "topics" are exclusive classifications, nor are they meant to be exhaustive with respect to *future* research in MKM. Another important point is that a *view*, like "document", should not be misunderstood as labeling papers which are about mathematical documents, but rather papers which focus on document-level issues. While documents and libraries are clearly inter-related, document-level issues and library-level issues and concerns differ significantly.

Some topics may seem unbalanced—like representation spanning from data-structures to formal theories. However, the papers on this topic all had one thing in common: how to encode (via some representation) some important piece of mathematical knowledge. As the knowledge being represented varied across different scales, so do the tools used. Some topics, like representation and presentation, are in some sense dual to each other; nevertheless, many papers deal exclusively with one of these topics, while others focus on the highly non-trivial relationship between presentation and representation. To muddy things further, some mechanisms (like diagrams) are used to denote both syntax and semantics, i.e. presentation and representation, often simultaneously.

Some topics may in fact overlap sufficiently that, if the community agrees, they should be merged. While there are significant differences in the papers that

---

[5] Even though mathematicians routinely use web 1.0 mechanisms as part of the social fabric of creating mathematics.

[6] A similar data-gathering effort was done on these sources, but that data was not included in our results.

deal with "organization" versus those dealing with "frameworks", perhaps that has more to do with the points of view than the topic. Similarly, the two authors absolutely agree with one reviewer who mentioned that "communication" and "translation" ought to be the same topic—however this is not (yet?) the community view, and so we did not feel like we should impose our view onto the topics in that way. Some topics rely on another (like extraction on document), but as topics of concern for a paper, they center of very different issues. Similarly, translation can be seen as the strongest possible form of extraction, but in practice these two topics are treated by very different techniques, and thus seemed to deserve separate classifications.

7   Conclusion

Our review of the MKM literature has produced a two-dimensional framework based on views and topics for classifying and analyzing MKM research. Although some bias on our part has certainly crept into our analysis, we have made a concerted effort to let the literature speak for itself. Our results show that the MKM community is pursuing a wide range of topics from a reasonably balanced set of view points. Our analysis shows that some trends and correlations are clearly evident such as the persistent interest in the formal view and the strong correlation between the formal view and the representation topic.

What stands out most in this work are the views. MKM researchers take different points of view when they do their research and write their results. The six views we have identified appear to cover, either individually or in combination, the views exhibited in the MKM literature. The views embody the different ways people see mathematical knowledge as well as the different ways people see mathematics itself. Like Parnas [38], we refuse to oversimplify MKM and shoehorn it into a hierarchy. More productive is to frankly embrace its complexity, and try to tame it with tools appropriate for a complex field rather than to do forensics on a carcass.

References

1. Adams, A.A.: Digitisation, representation, and formalisation. In: Asperti, et al. (eds.) [4], pp. 1–16
2. Adams, A.A., Davenport, J.H.: Copyright issues for MKM. In: Asperti, et al. (eds.) [3], pp. 1–16
3. Asperti, A., Bancerek, G., Trybulec, A. (eds.): MKM 2004. LNCS, vol. 3119. Springer, Heidelberg (2004)
4. Asperti, A., Buchberger, B., Davenport, J.H. (eds.): MKM 2003. LNCS, vol. 2594. Springer, Heidelberg (2003)
5. Aspinall, D., Lüth, C., Winterstein, D.: A framework for interactive proof. In: Kauers, et al. (eds.) [27], pp. 161–175
6. Autexier, S., Campbell, J., Rubio, J., Sorge, V., Suzuki, M., Wiedijk, F. (eds.): AISC 2008, Calculemus 2008, and MKM 2008. LNCS, vol. 5144. Springer, Heidelberg (2008)

7. Borwein, J.M., Farmer, W.M. (eds.): MKM 2006. LNCS, vol. 4108. Springer, Heidelberg (2006)
8. Borwein, J.M., Stanway, T.: Managing digital mathematical discourse. In: Asperti, et al. (eds.) [4], pp. 45–55
9. Boyer, R.: The QED Manifesto. In: Bundy, A. (ed.) Proceedings CADE 12, pp. 238–251 (1994)
10. Buchberger, B., Caprotti, O. (eds.): Electronic Proceedings of the First International Workshop on Mathematical Knowledge Management: MKM 2001. RISC-Linz (2001),
    http://www.risc.uni-linz.ac.at/about/conferences/MKM2001/Proceedings
11. Buchberger, B., Gonnet, G.H., Hazewinkel, M. (eds.): Mathematical Knowledge Management. Ann. Math. Artif. Intell, vol. 38. Kluwer, Dordrecht (2003)
12. Bundy, A.: Automating signature evolution in logical theories. In: Autexier, et al. (eds.) [6], pp. 333–338
13. Cairns, P.A., Gow, J., Collins, P.: On dynamically presenting a topology course. Ann. Math. Artif. Intell. 38(1-3), 91–104 (2003)
14. Cajori, F.: A History of Mathematical Notations. Dover, New York (reprint) (1929)
15. Coen, C.S.: From proof-assistants to distributed libraries of mathematics: Tips and pitfalls. In: Asperti, et al. (eds.) [4], pp. 30–44
16. Davis, M.: The Universal Computer: The Road from Leibniz to Turing. W. W. Norton & Co., Inc., New York (2000)
17. de Bruijn, N.G.: A Survey of the Project Automath. To H.B. Curry: Essays on Combinatory Logic, 579–606 (1980)
18. Encyclopédie ou dictionnaire raisonné des sciences, des arts et des métiers, pp. 1751–1772
19. Engelbart, D.: A research center for augmenting human intellect (demo) (December 1968), http://en.wikipedia.org/wiki/The_Mother_of_All_Demos
20. Farmer, W.M.: MKM: A new interdisciplinary field of research. ACM SIGSAM Bulletin 38, 47–52 (2004)
21. Farmer, W.M.: Mathematical knowledge management. In: Schwartz, D.G. (ed.) Encyclopedia of Knowledge Management. Information Science Reference, pp. 599–604 (2005)
22. Frege, G.: Begriffsschrift: eine der arithmetische nachgebildete Formelsprache des reinen Denkens. L. Nebert, Halle a/S (1879/1997)
23. Gödel, K.: Über formal unentscheidbare Sätze der Principia Mathematica und verwandter Systeme. Monatshefte für Mathematik und Physik 38(1), 173–198 (1931)
24. Guidi, F., Schena, I.: A query language for a metadata framework about mathematical resources. In: Asperti, et al. (eds.) [4], pp. 105–118
25. Heeren, B., Jeuring, J., van Leeuwen, A., Gerdes, A.: Specifying strategies for exercises. In: Autexier, et al. (eds.) [6], pp. 430–445
26. Horozal, F.F., Brown, C.E.: Formal representation of mathematics in a dependently typed set theory. In: Kauers, et al. (eds.) [27], pp. 265–279
27. Kauers, M., Kerber, M., Miner, R., Windsteiger, W. (eds.): MKM/CALCULEMUS 2007. LNCS, vol. 4573. Springer, Heidelberg (2007)
28. Kohlhase, A., Kohlhase, M.: An exploration in the space of mathematical knowledge. In: Kohlhase (ed.) [29], pp. 17–32
29. Kohlhase, M. (ed.): MKM 2005. LNCS, vol. 3863. Springer, Heidelberg (2006)
30. Libbrecht, P., Desmoulins, C., Mercat, C., Laborde, C., Dietrich, M., Hendriks, M.: Cross-curriculum search for Intergeo. In: Autexier, et al. (eds.) [6], pp. 520–535

31. Mathematical Knowledge Management Symposium,
    http://www.macs.hw.ac.uk/~fairouz/mkm-symposium03/
32. Melis, E., Büdenbender, J., Goguadze, G., Libbrecht, P., Ullrich, C.: Knowledge
    representation and management in activemath. Ann. Math. Artif. Intell. 38(1-3),
    47–64 (2003)
33. A North American Workshop on Mathematical Knowledge Management
    (NA-MKM 2002) (2002), http://imps.mcmaster.ca/na-mkm-2002/
34. Second North American Workshop on Mathematical Knowledge Management
    (NA-MKM 2004) (2004), http://imps.mcmaster.ca/na-mkm-2004/
35. Workshop on Mathematical Knowledge Management: Sustainability, Scalability
    and Interoperability, http://projects.cs.dal.ca/ddrive/seminars/mkm.shtml
36. Nederpelt, R., Geuvers, J., de Vrijer, R.: Selected Papers on Automath. Studies in
    Logic and the Foundations of Mathematics 133 (1994)
37. Padovani, L.: On the roles of LaTeX and MathML in encoding and processing
    mathematical expressions. In: Asperti, et al. (eds.) [4], pp. 66–79
38. Parnas, D.L.: On a "buzzword": hierarchical structure, pp. 429–440. Springer,
    Heidelberg (2002)
39. Turing, A.M.: Practical forms of type theory. Journal of Symbolic Logic 13, 80–94
    (1948)
40. Urban, J.: XML-izing Mizar: Making semantic processing and presentation of mml
    easy. In: Kohlhase (ed.) [29], pp. 346–360
41. Verchinine, K., Lyaletski, A.V., Paskevich, A., Anisimov, A.: On correctness of
    mathematical texts from a logical and practical point of view. In: Autexier, et al.
    (eds.) [6], pp. 583–598
42. Whitehead, A.N., Russell, B.: Principia Mathematica. Cambridge University Press,
    Cambridge (1910)
43. Wiedijk, F.: The QED manifesto revisited. Studies in Logic, Grammar and
    Rhetoric 10(23), 121–133 (2007)
44. Youssef, A.: Roles of math search in mathematics. In: Borwein, Farmer (eds.) [7],
    pp. 2–16
45. Youssef, A.: Methods of relevance ranking and hit-content generation in math
    search. In: Kauers, et al. (eds.) [27], pp. 393–406

A   View by Year Data

Percentage of weighted papers for each view, per year.

	2001	2003	2004	2005	2006	2007	2008
formal	39.4	35.3	37.8	28.2	54.5	19	32.5
document	12.1	16.7	23.1	26.3	2.27	9.52	20
digital	19.7	31.4	11.5	11.5	9.09	9.52	20
process	4.55	5.88	10.9	12.8	13.6	23.8	15
library	15.2	8.82	8.97	13.5	11.4	23.8	5
interactivity	9.09	1.96	7.69	7.69	9.09	14.3	7.5

B   View-Topic Data

Total number of weighted papers for each combination of view and topic.

	formal	document	digital	process	library	interactive
representation	22	7.3	3.8	6.5	4	3.8
case-study	16	2	4.5	5.3	6.3	3.5
mechanized	18	1.5	0	2.8	1.3	6
markup	6.2	6.2	7.7	1.8	3.3	2.8
presentation	6.3	7.3	3.5	0.83	3.5	4.5
extraction	7.5	3	6.5	1	3.5	0.5
search	2	0.5	7.5	2.5	6	0.5
practice	4	4.5	2.5	5.5	0.5	1
process	2.5	0.5	0.5	0	1	2.5
translation	0.5	1.5	0.5	0	0	1.5
usability	4.8	1	3	3.3	1.8	0
web	5.5	5	0.5	5	1.5	2.5
organization	2	2	3	1	0	0
natural-language	0.83	6.2	0.83	0.33	0.83	0
library	7	4.5	3	0	0.5	0
document	1.3	0	0.33	1	1.3	3
education	3.5	3.5	0.5	0	0.5	0
integrity	1	1.5	0.5	0.5	1.5	0
environment	5.5	1	1	2	0	1.5
maintenance	1	0	0	3	0	0
philosophy	1.3	1.2	6	1.5	2.7	3.3
communication	4.2	0.33	3.2	0.33	4	0
framework	1	2	0	1	2	3
publishing	0	1.3	0.33	0	1.3	0
interactivity	0	1	0	3	1.5	3.5

OpenMath Content Dictionaries for SI Quantities and Units

Joseph B. Collins

Naval Research Laboratory
4555 Overlook Ave, SW
Washington, DC 20375-5337
joseph.collins@nrl.navy.mil

Abstract. We document the creation of a new set of OpenMath content dictionaries to support the expression of quantities and units under the International System of Units (SI). While preserving many of the concepts embodied in the original content dictionaries, these new content dictionaries provide a foundation for quantities and units that is compliant with international standards. We respond to questions raised in prior efforts to create content dictionaries for units and dimensions by proposing and applying some rationalized criteria for the creation of content dictionaries in general. The results have been released and submitted to the OpenMath website as contributed content dictionaries.

1 Introduction

We are interested in the creation of a scientific markup for representing physics based models. In pursuing this objective, we have not found a sufficient body of markup for creating documents representing physics based models, rather we find we must develop further markup constructs in order to create such documents. Luckily, we find that we can build upon a developing body of work in mathematical markup. The first step in this endeavor, from a bottom-up perspective, is to properly address the representation of quantities and units.

OpenMath [1] represents a significant effort amongst the various attempts at representing mathematical knowledge, particularly in the problem area of representing mathematical semantics using web-oriented standards. Scientific knowledge is a mixture of mathematical representations and references to experiments and measurements. One of the fundamental intersections of this dual nature of scientific knowledge is in the representation of quantities and units. An initial attempt to capture some units and physical dimensions [2], [3] has resulted in several OpenMath content dictionaries (CDs). The prior OpenMath CDs we refer to are: dimensions1, units_metric1, units_imperial1, units_us1, units_time1, units_siprefix, units_ops1, and physical_consts1. In these efforts, some attention has been paid to observing conventions specified in the International System of Units, or Le Système International d'Unités, hereinafter simply referred to as SI, as expressed in [4], [5], and [6]. In these efforts there is an admitted incompleteness with respect to adherence to the SI in the implementation of the associated OpenMath CDs for units and dimensions.

J. Carette et al. (Eds.): Calculemus/MKM 2009, LNAI 5625, pp. 247–262, 2009.
© Springer-Verlag Berlin Heidelberg 2009

Our work here is an attempt to improve the representation of dimensions and units within OpenMath by building an new set of explicit SI CDs, following the current SI standard [7]. We do this in an attempt to close the gap between OpenMath CDs for units and dimensions and the expression of the formal standards of units and dimensions embodied in the SI. In this paper, we document a proposed restructuring of OpenMath CDs for units and dimensions, as well as attempt to provide answers to questions raised in prior efforts.

1.1   Guiding Principles

Prior to beginning this exposition, we review some of the guiding principles that we have developed and used. First, the SI represents a few things: a long-standing, slowly evolving international consensus; a well developed set of coherent standards for quantities and units; and a standard that dominates other, similar standards. Other standards, such as Imperial units and United States units, many of which predate the SI, have, over time, typically been redefined in terms of SI standards, and are largely in a process of being officially phased out. One of the purposes of the SI is to provide, from a few defined standard units, a way to repeatably measure all physical quantities. As such, the SI is, foremost, a standard based on physical measurement, not a mathematical standard. The definition of quantities, units, and their properties is essentially a posteriori, following from observation. While it is not generally anticipated that there would be major changes in these observations, in principle, if there were a distinct change, the meanings of what are defined in the SI would change. This being said, there is an observed mathematical structure to the concepts of quantities and units. This observed mathematical structure constitutes a physical theory, rather than mathematical truth: our acceptance of its truth rests primarily on consistency with observation and measurement. Being mathematical in nature, a physical theory's truth requires mathematical consistency, but the simplicity or complexity of its mathematical nature does not determine its validity. Neither can the mathematical nature of a physical theory be determined a priori. Much of the mathematical structure of quantities and units is described in the SI, and we intend to capture it as far as possible within the OpenMath framework.

While most non-specialists are familiar with the concept of units, the concept of quantities is perhaps somewhat more esoteric, particularly as the SI addresses it. Most importantly, some care is required in discussions of quantities and units, as colloquial usage may often be incorrect with respect to the SI. Many preconceptions may have to be abandoned for old habits to be replaced with SI compliant usage.

While embarking on this effort we were faced with several issues regarding how to make best use of OpenMath CDs. OpenMath and Content MathML [8] both embrace the concept that it is not only possible, but desirable, to separate the expression of mathematical semantics from the expression of the *presentation* of various mathematical symbols commonly in use. We advocate observing this distinction and avoiding the temptation to mix them more than might be necessary. In the development of content markup, we also suggest applying the following three criteria, as defined below: lack of ambiguity, convenience, and simplicity.

Ambiguity is anathema to content markup. Different from presentation markup, content markup loses utility if there is ambiguity in semantics. We suggest that ambiguity in mathematical semantics is tolerable, even desirable, in presentation markup. Presentation options of a given concept may overlap with presentation options of other concepts, potentially giving rise to ambiguity. In presentation, ambiguity is usually balanced with economy in the space allotted to symbols, relying on a context of implicit conventions. Presumably, a coherent set of presentation options can, and should, be documented in a style guide and perhaps implemented using a style sheet based translator operating on a document having content markup. Such an accompanying style guide can alleviate problems of ambiguity in the presentation markup.

By simplicity in content markup, one thing we mean is that we try to avoid implementing redundant constructs. While having multiple ways to express the semantics of a particular concept is not wrong, simplicity in a content markup language, i.e., having only one principal way to express a certain concept, seems desirable. Simplicity means fewer symbols to remember and will otherwise aid in the use of content markup, for example, in minimizing the job of writing translators for presentation of content. In Content MathML and OpenMath, operators and functions are currently represented in prefix form without much redundancy. For example, a prefix *divide* symbol represents the concept of division. While a presentation markup certainly does support alternative symbolic expressions of the operation of division, it is unnecessary for content markup to do so. Infix variations of division operators are not needed in content markup. Consequently, we do not insist that content markup include the various symbolic representations of division: an obelus symbol, a vinculum symbol, or a virgule or solidus symbol. By contrast, we suggest that this type of simplicity is not as important in presentation markup as it is in content markup. Presentation options of particular concepts need to be as numerous as the presentation conventions one intends to support, such as all of the above representations of the operation of division.

This does not mean there should never be any overlapping semantics: *divide* is redundant with exponentiation with negative integers. We find this redundancy acceptable because it is convenient to support both *divide* and *power* symbols, and it can be done without ambiguity. By convenience we mean that we may implement a possibly redundant set of concepts, usually because it is easier to do so than not to do so. Clearly, there are trade-offs between simplicity and convenience and the decision as to what is right and proper is subjective: the overarching consideration is overall economy of effort for the whole enterprise of implementation and use of the markup.

In sum, when faced with a question of what is the best way to represent a particular concept, we find we must first consider whether it relates to: a scientific or measurement issue; a presentation issue; or a mathematical semantic issue.

Secondly, in deciding whether to represent a particular concept, we must consider how to justify it based on principles of simplicity, convenience, and resolving ambiguity. We apply these guiding principles in the following sections.

2 SI Quantities

Quantities are a fundamental concept within the SI, where quantity is defined [7] as a "property of a phenomenon, body, or substance, where the property has a magnitude that can be expressed by means of a number and a reference". This *reference* means a reference amount of the same kind of quantity, called a unit. The magnitude is the number of unit amounts of the quantity required to be equal to the amount possessed by the object of interest.

The concept of quantities arises in multiple physical contexts, from the very specific to the abstract. For example, we can refer to the length of the specific piece of furniture standing in front of us, the wavelength of an arbitrary frequency of light, or the general concept of length. The objective is to be able to express all of these. From the SI definition, one can see that the concept of quantity is complex, and must have a combination of properties, which include a *dimension*, a *kind*, a *unit*, and a *magnitude*. The dimension, kind, and unit properties each have label, or name, values, and the dimension and unit properties also have SI symbol values, essentially abbreviated aliases of the names. A magnitude must have a numerical value expressed in some unit. These are discussed below.

2.1 Quantity Dimension

The most primitive concept of physics markup is that of *physical dimension*. With this property, mathematics is transformed to enable representation of physical quantities. We here define physical dimension, according to common usage, to include that which is defined as the *quantity dimension* in the SI standard. In this usage, the term physical dimension, such as is used in dimensional analysis, refers not only to the SI quantity dimension, but also to general SI quantities, including SI derived quantities.

The use of the term *dimension* in SI, as it relates to physical dimension, is, however, much more restrictive. The quantity dimensions in systems of quantities and units are given by the products of powers of a set of *base quantities*. As such, the base quantities form a basis for the space of quantity dimensions. Any system of quantities can, using its own set of base quantities, define a basis with which to span some space of physical dimensions. For the SI, there are seven base quantities: length, mass, time, electric current, thermodynamic temperature, amount of substance, and luminous intensity. The full set of SI quantity dimensions are generated by products of powers of these specific base quantities. Physical dimensions that are not in this set of SI quantity dimensions are not referred to as dimensions within SI. A different system of quantities might define a separate set of base quantities, consequently having a different set of quantity dimensions.

Derived quantities, as defined by SI, are quantities, in a system of quantities, which are defined in terms of the base quantities of that system. The SI requires that all quantities in the system be defined in terms of a product of powers of the base quantities. The SI introduces the concept of a mapping, dim, which maps a derived quantity to a quantity dimension. The requirement that all SI quantities be defined in terms of products of powers of base quantities essentially constitutes the definition of this mapping, dim. The SI does not bound the number of derived quantities that may be introduced.

Mathematical Structure of Quantity Dimensions. The mathematical structure of quantity dimensions in SI is summarized as follows: There are seven *base quantities*, length, mass, time, electric current, thermodynamic temperature, amount of substance, and luminous intensity. The *base quantities*, with the inclusion of an eighth, variously named *quantity of dimension one*, or *dimensionless quantity*, form an abelian generating set for the infinite abelian group of objects variously referred to as *quantity dimensions*, *dimensions of a quantity*, or just *dimensions*. The group multiplication operator is compatible with the multiplication operator for the field of real numbers. In addition to the base quantities, there is an unbounded set of *derived quantities*, which, in the SI, are defined by a name and a non-injective mapping to a quantity dimension,

$$\text{dim} : \textit{derived quantities} \rightarrow \textit{quantity dimensions}. \tag{1}$$

When two quantities, $Q1$ and $Q2$, are said to be *dimensionally equivalent*, this means that $\text{dim}(Q1) = \text{dim}(Q2)$. In addition to multiplication, each quantity dimension may be raised to any real power, though only rational powers ever seem necessary.

2.2   Kind of Quantity

The next property of a quantity is that of the *kind* of quantity. The kind of a quantity distinguishes between different quantities that may have the same quantity dimensions. The SI concedes that the concept of the kind of a quantity is to some extent arbitrary. Nevertheless, it is a necessary distinction. Perhaps the best illustration is by way of examples. The salinity of a solution is typically stated as a mass fraction, i.e. the mass of dissolved salt per unit of mass of solution. As such, salinity is a dimensionless quantity, i.e., mass/mass = 1. Angle is also dimensionless, given, for example, in radians as the ratio of the length of the subtended circular arc and the radius of the same circle. While dimensionally equivalent, one still considers salinity and angle to be distinct kinds of quantities. There are many dimensionless quantities distinguished by kind. Similarly, the quantities *torque* and *energy* have the same quantity dimension but are distinguished from each other by being different kinds of quantities.

Similar to the mapping, dim, we introduce the concept of another mapping, kind, implicit in the SI, which maps a quantity to a quantity kind. Complete equivalence of two quantities, $Q1$ and $Q2$, can only occur when both

$\dim(Q1) = \dim(Q2)$ and $\mathrm{kind}(Q1) = \mathrm{kind}(Q2)$ are true. While the SI recognizes that quantities may differ in kind, it does not standardize the definition of kind.

Distinction of quantities by kind can have rather fundamental consequences. The distinction between *inertial mass*, the resistance to acceleration due to an applied force in Newton's second law, and *gravitational mass*, the proportional factor by which an object influences other objects through the force of gravity in Newton's law of universal gravitation, was once debated. This debate was famously settled with Einstein's principle of equivalence and subsequent experimental measurements. We discuss later the import of differences in kinds of quantities.

2.3   Units

A *unit* is a defined reference amount of a given quantity. Having a unit, any quantity of the same kind may then be expressed as equivalent to some numerical amount of the reference quantity. For example, an arbitrary mass may be expressed as a numerical amount times a reference amount of mass. In the SI system of units, there are seven base units: the metre, kilogram, second, ampere, kelvin, mole, and candela, corresponding to the seven SI base quantities. There is a necessary one-to-one mapping between the base quantities and the base units. An additional unit, one, is added, corresponding to the dimensionless base quantity.

The essential reason for identifying the base units is that they serve as the measurement standards for most physical measurements. If, for example, we wanted to measure the length of something, we would need to calibrate our length-measuring device using a standard length. In SI, that standard length is the metre, and its definition is in terms of a measurement procedure. The same is true for all of the base units: each is defined in terms such that they may be readily used to calibrate measurement equipment for the corresponding physical quantity. This is referred to as the *practical realization* of the base units.

Similar to the dim mapping, the SI also posits a unit mapping with a range that is the set of products of powers of the SI base units, i.e., the set of *coherent derived units*. The domain of the unit mapping is the set of all quantities defined within the SI. As there is for each derived quantity a mapping to an SI quantity dimension, so too is there a mapping to a coherent derived unit. Because of this, the meaning of a quantity is defined in physically measurable terms, i.e., the base unit definitions. There can be no circular definitions, or any definitions of measurable quantities that are not rooted in definable measurements.

A limited number of coherent derived units are given special names. These named coherent derived units are: radian, steradian, hertz, newton, pascal, joule, watt, coulomb, volt, farad, ohm, siemens, weber, tesla, henry, degree Celsius, lumen, lux, becquerel, gray, sievert, and katal.

Mathematical Structure of Units: The mathematical structure of units in SI is summarized as follows: There are seven *base units*, metre, kilogram, second, ampere, kelvin, mole, and candela. The *base units*, with the inclusion of the unit named *one*, with symbol, 1, form an abelian generating set for the infinite abelian group of objects called the *coherent derived units*. The multiplication operator of this group of coherent derived units is compatible with the multiplication operator for the field of real numbers.

There is a one-to-one mapping of quantity dimensions to coherent derived units,

$$\text{unit} : \textit{quantity dimensions} \rightarrow \textit{coherent derived units}. \tag{2}$$

The unit of any quantity, Q, is given by $\text{unit}(\text{dim}(Q))$.

2.4   Magnitude of a Quantity

The magnitude of a specific scalar physical quantity represents the amount of that quantity, i.e., a mathematical product of a real number and a reference quantity, or unit. For example, the kilogram is the reference mass that resides at the International Bureau of Weights and Measures (BIPM) in Sèvres, France. In the SI system, the masses of all other physical objects are measured in proportion to that standard kilogram, where the proportion is expressed as a limited precision real number. There is also an accompanying error value, representing an estimate of the standard deviation, were an ensemble of such measurements to have been conducted. By default, when not specified, the error in a number is assumed to be half of the place value of the least significant digit expressed.

The SI introduces the concept of an operator that returns the numerical value of a quantity. We denote this mapping, num:

$$\text{num} : \textit{quantities} \rightarrow \textit{real numbers}. \tag{3}$$

Any quantity, Q, may be represented in the system by the unique product $\text{num}(Q) \cdot \text{unit}(Q)$. Clearly, while an arbitrary quantity is independent of unit system, num and unit are specific to a particular system of units, e.g., the SI.

Vector quantities, including complex numbers and tensors, are used in some situations. A full treatment of these is not possible here, partly because Open-Math currently expresses Cartesian vectors, but not vectors in other coordinates, e.g., spherical coordinates. When vector quantities are used, they may either be heterogeneous, where each component may be a distinct quantity, or homogeneous, where the components are pure numbers and the unit may be represented as a distinct factor. It should be noted that multiplication of base units with vector magnitudes is very much like scalar multiplication: when vector quantities fail to commute, it is the num parts of the quantities that fails to commute, not the unit parts. Extending the definition of the num mapping is straightforward for the much more common case of homogeneous Cartesian vectors of quantities, for example,

$$\text{num}((3\text{m}, 5\text{m}, 7\text{m})) = (3, 5, 7). \tag{4}$$

The representation of standard error with vector quantities is typically in terms of a covariance structure, a subject not elaborated in the SI, or here.

3   Operations on Quantities and Units

There are several operations that are defined on quantities and units. There are those that are specific to a system of quantities and units, namely dim, kind, unit, and num. The SI explicitly defines the dim, unit, and num mappings. In the SI, the unit mapping is symbolically represented with square brackets, and the num symbol is represented with curly braces. We add to this the kind mapping which provides some distinction between quantities which may be dimensionally equivalent. The equivalence of two quantities requires the equivalence of all four of these properties, i.e., two arbitrary quantities, $Q1$ and $Q2$, are equivalent if and only if $\dim(Q1) = \dim(Q2)$, $\text{kind}(Q1) = \text{kind}(Q2)$, and $\text{num}(Q1) = \text{num}(Q2)$. (If $\dim(Q1) = \dim(Q2)$, then $\text{unit}(Q1) = \text{unit}(Q2)$). The SI also specifies the multiplication, division, and raising of quantities to powers as required for the meaningful construction of quantities and units. If a quantity, say Qx, is expressed in a non-SI unit, say Ux, conversion may be effected by having defined $\text{unit}(Ux)$ when Ux itself is defined. Such conversions must be linear transformations: affine, or additive conversion, such as the conversion of thermodynamic temperature from kelvin to degree Fahrenheit, are not defined within this space of units and these operators. On the other hand, conversion of kelvin to degree Rankine is well defined.

3.1   Other Operations

There are several natural language usages which, with respect to dimensions and units, imply various mathematical operations. We partition these into the following categories: one, times, per, and plus; SI prefixes; cube, cubic, square, and squared.

One, Times, Per, and Plus: The dimensionless base quantity, one, and its corresponding unit, one, are essentially synonymous with the mathematical symbol of the same name. The presence of the unit or quantity dimension one in a quantity expression that we may want to represent in markup is generally implied, though rarely required explicitly. In text, multiplication is implied by juxtaposition of quantity names, using either a space (invisible times), a dot symbol, a hyphen, etc., and encoding such expressions into markup should follow those implications. As we have stated, multiplication of base quantities and base units is associative, commutative and otherwise compatible with multiplication of the real numbers, so the order of the encoded terms with respect to each other and numerical scalars is, in general, not semantically significant.

The term *per*, as in metre per second, implies an infix division operator, and could be encoded as such as long as it is unambiguously used.

The arbitrary addition of quantities, as pointed out for temperature, for example, in [3], is not always physically meaningful. In general, the SI does not discuss addition of quantities. We point out here that addition of quantities is undefined for quantities that are not of the same kind, without leaving out the possibility of additional constraints on addition of quantities. Similarly, the arithmetic relational operators equals, greater than, less than, etc., are only meaningful when quantities are of the same kind. The question as to whether quantities are of the same kind is really only answered by experimental validation of a law of physics. A law of physics is typically phrased as an arithmetic relation of quantities, implying the additivity of quantities involved. Arbitrary relations of quantities may be mathematically well defined without possessing physical meaning. Sometimes, as in the case of temperature, it is possible to identify different kinds of temperature, e.g., absolute (thermodynamic) temperature and relative temperature. The SI allows the definition, presumably by any user, of such derived quantities. The assignment of a value to the kind mapping in the definition of any such quantity, is also, presumably, left to the user, as is the task of ensuring the validity of the physical semantics.

SI Prefixes: SI prefixes are normally used by prepending a single prefix to an SI base unit, with the exception of the kilogram. For mass quantities, the prefix is prepended to the term *gram* as if it were the base unit, though gram is not presented without such a prefix. SI prefixes act like multiplicative numerical constants, each an integer power of 10. Normally, it is the user's choice to express, for example, 1000 microgram, 1 milligram, or 10^{-6} kilogram. Whichever choice is of no consequence for semantic content.

For capture of the semantic value of SI prefixes, we suggest the following: a) the use of an empty element for each occurrence of a named prefix; b) the allowance in content markup of multiple occurrences of prefixes (forbidden by SI in presentation); c) the abandonment of an OpenMath *prefix* symbol in the units_ops1 CD in preference to the multiplication operator, *times*. The enforcement of the SI-required, normal form of only using single prefixes is a presentation constraint, without mathematical meaning. It may be recommended practice in content markup, but should not be required.

Cube, Cubic, Square, Squared: The derived units *square metre, cubic metre, metre squared, metre cubed* are cases where the terms *square* and *cubic* act like prefix exponent operators, and *squared* and *cubed* act like postfix exponent operators. If used, a hyphen is interpreted as an infix multiplication operator. Similarly, *metre per second* and *metre/second* are cases where the symbols *per* and "/" act like infix division operators. In each of these cases, unless there is a compelling reason for supporting infix, postfix, or alternative, redundant prefix operators, simplicity and maintaining compatibility with the existing OpenMath mathematical CDs with *times, divide,* and *power* symbols suggests using the existing prefix forms of these same operations.

3.2   Appending of Units

While it may, in the expression of a quantity in markup, be good practice to append units to a numerical expression, this has no mathematical semantic value. The associativity and commutativity of multiplication of units with each other and with the numerical magnitudes makes the meaning clear regardless of order. The SI requirement to express the units of a quantity after the numerical value is strictly a prescription of a standard presentation form. Its practice may be recommended, but should not be required in content markup.

4 Number Representations

There are three requirements for numerical representations of quantities: range of absolute magnitudes, precision, and uncertainty. Magnitudes of physical constants and their uncertainties range from 10^{-72} to 10^{50} in SI units. The known material universe is estimated to be made up of approximately 10^{80} nucleons. Expressed in Planck units (not typically used for everyday physics) the overall size of the known universe is on the order of 10^{62} Planck lengths, the age is on the order of 10^{62} Planck times, and the mass of the order 10^{62} Planck masses. Any number representation should be able to support expression of numbers of these magnitudes. The current upper limit of precision of measurement for physical constants is about twelve to thirteen significant decimal digits, such as for the Rydberg constant.

While it may be ideal to have a format for arbitrary precision and arbitrary magnitude measurable values, IEEE double precision format [9] provides about 16 significant decimal digits and magnitudes spanning 10^{308} to 10^{-308}. While it is not inconceivable that physics-based computations may exceed the dynamic range the IEEE double precision format, for most practical purposes, that format appears adequate for the present.

Unambiguous representation of a limited precision real number requires the use of a scientific notation with significand of limited length and an exponent. Scientific notation is needed to express, for example, exactly two significant digits of 6.2×10^3, instead of writing 6200, the latter being ambiguous. Depending on how the significand of a double precision format is interpreted, the semantics of the precision limit of a measurement can be lost in the conversion to and from machine double precision. Machine formats do not typically support an inherent precision of a number which is distinct from the machine precision. From a semantic perspective, the machine precision is merely the maximum precision that may represented in a machine word, not the actual precision of the value that needs to be represented.

The IEEE double precision format does not in itself support the expression of measurement precision, which is really a form of uncertainty. Considering that in the conversion of a double precision literal to a machine double the limit of precision expressed in the literal is usually lost, it appears necessary to provide a separate mechanism for explicitly expressing the limit of precision, or for expressing the uncertainty. As such, we find it necessary to represent the concept of standard error for representing measured quantities, i.e., an estimate of the standard deviation, as a way to capture adequately the normal expression of uncertainty for scalar numerical magnitudes. Interpretation of uncertainties as bounded ranges, for example, is not standard practice. Proper treatment of uncertainty is sufficiently complex that we do not here provide a solution to this requirement, but suggest that more information, either an accompanying integer to represent the number of significant digits, or an accompanying float magnitude, will be needed to represent the standard error.

5 Physical Constants and Other Measured Quantities

Historically, the standardization of units began with units of practical interest, and as such are inherently anthropocentric. Over time, these have been supplanted with units having sizes that are still anthropocentric, but being defined in ways more amenable to increasingly precise scientific measurement. Nevertheless, any measurement device will have bounded precision. As a result, there are many scientifically measurable quantities that are difficult to capture adequately using only SI units.

The SI base units provides a set of units for representing the bulk of everyday quantities. It is possible, however, due to limits of precision in measurement apparatus, that macroscopic standards, such as the kilogram, are inadequate for measuring very small or very large quantities, such as the masses of quantities smaller than a microgram. It is known, for example, that the standard kilogram varies on the scale of micrograms, so it is not well defined to a corresponding precision. Electron and nucleon masses are significantly smaller than this. To adequately represent measurements of quantities having a combination magnitude and precision that fall outside the SI base units, and measurement equipment that uses them, the SI provides some other, off-system, measured units. The values of these in terms of SI units are obtained experimentally. These units are the electronvolt, the dalton, and the astronomical unit as units of energy, mass, and length, respectively.

An alternative system of units that is frequently used by physicists, for example, in studying cosmology and quantum gravity, are Planck units. The magnitudes of Planck units and the measurement apparatus required to gauge them make them inappropriate for use in anthropocentric applied physics. Planck units are sometimes called "God's units", as they comprise a natural or intergalactic standard set of units, completely defined using universal constants. The five Planck units are the Planck length, Planck mass, Planck time, Planck charge, and Planck temperature. They are completely defined in terms of the following five universal physical constants: the Newton's gravitational constant, the reduced Planck constant, the speed of light, the Coulomb force constant, and the Boltzmann constant. They are largely defined by experimental measurement, and updated measured values are published periodically by, for example, the international organization CODATA. These Planck units and the constants that define them are intrinsic to the physical laws that appear to describe the origins of the universe and physics as we know it. Consequently, other unit systems and fundamental constants are seen as derivative with respect to these.

6 Proposed Modifications to the OpenMath CD Library

We agree with prior work on the definition of quantity dimensions and units as empty XML elements. This appears to be their most natural representation.

We do, however, recommend the reorganization of the dimension and unit CDs appropriate to the properties of the SI quantities and units.

We believe that SI quantity and unit CDs should be specifically labeled as such, and the base units and quantities separately identified. Other systems having their own base quantities and units should similarly be distinctly labeled. Non-SI units and quantities should not be freely intermixed with SI quantities and units, but should be coupled in some way so as to distinguish them from units defined with respect to a different system. For example, while the contents of the dimensions1 CD may all be thought of as dimensions, in SI there are only seven base quantities serving as *quantity dimensions*. Other quantity dimensions are products of powers of these base quantities, i.e., derived via mathematical rules of construction, and generally need not be individually defined. For these reasons, we introduce the SI_BaseQuantities1 CD and suggest the deprecation of any SI base quantities from the dimensions1 CD. The SI_BaseQuantities1 CD defines length, mass, time, current, temperature, amount of substance, and luminous intensity. We additionally introduce the corresponding SI_BaseUnits1 CD and suggest the deprecation of SI base units from the existing CDs where they occur. The SI_BaseUnits1 CD defines metre, kilogram, second, ampere, kelvin, mole, and candela. For completeness, in both the SI_BaseQuantities1 and SI_BaseUnits1 CDs we include the symbol *one*, equating it to the symbol of the same name found in the OpenMath alg1 CD.

As the SI base quantities and units are defined as standards, the definition and a reference to the defining documents should be explicitly cited within those CDs. These definitions are generally sufficiently brief to be described within the OpenMath Description elements. Typically, new unit standards are ratified by a standards body at a periodically held conference, so a citation should, for example, name the conference and year. A change in SI definitions should prompt new versions of these CDs.

Derived quantities are those that are defined in terms of the base quantities. Some derived quantities are defined in the SI, particularly when they have names that are not mathematically constructed, or have corresponding specially named units. While the SI admits an unlimited number of derived quantities, and can only specify rules for their creation, it is reasonable to identify as SI quantities and units those that are specifically mentioned in the standard. For this reason we introduce SI_DerivedQuantities1, which defines angle, solid angle, frequency, force, pressure, energy, power, charge, voltage, capacitance, resistance, conductance, magnetic flux, magnetic flux density, inductance, Celsius temperature, luminous flux, illuminance, radioactivity, absorbed dose, equivalent dose and catalytic activity which all have named SI units.

Correspondingly, we define in SI_NamedDerivedUnits1 the radian, steradian, hertz, newton, pascal, joule, watt, coulomb, volt, farad, ohm, siemens, weber, tesla, henry, degree Celsius, lumen, lux, becquerel, gray, sievert, and katal. We also include in SI_NamedDerivedUnits1 the exceptional implied unit, gram. We include the gram because, even though its SI compliant presentation requires use of a prefix, the gram is the semantic root of all SI mass units. In the interest of simplicity, we believe that unit names that reflect mathematical construction, such as metre_per_second, or metre_squared, should not be included in CDs.

SI_DerivedQuantities1 may, in principle, include all derived quantities that are compliant with SI. There is no mathematical bound in the number of such quantities that may be defined. OpenMath custom is to not make content dictionaries arbitrarily large. No clear means of restricting the size in general suggests itself. We include in SI_DerivedQuantities1, somewhat arbitrarily, the quantities area, volume, speed, momentum, moment of force, density, concentration, heat, and entropy along with those quantities that have their own SI units, simply because these additional quantities are specifically mentioned within the SI.

6.1   SI Symbols

Each SI base quantity and base unit, and the named units corresponding to some derived quantities, have names with prescribed spellings, as well as associated SI symbols which act as abbreviated names, such as "m" for metre. In the spirit of first identifying what is necessary for semantic capture and in the spirit of simplicity, we use the SI names of quantities and units for their OpenMath symbol names in the CDs, choosing those over the short, SI symbols as being less ambiguous and less subject to errors in usage. In OpenMath, the presentation of many symbols is not emphasized. For example, the gradient operator does not reference the nabla symbol. Accordingly, it seems that we can, in general, neglect the representation of the semantically redundant SI symbols in OpenMath CDs. As a compromise, we suggest that symbols be identified in the *Description* element of the *CDDefinition* for reference purposes.

6.2   Capitalization and Abbreviation

Persons names are not capitalized in SI unit names, with the single exception of "degree Celsius". Person's names are capitalized when used in quantity names. Otherwise capitals are only used when the dimension and unit names begin a sentence. The SI symbols for units may have capitals: one should consult the standard for specific values. Abbreviations of unit and quantity names are explicitly barred. While these are principally presentation considerations, we follow these conventions in the symbol definitions within the CDs.

6.3   Non-SI Units

There are many non-SI units in use, with varying degrees of status with respect to the SI. There are the following four categories of units: coherent SI units; non-SI units accepted for use with the SI; non-SI units that have been deprecated; other non-SI units. Off-system, or non-SI units are those that are not coherent SI units. For example, minute, hour, and day are off-system units, defined in terms of the second of time and retained for use with the SI. The degree, minute, second, and gon of arc, and the litre are all defined in terms of SI units and retained for use with SI units. Similarly, the measured units mentioned earlier, the electronvolt, the dalton, and the astronomical unit, are off-system units, but used with SI. We propose two CDs to represent the category of non-SI units accepted for use with the SI: SIUsed_OffSystemUnits1 and SIUsed_OffSystemMeasuredUnits1. We choose to distinguish the Planck units and the physical constants that define them in their own CD, called Universal-PhysicalConstants.

We decline to propose CDs for other non-SI units at this time. Instead, we suggest some criteria for their construction. Many, if not most, non-SI units have been redefined in terms of SI units. Units not defined or mentioned in the SI should appear in non-SI CDs with their corresponding definition in terms of SI units. For example, the most common *foot* is the international foot. Since the international foot overrides other foot standards, we would recommend its Open-Math symbol be merely named *foot*, not *international foot*. In 1958 the United States and countries of the Commonwealth of Nations defined the length of the international foot to be equal to 0.3048 metre. This should be defined using formal mathematical property statements, and both the unit and num symbols, in this case, unit(foot) = metre and num(foot) = 0.3048. (Following this convention of defining non-SI units will allow straightforward unit conversions). Other, less common, foot units may have an OpenMath symbol distinguished in name, such as the *United States survey foot*. In any case, an OpenMath Description element should make it unambiguous which unit is being identified by citing an appropriate standards document, as described above. Other units that are both non-SI and undefined in terms of SI units must, of metrological necessity, be defined within other unit systems. Other unit systems may be rooted in their own content dictionaries. Units defined within other unit systems in general will have no exact, mathematical conversions to SI units, only approximate, metrological conversions.

6.4 Interaction with Other OpenMath Content Dictionaries

OpenMath symbols applicable to quantities and units from existing OpenMath CDs include: zero, one, divide, minus, plus, power, product, root, sum, times, unary_minus, eq, eqs, lt, gt, neq, leq, geq, and approx.

In [2] an interesting proposal was made for using the names of physical dimensions as types for units. Due to questions, we do not here provide the Small Type System (STS) CDs for the SI units and dimensions. Certainly the SI dimensions would appear to qualify as being of type PhysicalDimension. One could possibly create a new, more restrictive sub-type of PhysicalDimension, called SI_Dimension. As for an STS type for units, if we were to follow the suggestion of using SI dimensions as types for units, there would be an infinite number of possible types, i.e., certainly all of the possible quantity dimensions. Certainly such a type system would be well structured. We have the means to construct any of them, but not to list them all. It is not clear why a single type, say *Unit*, would not suffice. One of the purposes for having each physical dimension, or quantity dimension, be a type for units, is to perform type checking. This purpose, as well as unit checking and unit conversions for different units of the same quantity dimension, would seem to be adequately served with the use of the mappings, kind, dim, unit, and num.

7 Summary

In total, we propose eight new OpenMath CDs, which have been released and submitted to the OpenMath website as contributed content dictionaries. Our proposed SI CDs for quantities, SI_BaseQuantities1 and SI_DerivedQuantities1, and the proposed criteria for accepting dimension definitions, largely make obsolete the prior dimensions1 OpenMath CD. Our proposed SI CDs for units, SI_BaseUnits1, and SI_NamedDerivedUnits1, and the proposed criteria for accepting unit definitions, largely make obsolete the prior units_metric1 OpenMath CD. We do not suggest replacement of the units_siprefix CD, though we suggest deprecation of the units_ops1 CD. We incorporate the above described dim, kind, unit, and num symbols into a CD named SI_Functions1. Our proposed FundamentalPhysicalConstants1 is redundant with some symbols defined in physical_consts1, but also introduces the non-SI Planck units. SIUsed_OffSystemUnits1 replaces minute, hour, day of time in units_time1. SIUsed_OffSystemMeasuredUnits1, does not affect existing CDs.

We do not comment on any particulars regarding the units_imperial1 and units_us1 CDs, at this time, other than to say that since they have not been explicitly defined in terms of SI unit CDs, we may want to redefine them in CDs, with appropriate citations, in terms of SI units, if appropriate standards organizations have done so.

Overall, these new CDs, as proposed, isolate the essentially physics-based SI content into two CDs: the SI_BaseUnits1, and the SIUsed_OffSystem MeasuredUnits1. The additional FundamentalPhysicalConstants1 is also essentially physics-based, as would be any definition of measured constants or non-SI base units of other systems. The other CDs are principally mathematical in nature, where any of their properties attributable to the measurement process is derived by association to symbols defined in the physics-based CDs.

Finally, we have, for lack of space and time, neglected to elaborate on the general representation of vector quantities and the uncertainties of quantities.

References

1. Buswell, S., Caprotti, O., Carlisle, D.P., Dewar, M.C., Gäetano, M., Kohlhase, M.: The OpenMath Standard 2.0 (2004), http://www.openmath.org
2. Davenport, J.H., Naylor, W.A.: Units and Dimensions in OpenMath (2003), http://www.openmath.org/documents/Units.pdf
3. Stratford, J., Davenport, J.H.: Unit Knowledge Management. In: Autexier, S., Campbell, J., Rubio, J., Sorge, V., Suzuki, M., Wiedijk, F. (eds.) AISC 2008, Calculemus 2008, and MKM 2008. LNCS, vol. 5144, pp. 382–397. Springer, Heidelberg (2008)

4. Organisation Intergouvernementale de la Convention du Mètre. The International System of Units (SI), 8th edn.,
   http://www.bipm.org/utils/common/pdf/si_brochure_8_en.pdf
5. IEEE/ASTM. SI 10-1997 Standard for the Use of the International System of Units (SI): The Modern Metric System. IEEE Inc. (1997)
6. Taylor, B.N.: Guide for the Use of the International System of Units (SI) (May 2007), http://physics.nist.gov/Pubs/SP811
7. Technical Committee ISO/TC 12, Quantities and units. DRAFT INTERNA-TIONAL STANDARD ISO/DIS 80000-1, International Organization for Standard-ization (2008)
8. Content MathML Version 2, http://www.w3.org/TR/MathML2/chapter4.html, Content MathML Version 3 (Draft),
   http://www.w3.org/TR/MathML3/chapter4.html
9. IEEE 754-1985, see for example, http://en.wikipedia.org/wiki/IEEE_754-1985

Unifying Math Ontologies: A Tale of Two Standards

James H. Davenport[1] and Michael Kohlhase[2]

[1] Department of Computer Science
University of Bath, Bath BA2 7AY, United Kingdom
J.H.Davenport@bath.ac.uk
[2] School of Engineering & Science, Jacobs University Bremen
Campus Ring 12, D-28759 Bremen, Germany
m.kohlhase@jacobs-university.de

Abstract. One of the fundamental and seemingly simple aims of mathematical knowledge management (MKM) is to develop and standardize formats that allow to "represent the meaning of the objects of mathematics". The open formats OpenMath and MathML address this, but differ subtly in syntax, rigor, and structural viewpoints (notably over calculus). To avoid fragmentation and smooth out interoperability obstacles, effort is under way to align them into a joint format OpenMath/MathML 3. We illustrate the issues that come up in such an alignment by looking at three main areas: bound variables and conditions, calculus (which relates to the previous) and "lifted" n-ary operators.

Whenever anyone says "you know what I mean", you can be pretty sure that he does not know what he means, for if he did, he would tell you.   — H. Davenport (1907–1969)

1 Introduction

One of the fundamental and seemingly simple aims of mathematical knowledge management (MKM) is to develop and standardize representation formats that allow one to specify the meaning of the objects and documents of mathematics. The open formats OpenMath and MathML address the key sub-problem of representing mathematical objects from a content markup perspective: mathematical objects are represented as expression trees. As the formats were developed by different communities, they differ subtly in syntax, rigor, and structural viewpoints (notably over calculus). The efforts to mitigate the interoperability problem by establishing translations between the formats have done more to unearth subtle problems than to completely solve them in the past.

Both efforts shared the goal of representing mathematics "as it is", rather than "as it ought to be". A relevant example of the difference is given by [12], where the original text is

$$\text{The function } \sqrt{|x|} \text{ is not differentiable at } 0 \qquad (1)$$

J. Carette et al. (Eds.): Calculemus/MKM 2009, LNAI 5625, pp. 263–278, 2009.

while its formalised equivalent is

$$\neg(\lambda_{x:\mathbb{R}}(\sqrt{|x|}) \text{ is differentiable at } 0). \tag{2}$$

The key features are the typing of x as being in $\mathbb{R}$, and the conversion of $\sqrt{|x|}$ from an expression to a function. Both OpenMath and MathML, the latter explicitly as one of its design goals

> "Encode mathematical material suitable for teaching and scientific communication at all levels" [5, 1.2.4],

wish to encode both styles, or levels of formality, of mathematics. This is a particular problem for calculus. MathML and OpenMath have rather different views of calculus, which goes back to a fundamental duality in mathematics. These views can, simplistically, be regarded as:

- what one learned in calculus/analysis about *functions*, which *we* will write as $D_{\epsilon\delta}$: the "differentiation of ϵ–δ analysis" (similarly $\frac{d}{d_{\epsilon\delta}x}$, and its inverse $_{\epsilon\delta}\int$);
- what is taught in differential algebra about (*expressions* in) differential fields, which *we* will write as D_{DA}: the "differentiation of differential algebra" (similarly $\frac{d}{d_{\mathrm{DA}}x}$, and its inverse $_{\mathrm{DA}}\int$).

(2) is unashamedly the former, while (1) talks about a function, but actually gives an expression. This duality shows up whenever one talks about variables: while

$$2x \neq 2y, \tag{3}$$

$$(\lambda x.2x) =, \text{ or at least } \equiv_\alpha, (\lambda y.2y). \tag{4}$$

So does

$$\frac{dx^2}{dx} = \frac{dy^2}{dy}? \tag{5}$$

The variables are clearly free in (3) and bound in (4). Any system which attempted to *force* either interpretation on (5) would not meet the goal stated above.

In this paper we report on an ongoing effort of the W3C MathML Working group and members of the OpenMath Society to merge the ontologies[1] on which the OpenMath and MathML formats are based and thus align the formats, so that they only differ in their concrete XML encodings. This task proves to be harder than might initially be expected. We explain why, motivated by a study of four areas (which in fact turn out to be inter-related):

---

[1] Here we use the word "ontology" in its general, philosophical meaning as the study of the existence of objects, their categories and relations amongst each other, and not in the Semantic Web usage, where it is restricted to formal systems with tractable inference properties (description logics). Note furthermore that we are speaking as much about a "meta-ontology" of mathematical representation concepts as about "domain ontologies" that describe the mathematical concepts themselves. Having made this distinction, we will conveniently gloss over it in the rest of the paper.

1. constructions with bound variables;
2. the <condition> element of MathML;
3. the different handling of calculus-related operations in the two;
4. the "lifting" of n-ary operators, such as $+$ to $\sum$.

This paper is a short version of [10], which contains the details of the constructions. OpenMath-specific details of the proposals are in [9,8].

2   OpenMath and MathML

We will now recap the two formats focusing on their provenance and representational assumptions and then sketch the measures taken for aligning the languages. Sections 3, 4, 6, and 7 will detail the problem areas identified above. The first two leading to an extension proposal for OpenMath Objects and strict content MathML in Section 5, which is evaluated in the latter two. Section 7 concludes the paper.

2.1   MathML

MathML is an XML-based language for capturing mathematical the presentation, structure and content of mathematical formulae, so that they can be served, received, and processed on the World Wide Web. Thus the goal of MathML is to provide a similar functionality that HTML has for text. The present recommended version of MathML format is MathML 2 (second edition) of October 2003 [5]. MathML 1 had been recommended in April 1998 and revised as MathML 1.01 in July 1999.

MathML, starting from version 1.0, had a split into **presentation MathML**, describing what mathematics "looked like"[2], and **content MathML**, describing what it "meant". In this paper we will concentrate on content MathML, since the role of presentation MathML as a high-level presentation format for Math on the Web is (largely) uncontested. MathML's Content markup has ambitious goals:

> *The intent of the content markup in the Mathematical Markup Language is to provide an explicit encoding of the* underlying mathematical structure *of an expression, rather than any particular rendering for the expression.*   [5, section 4.1.1]

This mandate is met in MathML 1/2 by representing mathematical formulae as XML expression trees that follow the applicative structure of operators and their arguments: function application is represented by the apply elements where the first child is interpreted as the operator and the remaining children as their arguments. MathML 2 supplies about 90 built-in elements for mathematical operators, and the csymbol extension mechanism described later. The language has a fairly limited vision of what might be in "content":

---

[2] Which could include "sounded like" (for aural rendering) or "felt like" (e.g. for Braille), and MathML included a range of symbols such as ⁢ to help with this task.

The base set of content elements are chosen to be adequate for simple coding of most of the formulas used from kindergarten to the end of high school in the United States, and probably beyond through the first two years of college, that is up to A-Level or Baccalaureate level in Europe. [5, 4.1.2]

This is often referred to as the **K-14 fragment** of mathematics, by analogy with some countries use of "K–12" for the range of school mathematics. Since Version 2, MathML does have an extension mechanism via the `csymbol` elements and their `definitionURL` attributes, but this was rarely used except to achieve some form of OpenMath interoperability, or for proprietary extensions (e.g. Maple).

MathML tries to cater to the prevalent representational practices of mathematicians, and provides a good dozen structural XML elements for special constructions, e.g. set, interval and matrix constructors, and allows to "lift" various associative operators to "big operators" acting on sets and sequences simply by associating them by bound variables and possibly qualifier elements to specify the domain of application.

The MathML approach to specifying the "meaning" of expression trees largely follows a "you know what I mean" approach that alludes to a perceived consensus among mathematical practitioners on the K-14 fragment. The meaning of a construction is alluded to via examples rather than defined rigorously, intending to be "formal enough" to cover "*a large number of applications*" [5, 4.1.2], while remaining flexible enough not to preclude too many.

2.2 OpenMath

OpenMath [4] is a standard for the representation and communication of mathematical objects. It has similar goals to content MathML and focuses on encoding the meaning of objects rather than visual representations to allow the free exchange of mathematical objects between software systems and human beings. OpenMath has been developed in a long series of workshops and (mostly European) research projects that began in 1993 and continues through today. The OpenMath 1.0 and 2.0 Standards were released by the OpenMath Society in February 2000 and June 2004. OpenMath 1 fixed the basic language architecture, while OpenMath2 brought better XML integration, structure sharing and separated the notion of OpenMath Content Dictionaries from their encoding.

Like content MathML, OpenMath represents mathematical formulae as expression trees, but concentrates on an extensible framework built on a minimal structural core language with a well-defined extension mechanism. Where MathML supplies more than a dozen elements for special constructions, OpenMath only supplies concepts for function application (`OMA`), binding constructions (`OMBIND`; MathML 2 lacks an analogous element and simply uses `apply` with bound variables, hence the (inferred) Rule 1 below). Where MathML provides close to 100 elements for the K-14 fragment, OpenMath gets by with only an `OMS` element that identifies symbols by pointing to declarations in an open-ended set of Content Dictionaries (see below).

An OpenMath Content Dictionary (CD) is a document that declares names (OpenMath "symbols") for basic mathematical concepts and objects. CDs act as the unique points of reference for OpenMath symbols (and their encodings the OMS elements) and thus supply a notion of context that situates and disambiguates OpenMath expression trees. To maximize modularity and reuse, a CD typically contains a relatively small collection of definitions for closely related concepts. The OpenMath Society maintains a large set of public CDs, including CDs for all pre-defined symbols in MathML 2. There is a process for contributing privately developed CDs to the OpenMath Society repository to facilitate discovery and reuse. OpenMath does not require CDs be publicly available, though in most situations the goals of semantic markup will be best served by referencing public CDs available to all user agents.

The fundamental difference to MathML is in terms of establishing meaning for mathematical objects. Rather than appealing to mathematical intuition, OpenMath defines a free algebra $\mathcal{O}$ of "OpenMath Objects" which acts as (initial) model for encodings of mathematical formulae. OpenMath Objects are essentially labeled trees, with α-conversion for binding structures and Currying for nested semantic annotations. Note that since $\mathcal{O}$ is initial it is essentially unique and identifies (in the sense of "declares to be the same") fewer objects than any other model. As a consequence two mathematical objects must be identical, if their OpenMath representations are, but may coincide, even if their representations are different. The OpenMath standard therefore considers OpenMath objects as primary citizens and views the "OpenMath XML encoding" as just an incidental design choice for an XML-based markup language. In fact OpenMath specifies another encoding: the "binary encoding" designed to be more space efficient at the cost of being less human-readable. "OpenMath XML encoding" as just an incidental design choice for an XML-based[3] markup language.

The initial algebra semantics of OpenMath objects is intentionally weak to make the OpenMath format ontologically unconstrained and thus universally applicable. It basically represents the accepted design choice of representing objects as formulae. Any further (meaning-giving) properties of an object o are relegated to the content dictionaries referenced in o, where they can be specified formally ("Formal Mathematical Properties" as FMP elements which are themselves OpenMath objects) or informally ("Commented Mathematical Properties" as CMP elements containing text). Thus the precision of OpenMath as a representation language can be adapted by allowing CDs to range from fully formal (by providing CDs based on some logical system) to fully informal (where CDs are essentially empty). While this can be seen as a failure of OpenMath to supply semantics ("OpenMath is only syntax"), we see it as being as flexible as mathematical vernacular that gives the same freedom.

The question "does this OpenMath object o have formal semantics?" does not have an unambiguous answer. Rather, o has a meaning *for the system S* if each OpenMath symbol in o **either**:

---

[3] OpenMath also has a more space-efficient binary encoding.

1. is built into the OpenMath $\leftrightarrow$ S phrasebook **or**
2. has enough semantics deducible in S from the FMPs (which may be a recursive process).

Here S might be either a software system, or a logical system such as ZF.

2.3   The OpenMath/MathML 3 Alignment Process

Most of these differences between MathML and OpenMath can be traced to the different communities who developed these representation formats. MathML came out of the "HTML Math Module", an attempt to develop LaTeX-quality presentation of mathematical on the Web, something sorely missing from the otherwise very successful HTML. The guiding goal for OpenMath on the other hand was to develop an open interchange format among computer algebra systems, which resulted in a much stronger emphasis on the meaning of objects to make the exchange of sub-problems safe.

Even though interoperability between OpenMath and and MathML was always a strong desideratum for both communities, the two representation formats evolved independently and in line with the fundamental assumptions outlined in the two previous sections. Interoperability was attempted from the MathML side by integrating the `csymbol` element in MathML 2 and specifying parallel markup, i.e. allowing OpenMath representations to be embedded into MathML with fine-grained cross-referencing. The OpenMath Society developed CDs with analogues for "all predefined operators" and specified the correspondence between expression trees in [3]. Although 30 pages long, the fact that this document is still incomplete may serve as an indication that the problem is not trivial. As we will see below, mapping the MathML operators is not enough in the presence of different structural elements in the formats.

In June 2006 the W3C rechartered the MathML Working Group to produce a MathML 3 Recommendation, and the group identified the lack of regularity and specified meaning as a problem to be remedied in the charter period. The group decided to establish meaning for content MathML expressions based on Open-Math objects without losing backwards compatibility to content MathML 2. In the end, content MathML was extended to incorporate concepts like binding structures and full semantic annotations from OpenMath and a structurally regular subset of the extended content MathML was identified that is isomorphic to OpenMath objects. This subset is called **strict content MathML** to contrast it to full content MathML that was seen to strike a more pragmatic balance between regularity and human readability. Full content MathML borrows the semantics from strict MathML by a mapping specified in the MathML 3 specification that defines the meaning of non-strict (**pragmatic**) MathML expressions in terms of strict MathML equivalents. The division into two sub-languages serves a very important goal in standardization: to clarify and codify best (engineering) practices without breaking legitimate uses in legacy documents. In the current third version of MathML, the latter is a primary concern.

In June 2007, the OpenMath society chartered a group of members which includes the authors of this paper to work on version 3 of the OpenMath standard which would recognize content MathML 3 as a legitimate OpenMath encoding, to help define the pragmatic to strict mapping MathML, and to provide the necessary CDs, which would be endorsed by the W3C Math Group and the OpenMath Society. The discussions and the resulting CDs are online in the SWiM Wiki [16] [15]

Subsequent sections describe the problem areas that came up during the work and needed to be circumnavigated.

3   Set Constructors in MathML

With the K-14 scope discussed above, MathML found that it needed more sophisticated concepts, such as bound variables, to express the concepts that are manipulated *informally* at that level. One conspicuous example from K-14 is that of sets constructed by rules [5, 4.2.1.8].

> *A typical use of a qualifier is to identify a bound variable through use of the bvar element [...] The* condition *element is used to place conditions on bound variables in other expressions. This allows MathML to define sets by rule, rather than enumeration, for example. The following markup, for instance, encodes the set $\{x \mid x < 1\}$:*

```
1    <set>
        <bvar><ci>x</ci></bvar>
        <condition>
            <apply><lt/><ci>x</ci><cn>1</cn></apply>
        </condition>
6    </set>
```

Here (with the benefit of a great deal of hindsight, it should be pointed out) we can see the start of the problem. What would we have meant if we had changed the second[4] x to y? We would, of course, have written the MathML equivalent of $\{x \mid y < 1\}$, and the MathML would be as eccentric as that set of symbols. We therefore deduce the following (undocumented) rule, which corresponds to OpenMath's formal rules for OMBIND.

Rule 1 (MathML). *Variables in* bvar *constructions 'bind' the corresponding variable occurrences in the scope of the parent of the* bvar. *However, the variable may (e.g. $\forall$) or may not (e.g. $\frac{d}{dx}$) be bound in the sense of α-convertibility.*

Here the first problem of interpreting pragmatic MathML elements raises its ugly head. In OpenMath, we can represent the set[5] $\{x \in \mathbb{R} \mid x < 1\}$ by the representation

---

[4] Changing both of them would have been an α-conversion.

[5] Note that the OpenMath CDs require a larger set to be specified (to avoid Russell's paradox). It would not be a problem to provide a CD for what is often called "naïve set theory" that leaves out this safety device. However, such a system would have the same difficulties that the MathML above has: do we mean $(-\infty, 1)$ or $[0, 1)$, and is this a subset of $\mathbb{Z}$ or $\mathbb{R}$?

```
<OMOBJ version="2.0">
 <OMA>
   <OMS cd="set1" name="suchthat"/>
   <OMS cd="setname1" name="R"/>
   <OMBIND>
     <OMS cd="fns1" name="lambda"/>
     <OMBVAR><OMV name="x"/></OMBVAR>
     <OMA>
       <OMS cd="relation1" name="lt"/>
       <OMV name="x"/>
       <OMI> 1 </OMI>
     </OMA>
   </OMBIND>
 </OMA>
</OMOBJ>
```

This makes use of a binding construction (`OMBIND`) with a λ operator that constructs functions[6] from an expression with a bound variable. This kind of construction is standard in logical systems and λ-calculus, for which is is motivated as follows in a standard introductory textbook (our emphasis):

> To motivate the λ-notation, consider the everyday mathematical expression '$x - y$'. This can be thought of as defining either a function f of x or g of y ... And there is need for a notation that gives f and g different names in some systematic way. *In practice* mathematicians usually avoid this need by various 'ad hoc' special notations, but these can get very clumsy when higher-order functions are involved.   [11, p. 1]

To achieve interoperability with OpenMath objects, MathML 3 introduces the `bind` element in analogy to the OpenMath `OMBIND`. It could be argued that the "K–14" brief of MathML rules out higher-order functions, but in the example above we can see here the need, in a purely first-order case, to resort to "well, you know what I mean" without it. Extending MathML 3 with a `bind` element that encodes an *OpenMath binding object* takes the guessing of Rule 1 out of MathML and makes the meaning unambiguous. The MathML 3 specification does however need to specify the strict content MathML equivalent for the MathML 2 example above in order to give it an OpenMath Object semantics.

4   Calculus Issues

MathML and OpenMath have rather different views of calculus, which goes back to the fundamental duality in mathematics mentioned earlier.

Roughly speaking, the MathML encoding corresponds more closely to $D_{\epsilon\delta}$ and the OpenMath one to D_{DA}. If we were to look at the derivative of x^2 as in Figure 1, we might be tempted to see only trivial syntactic differences: in the MathML encoding we see a differential operator that *constructs a function*

---

[6] Here we also make use of the duality between sets and Boolean-valued functions that are their characteristic functions

```<apply>```
  ```<diff/>```
 	
  ```<bvar><ci>x</ci></bvar>```
  ```<apply>```
   ```<power/>```
   ```<ci>x</ci>```
   ```<cn>2</cn>```
  ```</apply>```
 	
```</apply>```
	```<OMA>```
	  ```<OMS cd="calculus1" "name="diff"/>```
	  ```<OMBIND>```
	   ```<OMS cd="fns1" name="lambda"/>```
	   ```<OMBVAR><OMV name="x"/></OMBVAR>```
	   ```<OMA>```
	    ```<OMS cd="arith1" name="power"/>```
	    ```<OMV name="x"/>```
	    ```<OMI>2</OMI>```
	   ```</OMA>```
	  ```</OMBIND>```
	```</OMA>```

Fig. 1. MathML 2 and OpenMath2 differentiation compared

from an expression with a bound variable[7] declared by a **bvar** element. The OpenMath encoding sees the differential operator as a functional that transforms one function (the square function) into another (its derivative). It is possible to do this without any variables, as in $\sin' = \cos$. Given the history of the two standards, this difference of encoding is not surprising, since D_{DA} is what computer algebra systems do (and what humans do, most of the time, even while interpreting the symbols as $D_{\epsilon\delta}$), whereas human beings generally *think* they are doing $D_{\epsilon\delta}$ and communicate mathematics that way.

For partial differentiation we see the same general picture, but the concrete representations drift further apart: For $\frac{d^{m+n}}{dx^m dy^n} f(x,y)$, MathML would use

```
<apply>
  <partialdiff/>
  <bvar><ci>x</ci><degree><ci>m</ci></degree></bvar>
  <bvar><ci>y</ci><degree><ci>n</ci></degree></bvar>
  <degree><apply><plus/><ci>m</ci><ci>n</ci></apply></degree>
  <apply><ci type="function">f</ci><ci>x</ci><ci>y</ci></apply>
</apply>
```

using **degree** qualifiers inside the **bvar** elements for the orders of partial differentiations and a **degree** qualifier outside for the total degree. The following representation is proposed in [3]:

```
<OMA>
  <OMS cd="calculus1" name="partialdiff"/>
  <OMA>
    <OMS cd="list1" name="list">
    <OMV name="m"/>
    <OMV name="n"/>
  </OMA>
  <OMBIND>
    <OMS cd="fns1" name="lambda"/>
    <OMBVAR><OMV name="x"/><OMV name="y"/></OMBVAR>
    <OMA><OMV name="f"><OMV name="x"/><OMV name="y"/></OMA>
  </OMBIND>
</OMA>
```

For the problems caused by wishing to represent $\frac{d^k}{dx^m dy^n} f(x,y)$, see [13] and the proposed solution in [8].

---

[7] With the insights from the last section, MathML 3 would probably use a **bind** element, emphasizing the role of the differentiation operator as a function constructor.

Integration is even more problematic than differentiation. MathML interprets integration as an operator on expressions in one bound variable and presents as paradigmatic examples the three expressions below, which differ in which ways the bound variables are handled.

a: $\int_0^a f(x)dx$	b: $\int_{x \in D} f(x)dx$	c: $\int_D f(x)dx$

```<apply>```
``` <int/>```
``` <bvar>```
```   <ci>x</ci>```
``` </bvar>```
``` <lowlimit>```
```   <cn>0</cn>```
``` </lowlimit>```
``` <uplimit>```
```   <ci>a</ci>```
``` </uplimit>```
``` <apply><ci>f</ci>```
```   <ci>x</ci>```
``` </apply>```
```</apply>```
	```<apply>```
	``` <int/>```
	``` <bvar>```
	``` <ci>x</ci>```
	``` </bvar>```
	``` <condition>```
	``` <apply><in/>```
	``` <ci>x</ci>```
	``` <ci>D</ci>```
	``` </apply>```
	``` </condition>```
	``` <apply><ci>f</ci>```
	``` <ci>x</ci>```
	``` </apply>```
	```</apply>```
		```<apply>```
		``` <int/>```
		``` <bvar>```
		``` <ci>x</ci>```
		``` </bvar>```
		``` <domainofapplication>```
		``` <ci>D</ci>```
		``` </domainofapplication>```
		``` <apply><ci>f</ci>```
		``` <ci>x</ci>```
		``` </apply>```
		```</apply>```

OpenMath can model usages (a) and (c) easily enough, via its **defint** operator: in fact usage (a) is modeled on the lines of (c), as $\int_{[0,a]} f(x)dx$, which means that we need to give an eccentric[8] meaning to 'backwards' intervals in order to encode the traditional mathematical statement

$$\int_a^b f(x)dx = -\int_b^a f(x)dx. \tag{6}$$

A more logical view is to regard the two notations as different, and define $_{\epsilon\delta} \int_{[a,b]}$ (via limits of Riemann sums, or whatever other definition is appropriate), and then

$$_{\epsilon\delta} \int_a^b f = \begin{cases} _{\epsilon\delta} \int_{[a,b]} f & a \le b \\ -_{\epsilon\delta} \int_{[b,a]} f & a > b \end{cases}, \tag{7}$$

whereas

$$_{\text{DA}} \int_a^b f = \left(_{\text{DA}} \int f \right)(b) - \left(_{\text{DA}} \int f \right)(a) \tag{8}$$

by definition.

Usage (b) might not worry us too much at first, since it is apparently only a variant of (c). The challenge comes when we move to multidimensional integration (in the $_{\epsilon\delta} \int$ sense). [2, p. 189] has a real integral over a curve in the complex plane,

$$\frac{1}{2\pi} \int_{|t|=R} \left| \frac{f(t)}{t^{n+1}} \right| |dt| \tag{9}$$

---

[8] Along the lines of "the set $[b, a]$ is the same as $[a, b]$ except that, where it appears as a range of integration, we should negate the value of the integral"! [13]. It is possible to regard 'backwards integration' as an "idiom" and (6) as the explanation of that idiom, but this seems circular.

whereas [1, p. 413, exercise 4, slightly recast] has an integral where we clearly want to connect the variables in the integrand to the variables defining the set:

$$\int\int\int_{\left\{\frac{x^2}{a^2}+\frac{y^2}{b^2}+\frac{z^2}{c^2}\leq 1\right\}} \left(\frac{x^2}{a^2}+\frac{y^2}{b^2}+\frac{z^2}{c^2}\right) dx dy dz \tag{10}$$

5   A Radical Proposal: Enhanced Binding Operators

The multiple points of view in the $\epsilon\delta$ vs. DA discussion can be seen in other situations, as witnessed by the difference between the OpenMath and MathML representations of the set $\{x|x < 1\}$ above. There seem to be two styles of thinking about mathematical objects. The first one — we will call it the **first-order style** — manifests itself as the $\epsilon\delta$-style in calculus. This style avoids passing around functions and sets as arguments to operators and uses expressions with bound variables instead. The second style — which we will call the **higher-order style** — allows functions and sets as arguments and relies heavily on this feature for conceptual clarity. It can be argued that the higher-order style is more modern[9], but arguably the first-order style still permeates much of mathematical practice. And if we take the use of mathematics in the Sciences and Engineering into account probably accounts for the vast majority of mathematical communication. Therefore we argue that both representational styles must be supported by MathML and OpenMath (and strict content MathML)

Examples like (9) and (10) show that the binding objects in OpenMath are too weak representationally to accomodate the first-order style of representation faithfully, and so force the reader into a higher-order style: we want the triple integration operator in (10) to range over a restricted domain of integration, and we want to give this domain as an *expression over the integration variables*[10], at least in $\epsilon\delta$ variant of integration. Moreover, given the discussion in Section 3 we need these variables to participate in α-conversion. How might we encode this in OpenMath? Figure 2 shows 4 alternatives[11]:

1. **In the binder** We can interpret $\int\int\int_{\left\{\frac{x^2}{a^2}+\frac{y^2}{b^2}+\frac{z^2}{c^2}\leq 1\right\}}$ as a complex binding operator, as in `forallin` and try to use that in a binding object. But this

---

[9] It has gained traction in the second half of the 20th century with the advent of category theory in Math and type theories in Logic.

[10] The original formulation in [1], which was "$\int\int\int_S\dots$ where $S = \{\cdots\}$", transcends the scope of both MathML and OpenMath, which restrict themselves to mathematical formulae. In fact MathML 2 had limited support for inter-formula effects with the `declare` element, but deprecates this element in MathML 3 since it cannot be defined on an intra-formula level. Thus the (important) issue of connecting bindings between different formula must be relegated to representation formats that transcend individual formulae, such as the OMDoc format [14].

[11] We use boxed formulae as placeholders for their (straightforward but lengthy) Open-Math2 encodings.

runs foul of the OpenMath2 dictum that the binding operator is not subject to α-conversion by its own variables; so this avenue is closed.

2. **In the body** On the other hand we can interpret the domain restriction as part of the binding object, and represent (10) as (2) in Figure 2. But this is impossible in OpenMath2, since only one OpenMath object after the `OMBVAR` element is allowed.

3. **In the body (2)** We can solve this problem by inventing a mathematically meaningless "gluing" operator

4. **separately** It is possible to represent an integration formula in OpenMath2 that is supposedly equivalent mathematically to (10) using the Differential Algebra approach: but this is, from the $\epsilon\delta$ point of view, totally unnatural, since it is α-equivalent to the expression in Figure 3 which is unreadable for a human, and also destroys commonality of formulae.

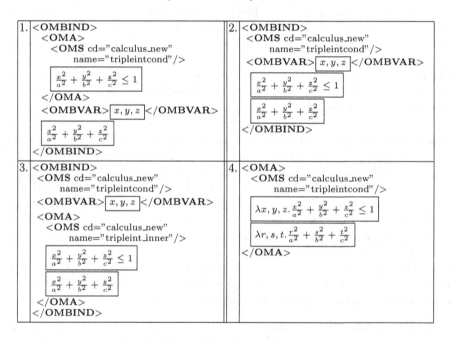

Fig. 2. The Alternatives

Solution 1 makes bound variables have an unusual, to say the least, scope, and solution 4 is higher-order style, so we are left with the other two. They have quite a lot in common, since they both achieve the fundamental goal of making both the region and the integrand subject to the *same* binding operation. We can summarise the points as follows.

```
<OMA>
  <OMS cd="calculus_new"
       name="tripleintcond"/>
```
$\lambda x, y, z. \frac{x^2}{a^2} + \frac{y^2}{b^2} + \frac{z^2}{c^2} \leq 1$

$\lambda z, y, x. \frac{z^2}{a^2} + \frac{y^2}{b^2} + \frac{x^2}{c^2}$
```
</OMA>
```

Fig. 3. α-equivalent of 4 above

2: pro: Mathematically elegant; fits into both the XML and binary encodings of OpenMath.

2: con: Requires a change to the abstract description of the OpenMath standard.

3: pro: No change to the OpenMath standard.

3: con: Needs a new, mathematically meaningless, symbol such as `tripleint_inner` for each symbol such as `tripleintcond`.

Option 2 is our preferred route, and the rest of this paper assumes that, but the changes to adopt option 3 should be obvious. The changes to the OpenMath standard to adopt option 2 are in the Appendix of the full paper [10].

6   Conditions in MathML

Our proposal above still leaves us with the problem to figure out the meaning of the `condition` from the examples and to specify their meaning in terms of OpenMath3 objects. MathML 2 introduces 23 examples of its usage, described in Table 1 of [10], and a further 31 in Appendix C, described in Table 2 of [10]. These can be roughly categorised as follows (where $a + b$ means "a in Chapter 4 and b in Appendix C").

5+14 are used to encode $\exists n \in \mathbb{N}$ or $\forall n \in \mathbb{N}$ (or equivalents). Strictly speaking, these usages are not necessary, because of the equivalences below.

$$\exists v \in S \quad p(v) \Leftrightarrow \exists v \quad (v \in S) \wedge p(v) \tag{11}$$

$$\forall v \in S \quad p(v) \Leftrightarrow \forall v \quad (v \in S) \Rightarrow p(v) \tag{12}$$

However, in practice, it would be better to have a convenient shorthand for these, hence the proposal in [9] for OpenMath symbols `existsin` and `forallin`, which are constructors for complex binding operators that include restricting the domain of quantification.

6+4 can be replaced by the OpenMath `suchthat` construct [10, 10, 1]

2+2 are solved by the use of `map` in OpenMath.

So we see that for all concrete operators, we have a natural strict content MathML/Open-Math equivalent. In the other cases we use the translation in Figure 4

Pragmatic MathML	Strict MathML

`<apply>` W
  `<bvar>`X`</bvar>`
  `<condition>`Y`</condition>`
  Z
`<apply>`
	`<bind>` W'
	  `<bvar>`X`</bvar>`
	  Z
	  Y
	`<bind>`

Fig. 4. Translating MathML with `condition`

afforded by OpenMath-/strict MathML extended according to our proposal. Here W is a binding operator and X stands for any number of variables in the `bvar` construct and Y, Z are arbitrary MathML expressions. Since we have treated all concrete operators, W must be either a `ci`, `cn`, a complex MathML expression, or a `csymbol` element. We believe the first two cases have not been used, since there is no plausible way to give them meaning; we propose to deprecate such usages in MathML 3. In contrast to that, the `csymbol` case is an eminently legitimate use, and therefore have to provide a W' in the rule above. But in MathML 2, a `csymbol` element only has a discernible meaning, if it carries a `definitionURL` attribute that points to a description D of the symbols' meaning, which will specify the meaning of the expression in terms of X, Y and Z. This description can be counted as (or turned into) a CD D' that declares a binary binding operator that can be referenced by a `csymbol` element W' which points to this declaration. Note that if D described a usage of the operator W without a `condition` qualifier, then D' must also declare the unary binding operator W; this must be different from W', since OpenMath operators have fixed arities. Finally, note that the case where W is a complex expression is analogous to the previous cases depending on the head symbol of W.

7   Lifting Associative Operators

Binary associative operators have notational peculiarities of their own. While we tend to write then as binary, as "$a + b + c$", we recognise that this is "really" one addition of three numbers, and both MathML-Content and OpenMath would represent this as a `plus` with three arguments. Mathematica distinguishes such operators as `Flat` and OpenMath's Simple Type System [6] as `nassoc`. It therefore makes sense to think of applying them to collections of arguments, and mathematical notation does this all the time (see table 5).

"small"	$a_1 + a_2 + a_3$	$a_1 a_2 a_3$	$a_1 \cap a_2 \cap a_3$	$a_1 \cup a_2 \cup a_3$	$a_1 \otimes a_2 \otimes a_3$	$a_1 \vee a_2 \vee a_3$
small Unicode			225C	225B	220A	225F
"big"	$\sum_{i=1}^{3} a_i$	$\prod_{i=1}^{3} a_i$	$\bigcap_{i=1}^{3} a_i$	$\bigcup_{i=1}^{3} a_i$	$\bigotimes_{i=1}^{3} a_i$	$\bigvee_{i=1}^{3} a_i$
big Unicode	1350	1351	1354	1353	134E	1357

Fig. 5. "Big" operators

With the exception of $\sum$ and $\prod$, which [7] regarded as being among the "irregular verbs" of mathematical notation, we can see a familiar pattern: the operator that applies to a collection of argument is "bigger" than its infix binary equivalent. The designers of Unicode have done as well as might be hoped for in mapping these symbols to 'related' code points in Unicode space for the corresponding glyphs.

How are these "big" operators going to be represented? For those it "knows" about [5, 4.2.3.2] (the list is, with our decorations. given in Figure 6: the ones marked $^\mathbb{P}$ are no longer n-ary in *strict* MathML 3),

---

`plus`, `times`, `max`[*], `min`[*], `gcd`[*], `lcm`[*], `mean`[‡], `sdev`[‡], `variance`[‡], `median`[‡], `mode`[‡], `and`[*], `or`[*], `xor`[†], `union`[*], `intersect`[*], `cartesianproduct`[†], `compose`[†], `eq`[$^\mathbb{P}$], `leq`[$^\mathbb{P}$], `lt`[$^\mathbb{P}$], `geq`[$^\mathbb{P}$], `gt`[$^\mathbb{P}$]

Fig. 6. MathML 2's n-ary operators

| `<apply>`
  `<or/>`
  `<bvar><ci>i</ci></bvar>`
    `<lowlimit><cn>1</cn></lowlimit>`
    `<uplimit><cn>3</cn></uplimit>`
  $\boxed{a_i}$
`</apply>` | `<OMA>`
  `<OMS name="apply_to_list" cd="fns2"/>`
  `<OMS name="or" cd="logic1"/>`
  `<OMA>`
    `<OMS cd="list1" name="make_list"/>`
    $\boxed{1}\;\boxed{3}\;\boxed{\lambda i.a_i}$
  `</OMA>`
`</OMA>` |

Fig. 7. $\bigvee$ in OpenMath and MathML

MathML can use bound variables and conditions, so the last item from Figure 5 would be shown on the left in Figure 7. It is not clear from [5] whether the same construct can be applied to a user-defined operator, but it would be reasonable. OpenMath, on the other hand, has an explicit lifting operator `apply_to_list`, see Figure 7 right.

Many of the operators $\oplus$ listed in Figure 6, those we have marked *, have two additional properties:

idempotence $\forall f \; f \oplus f = f$;
monotonicity There is some discrete order $\succ$ such that $\forall f, g \; f \oplus g \succ g$.

The first means that it make sense to apply $\oplus$ to a *set*, i.e. $\bigoplus S$. The second means that it makes sense to talk about $\bigoplus_{i=1}^{\infty} s_i$, as being the point where the construct stabilises under $\succ$, or some kind of infinite object otherwise. OpenMath's construction has no problem with, say, $\bigvee F$, but MathML has to write this as $\bigvee_{p \in F} p$ and use `condition` to represent the $p \in F$.

The statistical operators (marked ‡), when applied to discrete sets, and those marked †, only make sense over finite collections, but $\sum$ and $\prod$, as well as being lexically irregular in not being the infix operators writ large, are different in that they *can* have a calculus connotation. Here neither OpenMath nor MathML 3 make any clear distinctions, nor, in their defence, do the vast majority of mathematics texts. Is that sum meant to be absolutely convergent or only conditionally convergent? Only a careful analysis of the surrounding text will show, if then.

To help those authors who wish to make such distinctions, OpenMath probably *ought* to have a CD of symbols with finer distinctions, just as it should for the various kinds of integrals such as Cauchy Principal Value.

8 Conclusion

We have listed four areas where MathML (1–2) and OpenMath have taken different routes to the expressivity of mathematical meaning. In the case of MathML's. `condition`, we have seen one very general concept that does not have a single formalisation, and this led to the pragmatic/strict distinction in MathML 3. We have seen the utility of "restricted" quantifiers, even though they are not logically necessary, and [9] proposes their addition to OpenMath.

In the case of the calculus operations, this reflected a genuine split in the approaches to the calculus operations, whether one viewed them as algebraic or analytic operations. Since neither is 'wrong', but the two *are* different (for example the "Fundamental Theorem of Calculus" is a theorem from the analytic point of view, but a definition in the algebraic view), a converged view at MathML/OpenMath 3 should incorporate both.

Acknowledgements

The unification effort described here has benefited from the input of many people, notably Olga Caprotti, David Carlisle, Sam Dooley, Christoph Lange, Paul Libbrecht, Bruce Miller, Robert Miner, Florian Rabe, Chris Rowley. The authors are indebted to David Carlisle for comments on an earlier version of the paper.

References

1. Apostol, T.M.: Calculus, 2nd edn., vol. II. Blaisdell (1967)
2. Borwein, P., Erdélyi, T.: Polynomials and Polynomial Inequalities. Springer Graduate Texts in Mathematics, vol. 161 (1995)
3. Carlisle, D., Davenport, J., Dewar, M., Hur, N., Naylor, W.: Conversion between MathML and OpenMath. Technical report, The OpenMath Society (2001)
4. The OpenMath Consortium. OpenMath Standard 2.0 (2004),
   http://www.openmath.org/standard/om20-2004-06-30/omstd20.pdf
5. World-Wide Web Consortium. Mathematical Markup Language (MathML) Version 2.0 (Second Edition): W3C Recommendation, October 21 (2003),
   http://www.w3.org/TR/MathML2/
6. Davenport, J.H.: A Small OpenMath Type System. ACM SIGSAM Bulletin 2, 34, 16–21 (2000)
7. Davenport, J.H.: OpenMath in a (Semantic) Web (2008),
   http://www.jem-thematic.net/node/592
8. Davenport, J.H., Kohlhase, M.: Calculus in OpenMath. In: 22nd OpenMath Workshop (submitted, 2009)
9. Davenport, J.H., Kohlhase, M.: Quantifiers in OpenMath. In: 22nd OpenMath Workshop (submitted, 2009)
10. Davenport, J.H., Kohlhase, M.: Unifying Math Ontologies: A tale of two standards (full paper) (2009), http://opus.bath.ac.uk/13079
11. Hindley, J.R., Seldin, J.P.: Lambda-Calculus and Combinators. Cambridge University Press, Cambridge (2008)
12. Kamareddine, F., Nederpelt, R.: A Refinement of de Bruijn's Formal Language of Mathematics. J. Logic, Language & Information 13, 287–340 (2004)
13. Kohlhase, M.: OpenMath3 without conditions: A Proposal for a MathML3/OM3 Calculus Content Dictionary (2008),
    https://svn.openmath.org/OpenMath3/doc/blue/noconds/note.pdf
14. Kohlhase, M.: OMDoc – An Open Markup Format for Mathematical Documents [version 1.2]. LNCS (LNAI), vol. 4180, pp. 25–32. Springer, Heidelberg (2006)
15. Christoph Lange. OpenMath wiki (2009), http://wiki.openmath.org
16. Lange, C., Palomo, A.G.: Easily editing and browsing complex OpenMath markup with SWiM. In: Libbrecht, P. (ed.) Mathematical User Interfaces Workshop 2008 (2008)

Integrating Web Services into Active Mathematical Documents

Jana Giceva, Christoph Lange, and Florian Rabe

Computer Science, Jacobs University Bremen
{j.giceva,ch.lange,f.rabe}@jacobs-university.de

Abstract. Active mathematical documents are distinguished from traditional paper-oriented ones by their ability to interactively adapt to a reader's inputs. This includes changes in the presentation of the content of the document as well as changes of that content itself.

We have developed the JOBAD architecture, a client/server infrastructure for active mathematical documents. A server-side module generates active documents, which a client-side JavaScript library makes accessible for user interaction. Further server-side modules – in the same backend, or distributed web services – dynamically respond to callbacks invoked when the user interacts with the client. These three components are tied together by the JOBAD active document format, which backwards-compatibly enhances MathML by information about interactivity.

JOBAD is designed to be modular in the specific web services offered. As examples, we present folding and elision in mathematical expressions, type and definition lookup of symbols, as well as conversion of physical units. We evaluate our framework with a case study where a large collection of lecture notes is served as an active document.

1 Introduction and Related Work

Documents are an important interface for distributing mathematical knowledge. Recently, the technological development has shifted attention more and more towards digital documents and the added-value they can offer. This has led to a number of research efforts on interactive mathematical documents involving features such as adapting mathematical documents, interactive exercises, and connecting to mathematical web services.

The ActiveMath project investigates how to *aggregate documents* from a knowledge base such that the resulting document contains exactly the topics that the reader wants to learn and their prerequisites [Act08]. *Interactive Exercises* have been realized in ActiveMath and MathDox [GM08, CCK+08, CCJS05]. Here, the user enters the result into a form and then gets feedback from a solution checker in the server backend. ActiveMath comes with its own web services [MGH+06], and MathDox has originally been designed for talking to computer algebra systems but can also connect to other services via MONET (see below). Gerdes et al. have developed a reusable exercise feedback service for

J. Carette et al. (Eds.): Calculemus/MKM 2009, LNAI 5625, pp. 279–293, 2009.
© Springer-Verlag Berlin Heidelberg 2009

exercises that has also been integrated with MathDox [GHJS08]. Besides supporting MathDox's own communication protocol, Gerdes's service also complies to the XML-RPC and JSON data exchange standards [GHJS08]. The services developed within the MathDox and ActiveMath projects, such as the ActiveMath course generator, are potentially open to any client, but have not been used with any frontend other than their primary one so far [Ull08, MGH+06].

There are also elaborate *web service architectures* for mathematical web services that have been designed for integration with many systems, such as the ones developed in the SCIEnce [SCI09] and MONET projects [Mon05]. SCIEnce explicitly targets symbolic computation and grid computing and does not consider documents as user interfaces. MONET is an architecture that, in principle, allows for any kind of mathematical web service. Still, mainly computational web services have been developed and evaluated within that framework. Web services can register with a central MONET broker that accepts requests, which do not directly call a web service but consist of a problem description (e. g., solve an equation, given as an OpenMath expression). The broker then forwards the request to the best-matching service. The above-mentioned MathDox allows for access to MONET web services via a document interface.

Asynchronous communication with a server backend (AJAX) allows for client/-server interaction without submitting forms. It is a prerequisite for responsive browser-based applications: A client-side script can exchange small data packets with a server backend and insert responses from the server into the current page. This technique is employed by MathDox [CCK+08] and Gerdes's frontend to their feedback service [GHJS08].

Despite the efforts mentioned above, there is still a lot of static mathematical content on the web. Where documents act as frontends to web services, as in the above-mentioned systems, they have usually been designed to give access to a small selection of web services performing very specific tasks (mostly giving feedback to exercises and symbolic computation) – as is the case with ActiveMath and MathDox.

Our goal is to facilitate the integration of diverse web services into mathematical documents – inspired by the Web 2.0 technology of *mashups* [O'R05, AKTV08]. Originally, mashups were handcrafted JavaScripts pulling together web services from different sites. Since then several mashup development kits have been developed, e.g., Yahoo! Pipes [Yah09] or Ubiquity [Moz09]. We aim at a similar development kit for mathematical applications. Our vision of an *interactive document* is a document that the user can not just read, but adapt according to his preferences and interests *while* reading it – not only by customizing the display of the rendered document in the browser, but also by changing notations (which requires re-rendering) or retrieving additional information from services on the web. Consider a student reading lecture notes: Whenever he is not familiar with a mathematical symbol occurring in some formula, he should be able to look up its definition without opening another document, but right in his current reading context. Or consider the problem of converting between physical units (e. g., imperial vs. SI). There are lots of unit converters on the web (see [Str08] for a survey), but instead of manually opening one and copying numbers into its entry form, we want to enable an in-place conversion.

In [KMR08, KMM07], we investigated how OMDoc documents containing content markup can be rendered as XHTML with embedded presentation MathML. The rendering was relative to context dimensions such as the native language of the reader or the field of knowledge. This approach used *notation definitions* to translate content markup into presentation markup. It focused on generating documents *before* the user gets to read them. In the present paper, we continue this line of research and introduce JOBAD as the client-side counterpart. JOBAD is a JavaScript API for OMDoc-based Active Documents. While its primary intended application is to be part of the active documents served by our server, it is independent of the server and can be flexibly reused to enable any mathematical document to interact with the reader or web services. Our contribution includes the JOBAD interactive document format, an XHTML+MathML-based interface language between server- and client-side computation.

In Section 2, we present the main component of JOBAD, a collection of small JavaScript modules that add interactive services to a mathematical document. Here we assume a broad notion of "service" including local interactive functionality as well as any service with an HTTP interface, regardless of whether it complies with a "heavyweight" web service standard like XML-RPC or MONET. In Section 3, we present several web services that we have implemented and describe how to integrate third-party services. In Section 4, we briefly describe a first JOBAD case study, and we conclude in Section 5.

2   An Architecture for Active Documents

The JOBAD architecture is divided into the actual mathematical services, the user interface elements, and generic communication and document manipulation functions (see figure 1). On the client side, JOBAD consists of a JavaScript main

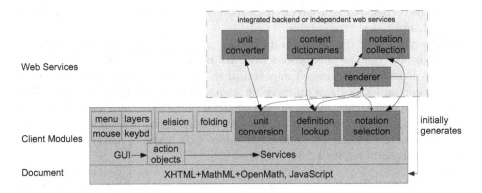

Fig. 1. JOBAD Architecture. Note the central role of the rendering service, which both generates JOBAD-compliant documents and is needed for many other services.

module and one independent module for each service. The server controls the available functionality by loading a collection of service modules into the document. We distinguish three kinds of interactive services by the amount of data they exchange with the web service backend: 1. services that merely draw on data embedded into the document, 2. services that send a symbol identifier and an action verb to the backend in order to retrieve additional information, 3. and services that send complex mathematical content expressions to the backend. The decision which kind of service should be used for a particular functionality depends considerably on the format used for interactive documents: Some meta-information about the current appearance may be embedded into the document to permit local interaction, for example, alternative presentations or parallel content markup. If this is not feasible, further information must be embedded that instructs the JOBAD client how to retrieve the external document fragments.

2.1   A Document Format That Enables Services

The MathML specification leaves some details about the structure of a document to application developers [W3C]. Therefore, for JOBAD, we pose some additional requirements:

1. Alternative displays, among which the user can switch, SHOULD[1] be realized by `maction` elements and an `@actiontype` attribute that indicates the type of service [W3C, section 3.6.1]. Particularly, services that rewrite a formula SHOULD retain the previous state of the formula as an alternative to which the user can switch back.
2. Unless subterms are already part of a grouping operator (e. g. radicals, or super-/subscripts, or fractions; see [W3C, table 3.1.3.2] for a complete list), they MUST be grouped using the invisible `mrow` element (which is optional in MathML).
3. For services that need access to the semantics of mathematical expressions, the latter MUST be provided as parallel content markup [W3C, chapter 5.4]. There MUST be cross-links from the subterm-grouping elements of Item 2 to the corresponding content elements. Content elements MAY be annotated with additional attributions.
4. If services that directly operate on the presentation markup to customize its display require annotations that will be interpreted in a service-specific way, it may be considered impractical to add them to the parallel content markup and look them up there. In such a case, annotations MAY be added directly to presentation markup elements as attributes from another namespace (see [W3C, section 2.4]). It is RECOMMENDED that such annotations require considerably less additional space than parallel markup; otherwise parallel markup SHOULD be preferred.

Requirements 2 and 4 lead to a linear overhead in the size of the formulas (with a small factor), requirement 3 as well but with a factor of 2 to 4, as content

---

[1] We use these capitalized keywords in accordance with RFC 2119 [Bra97].

markup, element IDs, and references to those IDs are added. Requirement 1 can lead to an exponential overhead when alternative displays are nested. This could be avoided if Presentation MathML allowed for structure sharing between expressions (which only Content MathML allows so far).

As these requirements are hard to satisfy by manual authoring, this markup is generated by our rendering service that generates presentation from content markup using pattern matching-based notation definitions for all symbols (cf. section 3.1). In the following, we will describe these requirements in more detail by discussing how they are used in a variety of services we implemented.

Switching between Alternative Displays. An `maction` element can have multiple children, one of which is displayed at a time, controlled by the `@selection` integer attribute – whose value can be changed at runtime. The MathML specification suggests possible `@actiontype` values but does not prescribe a fixed semantics for them. We have introduced several values for that attribute and clearly specified the desired behavior of the services using them.

In particular, we use `maction`, with the action type `abbrev`, for author-defined abbreviations of complex expressions. Consider a physics document, where the author wants to provide $W_{\mathrm{pot}}(R)$ (potential energy) as an instructive abbreviation of the complex term $\frac{-e^2}{4\pi\epsilon_0 R/2}$. We introduce the OpenMath symbol `folding#abbrev` that serves as an attribution key. Our rendering algorithm uses the value of such an attribution as an alternative display. For example,

```
<OMATTR>
   <OMATP>
      <OMS cd="folding" name="abbrev"/>
      W_pot(R)
   </OMATP>
   -e²/4πε₀R/2
</OMATTR>
```

$$\frac{-e^2}{4\pi\epsilon_0 R/2}$$

is rendered as

```
<maction actiontype="abbrev" selection="1">
   -e²/4πε₀R/2
   W_pot(R)
</maction>
```

$$\frac{-e^2}{4\pi\epsilon_0 R/2}$$
$$W_{\mathrm{pot}}(R)$$

Other JOBAD services create `maction`s on the fly (using the name of the service as the value of the `@actiontype` attribute) whenever they rewrite a term t into t' (e. g. by converting units; see Sect. 3.3). This allows for making interactions *undoable*: The previous state t of a term is preserved in the second, hidden child of the `maction`, and the user can switch back to it. Not only do interactions become *un*doable locally, but they also become *re*doable: When information from a remote web service has been used to rewrite $t \rightsquigarrow t'$, as is the case with unit conversion, e. g., and the user switches back to t, he can recover t'

without causing the information to be retrieved from the web once more, as it is still cached in the `maction`.

Grouping Subterms for Interactive Folding. Besides author-defined abbreviations, we have implemented interactive folding of *arbitrary* subterms, so that a reader can hide them if he feels distracted. Any subterm that is properly grouped (see requirement 2 above) is eligible for folding. When the user requests folding of a subexpression for the first time, we put both the original subterm and its folded version into an `maction` element with `actiontype folding` for making the action undoable (see above).

As an example, consider the expression $[1 + [2 \cdot x]]$, where square brackets denote `mrow`s. Suppose the user selects [part of] the subterm $2 \cdot x$ or right-clicks somewhere in that term and requests it to be folded. Then, the formula will display as $[1 + \ldots]$. Clicking on the dots and selecting the "unfold" action from the user interface (e.g. the context menu) will restore the original appearance.

Cross-Linked Parallel Markup. MathML provides the ability to attach semantic annotations to presentation markup as so-called *parallel markup* [W3C, chapter 5.4]. We use this to obtain the content counterpart of presentation expressions selected by the user. The direction of such cross-links is generally unspecified in MathML, but we fix it to "presentation→content" as that is the most natural direction for looking up a formal representation of the user's selection. Moreover, it is the only direction in which associative infix operators can be cross-linked, as multiple presentation operators have to link to one content symbol (cf. Fig. 2).

Therefore, we require links from symbols, numbers, identifiers, and subterm-grouping presentation elements to corresponding content elements, given by `@xref` attributes. If a content expression is to be looked up for a given selection of presentation elements, we locate the closest common ancestor of all selected XML elements that carries an `@xref` attribute and dereference it to obtain the corresponding content expression. An example is given in Figure 2.

Since our rendering algorithm supports pattern-matching-based and thus non-compositional translations from content to presentation markup, not every content subexpression corresponds to a presentation expression. For example, in the presentation element corresponding to $\sin^2 x$, there will be no subexpression pointing to the content expression $\sin x$.

Lightweight Annotations for Flexible Elisions. The rendering algorithm that we introduced in [KMR08] enables a flexible elision of redundant brackets. We now describe how to utilize the output generated by that algorithm – and thus our rendering service, cf. Sect. 3.1 – for deferring the decision which brackets to elide until the document is read.

When rendering a content expression $f(t_1, \ldots, t_m)$, brackets around the rendering of $t_i = g(s_1, \ldots, s_n)$ are redundant if the operator f binds weaker than g. Binding strength is determined by comparing the i-th input precedence of f

```
<semantics>                          <annotation-xml encoding="OpenMath">
 <!-- a+b  2  c  -->                  <OMA id="E">
 <mrow xref="#E">                      <OMS cd="arith1" name="plus"
  <mi xref="#E.1">a</mi>                id="E.0"/>
  <mo xref="#E.0">+</mo>               <OMV name="a" id="E.1"/>
  <mrow xref="#E.2">                   <OMA id="E.2">
   <msup xref="#E.2.1">                  <OMS cd="arith1" name="times"
    <mi xref="#E.2.1.1">b</mi>           id="E.2.0"/>
    <mn xref="#E.2.1.2">2</mn>          <OMA id="E.2.1">
   </msup>                               <OMS cd="arith1" name="power"
   <mo xref="#E.2.0">&#x2062;            id="E.2.1.0"/>
   <!-- INVISIBLE TIMES -->              <OMV name="b" id="E.2.1.1"/>
   </mo>                                 <OMI id="E.2.1.2">2</OMI>
   <mi xref="#E.2.2">c</mi>             </OMA>
  </mrow>                               <OMV name="c" id="E.2.2"/>
  <mo xref="#E.0">+</mo>               </OMA>
  <mi xref="#E.3">d</mi>               <OMV name="d" id="E.3"/>
 </mrow>                               </OMA>
                                      </annotation-xml>
                                     </semantics>
```

Fig. 2. Parallel markup: Presentation markup elements point to content markup elements. The light gray range is the user's selection, with the start and end node in bold face. We first look up their closest common ancestor that points to content markup, and then look up its corresponding content markup – here: *E.2*.

with the output precedence of g. Redundant brackets are retained in the output document, but annotated with the difference between these precedences as the *elision level*. Besides brackets, our rendering service supports other *elision groups*, e.g., for type annotations, some of which are essential whereas others can be inferred. Then brackets are the special case of the elision group fence.

All visible Presentation MathML elements and all grouping elements can carry the attributes @egroup and @elevel from the OMDoc namespace. The value of the former can be any string with the value fence being reserved for brackets. The value of the latter can be any integer or the strings infinity and -infinity. In case an element is member of multiple elision groups, the @egroup and @elevel attributes can contain a space-separated list. For example, a left bracket, annotated with elision information, looks as follows:

```
<mo fence="true" omdoc:egroup="fence" omdoc:elevel="100">(</mo>
```

Our elision service allows the user to choose one *visibility threshold* for each elision group. If T_g is the threshold of elision group g, then all elements of group g whose elision level is above T_g are invisible. This is realized by using maction elements with action type elision that switch between an expression and an invisible mspace element. This also permits a document to provide initial visibility status for its elements.

2.2   User Interface

JOBAD offers various user interface elements for input and output. By right-clicking, a context menu can be requested for the object under the cursor or for the range of selected objects. A selection can be made in the usual way of dragging the mouse, or by repeated clicking on any part of a formula. In the latter case, the selection is extended step by step, always advancing to the parent grouping element. While performing actions on the current selection makes sense for services like folding or definition lookup (cf. Sect. 3.2), other services, such as elision, are also made available globally: If desired, bracket elision can be controlled document-wide by hotkeys from 0 (no redundant brackets displayed) to 9 (all redundant brackets displayed), and a collapsible toolbar placed next to each formula offers one slider per elision group for controlling the thresholds locally.

Calls to services are represented by generic action objects, which allows for providing diverse access to them. The same elision action can, e. g., be triggered via a local context menu, from a formula-local toolbar, and via a global keyboard shortcut.

Besides rewriting formulae, JOBAD offers tooltip-like popups for displaying information on demand. These can be annotations hidden in the document, but we mainly use it for displaying responses from web services, such as the definition of a symbol that the user wanted to look up (cf. Sect. 3.2).

3   Web Services and Their Integration

The easiest way of realizing a mathematical web service is to expose functionality via an HTTP interface. When adopting the REST pattern [Fie00], URLs directly represent mathematical resources (e. g., OpenMath symbols). This can be used, for example, to retrieve the definition of a symbol. We call such services *symbol-based*. More complex services act on the selected expression. In those cases, we use parallel markup to obtain the corresponding content expression and include it in the body of the HTTP request. We call such services *expression-based*.

In the JOBAD framework, we do not commit to a fixed set of web services. Rather do we specify a way of how a JOBAD-compliant document server *advertises* available web services. For each service (client-side service or web service) the document server chooses to offer in the active documents it serves, it MUST serve the corresponding JOBAD JavaScript module to the client. In the head of an active document, each JavaScript module MUST be initialized. To modules that access web services, a description of a web service backend they can connect to MUST be provided. In the case of an integrated backend, these can be components of that backend, but it can also be remote web services that the document server is aware of.

The description of a symbol-based service MUST consist of a name that is displayed to the user and a URL that invokes the service. The URL MAY contain placeholders for cdbase, cd, and name of the symbol. Similarly, the description of an expression-based service MUST consist of a name and a URL.

Listing 1.1. Service initialization code in a document

```
<!-- utility functions (module loading, document manipulation,
    client/server communication) -->
<script src="../scripts/jobad.js"/>
<!-- our own initialization follows -->
<script type="text/javascript">
// GUI elements to be enabled
jobadInit("contextmenu"); // loads the context menu
// In-document services
jobadInit("elision");
// Web services
jobadInit("definition-lookup", "Look_up_definition",
  "http://jobad.mathweb.org/backend?action=definition-lookup
_&cdbase=$cdbase&cd=$cd&name=$name");
</script>
```

3.1 Rendering

The rendering service is a prerequisite for making output from other services human-readable. In its simplest form, it accepts as input (in the body of an HTTP POST request) a fragment of OpenMath and returns the result of rendering it to JOBAD-enriched Presentation MathML. In the following, we will use render(c) for the result of rendering a content markup fragment c.

Our rendering service is implemented using the JOMDoc library, which implements the rendering algorithm described in [KLM+09, KMR08]. It has access to a collection of *notation definitions*, which map content markup patterns to presentation markup templates [KMR08].

3.2 Definition and Type Lookup

The definition lookup service sends the ID of a symbol σ to the server and expects as a response a content-markup formula containing a term that defines σ. The type of a symbol can be looked up analogously. Our implementation uses the RESTful URI format introduced at the beginning of this section. The information that we want to look up is encoded by the value of the *action* parameter, either definition-lookup or type-lookup. In the following, we will use def(σ) and type(σ) for the definition of a symbol, or the type, resp., as looked up by this service.

On the server side, the lookup is enabled by representing content dictionaries (CDs) in OMDoc [Koh06]. There, a symbol with type declaration and definition is represented as shown below, which allows for easy retrieval, e. g., using XPath. The situation in an OpenMath CD is similar, but as "definitional mathematical properties" (DefMPs) have not yet been specified, there is no way of identifying a definition of a symbol among its various mathematical properties.

```
<!-- Content markup omitted here to save space -->
<symbol name="sin">
```

```
<type><OMOBJ>C → C</OMOBJ></type>
</symbol>
<definition for="#sin" type="simple">
  <OMOBJ>sin z = 1/2i (e^{iz} − e^{−iz})</OMOBJ>
</definition>
```

Our current client-side implementation displays render(def(σ)) in a tooltip overlay at the cursor position. Alternatively, *definition expansion* is possible, which replaces the selected occurrence of a symbol with its definition and re-renders the formula. The original formula before definition expansion is kept as an `maction` alternative using the `actiontype definition-expansion`. In contrast to definition expansion, which can only be offered for simple ($\sigma := \mathrm{expr}$) or pattern-based ($f(x) := \mathrm{expr}(x)$) definitions, and for inductive definitions, if applied for a single step, definition lookup can even be offered for implicit definition. (See [Koh06, chapter 15.2] for definition types supported by OMDoc.)

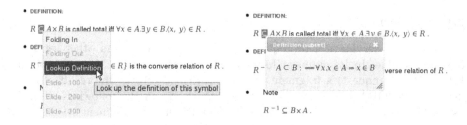

Fig. 3. Looking up a definition (left: selecting the action, right: the result); example taken from our lecture notes; cf. Sect. 4

As the desired MIME type of the response can be given in the HTTP request header for so-called *content negotiation*, we can distinguish requests for content markup from requests for a rendered formula while still using the same URL:

```
GET /backend?action=lookup-definition
 &cdbase=...&cd=transc1&name=sin HTTP/1.1
Host: jobad.mathweb.org
Accept: application/openmath+xml
```

Analogously, the MIME type `application/xhtml+xml` would be used to obtain a response rendered in XHTML with Presentation MathML.

3.3 Unit Conversion

The unit conversion service assumes the OpenMath encoding for units as specified in [DN03]: Base units are symbols in special CDs; derived units can be formed by multiplication or division of base units with numeric factors or other base units. For example:

```
<OMA>
  <OMS cd="arith1" name="times"/>
  <OMI>1</OMI>
  <OMS cd="units_metric1" name="metre"/>
</OMA>
```

The unit conversion service accepts one such expression o, plus a target unit u. If a conversion is possible, the result is returned as an OpenMath expression, which we denote by $\mathrm{uc}(o, u)$. On the client side, this result has to be integrated into the current formula. Let p with $o = c(p)$ be the presentation markup that the user selected; then we add $p' = \mathrm{render}(\mathrm{uc}(o, u))$ as an `maction`-alternative for p to the document.

We have not implemented our own unit converter but use the one developed by Stratford and Davenport [SD08, Str08], which performs conversions according to the OpenMath FMPs of the unit symbols involved. In its current version, their web service does not talk OpenMath but uses string input/output, so we have to convert values between their OpenMath and string representation (e. g. "1 metre").

3.4   Integrated Backends and Environments

For a clean conceptual model, we have treated our web services separately. From an efficiency point of view it does, however, make sense to arrange multiple services in an integrated backend. Consider unit conversion: Stratford's unit conversion web services internally relies on OpenMath CDs that declare one symbol per unit and define conversion rules for obtaining derived units [Str08]. Definitions of symbols are looked up from CDs as well. Last but not least, the rendering service needs notation definitions for the unit symbols, and CD authors often provide default notations for their symbols. Thus, offering those three services independently requires redundantly storing knowledge about symbols in three places. An integrated backend also saves time, as can be seen for definition lookup: With separate lookup and rendering services, the client-side active document has to connect to two web services in succession. An integrated backend could, however, offer readily rendered definitions by composing two of its internal functions and only minimally extending its external HTTP interface (cf. Sect. 3.2). We have implemented an integrated backend that performs rendering and definition lookup, and acts as a proxy for communicating with third-party web services on different domains to avoid security problems due to cross-site scripting.

4   Evaluation

Proof-of-concept demonstrations of individual JOBAD services can be tried at the JOBAD web site [JOB09]. Besides that, we conducted two evaluations to analyze the feasibility and scalability of our framework: 1. We loaded a large OMDoc document into our server and activated the elision, subterm folding, and definition lookup services. 2. We integrated an external unit conversion service, which was added after the main phase of the development, to get an understanding of the investment needed to integrate further services.

The former involved the complete lecture notes of a first-year undergraduate computer science course. These lecture notes are originally maintained in LaTeX with semantic annotations, which can be automatically converted to OM-Doc [Koh08]. The annotations in the source documents comprise content-markup formulae, informal definitions of symbols, and notation definitions. The OMDoc representation is then rendered into the JOBAD format, which is viewed using the JOBAD client, which offers flexible bracket elision, subterm folding, and definition lookup [JOB09]. So far, we have used this as a stress test, but for the Fall lecture we plan an evaluation where one group of our students will work with the static XHTML version of the lecture notes and a second group with the JOBAD-enriched active document.

The most complicated step in the latter evaluation was adapting the string-oriented interface of Stratford's unit conversion web service to our OpenMath interface. Most of the other required functionality turned out to be already available and just had to be composed. We chose the context menu interface and added a submenu containing the target units[2]. Checking whether the selected term was a quantity with a unit reduced to looking up its corresponding content markup (cf. Sect. 2.1) and performing a simple XPath node test on the latter. Sending a string to the web service and waiting for the response is a standard JavaScript function. Rendering the result of conversion (after converting it back to OpenMath) is done by another service. Finally, replacing two XML subtrees in a formula (both in the presentation and the content markup) and hiding the previous presentation tree in an `maction`, is a utility function provided by the JOBAD core and also used by other services.

5   Conclusion and Future Work

We presented JOBAD, our architecture for active mathematical documents. Our documents are generated dynamically from content-markup and viewed in a web browser, via which the reader can change interactively both content and form of the document. JOBAD constitutes the reader interaction component of our research group's framework for mathematical documents. As such, it is fully integrated into the authoring [Koh08, Lan08], notation management [KMR08], and storage [TNT09] work flows developed by our group.

We gently extended the Presentation MathML format to create an interface language, in which the document server can embed into the served document information about interactivity or instructions on how to retrieve that information. This extension is backwards compatible in the sense that the markup is still valid MathML, and switching off JavaScript yields the same static documents as before.

---

[2] A static list at the moment; obtaining admissible target units for a given input is neither supported by our client-side implementation nor by the unit conversion web service at the moment.

We have implemented and evaluated an initial set of services that constitute a representative selection of the possibilities we envision. Folding and flexible elision work locally, type and definition lookup retrieve additional information based on a symbol URI, unit conversion sends a content markup object as an argument to a web service. The former service is based on presentation markup generated by the server a priori. For the latter two, the JOBAD modules are passed initialization parameters that instruct them about the server and its URL format. For the latter one, parallel markup is utilized to obtain the content representation of a presentation expression.

A specific design feature of JOBAD is its extensibility. Offering new services for documents in the JOBAD interactive document can be achieved by adding very little new JavaScript code. Adding new user interaction components and binding them to JOBAD services is possible with minimal effort. Finally, the JOBAD client code requires only very few properties of the specific server backends, so that the same client can be easily used with different web services even in the same document.

Future work will be based on this, and we intend to rapidly develop more services, but also to invite contributions from external developers. Due to the modularity of our framework, we expect that this work load can be divided into small and manageable units that can be handled efficiently by students. In particular, we intend to approach the following services:

Notation selection: Our rendering service can already annotate every rendered symbol with a reference to the notation definition in the backend that was used for rendering it [KLM+09]. This information can be used to ask the backend for alternative notations, to allow the user to select from them, and have the current formula re-rendered accordingly.

Guided tour (extension of lookup): This service generates a linear tutorial containing an explanation of every symbol in the current selection, and of every symbol occurring in these explanations, and so on, until some foundational theory is reached.

Flattening: Many documents consist of components that are combined by a module system (see [RK08]). A flattening service replaces import links with the (possibly translated) copy of the imported document.

Search: Our group has developed a semantic search engine for mathematical formulae [KAJ+08]. Therefore, a service that searches the web (or the server database) for the selected expression will be easy to realize.

Links to web resources: The OpenMath wiki [Lan09] not only provides symbol definitions, but also hosts discussions about them. Its architecture allows for linking symbols to further web resources, e.g. Wikipedia articles about mathematical concepts, which can then be made available in a document.

Adaptive display of statement-level structures: On the level of definitions, theorems, and proofs, we generate a different kind of parallel markup from OM-Doc sources, namely XHTML+RDFa [ABMP08]. We have already used this for visualizing rhetorical structures in mathematical documents (cf. [Gic08]; demo available at [JOB09]) and plan to extend it to structured proofs.

Editing: Our group has developed the Sentido formula editor [LGP08]. An edit service will pass the selected term to a Sentido popup window and eventually replace it in the current document.

Saving: After a user has adapted a document, it is desirable to upload its configuration to the database.

Furthermore, we will integrate the JOBAD architecture into our various integrated document management systems, such as the semantic wiki SWiM [LGP08], and the panta rhei document browser and community tool [pan].

Acknowledgments. The authors would like Jonathan Stratford for help with his unit converter, Christine Müller for fruitful discussions on notation selection, Jan Willem Knopper for hints on designing a modular JavaScript library, as well as David Carlisle for clarifications on MathML. This work was supported by JEM-Thematic-Network ECP-038208.

References

[ABMP08] Adida, B., Birbeck, M., McCarron, S., Pemberton, S.: RDFa in XHTML: Syntax and processing. Recommendation, W3C (2008)

[ACR⁺08] Autexier, S., Campbell, J., Rubio, J., Sorge, V., Suzuki, M., Wiedijk, F. (eds.): AISC 2008, Calculemus 2008, and MKM 2008. LNCS (LNAI), vol. 5144. Springer, Heidelberg (2008)

[Act08] ACTIVEMATH (2008), http://www.activemath.org/

[AKTV08] Ankolekar, A., Krötzsch, M., Tran, T., Vrandečić, D.: The two cultures: Mashing up Web 2.0 and the Semantic Web. Web Semantics 6(1) (2008)

[Bra97] Bradner, S.: Key words for use in RFCs to indicate requirement levels. RFC 2119, Internet Engineering Task Force (1997)

[CCJS05] Cohen, A.M., Cuypers, H., Jibetean, D., Spanbroek, M.: Interactive learning and mathematical calculus. In: Kohlhase, M. (ed.) MKM 2005. LNCS (LNAI), vol. 3863, pp. 330–345. Springer, Heidelberg (2006)

[CCK⁺08] Cuypers, H., Cohen, A.M., Knopper, J.W., Verrijzer, R., Spanbroek, M.: MathDox, a system for interactive Mathematics. In: Proceedings of World Conference on Educational Multimedia, Hypermedia and Telecommunications. AACE (2008)

[DN03] Davenport, J.H., Naylor, W.A.: Units and dimensions in OpenMath (2003), http://www.openmath.org/documents/Units.pdf

[Fie00] Fielding, R.T.: Architectural Styles and the Design of Network-based Software Architectures. PhD thesis, University of California, Irvine (2000)

[GHJS08] Gerdes, A., Heeren, B., Jeuring, J., Stuurman, S.: Feedback services for exercise assistants. Technical Report UU-CS-2008-018, Utrecht University (2008)

[Gic08] Giceva, J.: Capturing Rhetorical Aspects in Mathematical Documents using OMDoc and SALT. Technical report, Jacobs University, DERI Galway (2008), https://svn.kwarc.info/repos/supervision/intern/2008/giceva_jana/project/internship%20report.pdf

[GM08] Goguadze, G., Melis, E.: Feedback in ActiveMath exercises. In: International Conference on Mathematics Education (ICME) (2008)

[JOB09]   https://jomdoc.omdoc.org/wiki/JOBAD (2009)
[KAJ+08]   Kohlhase, M., Anca, Ş., Jucovschi, C., González Palomo, A., Şucan, I.A.:
             MathWebSearch 0.4, a semantic search engine for mathematics (manuscript
             2008),
             http://mathweb.org/projects/mws/pubs/mkm08.pdf
[KLM+09]   Kohlhase, M., Lange, C., Müller, C., Müller, N., Rabe, F.: Notations
             for active documents. KWARC Report 2009-1, Jacobs University (2009),
             http://kwarc.info/publications/papers/KLMMR_NfAD.pdf
[KMM07]   Kohlhase, M., Müller, C., Müller, N.: Documents with flexible notation
             contexts as interfaces to mathematical knowledge. In: Libbrecht, P. (ed.)
             Mathematical User Interfaces Workshop (2007)
[KMR08]   Kohlhase, M., Müller, C., Rabe, F.: Notations for Living Mathematical
             Documents. In: Autexier, et al. (eds.) [ACR+08]
[Koh06]   Kohlhase, M.: OMDoc – An Open Markup Format for Mathematical Doc-
             uments [version 1.2]. LNCS (LNAI), vol. 4180. Springer, Heidelberg (2006)
[Koh08]   Kohlhase, M.: Using LaTeX as a semantic markup format. Mathematics in
             Computer Science (2008)
[Lan08]   Lange, C.: SWiM – a semantic wiki for mathematical knowledge manage-
             ment. In: Bechhofer, S., Hauswirth, M., Hoffmann, J., Koubarakis, M. (eds.)
             ESWC 2008. LNCS, vol. 5021, pp. 832–837. Springer, Heidelberg (2008)
[Lan09]   Lange, C.: OpenMath wiki (2009), http://wiki.openmath.org
[LGP08]   Lange, C., González Palomo, A.: Easily editing and browsing complex
             OpenMath markup with SWiM. In: Libbrecht, P. (ed.) Mathematical User
             Interfaces Workshop (2008),
             http://www.activemath.org/~paul/MathUI08
[MGH+06]   Melis, E., Goguadze, G., Homik, M., Libbrecht, P., Ullrich, C., Winterstein,
             S.: Semantic-aware components and services of activemath. British Journal
             of Educational Technology 37(3) (2006)
[Mon05]   MONET (March 2005), http://monet.nag.co.uk/mkm
[Moz09]   Mozilla Labs. Ubiquity (2009), http://ubiquity.mozilla.com
[O'R05]   O'Reilly, T.: What is Web 2.0 (2005),
             http://www.oreillynet.com/pub/a/oreilly/tim/news/
             2005/09/30/what-is-web-20.html
[pan]   The panta rhei Project (March 2009),
             http://trac.kwarc.info/panta-rhei
[RK08]   Rabe, F., Kohlhase, M.: An exchange format for modular knowledge. In:
             Rudnicki, P., Sutcliffe, G. (eds.) Knowledge Exchange: Automated Provers
             and Proof Assistants (KEAPPA) (November 2008)
[SCI09]   The SCIEnce project – symbolic computation infrastructure for europe
             [online] (2009)
[SD08]   Stratford, J., Davenport, J.H.: Unit knowledge management. In: Autexier,
             et al. (eds.) [ACR+08]
[Str08]   Stratford, J.: Creating an extensible unit converter using openmath as the
             representation of the semantics of the units. Technical Report 2008-02,
             University of Bath (2008),
             http://www.cs.bath.ac.uk/pubdb/download.php?resID=290
[TNT09]   TNTBase Project (2009), https://trac.mathweb.org/tntbase/
[Ull08]   Ullrich, C.: Courseware Generation for Web-Based Learning. LNCS,
             vol. 5260. Springer, Heidelberg (2008)
[W3C]   W3C. Mathematical Markup Language (MathML) 3.0, 3rd edn.
[Yah09]   Yahoo! Pipes [online] (2009)

Representation for Interactive Exercises

George Goguadze

Faculty of Computer Science, University of Saarland
george@activemath.org

Abstract. Interactive exercises play a major role in an adaptive learning environment ACTIVEMATH. They serve two major purposes: training the student and assessing his current mastery, which provides a basis for further adaptivity. We present the current state of the knowledge representation format for interactive exercises in ACTIVEMATH. This format allows for representing multi-step exercises, that contain different interactive elements. The answer of the learner can be evaluated semantically. Various types of feedback and hint hierarchies can be represented. Exercise language possesses a construction for specifying additional components generating (parts of) the exercise. One example of such component is a Randomizer, which allows for authoring parametrized exercises. Another example is so-called Domain Reasoner Generator, that automatically generates exercise steps and refined diagnosis upon the learner's answer. This turns ActiveMath system into an ITS as soon as some Domain Reasoner is connected to it. Finally, several tutorial strategies can be applied to the same exercise. This strategies control feedback and the way the exercise is navigated by the learner, and can adapt to the learner.

1 Introduction

This paper describes knowledge representation for interactive exercises in the ACTIVEMATH learning environment and shows how this knowledge representation allows for encoding exercises that are reusable with different tutorial and presentation strategies. This knowledge representation is a refinement of the format described in [4] as explained in the section 2. A need for such a refinement arose from the requirements of several research projects using ACTIVE-MATH system for exploring different tutorial strategies for interactive exercises. Such strategies have been developed in ACTIVEMATH for the needs of the German research projects ATuF (Adaptive Tutorial Feedback) and ALOE (Adaptive Learning On Errors). We describe the tutorial and presentation strategies in the section 5.

ACTIVEMATH is a web-based learning platform for mathematics. It incorporates the set of tools for learner and teacher that facilitate learning process. In addition to its adaptive course generation [12] and student model [2] a central component is its subsystem for interactive exercises that features ACTIVEMATH's main functionalities of 'intelligent tutoring' including the student input's diagnosis. Exercise subsystem serves two fundamental purposes. On one hand, it

J. Carette et al. (Eds.): Calculemus/MKM 2009, LNAI 5625, pp. 294–309, 2009.
© Springer-Verlag Berlin Heidelberg 2009

provides the student model with an assessment of the learner's knowledge, that serves as a basis for further adaptivity. On the other hand, exercises train the learner in order to increase his mastery of the domain concepts.

ACTIVEMATH is not specific to any particular mathematical domain. Among the current courses are e.g. calculus, fraction arithmetics, statistics, analytic and algebraic geometry, optimization and operation research, etc. For diagnosis in multiple mathematical domains ACTIVEMATH needs services of Computer Algebra Systems (CAS) and multiple domain reasoners. In the section 3.1 we briefly describe the semantic broker architecture for distributed mathematical diagnosis services developed in ACTIVEMATH.

OPENMATH[1] representation for mathematical formulas ensures interoperability with external CAS and other formal mathematical reasoning systems via using so-called phrasebooks, that transform OPENMATH representation into the syntax of a particular reasoning service and then translate the results back to OPENMATH. A specific generic query language has been defined in ACTIVE-MATH for querying external services for diagnosis upon the learner's answer and retrieving (parts of) correct solution.

ACTIVEMATH uses OMDOC standard for mathematical documents[2] for its course contents. Knowledge representation of interactive exercises extends OM-DOC. It reuses the micro-structure elements `CMP` (Commented Mathematical Property) of OMDOC and their contents for representing static content of exercise steps. This includes text, formulas, images, formatting elements. By using several `CMP`s with different values for language attribute, we obtain multilingual exercises. Complex internal structure for multi-step interactive exercises with various types of interactivity and rich set of annotations for classifying feedback is defined. This representation can serve as a natural replacement of an OMDOC quiz module, which can represent only single-step multiple-choice exercises.

2   Knowledge Representation of Exercises

An exercise in ACTIVEMATH is represented as a finite state machine of states representing tutorial actions of the system and transitions between those states representing reactions upon learner's answers to the tasks in the exercise states. Transitions are triggered by some conditions satisfied upon the learner's input. This representation resembles a tutorial dialogue consisting of consequent actions of a student and tutor.

Exercises representation in ACTIVEMATH is designed to be domain independent, and low level enough to be suitable for automatic generation, descriptive in its nature and suitable for authoring.

Exercise states and transitions have themselves complex internal structure. Several types of interactive elements such as fill-in blanks, multiple-choice questions, mappings and re-orderings are supported. Transitions can represent complex constraints upon single or multiple answers and their combinations. Each

---

[1] see http://www.openmath.org

[2] see http://www.omdoc.org

state is annotated with metadata, and each transition is annotated with diagnosis information.

In comparison, the quiz module of OMDOC represents only single step exercises that consist of the block of multiple-choice questions with answers attached to each choice.

In ACTIVEMATH each exercise state either represents a task for a learner, feedback to the previous action of the learner, or interaction in which the learner provides the solution to the task and submits it into the system. Each state apart from the terminal one issues one or more transitions to following states.

A transition represents a move from the current state to the next one, in conditional transitions the learner's answer should satisfy a particular condition that triggers the transition. Each transition contains a pointer to the target state.

Schematically, nodes and transitions of an exercise are shown in the Figure 1. The feedback loop consists of:

- presenting a task to the learner
- interaction, in which the learner enters the answer to the current task
- presenting feedback to the learner

In comparison to the previous version of exercise format ([4]), each component of a feedback loop is represented as a separate node. This separation facilitates application of tutorial strategies to the exercise.

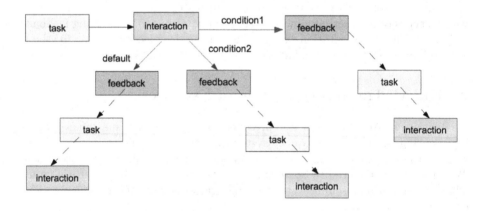

Fig. 1. Structure of an exercise

Consider a simple exercise, in which the learner has to differentiate the function $f(x) = 2 \cdot x$. The finite state automaton for this exercise is shown in the Figure 2. The first state (task1) of our finite state machine defines a task to calculate $(2x)'$ and forwards the learner to the interactive step interaction1. If the learner's answer is equal to $2 \cdot (x)'$ the next state will be feedback1, followed by the new task (task2) via unconditional transition. After this the next interaction interaction2 is presented. In case the learner entered the final correct answer 2 he is forwarded directly to the state `feedback3`. In all other cases, the default transition is triggered and the learner is forwarded to the `feedback2`. The states that do not issue any transitions are terminal. As soon as the terminal state is reached the exercise is finished. In our example `feedback2` and `feedback3` are terminal.

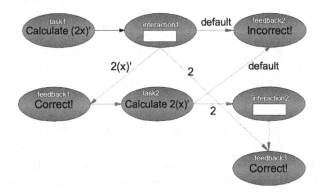

Fig. 2. A simple exercise graph

2.1 Types of Exercise States

Exercise states can be static – representing problem statements for tasks and feedback to the learner, and interactive – representing different types of interaction with learner such as fill-in-blanks, multiple choice questions, mappings and orderings. Figure 3 shows examples of interactive elements in ACTIVEMATH. Different styles of fill-in-blanks are shown in the screen shots. When entering mathematical formulas, complex palettes as well as simple input fields can be used. Multiple choice questions can be presented as radio buttons or drop-down menus and so on.

Static states issue an unconditional transition to the following static or interactive state. Interactive states can contain more then one conditional transition and a default transition for the case if none of the defined conditional transitions are triggered.

2.2 Metadata of Exercise States

In order to achieve accurate assessment of the learner's knowledge and enable automatic application of tutorial strategies to the exercise representation exercise states have to be annotated with metadata. Depending on the type of the state it can be assigned different metadata.

Metadata of Tasks. Each task is training or assessing mastery of a particular cognitive processes w.r.t. the focus concepts of the task at a certain difficulty level. These are domain concepts represented as items in the knowledge base of the ACTIVEMATH system.

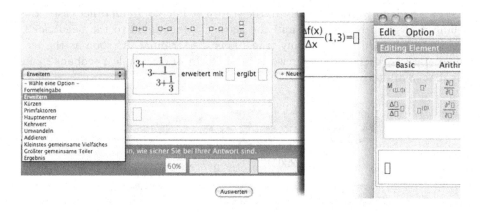

Fig. 3. Interactivity in ACTIVEMATH exercises

There are two main reasons for annotating the tasks of the exercise states with these metadata. Firstly, the Learner Model needs information about the the concepts and cognitive processes that are mastered in the task together with the diagnosis upon the learner's answer to this task in order to update the learner's mastery values. The quality of this updating defines the whole estimation of learner's current literacy in the domain that in its turn defines the further tutorial actions of the system.

Secondly, in an automated tutorial strategy that needs to generate feedback for the current state the metadata can be used in various ways. For example, in order to generate a conceptual hint it is necessary to know which concept is trained in the task.

Therefore, exercise tasks are annotated with the following metadata:

- `relation` to the identifier of the domain concept
- `competency` describing the cognitive process trained/assessed by the task
- `competency level` describing the degree of the resource within the set of competencies
- `difficulty` describing the difficulty of the task (relative to learning context of the student)

The value set for `competency` has been the subject of constant refactoring in ACTIVEMATH. The first `competency` scheme that was used in ACTIVEMATH was derived from a Bloom's taxonomy of learning goal levels. This scheme was exchanged by the adoption of PISA standard methodology (see e.g. [8]) within LEACTIVEMATH project (see [6]), and finally the new scheme was derived in collaboration with TU Dresden, as described in [9].

Feedback Metadata. Within a research project ATuF, we classify the feedback content, deriving our classification from the previous feedback classification of feedback in the LEACTIVEMATH[3] project enriched with feedback components

---

[3] see http://www.leactivemath.org

from Narciss's conceptual framework for feedback in interactive instruction (see [10,11]). Also the form, and timing of feedback presentation is varied by AC-TIVEMATH's exercise presentation strategies (see section 5.3).

We differentiate the following types of feedback:

- `conceptual` feedback provides information about the domain concepts, their interrelations, information about task constraints and so on
- `procedural` feedback is informative and can suggest how to proceed in the solution (e.g. which rules to apply, how to simplify the task and so on)
- `product` feedback is giving out (parts of) correct solution
- `meta-cognitive` feedback informs of or suggests the meta-cognitive behavior of the learner, it refers to aspects such as motivation, help seeking etc.

Each of these dimensions of feedback is further subdivided into `components` which are adopted from Narciss' content-related classification of feedback components (see, e.g. [10]). This subdivision is shown in the table 1.

Table 1. Components of feedback types

type	component	definition
procedural	KR	knowledge of result (correct/incorrect)
	KH	know how to proceed
	KC	knoweldge about concepts
conceptual	KCC	knowledge about concept connections
	KTC	knowledge of task constraints
product	KM	knowledge of mistakes
	KCR	knowledge of correct result
meta-cognitive	KMC	knowledge on meta-cognition
	KP	knowledge of performance

Another dimension of feedback that ACTIVEMATH currently uses is the type of `instruction` that the feedback represents. The possible values of `instruction` are *feedback, hint, explanation*, and *guiding_question*.

Such annotations provide a basis for automated application of tutorial strategies, defining algorithms for feedback selection within an internal feedback loop in the exercise, as we show in the section 5.

2.3   Structure of Transitions

There are two types of transitions – conditional and unconditional. Unconditional transitons connect static nodes to each other or to the following interactive nodes. Conditional transitions represent conditions that have to be satisfied upon the learner's answer for this transition to be triggered. Conditions represent three basic types of comparisons, that are *syntactic*, *numeric* and *semantic* equivalences. Syntactic equivalence represents a literal comparison of the learner's answer to the given value. Numeric equivalence means equivalence of real numbers modulo given ϵ. For more complex comparisons a generic query *compare* can be used, as defined in the section 3.2. Finally, a default transition is triggered when none of the defined transitions fire.

Each conditional (and default) transition contains a `diagnosis` element. Diagnosis represents information on learner's success w.r.t. the task in case the current transition fires, represented using `achievement` element. In case of wrong answer it can also contain references to misconceptions if any diagnosed in the condition.

```
<exercise id="deriv_2x" for="mbase://openmath-cds/calculus1/diff">
 <metadata>...</metadata>
 <task id="task1">
  <taskmetadata>
    <relation type="for">
      <ref xref="mbase://calculus/diff_rules/const_mult"/>
    </relation>
    <competency value="solve" subvalue="apply_algorithms"/>
  </taskmetadata>
  <CMP xml:lang="en">Calculate $(2x)'$</CMP>
  <transition xref="interaction1"/>
 </task>
 <interaction id="interaction1">
  <CMP xml:lang="en"><with for="blank1"/></CMP>
  <interaction_map><blank id="blank1"/></interaction_map>
  <transition_map>
    <transition xref="feedback1">
      <diagnosis><achievement value="1.0"/></diagnosis>
      <condition><syn_eq>$2(x)'$</syn_eq></condition>
    </transition>
    <transition xref="feedback3">
      <diagnosis><achievement value="1.0"/></diagnosis>
      <condition><syn_eq>$2$</syn_eq></condition>
    </transition>
    <default xref="feedback2">
      <diagnosis><achievement value="0.0"/></diagnosis>
    </default>
  </transition_map>
 </interaction>
 <feedback id="feedback1" type="procedural" component="kr">
 <CMP xml:lang="en">Correct!</CMP>
 <transition xref="task2"/> ...
</exercise>
```

Fig. 4. Excerpt from exercise encoding

2.4   Sample Exercise Representation

Consider again the sample exercise from the Figure 2. An excerpt from the representation of this exercise is shown in the Figure 4 (OPENMATH formulas are represented using latex syntax to save space). The task metadata defines relation to the focus concept of the step - constant multiplication rule and a competency trained to be 'solve/apply_algorithms'. In the interacion the interactive elements are separated from the rest content and grouped separately within an element `interaction_map`. This is needed for exercises in multi-lingual exercises, in which placeholders from CMPs with different languages have to be assigned to the same interactive element using relation *for*.

The states either contain one unconditional transition or a list of conditional transitions within a container element `transition_map`.

Feedback has attributes *type* and *component* assigned, *instruction* attribute is not assigned and takes the default value 'feedback'.

3   Exercise Subsystem Architecture

Figure 5 shows components of the exercise system architecture. The central component of the exercise subsystem is the `Exercise Manager`, which coordinates the other components and controls the exercise process. Another important component is the `Exercise Generator` which is responsible for generating the nodes of the exercise.

This nodes can either be obtained from an authored part of the exercise representation, or generated automatically. The `Diagnoser` remotely connects via a `Query Broker` to external services capable of intelligent diagnosis. Based on the diagnosis of the user action provided by a CAS or by a domain reasoner, the `Feedback Generator` component generates feedback automatically or selects appropriate authored feedback from the exercise representation. Various `Tutorial Strategies` can be applied to the exercises that define, e.g., what type of feedback has to be generated depending on the situation of the student, his previous activities and the strategy's pedagogical approach.

3.1   Diagnosis of the Learner's Answer

Conditional transitions represent conditions to be satisfied upon the learner's answer for this transition to be triggered. ACTIVEMATH possesses a framework for distributed semantic evaluation services, realized via a `Query Broker` shown in the Figure 5.

This broker receives a query from the `Diagnoser` component that is matched against the list of available semantic services and the service that can answer this query is automatically selected based on the semantic context of the query. After selecting appropriate service, the OPENMATH representation of the expression to be evaluated is translated to the language of the corresponding external evaluation service via so-called phrasebooks. The evaluated expression returned to ACTIVEMATH is then translated back into OPENMATH.

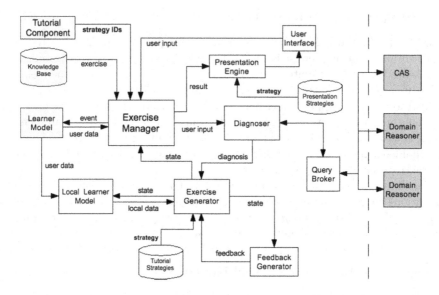

Fig. 5. Exercise subsystem architecture

Currently, ACTIVEMATH integrates and communicates with the following CASs: YACAS[4], Maxima[5], and WIRIS[6]; phrasebooks for Maple[7] and Mathematica[8] are available too.

3.2 Query Language

In ACTIVEMATH we defined a generic format for queries used to access a diagnosis service. A *context* included into queries defines (sub-)sets of rules and functions that a domain reasoner or a CAS is allowed to use for the diagnosis. The background for this restriction is that depending on the learner's activity history, situation and on the pedagogical approach, different rules (and functions) should be usable. The query representation includes:

- **action** of the query (e.g. `getResults`, `getUserSolutionPaths`, `compare`)
- **(list of) expressions** e.g., task, user answer, correct answer
- **context** of action identifying the set of applicable rules
- **number** of iterations domain reasoner should perform in the given context

In the following, e, e_1, e_2, are OPENMATH expressions, C is a context of a query, N is the number of iterations. A solution path is a list of results of consecutive rule applications, which are annotated with rule identifiers.

---

[4] see http://yacas.sourceforge.net
[5] see http://maxima.sourceforge.net
[6] see http://www.wiris.com
[7] see http://www.maplesoft.com
[8] see http://www.wolfram.com

Currently the following queries to diagnostic services are used in ACTIVEMATH:

- query(getResults, e, C, N) - returns the list of final nodes of all paths of length N starting at e in the context C
- query(compare, e_1,e_2, C, N) - returns true if there exists a path of the length N from e_1 to e_2 in the context C, false otherwise
- query(getRules, e, C) - returns the list of the identifiers of expert rules applicable to e in context C
- query(getBuggyRules, e_1, e_2, C, N) - returns the list of the identifiers of all buggy rules that belong to a path from e_1 to e_2 in the context C. This query is possible for those domain reasoners that can reason with (typical) buggy rules and some CASs, which can be extended to do so.
- query(getUserSolutionPaths, e_1, e_2, C, N) - returns the list of all paths of length N from e_1 to e_2 in the context C
- query(getExpertSolutionPaths, e, C, N) - returns the list of all paths of length N starting at e in the context C. In this query C can consist of expert rules only.
- query(getNumberOfStepsLeft, e, C) - returns the number of steps left to reach the final node of the shortest expert solution path in context C
- query(getRelevance, e_1, e_2, C) - returns 'true' if the expression e_2 is closer than e_1 to the actual solution in the context C,

Exercises using CAS evaluation mostly use the query compare and its special cases syn_eq, num_eq, and sem_eval to compare the learner's answer to the correct solution and the query getResults to get the final correct result. In the section 4 we give usage examples of other queries.

4   Generating Exercises

Exercise Generator provides a generic interface for other custom generators that can generate the exercise knowledge representation in their own way.

Each generator can define a custom diagnosis service which is a subclass of a standard Diagnoser component. The custom Diagnoser can send more advanced queries to domain reasoner services, that helps to diagnose learner's errors more deeply and generate error related and remedial feedback.

Consider some examples of automatically generated exercises and other tools reusing exercise representation and running engine.

4.1   Randomized Exercises

Using a custom instance of an Exercise Generator, so-called Randomizer component of ACTIVEMATH and an extension mechanism present in the knowledge representation of exercises, one can author whole classes of parametrized exercises. Such an extension mechanism offers specifying the name of the Exercise Generator and the parameters it uses together with the set of their possible values. The Randomizer generator receives as input an exercise representation, parametrized using a set of meta-variables. This variables can be used for problem definition and the solution of the exercise. They can be also used in the conditions, comparing the learner's answer to the correct solution. Randomizer also receives the set of possible values of these meta-variables which randomly substitutes each time the new instance of the exercise is started.

This approach has its limitations, authors have to carefully design their exercises, since although Randomizer provides basic functionality for avoiding semantic inconsistencies such as singularities in the process of substitution, still it is not easy to avoid all possible inconsistencies that might appear in the process of solution. See [1] for more advanced features of the Randomizer.

4.2   CAS Console and Domain Reasoner Console

Computer Algebra Console is a small tool that is realized via a custom Exercise Generator in which the learner can just type a mathematical expression that is evaluated by the exercise subsystem and the result is presented to the learner in form of feedback. This tool allows the learner to use any back-end Computer Algebra System connected to ACTIVEMATH using a single user interface.

Similar to a CAS console, a Domain Reasoner Console allows the learner to define a problem he wants to tackle and then solve it receiving feedback powered by the back-end domain reasoner system. Currently, ACTIVEMATH defines a Domain Reasoner Console for solving problems for symbolic differentiation. For more information about the domain reasoner currently connected to ACTIVEMATH, see [3].

In order to create such an explorative 'multi'-exercise, a specific Exercise Generator is programmed. This generator, similarly to the CAS console, sends queries to the Query Broker that forwards them to the appropriate Domain Reasoner service. These queries request diagnosis upon learner's answers, and (parts of) following correct solution paths. Based on this information, the new states of the exercise are generated.

If the domain reasoner is capable of diagnosing typical errors, such diagnosis can be used for generating informative or remedial error-related feedback. For example, the query getBuggyRules returns the buggy rules that have been diagnosed for the learner's answer, the query getUserSolutionPaths returns the full erroneous paths the user might have taken.

The query getRules returns the set of applicable rules that can be suggested to the learner in a procedural hint.

Queries getResults and getExpertSolutionPaths can give out (parts of) correct solutions and solution paths that can be given to the learner in a bottom-out hint.

5   Tutorial and Presentation Strategies

We define a *solution space* of an exercise to be an (authored or generated) collection of all possible solution paths of this exercise, possibly including erroneous paths.

A tutorial strategy for a multi-step interactive exercise defines the way the learner navigates through the solution space of an exercise together with form and content of feedback presented to the learner in the solution process.

Tutorial strategy specifies, e.g. how many times the learner is allowed to repeat the current step in case of wrong answer, what kind of feedback the learner receives and in which sequence. Tutorial strategy can decide to reveal (parts of) correct problem solving strategy.

Each domain reasoner defines its own problem solving strategy for the given domain. Recent research, described in [7] allowed to realize different problem solving strategies. It is ongoing work to integrate the domain reasoners mentioned in [7] and combine tutorial strategies of ACTIVEMATH with different problem-solving strategies in the domains of propositional logic, linear algebra and fraction arithmetics.

A presentation strategy defines graphical presentation of the exercise, the placement of feedback, highlighting, folding, and other aspects that might seem to be secondary, but play a big role in the perception of the exercise content.

5.1   Applying Tutorial Strategies

Consider in more detail the mechanism for applying tutorial strategies. The `Exercise Generator` consults the strategy before producing the current state of the exercise. The `Local Learner Model` contains procedural information such as how many trials the learner already had to tackle the current task, how many and what type of hints have been already presented to the learner, what is the current mastery level of the learner w.r.t. the task and so on. This information is received by the `Exercise Generator` and matched against the current tutorial strategy. Based on this information the next step is created in which feedbacks of appropriate `type` and `component` are generated or extracted from the knowledge representation of the exercise. The assembled exercise step is passed to the `Exercise Manager`. The latter forwards the processed exercise step to the `Presentation Engine` which applies the current `presentation strategy`.

Tutorial and presentation strategies are not attached to concrete exercises, but can be freely assigned to exercises by the configuration of the system, or by any component of ACTIVEMATH that calls the exercise. In this sense, the exercises are *reusable* within different educational scenarios, using various tutorial and presentation strategies.

5.2   Sample Tutorial Strategies

In the Figure 6 two different feedback strategies are shown, that are used in the ATuF project, mentioned above.

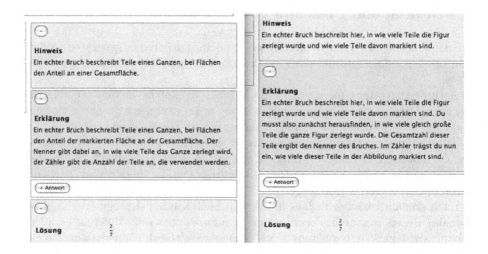

Fig. 6. ATuF Feedback Strategies

The first strategy is providing the learner with sequenced feedback - first a conceptual hint, then a conceptual explanation and then the correct solution. The second strategy is providing a sequence - procedural hint, procedural explanation and the correct solution. Using different variations these feedback sequences in the ongoing school experiment the researchers of TU-Dresden are collecting empiric results aiming to analyze the effect of procedural and conceptual feedback components in the problems with fraction addition.

Other tutorial strategy partially shown in the Figure 8 is used within the project ALOE in which the effect of erroneous examples on is investigated - classroom studies are running in several schools in Germany. In this strategy the learner has to find an error in a given solution and correct it. In the interaction the learner has a possibility to modify any step, not only the erroneous one.

For more tutorial strategies in ACTIVEMATH exercises, see [5].

5.3   Presentation Strategies

In the Figure 7 different presentation strategies for the same exercise are shown. Both of the strategies allow the learner to add or delete intermediate steps in his calculation, but the first strategy provides a graphical palette and pre-defined templates for kinds of steps he might want to perform.

Another two strategies that are aiming at immediate feedback are shown in the Figure 8 as opposed to the delayed feedback strategies in the Figure 7. In the Figure 8, the screen shot at the left hand side shows similar interface to the left hand side image in the Figure 7, but here the learner can only make one step at a time. The image on the right hand side of the second Figure shows several steps at once, but the learner can not add or remove steps here and has to edit and evaluate all steps in one iteration.

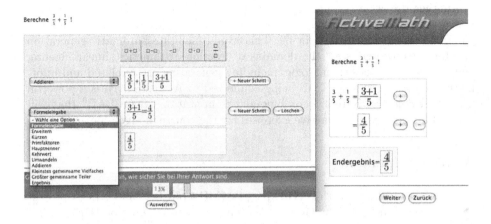

Fig. 7. Different Presentation Strategies for the same exercise

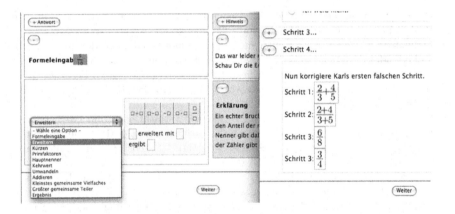

Fig. 8. Different Presentation Strategies for the same exercise

6   Conclusion and Future Work

In this paper we gave an overview of the current updates of knowledge representation for interactive exercises in ACTIVEMATH learning environment, discussed its architecture, and tutorial and presentation strategies that can be automatically applied to exercises.

Currently the tutorial strategies are realized as programs extending the standard `Exercise Generator`. One interesting direction of future work is finding a generic representation for the tutorial strategies instead of programming those.

We hope to be able to derive such representation from the results of currently running research projects such as ATuF and ALOE, aiming to explore how different tutorial strategies for interactive exercises can improve learning.

Another direction of ongoing research is establishing connection to several new domain reasoners developed at Open University of Netherlands, which implement strategies described in [7]. This would facilitate automatic generation of interactive exercises for the domains of fraction arithmetics, linear algebra, and first order propositional logic, including generating error-related and informative feedback. Another CAS being connected to ActiveMath is a powerful Computer Algebra System for polynomial computation SINGULAR[9] that will be used for powering interactive exercises for the course of algebraic geometry.

Acknowledgement

This article results from the ATuF project (ME 1136/5-1) and ALOE project (ME 1136/7) funded by the German National Science Foundation (DFG).

References

1. Dudev, M., González Palomo, A.: Generating Parametrized Exercises, Student Project at the University of Saarland (2007)
2. Faulhaber, A., Melis, E.: An efficient student model based on student performance and metadata. In: Fakotakis, N., Ghallab, M., Spyropoulos, C.D., Avouris, N. (eds.) 18th European Conference on Artificial Intelligence (ECAI 2008). Frontiers in Artificial Intelligence and Applications, vol. 178, pp. 276–280. IOS Press, Amsterdam (2008)
3. Grabowski, B., Gäng, S., Herter, J., Köppen, T.: MathCoach und LaplaceSkript: Ein programmierbarer interaktiver Mathematiktutor mit XML-basierter Skriptsprache. In: Jantke, K.P., Fähnrich, K.-P., Wittig, W.S. (eds.) Leipziger Informatik-Tage. LNI, vol. 72, pp. 211–218. GI (2005)
4. Goguadze, G., González Palomo, A., Melis, E.: Interactivity of Exercises in ActiveMath. In: Looi, C.-K., Jonassen, D., Ikeda, M. (eds.) Towards Sustainable and Scalable Educational Innovations Informed by the Learning Sciences - Sharing Good Practices of Research, Experimentation and Innovation. Frontiers in Artificial Intelligence and Applications, vol. 133, pp. 109–115 (2005) LNCS (2008)
5. Goguadze, G., Melis, E.: One Exercise - Various Tutorial Strategies. In: Woolf, B.P., Aïmeur, E., Nkambou, R., Lajoie, S. (eds.) ITS 2008. LNCS, vol. 5091, pp. 755–757. Springer, Heidelberg (2008)
6. Goguadze, G., Ullrich, C., Melis, E., Siekmann, J., Groß, Ch.: Structure and Metadata Model, LeActiveMathDeliverable D6, LeActiveMathConsortium (2004)
7. Heeren, B., Jeuring, J., van Leeuwen, A., Gerdes, A.: Specifying Strategies for Exercises. In: Autexier, S., Campbell, J., Rubio, J., Sorge, V., Suzuki, M., Wiedijk, F. (eds.) AISC 2008, Calculemus 2008, and MKM 2008. LNCS, vol. 5144, pp. 430–445. Springer, Heidelberg (2008)
8. Klieme, E., Avenarius, H., Blum, W., Döbrich, P., Gruber, H., Prenzel, M., Reiss, K., Riquarts, K., Rost, J., Tenorth, H., Vollmer, H.J.: The development of national educational standards - an expertise. Technical report, Bundesministerium für Bildung und Forschung (2004)

---

[9] see http://www.singular.uni-kl.de

9. Melis, E., Faulhaber, A., Eichelmann, A., Narciss, S.: Interoperable competencies characterizing learning objects. In: Woolf, B.P., Aïmeur, E., Nkambou, R., Lajoie, S. (eds.) ITS 2008. LNCS, vol. 5091, pp. 416–425. Springer, Heidelberg (2008)
10. Narciss, S.: Feedback strategies for interactive learning tasks. In: Spector, J.M., Merrill, M.D., van Merrie"nboer, J.J.G., Driscoll, M.P. (eds.) Handbook of Research on Educational Communications and Technology, 3rd edn. Lawrence Erlbaum, Mahwah
11. Narciss, S., Huth, K.: How to design informative tutoring feedback for multi-media learning. In: Leutner, D., Niegemann, H., Brünken, R. (eds.) Instructional Design for Multimedia Learning, pp. 181–195. Waxmann, Münster (2004)
12. Ullrich, C.: Pedagogically Founded Courseware Generation for Web-Based Learning. LNCS (LNAI), vol. 5260. Springer, Heidelberg (2008)

The Characteristics of Writing Environments for Mathematics: Behavioral Consequences and Implications for Software Design and Usability

Davood G. Gozli[1], Marco Pollanen[2], and Michael Reynolds[1,*]

[1] Dept. of Psychology, Trent University, Peterborough, ON, K9J 7B8, Canada
[2] Dept. of Mathematics, Trent University, Peterborough, ON, K9J 7B8, Canada
marcopollanen@trentu.ca or michaelchanreynolds@trentu.ca

Abstract. Effective communication and collaboration of symbolic and quantitative knowledge requires the digitization of mathematical expressions. The multi-dimensionality of mathematical notation creates a challenge for mathematical software editors. There are two different approaches for handling the multi-dimensionality of mathematical notation: either using a two-dimensional writing environment in which symbols can be placed freely (unit-based) or using an environment in which single-dimensional structural elements can be nested (structure-based). The structure-based approach constrains how users write expressions. These constraints may conflict with how mathematics is normally written. A study is reported that examines how users write mathematical expressions using two graphic based editors: one that is structure-based and one that allows the free-form manipulation of selected symbols in a diagrammatic fashion (unit-based). The results are contrasted with how users handwrite mathematics in a physical medium and implications are drawn for future software design.

1 Introduction

Mathematical expressions are a fundamental tool for representing knowledge. The successful communication of mathematical expressions is heavily dependent on the use of visual representations. Indeed, even verbal communication of mathematics relies heavily on intermediary interfaces such as pen-and-paper or chalk-and-chalkboard. In order to facilitate knowledge transfer and the real-time communication and sharing of mathematical expressions, it is therefore critical that people be able to write mathematical expressions fluently and easily. The widespread use of digital communication technologies for knowledge dissemination, discussion and collaboration, suggests that achieving efficient communication of mathematics requires that mathematical expressions be easily digitized. The purpose of the present paper is to examine how people normally handwrite

---

* The present research was supported by a grant to MR and MP from the Social Sciences and Humanities Research Council of Canada (SSHRC). Authors are in alphabetical order.

J. Carette et al. (Eds.): Calculemus/MKM 2009, LNAI 5625, pp. 310–324, 2009.

mathematical expressions and examine how the characteristics of the writing environments affect how people write.

The digitization of mathematics is usually achieved using personal computers and relies on software programs to overcome limitations in the hardware interface. There are two major challenges for mathematics; (1) a large symbol set (the symbol problem) and (2) mathematical notation has a two-dimensional layout (the layout problem). Broadly speaking there are three different approaches to writing digital expressions; (1) use a text-based description, (2) use a digital pen, or (3) use a palette-based graphic editor. Next we briefly discuss how each of these approaches has solved the symbol problem and the layout problem.

The Symbol Problem

One solution to the symbol problem is to use a keyboard to write text-based synonyms of the graphic symbols. For example, {\cal F} \cap \Upsilon, is the TEX [5] representation for $\mathcal{F} \cap \Upsilon$. The keyboard interface is quick and efficient for this form of writing. However, the mapping between the text-based and the conventional graphic-based representation is not always transparent. Therefore, efficient use of this approach to writing mathematical expressions requires learning a potentially large lexicon of terms (e.g., MathML [1], provides access to over 2,000 symbols). Another solution is to use a keyboard and mouse in combination to select symbols from graphical palettes. This method allows for a transparent mapping between the visual representation of symbols written on paper and the ones written in the digital environment. One obstacle for this type of interface is that graphical symbols compete for a limited amount of space on the display, consequently (1) only a subset of mathematical symbols may be visible at any one time and (2) the symbols may be organized in a manner that is not immediately intuitive. Consequently, it can take a long time to enter even a simple expression if the appropriate symbols cannot be located right away. A third solution to the symbol problem is to use a digital pen so that the user can enter the symbols directly into the digital writing environment. This avoids many of the problems associated with the other two interfaces because the symbols are both transparent and are not hidden from the user. At present, these interfaces are still being refined so that they can correctly recognize the full range of user-drawn symbols (e.g., $\cdot, \bullet, O, o, 0, °, \circ, \odot, \otimes, \oplus, \oslash, \emptyset, \phi, \ominus, \theta, \Theta, \cdots,$), which is a formidable task in computing.

The Layout Problem

Text-only programs tend to use a writing environment that requires that expressions have a one-dimensional layout (i.e., a single string of characters). Consequently, they use nested grouping symbols (e.g., brackets & parentheses) to create sub-expressions so that a formula typically written with a two-dimensional layout can be written in a single dimension. For example,

```
\frac{\frac{a}{b}}{\frac{c}{d}+\frac{e}{f}}
```

is a TEX representation for $\frac{\frac{a}{b}}{\frac{c}{d}+\frac{e}{f}}$. The mapping between this one-dimensional and its traditional two-dimensional layout is not transparent. Thus, users are required to develop a certain level of expertise with the syntax of the language before being able to use the technology properly. In addition, given that the two-dimensional layout often conveys metaphorical properties of an expression, many of these will be lost in a one-dimensional representation [6].

A second solution to the layout problem is provided by digital pen-based technologies for writing mathematical expressions, such as FFES [17], which allow users to draw symbols anywhere on a virtual page. These interfaces have the promise of writing as easily as pen-and-paper with the additional benefit of having the software identify the expression. Great strides have been made in developing this interface, however, the potential for a robust interface for handwritten mathematics has yet to be achieved.

Two solutions to the layout problem have been developed for palette-based editors. The visual display of an expression in these editors has a two-dimensional layout. However, for the majority of palette-based editors the writing environment for the expression consists of nested one-dimensional structures (e.g., Microsoft Equation Editor and BrEdiMa [9]). For example, a fraction is typically created by selecting a fraction structure from the palette. The fraction structure usually inserts a fraction bar into the expression that is bound to empty one-dimensional writing slots above and below it (see [11]). Each slot may then be populated with their own nested sub-expressions. These slots are sub-divisions of the main writing space and are often indicated with outline boxes or background shading. While the graphical presentation of the expression may be two-dimensional for the reader, for the writer this approach can be thought of as a simple extension of the text-only method (i.e., indirect access to the two-dimensional layout) with grouping symbols replaced by slots. *This type of writing environment affords a structure-based writing style* because it constrains the order in which symbols are added to an expression by giving precedence to symbols that affect the expression's layout. Creating a correct visual representation of an expression, therefore, often requires understanding the deep layout structure of the expression and which symbols parse the physical layout of the expression, before writing.

A second solution that exists in palette-based editors is to allow users to "draw" their expressions by placing symbols on a virtual canvas with direct access to a two-dimensional space (i.e., XPRESS [12]). Once the expression is drawn a spatial analysis algorithm, similar to those from pen-based systems, is applied to identify the expression. As the symbols are chosen from palettes and "placed" by the user on a virtual canvas, there is little doubt about the identity of the symbols and their intended locations. This greatly reduces the complexity of expression identification as compared to handwritten mathematics. It also reduces the complexity of editing expressions in that in a two-dimensional space, items are directly accessible and users are not required to first choose among spatial units and then particular items. The two-dimensional writing environment allows users to select symbols from palettes in any order and place them anywhere they wish. *This type of writing environment affords a unit-based writing style* because it does not place any constraints on the order in which symbols are written.

In summary, at least three writing environments have been developed as solutions to the layout problem (1) a one-dimensional layout with multiple embedded substructures, (2) a two-dimensional layout, and (3) a two-dimensional layout constructed from nesting one-dimensional structures.

The Present Study

Although many studies have examined different solutions to writing digital expressions, it has not been possible to disentangle how writing is uniquely affected by solutions to the symbol and the layout problems. The recent development of XPRESS, a palette-based editor with a two-dimensional writing environment, provides a unique opportunity to examine how different solutions to the layout problem affect writing behaviour. The purpose of the present study, therefore, is to examine how the characteristics of a writing environment affect how people write mathematical expressions. Here we focus on two different writing environments that are palette-based.

Although writing environments have their own set of rules governing the types of actions that are permitted and how space is allocated, it is unclear whether this will result in *actual* behavioural differences in how mathematical expressions are written. Here we follow the suggestion that user behaviour can only be understood when (1) the types of behaviours permitted (i.e., affordance), actual use (i.e., effectivity), and goals, motives and perceptions (i.e., intentionality) are considered simultaneously [3,4]. To understand how the characteristics of a writing environment affect how people write expressions, people were observed as they wrote mathematical expressions by hand and using two different graphics based software environments. A handwriting condition was included to assess how people would write the expressions under natural conditions. The two software environments (BrEdiMa and XPRESS) were selected because they use similar interface technologies (i.e., keyboard and mouse) and representations (graphic symbols) to enter mathematical expressions, and therefore only the rules governing how symbols are arranged in the environments are qualitatively different. Novice users were examined to control for expertise. If the editors require that users change their writing style, then the use of novices will allow us to document some of the challenges that they encounter.

2   Method

2.1   Subjects

Seven members of the Cognitive Ethology Lab at Trent University participated in the present study, four of which were undergraduate students, two were graduate students, and one was a faculty member. One subject was left-handed. The subjects were all familiar with simple mathematical and logical notation, although none had previous experience using either of the software editors.

2.2   Stimuli

Eight different mathematical expressions were used (see Table 1). The stimuli were split into two sets, each set was typeset and printed a separate sheet of paper. The first set (Expressions 1–4) had fewer symbols (mean=7.5) than the second set (Expressions 5–8; mean=25.5).

Table 1. The two expression sets used in the study. Expression number is indicated in parenthesis.

Expression Set 1	Expression Set 2
(1) $\overline{A \wedge B \vee C}$	(5) $x = \dfrac{-b \pm \sqrt{b^2 - 4ac}}{2a}$
(2) $A \wedge \overline{\overline{B} \vee \overline{C}}$	(6) $r_{xy} = \dfrac{\sum (x_i - \bar{x})(y_i - \bar{y})}{(n-1)s_x s_y}$
(3) $\dfrac{\sqrt[3]{x^2 + 1}}{2x}$	(7) $r_{xy} = \dfrac{n \sum x_i y_i - \sum x_i \sum y_i}{\sqrt{n \sum x_i^2 - \left(\sum x_i\right)^2} \sqrt{n \sum y_i^2 - \left(\sum y_i\right)^2}}$
(4) $\displaystyle\sum_{i=1}^{100} i^2$	(8) $\displaystyle\lim_{h \to 0} \dfrac{f(x+h) - f(x)}{h}$

2.3   Apparatus

In the handwriting sessions, mathematical expressions were written using dry-erase markers on a 36" x 48" whiteboard affixed to the wall of a small conference room. Although the whiteboard environment is different than paper in a number of aspects (e.g., orientation of the writing surface, thickness of the writing instrument, and physical size of the written expression), the principles of the writing environments are assumed to be the same for the whiteboard and paper. The writing sessions were video recorded using a Canon HG10 video camera. The camera was operated by the experimenter and hand-held to allow for adequate observation of hand movements and written symbols.

In the software writing sessions, mathematical expressions were written in a small office using BrEdiMa and XPRESS. Comparing BrEdiMa and XPRESS allowed us to control many input and representation features and to isolate the difference between one- and two-dimensional writing environments. Both are standalone browser-based editors, and both have an AJAX-based front-end that allows users to create their expression and then submit their expression to a server which returns a LATEX preview. None of the subjects had used either interface before. Although the palettes in XPRESS contain more symbols than those in BrEdiMa (138 to 50), both are minimal editors relative to commercial alternatives.

Data was collected using an Acer personal computer with a core-2-quad processor, a 19" LG Flatron LCD screen and running Windows XP operating system. The editors were run inside the Mozilla Firefox Web browser (version 2.0.0.16). The writing sessions were recorded using in video format using SnagIt 9.0 by TechSmith (http://www.techsmith.com), which recorded all occurrences on the computer screen during a writing trial.

2.4   Procedure

Data collection was spread over six separate sessions (2 expression sets × 3 writing environments) that lasted approximately 20 minutes each. Thus each mathematical expression was written three times (once for each interface) and each interface was used on two separate occasions (once for each expression set). The conditions were always completed in the same order. The short expression set was used in sessions 1–3 and the long-expression set was used in sessions 4–6. Furthermore, the order of the writing environment was the same for each set, first with the whiteboard, then BrEdiMa, and finally XPRESS. On average, there was a 24-hour interval between each session, for each subject. The order of the sessions was chosen to control transfer between writing platforms. For instance, having the subjects use BrEdiMa before XPRESS enabled us to observe the way subjects prefer to write in the latter environment after being exposed to both writing methods.

At the beginning of each session, subjects were given a sheet containing the expressions to be written. Subjects were able to refer to the expressions before and during the writing process. Subjects were instructed to write the expressions one at a time and in the order on the sheet. They were encouraged to go quickly, but not to sacrifice accuracy for speed.

In the handwriting condition, subjects were instructed to signal the experimenter when they were starting an expression and when they were finished writing an expression. The whiteboard was erased between expressions, so that only one expression appeared on the whiteboard at any one time.

In the software editor conditions, the editors were loaded via the Internet before subjects arrived in the lab. Subjects were not given any instruction concerning how to use the interfaces, nor regarding the specific use of keyboard or mouse. However, they were asked to re-load the software using the browsers refresh button, after completing each expression, so that only one expression appears on the screen at any time.

3   Results

Data analysis began by coding over 4,800 discrete behavioural events in the video recordings. An event was defined as any action that had a direct effect on the mathematical symbols represented in the writing environment (e.g., adding a symbol). The timing of an event was linked to when changes occurred in the writing environment. An event did not need to be a correct step towards a successful

Table 2. Overall writing time (in seconds), number of events and time per event for each writing platform and as a function of each type of action event

	Whiteboard	BrEdiMa	XPRESS
Overall			
Total Time	168.9	1007.1	951.6
Number of Events	138.4	218.3	197.1
Time per event	1.2	4.5	4.8
Adding Units			
Total Time	164.9	852.3	684.0
Number of Events	135.9	196.0	152.3
Time per event	1.2	4.3	4.5
Deleting Units			
Total Time	1.4	154.9	37.4
Number of Events	0.7	22.3	10.7
Time per event	2.0	6.9	3.5
Modifiying Units			
Total Time	2.6	NA	230.1
Number of Events	1.9	NA	34.1
Time per event	1.4	NA	6.7

completion of a desired formula. Three broad classes of events were identified; Addition events consisted of all events directly required to add elements to the display; Deletion events included any action that was directly implicated in the removal of a unit; and Modification events included any action that directly changed either the spatial location or physical appearance of an element in the display. The data were analyzed using a repeated measures ANOVA with writing platform (Whiteboard, BrEdiMa, and XPRESS) as the repeated factor. Unless otherwise specified the degrees of freedom for all tests are 2 (treatment) and 12 (error) and significant findings are reliable at the .05 level.

As can be seen in Table 2, the writing environments differed dramatically in how long it took users, on average, to write the expressions (F= 36.4, MSE = 42285). Mathematical expressions were written fastest in the handwriting condition, followed by BrEdiMa and XPRESS, which were not reliably different (t(6) < 1). To better understand why users took longer to write the expressions using BrEdiMa and XPRESS we further examined the number of events that occurred during the writing process and how long users spent per event.

We assumed that changes in the number of events made while writing would indicate a change in the writing process. Consequently, we hypothesized that if the number of events changed across writing environments, then this would indicate that the properties of the different writing environments affected how people write the expressions. Consistent with users changing how they write expressions when using the software interfaces, they required substantially more events compared to the handwriting condition, $F = 21.4$, MSE $= 560$. Although BrEdiMa had 21 more events on average per subject than did XPRESS, the difference was not reliable ($t(6) = 1.5$, $p > .10$).

We also hypothesized that the average time per event would provide insight into how easily users were able to interact with the environment. Unsurprisingly, the average duration of an event was also substantially longer for the software editors compared to the handwriting condition, $F = 64.2$, MSE $= .429$. Given that BrEdiMa and XPRESS are both palette-based editors we anticipated that they would not differ on this dimension. Consistent with the difficulty of navigating the palettes being similar, the software editors did not differ in the average duration of an event, ($t(6) = 1.1$, $p > .10$).

Types of Events

In order to better understand how users were writing the expressions we examined performance as a function of the types of events that people engaged in. Three types of events were examined Additions, Deletions and Modifications.

Addition and Deletion Events

As can be seen in Table 2, users spent substantially more time adding and deleting symbols when using the software editors than they did when using the whiteboard ($F= 28.7$, MSE $= 31333$ and $F=14.0$, MSE $= 3223$, respectively). More time was spent adding and deleting symbols with the software editors because (1) the users made more addition and deletion events with the software editors, ($F = 12.4$, MSE $= 545$ and $F = 25.5$, MSE $= 31.9$, respectively), and (2) it took users more time to execute addition and deletion events with the software editors ($F = 50.0$, MSE $= .467$, and $F = 4.9$, MSE $= 22.6$, respectively).

In order to examine how the characteristics of the writing environments affected performance independent of interface type (pen vs. palette), we compared performance for BrEdiMa and XPRESS. We expected that XPRESS would require fewer addition events because the one-dimensional canvas used in BrEdiMa often requires the user to add new spatial locations for those elements that do not belong to the same structural domain (e.g., a suprascript location). Consistent with this hypothesis, 44 more addition events per subject were made with BrEdiMa than XPRESS ($t(6) = 2.8$, $p < .05$).

Furthermore, there were 12 more deletion events per subject when using BrEdiMa compared to XPRESS, consistent with giving precedence to structure symbols (for creating new one-dimensional writing canvases) increasing the difficulty of writing expressions ($t(6) = 3.3$, $p < .05$). Interestingly, despite taking twice as long to make a deletion event in BrEdiMa compared to XPRESS, the difference was not reliable ($t < 1$). The reason was a large amount of variability in duration of the deletion events in BrEdiMa. As it turns out, deleting or changing parts of an expression in a structure-based environment requires selecting the appropriate space. This can cause confusion since being in one space makes other spaces inaccessible for editing. In response to this inaccessibility, users tended to clear the writing environment (by refreshing the browser) and start over instead of removing an unwanted part of an expression. There were 23 re-starting events in BrEdiMa (more than 3 per subject), by comparison there were none in either XPRESS or handwriting conditions.

Writing Order

The nature of the writing environment can also affect a users writing style. As noted above, a two-dimensional writing space affords a unit-based writing style because symbols can be written in any order and placed at any location. In contrast, constructing a two-dimensional spatial layout using nested one-dimensional canvases affords a structure-based writing style because precedence must be given to symbols that affect the spatial layout of the expression. These predictions were assessed by examining how much variability in the order that symbols were added to the expression (writing order) was explained by the unit- and structure- based writing styles. The initial writing order was used as a measure of how people *attempted* to write the expression. The final writing order was used as a measure of how people ended up writing the expression. The variance explained by each writing style was calculated independently for each subject and each formula.

Our implementation of the unit-based writing style presumed writing order would be left-to-right and top-to-bottom (as opposed to random). Our rationale was that (1) equations are typically read left-to-right, top-to-bottom irrespective of the direction of a cultures text-based writing (e.g., Persian), and (2) this structure is argued to be linked to peoples understanding of the mathematical relationships (see [6]). Similarly, our implementation of the structure-based writing style presumed that precedence would be given to only those symbols that were required to add new one-dimensional writing slots (thus single line operators such as $*$, $+$, or $\div$ were not seen as having special priority). It was assumed that within the one-dimensional structures, a unit-based approach would be employed.

Handwriting. As can be seen in Table 3, the unit-based writing style best captured overall writing order, $F(1, 6) = 26.1$, MSE $= .211$. Indeed, only one person, a computer science major, wrote using the structure-based style. Given that the whiteboard interface very closely approximates writing with a pen and paper we expected that people would not change their writing style while writing an expression. Consistent with this prediction, the unit-based writing style captured both peoples initial- and final- attempts to write an expression equally well ($F(1, 6) = 3.7$, $p > .10$, MSE $= .004$). In order to assess whether people adjusted their writing style, but did so only once while writing the first expression, we examined performance for Expressions 1 and 2 more closely. The unit-based method accounted for 86% of the variability in Expression 1 and 87% in Expression 2, whereas the structure-based method explained only 16% of the variability in Expression 1 and 8% of the variability in Expression 2. Furthermore, there was difference between initial and final writing order ($F<1$), consistent with people not needing to adjust their writing style with this writing environment because it is similar to paper.

Table 3. Amount of variability (R^2) in initial- and final- symbol placement order as a function of writing style (unit-based vs. structure-based) and writing environment

Interface	Attempt	Unit-based	Structure-based
Whiteboard	Initial	.84	.53
	Final	.83	.51
BrEdiMa	Initial	.55	.78
	Final	.41	.92
XPRESS	Initial	.78	.57
	Final	.77	.52

An informal analysis of peoples writing behaviour revealed that deviations from the writing order predicted by the unit-based style primarily arose from violations of our left-to-right and top-to-bottom assumption. For instance, in Expression 3 people would often write the index of the radical after writing the radicand. Another violation is captured with Expression 4 (which was unique in that it was the only expression for which the unit and structure-based styles predicted the same writing order). Despite both writing styles making the same predictions, they only explained 74% of the variability in writing order. The reason: people added the "$\sum$" first and then were essentially random as to whether they would add the initial condition or the upper bound portions of the expression next.

BrEdiMa. As expected, writing order was best captured by the structure-based writing style, $F(1, 6) = 683.4$, $MSE = .011$. Unlike the handwriting condition, writing style tended to change as people wrote each expression. The unit-based writing style best captured peoples initial attempt to write an expression, whereas the final attempt was best captured by the structure-based style, $F(1, 6) = 66.3$, $MSE = .016$. This is consistent with people changing their writing style to give precedence to structure symbols that create the two-dimensional layout.

In order to better examine how people adjusted their writing style, we examined performance for Expressions 1 and 2 more closely. The unit-based writing style best captured the initial attempt at writing Expression 1, explaining 84% of the variability compared to 18% for the structure-based style. The final attempt at writing the expression was best explained by the structure-based writing style (100%) compared to the unit-based style (2%). This suggests that people quickly and efficiently adjusted to the demands of writing a two-dimensional expression using nested one-dimensional slots. This change in writing style persisted when users wrote Expression 2, which is of the same general form as Expression 1. This time, the unit-based style explained less than 1% of the variance in performance whereas the structure-based account explained 99% and the structure based account captured both the initial- and final- attempts. For each of the remaining expressions, however, (except for Expression 4, see below) there was a change between the initial and final writing order. In each case there was a reduction in the explanatory power of the unit-based writing style over time, with the structure-based method increasing in explanatory power. This suggests that subjects initial approach is to employ a unit-based writing style and adapt their approach as the situation warrants.

Deviations from the writing order predicted by the structure-based writing style primarily arose from two sources (1) when individual symbols need to be corrected (see below) and (2) when a symbol added more than one writing dimension and placed the cursor at an unpredicted location. One example of this is captured with Expression 3 in which the cursor was placed inside the radicand instead of at the index location.

XPRESS. Similar to the handwriting condition, writing order was best captured by the unit-based writing style, $F(1, 6) = 8.7$, MSE = .297. It is important to highlight that this was true despite carry over effects from having used the structure-based editor in the previous session. The amount of variability explained by both writing styles decreased from the initial attempt to the final attempt at writing an expression, $F(1, 6) = 17.4$, MSE = .024. This reduction in the explanatory power of both approaches was related to people modifying the appearance of symbols (see below).

Once again we examined performance for Expressions 1 and 2 separately. The initial writing order for Expression 1 was best explained by the unit-based writing style (70%) compared to the structure-based style (30%). The unit-based writing style captured even more variance in the final writing order (82%) compared to a decrease in the structure-based style (22%) suggesting that users were once again changing their writing style. This is consistent with some carry over from writing with a structure-based in BrEdiMa, but that people ultimately preferred the unit-based writing style. This change appears to have stabilized as early as Expression 2, were the unit-based writing style accounted for 82% of the variance in symbol placement order compared to 12% for the structure-based style. There was no difference between the initial and final writing order.

Deviations from the unit- and structure- based writing styles predominantly arose from violations of our assumption that people would write left-to-right and top-to-bottom and were very similar to the violations that occurred in the handwriting condition. Another source of error, above and beyond those observed with in the handwriting condition, concerned the modification of symbols. Sometimes people would replace a symbol if it was not an appropriate size.

Modification Events

Modifications accounted for approximately 1% of all events in the handwriting condition and 17% of all events for XPRESS. Modifications were not observed for BrEdiMa because the overall structure and the spatial relations among symbols are determined automatically by the structure-units that specify the spatial layout. Compared to the handwriting condition, substantially more time was spent modifying symbols in XPRESS, $(F(1, 6) = 41.9$, MSE = .362). This increase in time was a consequence both of more modification events, $(F(1, 6) = 77.7,$ MSE $= .004$), and more time being spent per modification event, $(F(1, 6) = 63.6,$ MSE $= 1.17$). Modification events within the handwriting condition typically consisted of extending the length of horizontal lines. In contrast, almost all types of symbols were resized in XPRESS. In XPRESS it was possible for users to change the structural position of a symbol by dragging it to a new location. Such changes in spatial location would require a deletion and an addition event when using either the whiteboard or BrEdiMa. If the majority of modification events were of this type, then this could have important implications for how we understand the consequences of having to change writing styles. The data were therefore reanalyzed to examine how many events involved changes in layout that were structural in nature and that could be conceptualized as a simple deletion-addition event when using the whiteboard, BrEdiMa and XPRESS (due a movement in the structural position of a symbol as opposed to subtle changes in relative spacing). This analysis revealed that no such events occurred with the whiteboard, 1.7 events with BrEdiMa and 2 events with XPRESS. Together, these data suggest that no change is required in how the addition, deletion, and modification data are understood.

4   General Discussion

The purpose of the present study was to examine how differences in the characteristics of writing environments affect how people write mathematical expressions. Handwriting and writing with the two-dimensional software environment were largely characterized by a unit-based writing style in which individual symbols were added in a left-to-right, top-to-bottom fashion. In contrast, writing with the one-dimensional software platform was characterized by a structure-based writing style in which precedence was given to symbols that created additional one-dimensional writing spaces. Thus, the indirect access to the two-dimensional writing space led users to change how they write. Although users were able to adjust to the demands of the structure-based writing style, there is evidence that it was less than intuitive. First, users seemed to adjust their writing style on an as-needed basis, after encountering problems. This suggests that they are able to remember specific instances were they have had to adjust, but have difficulty generalizing this knowledge to new contexts. Second, users found the environment difficult to navigate; this was most evident in the number of times symbols were deleted and the number of times users elected to rewrite an expression from scratch rather than fix an error. One concern is that the increased cognitive load that arises from having to operate in an unfamiliar writing environment may result in more dramatic performance deficits in time-pressured situations [16]. For instance it is well known that reading performance is dramatically impaired when having to perform a second task, even if it requires independent sensory and effector systems (e.g., [10,13]).

Although the two-dimensional software environment was more intuitive than the one-dimensional editor, users spent approximately the same amount of time using both editors. Users of the two-dimensional writing system spent a substantial amount of time modifying the display. The fact that 17% of the time using XPRESS was spent adjusting the cosmetic features of the users input, despite the software correctly recognizing the correct equations and reformatting the final output through LaTeX, suggests that sizing is an important issue that needs to be addressed in two-dimensional environments. The sizing issue is largely avoided when users handwrite an expression because they tend to draw symbols at an appropriate size. The size of a symbol is typically not an issue for most structure-based editors because sizing is accomplished automatically through the nested structure of the environment. One solution, therefore, to this problem in two-dimensional software environments might be to analyze structure on-the-fly to provide automated assistance with symbol sizing and association.

At present it is unclear how and whether using a pen-based digital interface will result in modification events. Given that handwriting seldom results in modifications (we did not observe any here), we anticipate that modifications would have to arise as a consequence of the handwriting recognition process. Failures recognizing written symbols will require users to clean-up or resize what they have written. Similarly, an error in layout analysis might require a user to manually resize, delete, and/or move a symbol in a written expression.

Although we discussed two violations of the left-to-right, top-to-bottom writing order, more occurred. Documenting violations of the "normal" reading and writing order may help develop a more intuitive structural interface and provide insight into the cognitive factors that influence mathematical writing. In some instances, violations occurred because the standard form of the expression violated the standard writing order (yet some people still apply the left-to-right and top-to-bottom order), as was true of the summation operator. In other instances, it is less clear why violations occurred. For instance, users may have waited to write the index in the third-root component of Expression 4 because (1) users understanding of the expression is incomplete, (2) there is forward momentum in the writing process, (3) users conceptualize the index as analogous to an accent, or (4) users are most familiar with writing the square root, which does not require an index. Presently there is insufficient evidence to discriminate among the many alternatives. One important objective for future research, therefore, is to more thoroughly document when violations of the left-to-right, top-to- bottom assumption occur.

Two additional lines of inquiry that are relevant for software development concern (1) writing expressions from memory and (2) how writing environments affect learning. For instance, in the handwriting condition people wrote brackets in the order they appeared. However, people may wait until all of the necessary symbols are written before using brackets to nest the symbols when writing from memory. With respect to learning, one-dimensional editors may improve fluency in mathematics because they require users to understand the deep structure of an expression. Consistent with this possibility, the nature of a writing technology has been shown to affect how people think about the material they are writing [2]. Interestingly, it may be possible to improve a user's understanding by implementing visual coding of structure given that the abstract meaning of an expression is cognitively related to the details of its actual representation [7,8]. Given that the size of symbols and their relative spacing are often used in mathematics to perceptually organize an expression it may be possible to examine the relative importance of these visual cues by examining how users make cosmetic changes to their written expressions using XPRESS (e.g., $4+4 \times 2+2$ vs. $4 + 4 \times 2 + 2$).

In the present study we examined how the writing environment affects writing behavior, and did not focus on the usability of particular interfaces. However, for future work, this does raise the issue that there are no standard methodologies and benchmarks for the scientific comparison of mathematical input interfaces. In comparison, the usability of text-input interfaces has been well studied. For example, there are widely used methodologies like the Roberts and Moran Methodology [14] that examine the usability of text editors in terms of time, error and learning. Our study does raise the issue about whether a methodology for comparing mathematical input interfaces should also consider if a mathematical interface forces changes in a users writing behavior, potentially increasing their cognitive load, irrespective of differences in performance measures such as time and accuracy.

5 Conclusions

Traditionally interfaces for mathematical expression entry were found mainly in document creation environments and computer algebra systems. The explosion of Web-based technologies has created a demand for new applications, such as online collaboration and assessment tools, many of which can be considered real-time applications. At the same time there have been several recent attempts (e.g., XPRESS, pen-computing) to develop two-dimensional mathematical writing environments. In this paper we have shown that two-dimensional environments allow users to write mathematical expressions in a more intuitive way than one-dimensional environments. Therefore, continued research into two-dimensional interfaces may have important implications for the development of future mathematical interfaces, especially real-time ones, where the main goal is to communicate quickly and effectively.

References

1. Carlisle, D., Ion, P., Miner, R., Poppelier, N., et al.: Mathematical Markup Language (MathML) Version 2.0. W3C Recommendation (February 21, 2001) http://www.w3.org/TR/2001/REC-MathML2-20010221/
2. Edwards, L.D.: Embodying mathematics and science: Microworlds as representations. Journal of mathematical Behavior 17, 53–78 (1998)
3. Gibson, J.: The Ecological Approach to Visual Perception. Lawrence Erblbaum Associates, Inc., Hillsdale (1986)

4. Kadar, E.E., Effken, J.: From discrete actors to goal-directed actions: Toward a process-based methodology for psychology. Philosophical Psychology 18, 353–382 (2005)
5. Knuth, D.E.: The TeX book. Addison-Wesley, Reading (1984)
6. Lakoff, G., Núñez, R.: Where Mathematics Comes From: How the Embodied Mind Brings Mathematics into Being. Basic Books, New York (2000)
7. Landy, D., Goldstone, R.L.: How abstract is symbolic thought? Journal of Experimental Psychology 33, 720–733 (2007)
8. Landy, D., Goldstone, R.L.: Formal notations are diagrams: Evidence from a production task. Memory and Cognition 35(8), 2033–2040 (2007)
9. Nakano, Y., Murao, H.: BrEdiMa: Yet Another Web-browser Tool for Editing Mathematical Expressions. In: Proceedings of MathUI 2006 (2006) (online)
10. O'Malley, S., Reynolds, M.G., Stolz, J.A., Besner, D.: Reading aloud is not automatic: Lexical and sub-lexical spelling to sound translation use central attention. Journal of Experimental Psychology Learning Memory and Cognition 34, 422–429 (2008)
11. Padovani, L., Solmi, R.: An Investigation on the Dynamics of Direct-Manipulation Editors for Mathematics. In: Asperti, A., Bancerek, G., Trybulec, A. (eds.) MKM 2004. LNCS, vol. 3119, pp. 302–316. Springer, Heidelberg (2004)
12. Pollanen, M., Wisniewski, T., Yu, X.: XPRESS: A Novice Interface for the Real-Time Communication of Mathematical Expressions. In: Proceedings of MathUI 2007 (2007) (online)
13. Reynolds, M., Besner, D.: Reading aloud is not automatic: Phonological recoding and lexical activation use central processing capacity. Journal of Experimental Psychology: Human Perception and Performance 32, 799–810 (2006)
14. Roberts, T.L., Moran, T.P.: The evaluation of text editors: methodology and empirical results. Communications of the ACM 26, 265–283 (1983)
15. Smithies, S., Novins, K., Arvo, J.: Handwriting-Based Equation Editor. In: Proceedings of Graphics Interface 1999, pp. 84–91 (1999)
16. Vicente, K.: The Human Factor: Revolutionizing the Way People Live with Technology. Knopf, Toronto (2003)
17. Zanibbi, R., Blostein, D., Cordy, J.R.: Directions in Recognizing Tabular Structures of Handwritten Mathematics Notation. In: Proc. Fourth Int'l IAPR Workshop on Graphics Recognition, pp. 493–499 (2001)

Canonical Forms in Interactive Exercise Assistants

Bastiaan Heeren[1] and Johan Jeuring[1,2]

[1] School of Computer Science, Open Universiteit Nederland
P.O. Box 2960, 6401 DL Heerlen, The Netherlands
{bhr,jje}@ou.nl
[2] Department of Information and Computing Sciences, Universiteit Utrecht

Abstract. Interactive exercise assistants support students in practicing exercises, and acquiring procedural skills. Many mathematical topics can be practiced in such assistants. Ideally, an interactive exercise assistant not only validates final answers, but also comments on intermediate steps submitted by a student, provides hints on how to proceed, and presents worked-out examples. For these purposes, fine control over the symbolic simplification procedures of the underlying mathematical machinery is needed.

In this paper, we introduce views for mathematical expressions. A view defines an equivalence relation by choosing a canonical form of mathematical expressions. We use views to track and recognize intermediate answers, to help in presenting expressions to a user, and to control the granularity of the steps in worked-out examples. We develop the concept of a view, discuss the laws it satisfies, and show how views are composed, which means that they can be used for multiple exercise classes.

1 Introduction

An interactive exercise assistant supports a student who stepwise solves an exercise. A student gets an exercise, for example about solving a system of linear equations, and takes steps towards the solution. Examples of interactive exercise assistants for mathematics are the Digital Mathematics Environment (DWO) of the Freudenthal Institute [5], MathDox [7], Aplusix [6], MathPert [3], WIMS [8], ActiveMath [9], and many more. Here is an example of a series of (correct) steps a student makes when solving a linear equation:

$$
\begin{array}{rcll}
1 - \dfrac{4x+2}{3} & = & 3x - \dfrac{5x-1}{4} & \\
\Longleftrightarrow \quad 12 - 4(4x+2) & = & 36x - 3(5x-1) & \text{times } 12 \\
\Longleftrightarrow \quad 12 - 16x - 8 & = & 36x - 3(5x-1) & \text{distribution} \\
\Longleftrightarrow \quad 12 - 16x - 8 & = & 36x - 15x + 3 & \text{distribution} \\
\Longleftrightarrow \quad 4 - 16x & = & 21x + 3 & \text{merging} \\
\Longleftrightarrow \quad 4 - 37x & = & 3 & \text{minus } 21x \\
\Longleftrightarrow \quad -37x & = & -1 & \text{minus } 4 \\
\Longleftrightarrow \quad x & = & \dfrac{1}{37} & \text{divide by } -37
\end{array}
$$

J. Carette et al. (Eds.): Calculemus/MKM 2009, LNAI 5625, pp. 325–340, 2009.

Most interactive exercise assistants would accept this derivation: they check that each step is correct by calculating that the solution of the equation has not changed. The comments on the right-hand side suggest that a single rewrite rule is applied at each step. However, simplification steps are silently performed at all these steps. For instance, unraveling the simplification of the left-hand side after the first step (multiply both sides by 12) gives:

$$(1 - \frac{4 \cdot x + 2}{3}) \cdot 12$$
$$\Longleftrightarrow \quad 1 \cdot 12 - \frac{4 \cdot x + 2}{3} \cdot 12 \qquad\qquad (a - b) \cdot c = a \cdot c - b \cdot c$$
$$\Longleftrightarrow \quad 12 - \frac{4 \cdot x + 2}{3} \cdot 12 \qquad\qquad \text{constant folding}$$
$$\Longleftrightarrow \quad 12 - \frac{(4 \cdot x + 2) \cdot 12}{3} \qquad\qquad \frac{a}{b} \cdot c = \frac{a \cdot c}{b}$$
$$\Longleftrightarrow \quad 12 - \frac{12 \cdot (4 \cdot x + 2)}{3} \qquad\qquad a \cdot b = b \cdot a$$
$$\Longleftrightarrow \quad 12 - \frac{12}{3} \cdot (4 \cdot x + 2) \qquad\qquad \frac{a \cdot c}{b} = \frac{a}{b} \cdot c$$
$$\Longleftrightarrow \quad 12 - 4 \cdot (4 \cdot x + 2) \qquad\qquad \text{constant folding}$$

The single step in the first derivation actually consists of around 15 basic rewrite steps. Expanding the steps in this derivation would make it very lengthy.

The first derivation shows a sequence of simplified terms that are in some canonical form. A *canonical form* of a mathematical expression is a standard way of (re)presenting that expression. These canonical forms play an important role in interactive exercise assistants, for instance for simplifying terms. The exercise assistants we have tested all have some notion of canonical forms, but their application is often rather subtle.

Most of the exercise assistants mentioned earlier can perform rewrite steps, followed by automatic simplification to some canonical form, and they can check that a student has not changed the solution of the exercise, which would indicate an error. These tools do not have explicit knowledge about strategies for solving the exercise, however. Therefore, they do not check whether the step made by the student is on the optimal path to the solution, whether the student makes progress, or give hints to students that are stuck. For these purposes we use strategies [10] in our feedback services. A strategy for an exercise describes exactly how to stepwise obtain a solution to an exercise. Strategies can be used to monitor progress, to check whether or not a step submitted by a student follows the strategy, to give hints, and to generate worked-out solutions.

Strategies have to include knowledge about canonical forms of expressions: we do not want to show the basic simplification steps in our hints or worked-out solutions, and we also do not want to force students to perform these simple rewriting steps. In this paper, we investigate the following research questions:

- *Economy of rules (Section 2).* How can we describe rewrite rules on a mathematical domain using a limited set of rules? For example, we want the rewrite rule $\frac{a}{b} + \frac{c}{b} = \frac{a + c}{b}$, but not also $-\frac{a}{b} + \frac{c}{b} = \frac{-a + c}{b}$ and $\frac{a}{b} - \frac{c}{b} = \frac{a - c}{b}$.
- *Canonical form (Section 2).* How can we ensure that we only show intuitive representations of expressions to users in worked-out examples? For example, $a + (-b)$ should be presented as $a - b$. And we should never show -0.

- *Granularity (Section 3).* How can we describe rewrite steps of different granularity, to mimic the typical steps users take? Users with different backgrounds will take steps of different granularity: a university student will usually take fewer steps in a calculation than a 10-year old.
- *Recognizing strategy steps (Section 4).* How can we determine that a student has performed a step that matches the step prescribed by the strategy? A user might have performed a step, but forgotten some of the simplification steps we assume. We want to accept automatic simplification, but we also want to accept partly simplified steps.

In this paper we present so-called *views* [18] to address these questions. Views are used to describe and calculate canonical forms, at each step. Our main contributions are the development of views, and the description of a derivation step in terms of a rewrite rule and a view in which the rule is applied. We use the functional programming language Haskell [14] to explain our ideas, and to show some actual code snippets of our implementation.

2 Views

In this section, we gradually explore the concepts of views and canonical forms. Our views are based on the views proposed by Wadler [18]. His views make it possible to combine pattern matching with abstract data types, and have their origin in research on programming languages. We use views for a very different purpose, namely for rewriting in the context of an interactive exercise assistant. Our views abstract over algebraic laws, and help to hide the underlying representation of mathematical objects.

We start by introducing a representation for mathematical expressions in Section 2.1, which we use in an exercise to perform some basic calculations with fractions. This will be our running example throughout this section. We discuss a number of definitions for matching expressions (Section 2.2), and show how these functions can be combined in Section 2.3. In the last two sections we make the concept of a view more precise with some definitions and properties, and we focus on choosing the canonical form of a view.

2.1 Abstract Syntax

We use the following abstract syntax to represent mathematical expressions. Abstract syntax is represented by a data type in Haskell, the programming language in which we have implemented our exercise assistants.

```
data Expr = Nat Integer   | Var String   | Negate Expr
        | Expr :+: Expr | Expr :*: Expr | Expr :-: Expr | Expr :/: Expr
```

Expressions are constructed from the natural numbers (*Nat*) and variables (*Var*), and can be combined into larger expressions using unary negation and the binary operators for addition, multiplication, subtraction, and division. The *Nat* constructor can only have a non-negative number, and we will maintain this invariant. Hence, the constant value -5 is represented by *Negate* (*Nat* 5). This data type is close to the concrete syntax of mathematical expressions, which makes it suitable for interactive exercise assistants since we can truthfully represent terms that are entered by users of the exercise assistants. The disadvantage of this representation is that it complicates the formulation of rules and strategies. We have to deal with atypical expressions, such as $x + (-2)$ or -0, and we want to avoid reporting these to our users.

In the remainder of the paper, we use the infix constructors surrounded by colons for the abstract representation of mathematical objects. Other representations, such as OpenMath [15] and MathML [17], are quite similar, be it more verbose.

2.2   Matching with Views

Consider the exercise of adding two fractions, targeted at primary school pupils. A first step would be to let the fractions have the same denominator, and for this one typically computes the lowest common denominator (*lcd*). Given an expression of type *Expr*, the following function returns its *lcd*:

> *lcd* :: *Expr* → *Maybe Integer*
> *lcd* ((*a* :/: *Nat b*) :+: (*c* :/: *Nat d*)) = *Just* (*lcm b d*)
> *lcd* _   = *Nothing*

where *lcm* is a predefined function which calculates the lowest common multiple of two integers. The function *lcd* is partial, which is reflected by the *Maybe* type constructor. The function only works for expressions of the same form as the left-hand side pattern: for all other values, the function fails in computing the *lcd* (that is, *Nothing* is returned). In fact, our intuitive definition of *lcd* is unsuitable for our *Expr* data type:

- Suppose we also want to use *lcd* when subtracting one fraction from another, e.g., $\frac{2}{3} - \frac{1}{4}$. This requires an extra case for our definition, in which we match on the constructor :−: at top-level.
- What if the first fraction is negative, as in $-\frac{1}{4} + \frac{2}{3}$? In combination with support for subtraction, this requires a substantial number of new cases.
- The denominator can also be negative ($\frac{1}{-4} + \frac{2}{3}$), leading to even more combinations that have to be considered.

In this scenario, pattern matching is not going to work because the number of cases will grow rapidly. Instead, we introduce *views* [18] to gain the flexibility we are searching for, without obscuring *lcd*'s definition. A view allows us to represent a collection of expressions by means of expressions of a particular canonical form. A view consists of two components: a function for mapping an expression to a canonical form, and a function mapping a canonical form back to an expression. We now introduce the former component, and defer the latter to Section 2.4.

Addition :
[A1]	$a + (b + c)$	$= (a + b) + c$
[A2]	$a + b$	$= b + a$
[A3]	$0 + a$	$= a$

Multiplication :
[M1]	$a \cdot (b \cdot c)$	$= (a \cdot b) \cdot c$
[M2]	$a \cdot b$	$= b \cdot a$
[M3]	$0 \cdot a$	$= 0$
[M4]	$1 \cdot a$	$= a$
[M5]	$a \cdot (b + c)$	$= (a \cdot b) + (a \cdot c)$

Equation :
[E1]	$(a = b)$	$=$	$(a + c = b + c)$
[E2]	$(a = b)$	$=$	$(a \cdot c = b \cdot c)$ $(c \neq 0)$

Negation :
[N1]	$-(-a)$	$= a$
[N2]	$a - a$	$= 0$
[N3]	$a - b$	$= a + (-b)$
[N4]	$-(a + b)$	$= (-a) + (-b)$
[N5]	$-(a \cdot b)$	$= (-a) \cdot b$
[N6]	$-(a / b)$	$= (-a) / b$

Division :
[D1]	a / a	$= 1$	$(a \neq 0)$
[D2]	$a / 1$	$= a$	
[D3]	$a / (b / c)$	$= a \cdot (c / b)$	$(c \neq 0)$
[D4]	$(a / b) / c$	$= a / (b \cdot c)$	
[D5]	$a \cdot (b / c)$	$= (a \cdot b) / c$	
[D6]	$(a + b) / c$	$= (a / c) + (b / c)$	

Fig. 1. Basic algebraic laws

Let the type *Match a b* be an abbreviation for a partial function from type *a* to type *b*:

type *Match a b* $= a \rightarrow Maybe\ b$

The intuition is that we view a value of type *a* in some specific way, and possibly as a value of a different type.

At top-level, *lcd* is expecting an addition, and we can apply some algebraic laws to put an expression into the expected form (if possible). Figure 1 lists a number of basic algebraic laws. The function *matchPlus* tries to match a plus at top-level, and uses laws [N3] and [N4] to do so. If it succeeds, it returns a pair containing the operands of the addition.

$matchPlus :: Match\ Expr\ (Expr, Expr)$
$matchPlus\ (a :+: b)\quad = Just\ (a, b)$
$matchPlus\ (a :-: b)\quad = Just\ (a, Negate\ b)$ -- law [N3]
$matchPlus\ (Negate\ a) = \textbf{do}\ (x, y) \leftarrow matchPlus\ a$
$\qquad\qquad\qquad\qquad Just\ (Negate\ x, Negate\ y)$ -- law [N4]
$matchPlus\ \_\qquad\qquad = Nothing$

In the case for negation, we call the function recursively on the negated term. If the call succeeds with a pair (x, y), both operands are negated. Preferably, a helper-function is used (instead of the constructor *Negate*) that removes double negations (law [N1]). More laws could be used in the above definition, such as the distribution rule [M5]. The challenge was to define *lcd* for adding fractions. Given our targeted audience, we want this distribution to be performed by the user prior to the addition. Therefore, we do not incorporate the law in *matchPlus*.

In the same fashion, we introduce a function to match a division. Here, we only push negations into the numerator.

$matchDiv :: Match\ Expr\ (Expr, Expr)$
$matchDiv\ (a :/: b)\quad = Just\ (a, b)$
$matchDiv\ (Negate\ a) = \mathbf{do}\ (x, y) \leftarrow matchDiv\ a$
$\qquad\qquad\qquad\qquad\quad Just\ (Negate\ x, y)\quad \text{-- law [N6]}$
$matchDiv\ \_\qquad\qquad = Nothing$

The third match-function alleviates the problems caused by the *Nat* constructor only accepting non-negative constants. This function matches a natural number preceded by one or more negations, and returns an integer value.

$matchCon :: Match\ Expr\ Integer$
$matchCon\ (Nat\ n)\quad = Just\ n$
$matchCon\ (Negate\ e) = \mathbf{do}\ c \leftarrow matchCon\ e$
$\qquad\qquad\qquad\qquad\quad Just\ (-c)\quad \text{-- constant folding}$
$matchCon\ \_\qquad\qquad = Nothing$

Note that $(-c)$ is the primitive negation operation applied to integer c.

2.3 Composing Match-Functions

With the helper-functions for matching expressions, we can define *lcd*. With some "plumbing" in the *Maybe* monad, this is not too difficult. However, we first present a number of combinators for composing match-functions, which will make it even more straightforward to write *lcd*.

The type constructor *Match* precisely fits the *Arrow* interface [13], which is a general interface to computation. In our case, we modeled partiality by introducing the *Maybe* monad, which turns *Match* into a Kleisli arrow: an arrow of type $a \to m\ b$ for some monad m. The advantage of turning *Match* into an arrow is that this gives us a set of combinators, without too much effort. The combinator ($\ggg$), for example, has type $Match\ a\ b \to Match\ b\ c \to Match\ a\ c$, and allows us to sequentially combine two matches: $m \ggg n$ first matches with m and then with n. Other arrow combinators are ($\ast\ast\ast$), which performs two matches in parallel, and *second*, which performs a match on the second component of a pair.

With the arrow combinators, we define *matchTwoFractions*, which views an expression as the sum of two fractions with constants in the denominators.

$matchTwoFractions :: Match\ Expr\ ((Expr, Integer), (Expr, Integer))$
$matchTwoFractions = matchPlus \ggg (matchFraction \ast\ast\ast matchFraction)$
$\quad\mathbf{where}$
$\qquad matchFraction :: Match\ Expr\ (Expr, Integer)$
$\qquad matchFraction = matchDiv \ggg second\ matchCon$

For each match-function we have made explicit the laws on which it is based. Therefore, it is easy to determine the laws involved in combinations of match-functions such as *matchTwoFractions*. We give an improved definition for *lcd*:

$$lcd :: Expr \rightarrow Maybe\ Integer$$
$$lcd\ e = \mathbf{do}\ ((a, b), (c, d)) \leftarrow matchTwoFractions\ e$$
$$Just\ (lcm\ b\ d)$$

2.4 Defining Views

This section defines views. We explain how views are used to calculate canonical forms, and which properties they satisfy. The definitions are given in Haskell.

So far, only functions for matching have been considered. With each partial function from a to b, we associate a *build* function which returns a value in the original domain. A view pairs a *match* and *build* function.

$$\mathbf{data}\ View\ a\ b = View\{match :: Match\ a\ b, build :: b \rightarrow a\}$$

For each view we assume that the two functions define a canonical form. We make this idea more precise in the definition of the function *canonical*, which returns the canonical form of an element under a given view:

$$canonical :: View\ a\ b \rightarrow a \rightarrow Maybe\ a$$
$$canonical\ view\ a = \mathbf{do}\ b \leftarrow match\ view\ a$$
$$Just\ (build\ view\ b)$$

We apply the *match* function of the view on an element, and on a successful match, we use the *build* function to return to the original domain. For convenience, we also define a simplification function, which returns the value at hand on a failing match:

$$simplify :: View\ a\ b \rightarrow a \rightarrow a$$
$$simplify\ view\ a = fromMaybe\ a\ (canonical\ view\ a)$$

The following properties of the *simplify* function should hold for all views, establishing a property for match and build pairs.

Property 1 (Idempotence). For every view v, *simplify* v is expected to be an idempotent function. If this is not the case, we say that view v is improper.

Property 2 (Soundness). Simplification with a view v should preserve the semantics of an object. Let a be some element in the domain of view v, and let $[\![\,\cdot\,]\!]$ denote the semantics of that domain. Then $[\![a]\!] = [\![simplify\ v\ a]\!]$.

Because each proper view defines a canonical form, it also defines an equivalence relation. Two elements can be tested for equivalence under a view by comparing their canonical forms. We use *simplify* to do the job:

$$viewEquivalent :: Eq\ a \Rightarrow View\ a\ b \rightarrow a \rightarrow a \rightarrow Bool$$
$$viewEquivalent\ view\ x\ y = simplify\ view\ x \equiv simplify\ view\ y$$

The overloaded equality operator $\equiv$ belongs to the *Eq* type class, and is normally implemented as equality on the abstract syntax. Hence, if *view* does not apply to x nor y, *viewEquivalent* tests for syntactic equality of x and y.

The functions for matching can be composed using the arrow interface, and likewise, we can compose views. In fact, we use the same interface for the *View* type constructor, which enables us to combine views. The *build* operation is also an arrow since it is an ordinary function, except in the opposite direction. As a consequence, we cannot implement the *pure* function for a view because we cannot automatically compute the inverse of a function. Views are closely related to the bidirectional arrows proposed by Alimarine et al. [1].

2.5 Choosing the Canonical Form

We continue the example of adding two fractions. Now that we can determine the lowest common denominator of two fractions (*lcd*), we need a rule to scale one of these fractions accordingly. For this purpose, we define *build* functions for *matchCon*, *matchPlus*, and *matchDiv*. The view for positive and negative constants (*conView*) pairs *matchCon* with a function that turns an integer value back into an *Expr* value.

> *conView* :: *View Expr Integer*
> *conView* = *View*{ *match* = *matchCon*, *build* = *buildCon* }

> *buildCon* :: *Integer* → *Expr*
> *buildCon* $n \mid n \geqslant 0$ = *Nat n*
> | *otherwise* = *Negate* (*Nat* (*abs n*))

This definition results in a proper view, and it respects the invariant imposed by the *Nat* constructor. When defining the builder for the plus view (the counterpart of *matchPlus*), we have another look at the algebraic laws in Figure 1. The matching function maps each of the expressions $3 + (-5)$, $3 - 5$, and $-(-3 + 5)$ to the pair $(3, -5)$. In the definition of the builder, we choose $3 - 5$ as the canonical representation for this pair.

> (.+.) :: *Expr* → *Expr* → *Expr*
> *Nat* 0 .+. *b*   = *b*   -- law [A3]
> *a*   .+. *Nat* 0   = *a*   -- law [A3] (and [A2])
> *a*   .+. *Negate b* = *a* :−: *b* -- law [N3]
> *a*   .+. *b*   = *a* :+: *b*

Here, we write the builder as the infix function .+., which should not be confused with the constructor function :+:. With the function *uncurry*, we turn .+. into a function of type (*Expr*, *Expr*) → *Expr*, which we use in the plus view.

> *plusView* :: *View Expr* (*Expr*, *Expr*)
> *plusView* = *View*{ *match* = *matchPlus*, *build* = *uncurry* (.+.) }

The builder function of the division view uses law [N6]: we omit its definition.

We conclude this section with a definition for the rule that scales a fraction to a certain denominator, in which we use a composed view both for matching and for building. For this occasion, we make a view that constrains the numerator and the denominator to be constant.

> $fractionView :: View\ Expr\ (Integer, Integer)$
> $fractionView = divView >>> (conView *** conView)$

The rule that scales a fraction can then be defined as follows:

> $scaleFraction :: Integer \rightarrow Expr \rightarrow Maybe\ Expr$
> $scaleFraction\ n\ e = \textbf{do}\ (a, b) \leftarrow match\ fractionView\ e$
> $\qquad\qquad\qquad\qquad \textbf{let}\ (c, zero) = n\ `divMod`\ b$
> $\qquad\qquad\qquad\qquad guard\ (zero \equiv 0)$
> $\qquad\qquad\qquad\qquad Just\ (build\ fractionView\ (c * a, n))$

We calculate the scale factor (c), and test whether the target value of the denominator (n) is a multiple of the old value (b). Then, we build an expression from the scaled fraction using the same view.

3   Granularity of Rewrite Steps

In this section, we return to the example of the introduction, and we take a closer look at the size (or granularity) of the rewrite steps in the derivation. For some exercises, the steps that a student is expected to take correspond exactly to the laws that are known for that domain. This is, for instance, the case in most exercise assistants in the area of logic, where propositions have to be manipulated using only a handful of rules, typically the ones appearing in textbooks on this subject. In such a scenario, the granularity of user steps is not an issue. In other cases, terms can be simplified automatically without an interest in intermediate values. For example, when performing Gaussian elimination, the focus of the student should be on applying the elementary row operations, not on simplifying the elements appearing in the matrix. It seems reasonable that a tool performs these simplifications automatically.

In the example of solving a linear equation, we are interested in intermediate results, but the steps should be at a conceptually higher level than the algebraic laws listed in Figure 1. Worked-out examples that are generated by the system should be at the right conceptual level (like the derivation in the introduction), just as hints about the direction to go. We start by making some of our assumptions explicit before we discuss the conceptual level of this exercise.

- Associativity of operators is implicit, meaning that a user cannot and should not distinguish $a + (b + c)$ from $(a + b) + c$. The system can thus minimize the use of parentheses in presenting terms. Commutativity, on the other hand, should be used with care. We want to respect the order in which terms appear as much as possible for a better user experience.

view	view type	description
plus View	$(Expr, Expr)$	match an addition (:+:) at top-level
div View	$(Expr, Expr)$	match a division (:/:) at top-level
con View	*Integer*	match a natural number, possibly preceded by some negations
sum View	$[Expr]$	order preserving summation $(e_1 + \ldots + e_n)$
product View	$(Bool, [Expr])$	order preserving multiplication $(e_1 \cdot \ldots \cdot e_n)$: the Bool indicates negation of the product
rational View	*Rational*	reduce by folding constants recursively
linear View	$(Rational, Rational)$	normalize a linear expression in x: use all laws to turn the expression into the form $a \cdot x + b$

Fig. 2. Summary of views on expressions

- Constant terms are normalized aggressively: the skills to manipulate fractions and integers are assumed to be present.
- The distribution of multiplication over addition (law [M5]) is an explicit step in the derivation. Laws to manipulate the sign of a term (laws [N1] up to [N6]) can be performed automatically.

Keeping the assumptions above in mind we define four operations to rewrite an equation until it is in a solved form. In an exercise assistant, these operations could be offered to a user as buttons, allowing the student to focus on the strategy, while the tool is doing the calculations. The operations are:

1. Add a term to both sides of the equation ([E1]). The term can be negative, in which case we are actually performing a subtraction.
2. Multiply both sides by a non-zero constant factor ([E2]): since this exercise is restricted to linear equations there is no point in allowing variables to appear in this factor. Division can be mimicked by multiplying by a fraction.
3. Remove parentheses, i.e., apply the distribution law ([M5]). In the remaining part of this section we make more precise where and how this is done.
4. Merge "similar" terms: this too will be made more precise.

3.1 Sum View and Product View

We define more views that help to implement the operations on an equation. Figure 2 gives a summary of the views on expressions in this paper. The sum view is similar to the plus view defined earlier, except that we now take associativity of the addition operator into account. The sum view converts an expression to a list of terms. Like the plus view, we push negations inside. For example, $3x - (1 - \frac{2x}{5})$ is viewed as a list of three elements, namely $[3x, -1, \frac{2x}{5}]$. The function for matching can be defined as[1]:

---

[1] Although intuitive, a more efficient definition would avoid having to concatenate lists (++) from recursive calls, especially for left-biased abstract syntax trees.

$matchSum :: Match\ Expr\ [Expr]$
$matchSum = Just \circ f\ False$ -- laws [A1], [N1], [N3], [N4]
where $f\ n\ (a :+: b)$ $= f\ n\ a \mathbin{+\!\!\!+} f\ n$ b
   $f\ n\ (a :-: b)$ $= f\ n\ a \mathbin{+\!\!\!+} f\ (\neg\ n)\ b$
   $f\ n\ (Negate\ a) = f\ (\neg\ n)\ a$
   $f\ n\ a$ $= [\textbf{if}\ n\ \textbf{then}\ Negate\ a\ \textbf{else}\ a]$

The first parameter of the helper-function f is a boolean indicating whether or not the expression has to be negated. The function $matchSum$ is total: for an expression without top-level additions, such as $3(x + 1)$, a singleton list is returned. For the builder of the sum view, we pass (.+.) and addition's neutral element to the *foldl* function, which constructs a left-biased tree.

$sumView :: View\ Expr\ [Expr]$
$sumView = View\ matchSum\ (foldl\ (.+.)\ (Nat\ 0))$ -- laws [A1], [A3]

A list is a natural data structure for viewing associative operators. If we also take commutativity into account, we can sort the list, or use the bag (multi-set) data structure. If the operator is also idempotent, such as logical conjunction, we can turn to sets.

We define the product view similarly. Contrary to the sum view, we propagate negations upwards such that we find negations that cancel each other out (law [N1]). The type signature of the product view is $View\ Expr\ (Bool, [Expr])$. The boolean in the pair indicates whether or not the product has to be negated: we omit its definition, but give some examples instead. Matching the expression $3 \cdot (-x \cdot \frac{1}{5})$ gives the pair $(True, [3, x, \frac{1}{5}])$. Although there is no special notation for the reciprocal function, we can also decompose divisors (but we don't have to), thereby also using law [D3] and taking care of its side-condition. The reciprocal function is its own inverse, and plays the same role as negation did for the sum view. The expression $(1 + 1) \cdot \frac{x}{4/7}$ could then be viewed as the pair $(False,$ $[1 + 1, x, \frac{1}{4}, 7])$. The builder of the product view takes care of neutral elements (law [M4]) and absorbing elements (law [M3]).

3.2   Normalizing Sums and Products

We discuss a normalization procedure for a list of expressions produced by the product view. Constant expressions (i.e., terms without variables) can be reduced to a rational number using constant folding techniques. Let us assume that the rational view (of type $View\ Expr\ Rational$) takes care of this. Products are normalized as follows: combine all constant rational numbers, even if they are not adjacent in the list. This operation is sound because multiplication is commutative. The order of the other, non-constant elements is left unchanged. The first occurrence of a constant rational number is replaced by the new, combined constant. Let this procedure be:

$normalizeProduct :: [Expr] \rightarrow [Expr]$

For instance, consider the list $[1 + 1, x, \frac{2}{8}, 7]$. The rational view is applied to each element, giving $[\textit{Just } 2, \textit{Nothing}, \textit{Just } (1 / 4), \textit{Just } 7]$. The product of the constants is $7 / 2$ of type $\textit{Rational}$. We use the rational view to turn this into an expression. This expression is placed in a list before the variable x.

When normalizing sums, we combine constants (using the rational view), but we also merge terms that are "similar". For example, $2x$ and $3x$ should be turned into $5x$ by using the commutative variant of [M5] (from right to left) and constant folding. Product normalization is used for finding similar terms.

Constant folding in sums and products seems straightforward, but preserving the order makes it more involved. We want to emphasize that this is necessary for a tool in order to react naturally on user requests. For example, adding 3 to both sides of the equation $1 + x = 2x - 1$ would (ideally) result in $4 + x = 2x + 2$, even though the constants appear at different sides of the addition operator.

3.3   A Strategy for Solving Linear Equations

We briefly sketch a strategy for solving linear equations using our strategy combinators [10]. A strategy prescribes the order of rewrite steps in a derivation. If both sides of the equation have the form $a \cdot x + b$, then we are done in three steps: move x to the left (law [E1]), move the constant to the right (again law [E1]), and finally scale the equation such that the a on the left-hand side becomes one (law [E2]). Each step can be skipped under certain circumstances. Let this be the basic strategy:

$$basicEquation = try\ varToLeft <*> try\ conToRight <*> try\ scaleToOne$$

Views are used to implement the rewrite steps of the strategy, i.e., $\textit{varToLeft}$, $\textit{conToRight}$, and $\textit{scaleToOne}$, in the same way as $\textit{scaleFraction}$ was defined in Section 2.5. For more involved equations, we first have to apply the distribution rule (law [M5]), after which we merge "similar" terms, and multiply both sides (law [E2]) to get rid of divisions. The overall strategy, which produces the derivation shown in the introduction, is:

$$\begin{aligned} solveEquation\ =\ &repeat\ (merge <|> distribute <|> removeDivision) \\ &<*> basicEquation \end{aligned}$$

4   Recognizing Strategy Steps

In this section, we briefly discuss how interactive exercise assistants can deal with formulas entered by a student. Such a submitted expression can be an intermediate answer in a larger derivation. We do not only want to validate that the intermediate term is correct, but we also want to recognize which rewrite rule has been applied. Three terms are involved in such a diagnosis: the term submitted by the student, the previous term in the derivation, and the term that was expected at this point. We use a strategy definition, such as $\textit{solveEquation}$, to predicted the expected term (possibly more than one).

On a submission, we use an equivalence relation to compare the submitted term with both the previous term and the expected term. We could use the semantic interpretation for checking equivalence, but this only establishes the soundness of the step, and ignores the direction in which the student continues the derivation. Using syntactic equality is also not an option since this test does not take minor differences in representation into account.

The equivalence relation should preferably be congruent, that is, compatible with the semantic interpretation of the symbols [2]. If not, it will be hard to predict whether or not two terms belong to the same equivalence class. The views we have seen operate at top-level: they are shallow. As a result, an equivalence relation that belongs to a view (defined in Section 2.4) is often not congruent. For example, the equivalence relations derived from the plus view and the div view are not congruent. To define a congruence relation using views, we have to recursively apply views.

We return to our running example of solving a linear equation. Each linear expression in x can be written as $a \cdot x + b$, where a and b are expressions in which x does not occur. In the remainder we assume that a and b are both constant rational numbers. We introduce two new views:

linearView   :: *View Expr*   (*Rational, Rational*)
equationView :: *View* (*Equation Expr*) *Rational*

The linear view returns a pair containing the a and b values. This view can easily be extended to the equation view, which first subtracts one side of the equation from the other, then applies the linear view, and finally divides the b value by $-a$. In fact, the equation view can be used as the semantic interpretation of our exercise. The view is not applicable to non-linear terms, or to terms that are not well-formed (e.g., division by zero).

With the equation view, we check whether or not a submitted term is correct. The linear view is used to test if the two sides of the equation still have the same meaning: if this is the case, we can exclude application of an equation rule ([E1] and [E2]). The derivation in the introduction is correct, and indeed, all equations in the derivation are equivalent under the equation view. In the middle part of the derivation, merging and distribution operations are performed. These operations work on expressions, not on equations. The left-hand sides of these equations ($12 - 4(4x + 2)$, $12 - 16x - 8$, and $4 - 16x$) and the right-hand sides ($36x - 3(5x - 1)$, $36x - 15x + 3$, and $21x + 3$) are equivalent under the linear view.

However, if we want to recognize distributions of multiplication over addition (law [M5]), then we need to distinguish $12 - 4(4x + 2)$ from $12 - 16x - 8$. These expressions are equivalent under the linear view. With the help of the sum and product views, and the normalization functions for sums and products, we can define a congruence relation that distinguishes these terms. The details of this relation are omitted from this paper. Merging alike terms, such as $12 - 16x - 8$ becoming $4 - 16x$, results in an expression from the same equivalence class.

We want some congruence relation for recognizing the steps of a user, but which? From a theoretic point of view, this relation should come from an equational theory based on a collection of laws or axioms. This way, we know which laws are available for testing equivalence. Unfortunately, it is not feasible to automatically derive an equivalence relation from a set of laws. Consequently, we have to restrict ourselves to certain collections.

5   Related Work

A popular approach in constructing computer aided assessment (CAA) systems is to delegate all calculations to a computer algebra system (CAS). This approach will give good instant results, since CAS typically have advanced built-in algorithms, and are very good in simplifying complex formulas. These systems are, however, not designed for interaction with a CAA system, and they cannot be configured easily for a finer control of the simplification procedure [11]. This becomes even more of a problem when dealing with interactive exercises.

The purpose of views is related to the design principles of MathPert [3,4]. We follow the guidelines for cognitive fidelity (the software solves the problem as a student does), glassbox computation (you can see how the software solves the problem), and customization of the software to the level of a user.

Beeson [4] claims that rewriting technology [2] is not enough to implement interactive systems that satisfy the above principles. He concludes that every operation has to be implemented as a function in the underlying programming language. We agree with his claim that rewriting alone is insufficient, however, we believe that a function implementing an operation can be given more structure: it is a rewriting step in the context of a view. The advantage of this separation is that we can still see operations as rewrite steps, but in the context of a view. Views can be reused for different exercise classes, and rewrite rules stay simple.

Interactive exercise assistants like the DWO [5] can be used to stepwise solve exercises. Most of these tools have no knowledge of strategies for solving exercises. As a consequence, intermediate steps are only compared against the final solution, and no hints or worked-out examples can be calculated. Most of these tools perform simplifications automatically, with similar results as we obtain. We have not found descriptions of how these tools implement canonical forms.

Proof planners that use computer algebra systems in their proofs run into similar problems as exercise assistants do: the form of the expression returned by the CAS might not coincide with the canonical form expected by the proof assistant. For example, Sorge [16] uses similar techniques as we do in the proof planner Ωmega. No concept of views is introduced though.

6   Conclusions and Future Work

In this paper, we have introduced views for specifying canonical forms. A view consists of a function for matching and one for reconstructing. Reconstruction after matching maps an expression to its canonical form, and matching after reconstruction is the identity function. The arrow combinators can be used to compose views, which makes them reusable for multiple exercise classes.

We have proposed views as a solution to the research questions posed in the introduction. A view defines a canonical form, which is used to show intuitive representations to users. It abstracts over a set of algebraic laws, which we have made explicit for the views introduced in this paper. This is helpful for determining the granularity of a rewrite step, which should correspond to the background of the student, but also for describing rewrite rules without having to worry about slightly different representations. Views help us to recognize expressions entered by students, provide helpful hints on how to proceed with the exercise, and generate worked-out examples with the right level of detail.

Views are useful in any situation where we need canonical forms of expressions: if for some reason $a + (-b)$ is to be preferred over $a - b$, we can define a view that calculates such a canonical form. In a strategy for solving an exercise, multiple views can coexist, for instance to show more detail at the start of a calculation.

The examples presented in this paper are exercises in calculating with fractions, and solving a linear equation. Views are also applicable to exercises outside the domain of mathematics. We are working on interactive exercises assistants for relation algebra and for an introductory programming course, and we believe that views will play a fundamental role within these exercise classes too.

We will proceed our research in the following directions. Multiple domain reasoners for classes of mathematical exercises are to be investigated for the upcoming European MathBridge project. Views will be used for implementing these reasoners. We have integrated our tools with the DWO, such that our step recognition technology can be used. The results are promising, and are expected to be used in mathematics courses in Dutch high schools next year. Further investigation is needed to understand how views can be incorporated in our generalized rewriting framework, in which we use generic programming techniques [12]. More information about our tools can be found on http://ideas.cs.uu.nl/.

Acknowledgements. We gratefully acknowledge discussions with Peter Boon and Wim van Velthoven, the authors of the DWO. Our introductory example was taken from the DWO applet on solving linear equations. We thank Alex Gerdes and the anonymous referees for their useful comments. This work was made possible by the support of the SURF Foundation, the higher education and research partnership organization for Information and Communications Technology (ICT), for the NKBW2 project. For more information about SURF, please visit http://www.surf.nl.

References

1. Alimarine, A., Smetsers, S., van Weelden, A., van Eekelen, M., Plasmeijer, R.: There and back again: arrows for invertible programming. In: Haskell 2005, pp. 86–97. ACM Press, New York (2005)
2. Baader, F., Nipkow, T.: Term Rewriting and All That. Cambridge University Press, Cambridge (1997)
3. Beeson, M.J.: A computerized environment for learning algebra, trigonometry, and calculus. Journal of Artificial Intelligence and Education 1, 65–76 (1990)

4. Beeson, M.J.: Design principles of MathPert: Software to support education in algebra and calculus. In: Kajler, N. (ed.) Computer-Human Interaction in Symbolic Computation, pp. 89–115. Springer, Heidelberg (1998)

5. Boon, P., Drijvers, P.: Algebra en applets, leren en onderwijzen (algebra and applets, learning and teaching, in Dutch) (2005), http://www.fi.uu.nl/publicaties/literatuur/6571.pdf

6. Chaachoua, H., et al.: Aplusix, a learning environment for algebra, actual use and benefits. In: ICME 2004: 10th International Congress on Mathematical Education (May 2008), http://www.itd.cnr.it/telma/papers.php

7. Cohen, A., Cuypers, H., Reinaldo Barreiro, E., Sterk, H.: Interactive mathematical documents on the web. In: Algebra, Geometry and Software Systems, pp. 289–306. Springer, Heidelberg (2003)

8. Gang, X.: WIMS: An interactive mathematics server. Journal of Online Mathematics and its Applications (2001)

9. Goguadze, G., González Palomo, A., Melis, E.: Interactivity of exercises in ActiveMath. In: International Conference on Computers in Education, ICCE 2005 (2005)

10. Heeren, B., Jeuring, J., van Leeuwen, A., Gerdes, A.: Specifying strategies for exercises. In: Autexier, S., Campbell, J., Rubio, J., Sorge, V., Suzuki, M., Wiedijk, F., et al. (eds.) AISC 2008, Calculemus 2008, and MKM 2008. LNCS (LNAI), vol. 5144, pp. 430–445. Springer, Heidelberg (2008)

11. Kyle, J., Sangwin, C.J.: To simplify or not to simplify: that is the question in the CAA of mathematics. In: 3rd International Conference on the Teaching of Mathematics (2006)

12. van Noort, T., Yakushev, A.R., Holdermans, S., Jeuring, J., Heeren, B.: A lightweight approach to datatype-generic rewriting. In: WGP 2008, pp. 13–24. ACM Press, New York (2008)

13. Paterson, R.: Arrows and computation. In: Gibbons, J., de Moor, O. (eds.) The Fun of Programming, Palgrave, pp. 201–222 (2003)

14. Jones, S.P., et al.: Haskell 98, Language and Libraries. The Revised Report. Cambridge University Press, Cambridge (2003); A special issue of the Journal of Functional Programming, http://www.haskell.org/

15. The OpenMath Society. The OpenMath Standard (2006), http://www.openmath.org/standard/index.html

16. Sorge, V.: Non-Trivial Computations in Proof Planning. In: Kirchner, H. (ed.) FroCos 2000. LNCS, vol. 1794, pp. 121–135. Springer, Heidelberg (2000)

17. W3C. MathML 2.0 (2001), http://www.w3.org/Math/

18. Wadler, P.: Views: A way for pattern matching to cohabit with data abstraction. In: POPL, pp. 307–313 (1987)

Spreadsheet Interaction with Frames: Exploring a Mathematical Practice

Andrea Kohlhase[1] and Michael Kohlhase[2]

[1] German Center for Artificial Intelligence (DFKI)
Andrea.Kohlhase@dfki.de
[2] Computer Science, Jacobs University Bremen
m.kohlhase@jacobs-university.de

Abstract. Since Mathematics really is about what mathematicians do, in this paper, we will look at the mathematical practice of *framing*, in which an object of interest is viewed in terms of well-understood mathematical structures. The new perspective not only allows to deepen the understanding of a resp. object, it also facilitates new insights. We propose a model for framing in the context of theory graphs, and show how framing can be exploited to enhance the interaction with MKM systems. We use the framing extension of our SACHS system — a semantic help system for MS Excel — as a concrete example.

1 Introduction

It has often been said that to understand mathematics one has to understand *what mathematicians do*, and in fact the value of a mathematical education is usually appraised more for the practices and abstract skills acquired with it, than for the concrete knowledge gained. For the field of Mathematical Knowledge Management (MKM) this suggests that we have to support mathematical practices in our systems and representation formats (see [KK06] for a call to arms to do just this).

A particular mathematical practice that comes to mind is to view objects of interest in terms of already understood structures and make creative use of this new perspective. For instance, we can understand certain point sets in three-dimensional space by viewing them as zeroes of polynomials. Then we may derive insights about these point sets by studying the algebraic properties of polynomials. For the purposes of this paper we will say that we are **framing** the point sets as algebraic varieties (sets of zeroes of polynomials). Other intuitive examples of framing in mathematics consist e.g. in equipping a differentiable manifold with a (differentiable) group operation (arriving at a Lie group), or interpreting a Boolean algebra as a field of sets via Stone's representation theorem. The practice of framing is so valuable, since it allows to transport insights between seemingly disparate fields. Indeed, in mathematics many of the most famous theorems earn their recognition *because* they establish profitable framings.

We adopt the term 'framing' for the mathematical practice we want to study because we want to highlight the particular approach to context mathematicians

J. Carette et al. (Eds.): Calculemus/MKM 2009, LNAI 5625, pp. 341–356, 2009.
© Springer-Verlag Berlin Heidelberg 2009

choose. In contrast to many MKM applications where 'to contextualize' means to manipulate appearances (presentations), we are interested here in the potential of manipulating substances (representations). We do not want to use the term 'view' since that is already taken in MKM and fails to address either the cognitive process aspect or its collaborative aspect. The term 'frame' has been used e.g. in Communication Research as *"schemata of interpretation that enable individuals to locate, perceive, identify and label occurrences within their life space and the world at large."* [SRWB86]; a frame is understood as a *scaffolding of concepts that influence the understanding of situations.* Therefore it seems to sit well with our demands.

In this paper, we will argue that framing in mathematics usually involves some kind of mapping or even isomorphism between the participating structures. We will propose a model for the mathematical practice of framing in the context of theory graphs, and we will show how framing can be exploited in the interaction design with MKM systems using our extension of the SACHS system — a semantic help system for MS Excel — as an example.

2   Modelling the Practice of Framing

We will set the mathematical practice of framing in the context of theory graphs following the "little theories approach" proposed in [FGT92], in which separate mathematical contexts are represented by separate theories. Structural relationships between contexts are represented as theory morphisms, which serve as conduits for passing information (e.g., axioms, definitions, and theorems) between theories (see [Far00]).

2.1   Semi-formal Theory Graphs and Framing

Theory graphs are one of the theoretical underpinnings of what is sometimes called **Formal Digital Libraries (FDL)**, which have been a focus of the MKM community. FDL have evolved from the libraries of theorem proving and verification systems, and the theory graph structure is used there for modularization by compartmentalizing knowledge about objects into modules (theories) and linking them by inheritance links (morphisms). This aspect seems to be an appealing starting point for modelling framing. But FDL are of rather limited use for mathematicians as most mathematics is not born formal. Indeed, formalization is a very specialized framing practice, which is more often than not at the very end of mathematical creative processes. Therefore, for our purposes we draw on **Semi-Formal Digital Libraries (SFDL)**, where axioms, definitions, theorems, and even theories can be given as annotated text fragments. As semi-formal representation formats like MathML, OpenMath, LaTeX, XHTML+MathML, Math-Lang [KWZ08], MathDox [CCB06] concentrate on mathematical formulae only or lack theory-level features, we will use our OMDoc format [Koh06], which generalizes the structural invariants of theory graphs to an informal level [RK08], but also accommodates fully formal representations. In the following, we will assume an OMDoc-based background SFDL with a fine-grained theory graph structure which acts as a content commons that contains our examples as theory subgraphs.

We will use the formal techniques and results about modular theory graphs from [MAH01, RK09] in an informal setting without checking the various category-theoretic prerequisites. This is generally justifiable by current practice in mathematics (see [BC00] for an extensive discussion): Arguments are presented informally and are considered *rigorous*, if they could in principle be elaborated into a formal system like first-order logic with set theory axioms which does meet the formal prerequisites. Even though such an elaboration is almost never done in practice, enough examples have been carried out that we can be confident that it is possible in all informal but rigorous arguments (in this paper). Thus we use the informal theory graph representation of the OM-Doc format [Koh06], which provides an infrastructure for theory morphisms and inter-theory reasoning without requiring formality.

Let us now briefly recap the salient features of semi-formal theory graphs to make the paper self-contained. A **theory** consists of a **signature** — i.e. a set of concepts or symbols — together with a set of **axioms** — i.e. distinguished members of the set of sentences induced by the signature — which act as basic assumptions of the theory. A signature mapping is called a **theory morphism**, iff all axioms of the source theory are consequences of the target theory's axioms. Thus we can use theory morphisms for the modularization of mathematics: Consider the diagram on the right where we have depicted theories as boxes consisting of the theory name, signature, axioms and theorem morphisms as arrows labeled with i:φ, where i is a name and φ a signature morphism. In our example we have a theory of monoids called Monoid (i.e. structures $\langle G, \circ \rangle$, where G is a set and $\circ: G \times G \to G$ an associative binary operation on G, such that there is an element e with $a \circ e = a$ and $e \circ a = a$). To extend this to a theory of commutative groups, we only have to add axioms for the existence of inverses and commu-

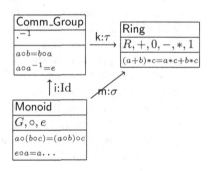

Fig. 1. A Theory Graph for Rings

tativity to the monoid axioms. So in a theory graph we only have to represent these *local* axioms and *import* the ones from Monoid. Note that the identity signature morphism induced by the import becomes a theory morphism by fiat. But we can do even more: to define the theory of rings called Ring, we can just import the Monoid axioms into the Ring theory via a signature morphism $\sigma := \{G \mapsto R^*, \circ \mapsto *, e \mapsto 1\}$ and the Comm_Group axioms via $\tau := \{G \mapsto R, \circ \mapsto +, e \mapsto 0, \cdot^{-1} \mapsto -\}$ and add the distributivity axiom.

The distinguishing property of theory morphisms is that they preserve theorems, i.e. after translation, all theorems from the source theory are theorems of the target theory. This is trivial for the definitional morphisms we have seen above, but also holds for **views**: theory morphisms, where we prove all the **proof obligations** (i.e. the translated axioms of the source) in the target theory. The representation theorems alluded to above give rise to views in this sense. We will need one more notion below: we will call a theory morphism $\sigma: S \to T$ **conservative**, iff $\sigma(s)$ is a theorem of T, iff s is one of S, i.e. the target theory does not introduce new knowledge about objects that can be expressed in terms of the source theory. Note that adding axioms to the target theory will usually render a theory morphism non-conservative; an exception are definitions like $G^*: = G \backslash \{e\}$ in Monoid. The significance of conservative theory morphisms is that any theorem that can be expressed only in terms of the source language can be transported back to the source theory.

Concretely, we *model the framing practice* by defining a **framing** to be the establishment (creating or choosing) of a theory morphism from a source theory (the **framing theory**) into the theory representing the problem (the **framed theory**). The theory morphism itself is called a **frame**. In situations where there is a unique morphism from a theory S to T, we will also say that S is a frame for T in a slight abuse of terminology. But note that in many situations we naturally have more than one morphism between two theories, for instance above we have the morphisms m and k∘i (theory morphisms compose naturally to theory morphisms). Mathematically, m frames Ring in terms of its multiplicative monoid structure and k ∘ i as the additive one. Note that for every theory S, the identity is a theory morphism, we call it the **natural frame** for S. Finally, we will say that frames $f_i: S \to T_i$ are **frame variants**, iff the T_i are pairwise inconsistent. In most practical cases the theories T_i add a single axiom each, e.g. specializing a parameter that was left unspecified in S in different ways. We will call these axioms the **loci** of the variants. We assume that frame variant relations (and their loci) are explicitly annotated in SFDL metadata; see [KMM07] for a proposal to integrate such data into the OMDoc format.

To strengthen our intuition about framing and the suggested model, we will have a closer look at three typical framing practices used in mathematics. From them we will draw more general conclusions concerning SFDL formalization and requirements for the interaction with frames.

2.2   Understanding Abstract Objects by Examples

A fine example of framing is the mathematical practice of supplying examples for abstract concepts. For instance most expositions of the concept of a monoid will give the natural numbers with addition as an example, and use it as a "near-miss" counterexample for being a group.

In [Koh06, section 15.4] we had argued that examples are triples $\langle o, P, A \rangle$, where o is a mathematical object ($\langle \mathbb{N}, + \rangle$ in our example), P is a property (being a monoid), and A is an assertion establishing $P(o)$. Re-interpreting examples as theory morphisms allows to package the same information much more plausibly. Consider the following theory graph fragment, where the theory $\mathbb{N}+$Ex builds on the natural numbers (specified e.g. by the Peano Axioms) and is connected by

a view e to Monoid.
Note that e carries
with it a set of proof
obligations, which to-
gether state the fact
that the structure
$\langle \mathbb{N}, + \rangle$ is a monoid.
To make $\mathbb{N}+\mathsf{Ex}$ into a
counterexample for a

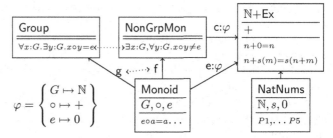

$$\varphi = \left\{ \begin{array}{l} G \mapsto \mathbb{N} \\ \circ \mapsto + \\ e \mapsto 0 \end{array} \right\}$$

group (not all natural numbers have additive inverses) we introduce a theory
NonGrpMon of non-group monoids with a local axiom that states that there is
an element x for which no y is an inverse (note that this is just the negation of
the group axiom). Then any framing c for $\mathbb{N}+\mathsf{Ex}$ naturally acts as a counterexample to the assumption that it is a group, since the local axioms of Group and
NonGrpMon are contradictory. In our terminology, the frames f and g are frame
variants and the local axioms are the variant loci — we show this by the dotted
bidirectional arrow.

2.3   Problem Solving

Another example of framing arises in word problems, i.e. mathematical problems
clothed in words. Problems like the following one appear in many high school
textbook on elementary trigonometry.

> **Problem 0.8.15:** *How can you measure the height of a tree you cannot
> climb, when you only have a protractor and a tape measure at hand.*

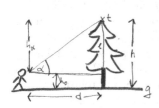

The standard solution is to assume that the tree
in question stands on flat ground, to mark the tree
at eye height and to use the protractor for sighting the top of the tree and the mark to determine
the angle α between the sightings. The tape measure can be used to determine the eye height (h_0)
and the distance d between sighting point and the
center of the tree. Then the height h of the tree is
$h = h_0 + h_\alpha = h_0 + d \tan(\alpha)$ according to the sketch on the right.

Even in this simple situation, framing is complex; consider what happens in
the solution process. The first step is to realize that certain concrete properties
of the problem do not matter, in this case the shape of the tree, its color, and
indeed that it is a tree at all; so in a first framing step, we map the problem to
a simpler one of determining the length of a mathematical line segment without
directly measuring it. The second step in solving the problem is to carefully add
further objects to the problem (e.g. the mark and the sighting point) so that a
solution can be found. And in a third step, the solution is mapped back to the
original problem and verified there.

In our example, we would posit a theory graph like the one on the right,
grounded in a theory **"Planar Geometry (PG)"**, which supplies knowledge

about right triangles, angles, and trigonometry. On this, we build a theory **"Planar Geometry for our Problem (PGP)"** that abstracts from all biological details of trees (they come from a Forestry theory) and only extends PG with two perpendicular line segments l and g and a point p at the end of l. This will be framing theory for our problem; the framing is given by the theory morphism p:φ, where φ maps p to the tree top, l to the center of the tree's trunk, and g to the ground the tree stands on. Since

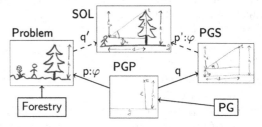

Fig. 2. Framing and Extending a Problem

our problem also inherits from the theory Forestry that contains assumptions like the one that the trunks of (fir) trees are straight and grow vertically, p is a view and thus constitutes a frame in our model. Note that all the theory morphisms in the graph up to now are conservative, so by [MAH01, Proposition 12] we can extend it by the union theory **"Solution (SOL)"** and the two dashed theory morphisms p′ and q′, which are again conservative. SOL contains the full information to understand the solution. As q′ is conservative over Problem, the computed height is the correct one for the problem.

2.4   Problem Transformation

In the third example, we study the contribution of framing to understanding and anchoring of mathematical structures using the well-studied **"Mutilated Checker Board Problem (MCBP)"** (see [Win01, KP06] and references there). The MCBP is based on a combinatorial problem, which we can formalize as a pair covering problem which (following a formulation of McCarthy [Win01]), we can pose as follows. Given a set S and a relation D on S, then we call a relation $R \subseteq D$ a **partial covering**, iff the pairs in R are pairwise disjoint, and a **covering of** S, iff the union of all pairs in R is S. Now the **"Pair Covering Problem (PCP)"** is to find a covering R for a given set S and relation D or to show that no covering exists. We are going to look at two special PCP.

In the **"Adjacent Fields Covering Problem (AFCP)"**, $S \subseteq \mathbb{N} \times \mathbb{N}$, and $\langle\langle i, j\rangle, \langle k, l\rangle\rangle \in D$, iff $|i - k| + |j - l| = 1$. In the **"Disjoint Set Covering Problem (DSCP)"**, S is the disjoint union of sets B and W and $D = B \times W$. In the MCBP S is a mutilated checker board (the squares of the board minus the black ones in the corners) and D is the adjacency relation. Finally, the **"Matchmaker Problem (MMP)"** is given as follows in [Sch09].

In a small but very proper Russian village, there were 32 bachelors and 32 unmarried women. Through tireless efforts, the village matchmaker succeeded in arranging 32 highly satisfactory marriages. The village was proud and happy. Then one drunken Saturday night, two bachelors in a test of strength, stuffed each other with pirogies and died. Can the matchmaker, through some quick arrangements come up with 31 satisfactory marriages among the 62 survivors?

Obviously, the DSCP and AFCP specialize the PCP, and the MMP specializes the DSCP, if we take the set B to be the set of village bachelors and W the set of unmarried women. Similarly, the MCBP specializes the AFCP if we identify checker board squares with their positions in $\mathbb{N} \times \mathbb{N}$, so we have the theory graph on the right. There are two crucial

$$
\begin{array}{ccc}
\text{MMP} & & \text{MCBP} \\
\uparrow c & \xrightarrow{\ e\ } & \uparrow d \\
\text{DSCP} & & \text{AFCP} \\
{}_a\diagdown & & \diagup_b \\
& \text{PCP} &
\end{array}
$$

insights that are important for solving the MCBP and that are driven by framing. The first one is that the DSCP is unsolvable unless $|B| = |W|$, as framing the problem as a matchmaking exercise will make clear even to non-mathematicians. The other insight is that by mapping the sets B and W in DSCP with the set of black and white squares respectively, then we obtain a view e into MCBP[1]. This allows to transport the insight that DSCP is unsolvable to the MCBP.

2.5   Conclusions for SFDL Formats and Interaction with Frames

Let us now see how the theory-morphism based model fares with respect to the different aspects of framing shown in the examples and which insights it provides for SFDL formats as well as interaction design for implementing framing.

The first example uses frames to specialize abstract objects into concrete examples, adding details by fixing the base set of the monoid to $\mathbb{N}$ and the operation to the addition function. At the same time the frame can be used to generalize $\langle \mathbb{N}, + \rangle$ by abstraction. If we want to exploit frames for user interaction in MKM systems, the user should be enabled to select and change frames. In a theory graph a **frame generalization** can be seen as an extension from a frame f to a frame f∘g. In this sense the framing e above can be seen as the generalization c∘f, where we have generalized N+Ex from an example for a non-group monoid to an example of a monoid. Here, the only possible *frame specialization* is taking back this generalization since we cannot change the framed theory. The example also shows that frame variants play an important role in understanding abstract mathematical objects and theories, and should therefore be supported by the interface (see for instance Figure 8 for a concrete example). If variant relations and loci are annotated in a SFDL, the explication of mathematical objects may become a simple planning exercise on theory graphs. Note that in our model of framing we can interpret the practice of giving examples as supplying the reader with a basic supply of prototypical framings (here Monoid is the framing theory for N+Ex) that the reader can later draw upon for problem solving.

In the second example we see that the framing morphism drives problem solving. It opens the real-world situation to methods from Planar Geometry, identifies the salient features, and pulls back the geometric solution into the original problem over a conservative morphism. We can also observe another effect: the opportunity to *model framing in the SFDL* allows us to (partially) tackle the formalization divide. The current practice in formal methods is that informal problem descriptions remain as unstructured texts outside the (formal)

---

[1] Note that even if we frame the set S in the AFCP as squares in a rectilinear grid, they are still uncolored, therefore the target theory of the frame e has to be MCBP.

system. As a consequence, the relation between the original and the formal representation of the problem remains unclear and has to be accepted by a leap of faith. If we view formalization as a framing process in the SFDL, we can support it by MKM systems and take the guessing out of formalization.

In the third example we have used framing in two facilities: for problem solving via conservative extensions, but also in the form of views of the problems into situations that appeal to the intuitions of the (human) reader. This allows to anchor the abstract, mathematical concepts in the real world and thus trigger insights that help problem solving. There is an interesting situation for user interaction: say the user started out with the natural frame for MCBP, which she then generalized to d to view it as an AFCP and then further generalized the frame to d∘b to consider the original problem as a Pair Covering Problem. In this situation she can *specialize* the PCP to a Disjoint Set Covering Problem via a. Formally, we call a frame change g ↦ g' a **frame specialization via f**, iff g' ∘ f = g. And indeed d ∘ b ↦ e is one in our example. But *in* the problem solving phase, framing is not safe, therefore in the envisioned user interface, we need to allow speculative frame specialization. In the example we might want to further specialize d ∘ b to c ∘ a (beyond what is known in the theory graph) to study the Mutilated Checker Board Problem as a Matchmaker Problem and possibly establish a suitable theory morphism that justifies the frame specialization a posteriori.

Note that in all three examples the different, salient aspects of framing could directly be tied to the existence of suitable theory morphisms in the underlying content commons. In the following we will present a first MKM system that illustrates how framing can extend the user interaction.

3   SACHS: A Semantic Help System for **MS Excel**

We will illustrate how framing can extend the user interaction in a semantic help system under development at the German Center for Artificial Intelligence (DFKI), Bremen. The SACHS system (Semantic Annotation for a Controlling Help System [KK08a]), aims to address usability problems in spreadsheet-based applications. For details about the ideas and design decisions behind the SACHS system we refer the reader to our paper *"Compensating the Computational Bias of Spreadsheets with MKM Techniques"* in this volume. We only recap those aspects here that are relevant for our framing extension — which we refer to as "framing-aware SACHS".

A controlling system is a means for the organization to control finances, i.e. to *understand* profits and losses and draw conclusions, thus a lack of overview hampers the process: if users are not sufficiently informed they cannot optimize the company outcome. Even though MS Excel spreadsheets have the potential to serve well as an interface for a financial controlling system, they are more often than not too complex in practice. Even longtime users cannot interpret all data and are not certain about their origins.

A key observation in SACHS is that spreadsheets are *active documents* whose surface structure can adapt to the environment and user input. For SACHS we take a foundational stance and analyze spreadsheets as semantic documents, where the formula representation is the computational part of the semantic relations about how values were obtained. To compensate the diagnosed computational bias we propose to augment the two existing semantic layers of a spreadsheet — the surface structure and the formulae by one that makes the intention of the spreadsheet author explicit. We encode this intention in SFDL and can use it as a basis to provide multi-layered, semantic help services. As we cannot disclose DFKI financial data, we will use the traditional spreadsheet from [Win06] as a running example (see Figure 3).

	A	B	C	D	E	F
1	**Profit and Loss Statement**					
2						
3	(in Millions)		Actual		Projected	
4		1984	1985	1986	1987	1988
5						
6	Revenues	3,865	4,992	5,803	6,022	6,481
7						
8	Expenses					
9	Salaries	0,285	0,337	0,506	0,617	0,705
10	Utilities	0,178	0,303	0,384	0,419	0,551
11	Materials	1,004	1,782	2,046	2,273	2,119
12	Administration	0,281	0,288	0,315	0,368	0,415
13	Other	0,455	0,541	0,674	0,772	0,783
14						
15	Total Expenses	2,203	3,251	3,925	4,449	4,573
16						
17	Profit (Loss)	1,662	1,741	1,878	1,573	1,908

Fig. 3. A Simple Spreadsheet after [Win06]

The central concept we establish is that of a **functional block** in a spreadsheet, i.e. a rectangular region in the grid where the cells can be interpreted as input/output pairs of a *function*. For instance, the cell range [E9:F9] (highlighted with the selection of [E9] by a borderline) is a functional block, since the cells represent projected salary costs as a function π of time; the pair $\langle 1987, 0.617 \rangle$ of values of the cells [E4] and [E9] is one of the pairs of π. The semantic help functionality of the SACHS system is based on an **interpretation**, i.e. a meaning-giving function that maps functional blocks to concepts in a SFDL. For instance our functional block [E9:F9] is interpreted to be the function of the projected salaries in a year for a business which we assume to be available as semantic background.

In [KK08a] we have presented the SACHS information and system architecture, and have shown how the semantic background can be used to give semantic help to the user on several levels like labels, explanations (as showcased in Figure 7) and dependency graphs (see Figure 4 on the right). For example, a user may not be aware that the spreadsheet concerns the profit statement of "SemAnteX Corp.", but can learn this from SACHS's dependency graph feature presented in Figure 4 by selecting cell [E9].

While the information about functional blocks and the meaning of their values (e.g. units), the provenance of data, and the meaning of formulae provided by the semantic background is a nice-to-have, in the development process it became painfully obvious that the interpretation (hence the information provided by the SACHS system to the user) is strongly dependent on the standpoint of the author — how she *frames* the data. In fact even the interpretation into a SFDL itself can be seen as a large frame. Therefore in the work reported in this paper we

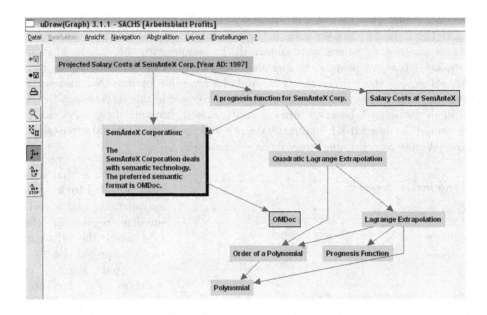

Fig. 4. Dependency Graph with 'uses'-Edges

went one step further and integrated framing as part of the interface to give the user of the spreadsheet more control over the interaction.

3.1   Framing in SACHS

Semantic help systems need various kinds of information: concepts of the system, its user interface, input/output data, etc. For all of these, we need to know a lot about the objects themselves and their relations, i.e. we need ontologies about them. Generally, when we talk about interacting with knowledge-based systems, we have to distinguish knowledge about the system itself from knowledge structures about the domain the system addresses. We consider the first kind of knowledge as part of the *system ontology* and the second kind part of the *domain ontology*.

To distinguish between the system and domain ontologies, the following test suggests itself: anything the system is parametric in must be part of the domain ontology, anything that is particular to the system belongs to the system ontology. For instance, in SACHS the system ontology contains information about concepts like spreadsheet cells, functional blocks, the interpretation, etc. whereas domain ontologies include knowledge about monetary systems, accounting concepts, or prognosis. If the SACHS system were applied to grading spreadsheets, the system ontology that is tied to the underlying spreadsheet application would remain fixed, but the domain ontologies would need to talk about grades, students, semesters,

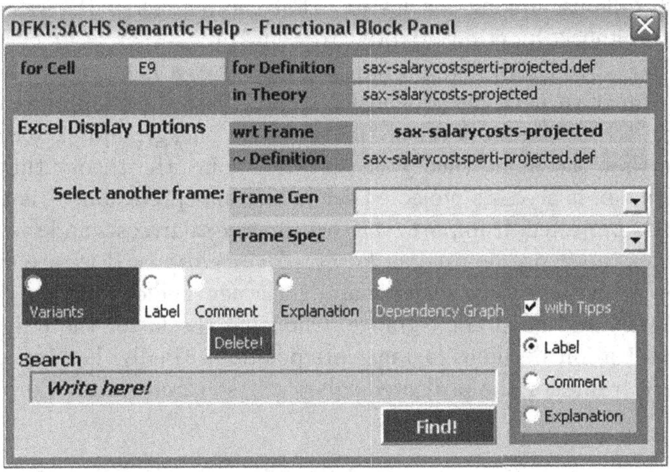

Fig. 5. SACHS's Functional Block Panel

courses, etc. Accordingly in semantics-based systems like SACHS, the domain-level functionality is driven by an explicit representation of the domain ontology — in the case of the SACHS system as an OMDoc-based SFDL.

As a consequence, we can also distinguish system- from domain-level framings in semantic help systems. Domain-level framings are triggered by theory morphisms in the SFDL, whereas the interaction design of the system must account for system-ontology level framings directly. In Figure 5 we find the SACHS panel extended by framing features. Once a cell is selected, the assigned definition in the SFDL with its home theory is shown as the *framed theory*. The natural framing theory determines the *framing theory* in the first step and all the background information is subsequently shown with respect to this frame. On the system-level the user is offered to change the frame via frame *generalization* or frame *specialization*.Moreover, in the field labeled "~Definition" the corresponding definition in the chosen framing theory is presented. The user might also choose to recover domain-dependent *variants* from the semantic background.

To get a better understanding of the role of framing in the interaction with the SACHS system, let us have a closer look at the more specific use example for cell [E9] bearing the dependency graph in Figure 4 in

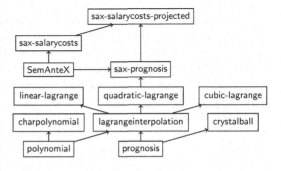

Fig. 6. A Fragment of the SACHS Domain-Ontology Theory Graph

mind, which tells us (among other stuff) that the number 0,617 was computed
a) using a prognosis function adapted to SemAnteX, that is *b)* based on the
Quadratic Lagrange Extrapolation function that is *c)* a Lagrange Extrapolation
that is *d)* a function used for prognosis. To illustrate the framing potential we
have to turn to the theory level of the semantic background sketched in Figure 6. Note that the home theory of cell [E9] — i.e. the theory that contains
the definition sax-salarycosts-projected.def in the interpretation — is the theory
sax-salarycosts-projected. It imports the theories sax-salarycosts and sax-prognosis.
These theories can hence be used as *frame generalizations*. If we are more interested in the latter theory, we select it and get a new choice of frame generalizations SemAnteX and quadratic-lagrange. Choosing the latter the only available
frame generalization becomes lagrangeinterpolation. Finally, here we can select
prognosis as a frame for the projected salary costs at SemAnteX Corp.

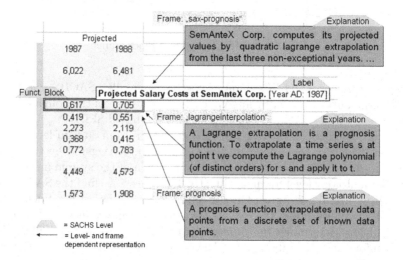

Fig. 7. Explanations within Distinct Frames

Importantly, with each change of frame the semantic information given
to the user changes. For instance, in Figure 7 we can see different explanations for
the same selected cell with respect to the resp. distinct frames. Note that usually
the user can only get the information with respect to the author's framing as the
resp. OMDoc document is fixed and consequently the imports-relation for the
home theory. Another author might have chosen to e.g. import the lagrangeinterpolation theory directly instead of importing the more specific sax-prognosis.
Here, the SACHS panel broadens the user's opportunities and takes back the
rigor and subjectivity of the author's choice of framing.

The set of *frame specializations* wrt. a certain framing theory consists of all
theories that import this framing theory. Frame specializations can supply the
user with surprising insights. For example, the theory prognosis is imported by
the theory crystallball, which offers the prognosis method of sitting in front of a

crystal ball and — disregarding the data set — coming up with a mapping from times to values. With this, the reader may realize that there are always worse possible prognosis functions.

Another interesting service a framing-aware SACHS can offer is the display of *variants*. That is, the concrete framing assumption reified in the MS Excel formula for a cell can be changed. The conventional way to deal with such variants in a spreadsheet is to just replace the formulae in the functional block with new ones and see what the result is; a destructive and error-prone process at best. Given enough background knowledge we can do better. In our example, we have three theories specializing lagrangeinterpolation with concrete Lagrange extrapolations of different order, from which we can derive spreadsheet formulae, which in turn can be entered into the spreadsheet. In the example in Figure 8, we are looking for variants for the '~Definition' lagrangeinterpolation.def in the framing theory for the definition sax-salarycostsperti-projected.def assigned by the author to cell [E9]. Concretely, selecting the option "Variants" in the SACHS panel shown in Figure 5 leads to the opening of the "Variants Panel" demonstrated in Figure 8. We see that there are three possible variants for the Lagrange

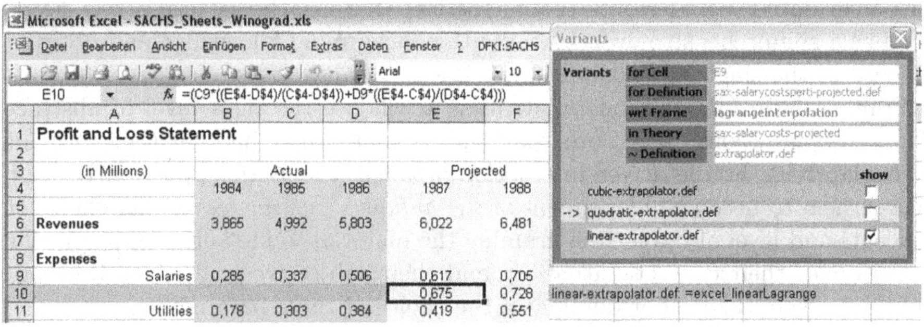

Fig. 8. Frame-based Variants

extrapolation function: the linear, the quadratic, and the cubic Lagrange extrapolations. Remember that the quadratic one was used as the SemAnteX prognosis function, this is marked by the arrow in front of this variant. In the example the user selected the variant linear-extrapolator.def. Once the check box is checked the SACHS system generates new space in the spreadsheet (the light grey row 10 in Figure 8) enabling the presentation of the variant values for the entire functional block. The according variant formula (in the MS Excel formula box at the top of Figure 8) is evaluated. Note that framing influences which concrete variants are available: if we have framed [E9] as the result of a Lagrange extrapolation, we should be allowed to vary the order k of the Lagrange Polynomial (if we have enough data points). If we have however framed [E9] only as the result of a general prognosis function then we should also have crystal ball prognosis at our disposition as a variant.

4   Conclusion

In this paper we have analyzed a common mathematical practice from an MKM perspective, i.e. with an eye towards finding the underlying knowledge structures and representing them in content markup formats so that they can be exploited to support the practices in mathematical software systems. We model this practice of framing a mathematical object as establishing a theory morphism into a theory describing it. We have shown that in many paradigmatic framing cases, the model is able to account for the salient aspects of framing. The theory graph based model is appealing for MKM, since it allows to leverage a large body of existing work.

To test the model further, we have applied it in a situation that is only loosely coupled with classical mathematics: a semantic help system based on spreadsheets. The connection to our model of framing is that the semantic facilities feed on a semiformal digital background library that is theory graph structured. We have shown that taking framings into account in the user interface allows users to find their subjective perspective in the semantic help system. The necessary framing possibilities were naturally present in the background theory graph for our example. We attribute this to the fact that the theory graph was developed as a comprehensive overview over the background knowledge and not just tailored to the single spreadsheet application at hand.

Framing-aware interactions allow users to choose the right level of abstraction of explanations. But note that this is more than just another form of user-adaptivity. Frame-driven interaction broadens the users' opportunities as it allows them to become independent of the *author's framing* — e.g. her choice of concepts and level of rigor — by framing the material to fit their own particular background, their concrete situation, and their subjective goals. In a framing-blind interface, the author dominates the choice of these parameters.

In [KK08b, Section 3.3] we have analyzed requirements for semantic formats to be used in educational technology. In particular, we distinguished three contexts in educational situations: a "content context", a "learner context" and an "interaction context". Usually only the first two are recognized and operationalized in systems. Here, the choice of frames and the navigation between framings are part of the interaction context made explicit in the SACHS user interface. Interestingly, theory graphs that have been thought of as exclusively belonging to the content context now enable a simple formulation of a complex aspect of the interaction context.

Incidentally various learning theories discuss the framing practice as the basis for abstraction processes and ultimately as 'causes' for learning. For example, KLAUS HOLZKAMP argues that every human being engages in an ever-present *"inner dialogue"* [Hol95, p. 25], the result of which turns into her specific actions. The dialogue entertains the idea of at least two distinct frames that inform the learning process. This suggests that framing is also an essential practice in any learning environment, hence the application of this MKM technology might reach much further than the application discussed in this paper.

Finally, framing-aware systems allow the user to explore variants afforded by the background knowledge. In controlling systems this seems to be especially useful to test variant modeling assumptions (like our prognosis functions), but testing variants is a central practice in the sciences and engineering as well.

To close the circle to our introduction, we believe that eventually, the MKM community should build systems that support *what mathematicians do*. In particular, they should exploit theory graphs to support the practice of framing in the mathematical domain proper as we strongly conjecture that such systems will be better suited to re-enliven reified mathematical knowledge.

References

[BC00]   Barendregt, H., Cohen, A.M.: Representing and handling mathematical concepts by humans and machines. In: ISSAC 2000: Proceedings of the 2000 international symposium on Symbolic and algebraic computation. ACM, New York (2000)

[BF06]   Borwein, J.M., Farmer, W.M. (eds.): MKM 2006. LNCS (LNAI), vol. 4108. Springer, Heidelberg (2006)

[CCB06]   Cohen, A.M., Cuypers, H., Barreiro, E.R.: Mathdox: Mathematical documents on the web. In: OMDoc – An open markup format for mathematical documents [Version 1.2] [Koh06], ch. 26.7, pp. 278–282

[Far00]   Farmer, W.: An infrastructure for intertheory reasoning. In: McAllester, D. (ed.) CADE 2000. LNCS (LNAI), vol. 1831, pp. 115–131. Springer, Heidelberg (2000)

[FGT92]   Farmer, W., Guttman, J., Thayer, X.: Little theories. In: Kapur, D. (ed.) CADE 1992. LNCS, vol. 607, pp. 467–581. Springer, Heidelberg (1992)

[Hol95]   Holzkamp, K.: Lernen: Subjektwissenschaftliche Grundlegung. Campus Verlag (1995)

[KK06]   Kohlhase, A., Kohlhase, M.: Communities of Practice in MKM: An Extensional Model. In: Borwein, Farmer (eds.) [BF06], pp. 179–193

[KK08a]   Kohlhase, A., Kohlhase, M.: Compensating the Computational Bias of Spreadsheets with MKM Techniques. In: Carette, J., Dixon, L., Sacerdoti Coen, C., Watt, S.M. (eds.) Calculemus/MKM 2009. LNCS (LNAI), vol. 5625, pp. 357–372. Springer, Heidelberg (2009)

[KK08b]   Kohlhase, A., Kohlhase, M.: Semantic knowledge management for education. Proceedings of the IEEE, Special Issue on Educational Technology 96(6), 970–989 (2008)

[KMM07]   Kohlhase, M., Mahnke, A., Müller, C.: Managing Variants in Document Content and Narrative Structures. In: Hinneburg, A. (ed.) Wissens- und Erfahrungsmanagement LWA (Lernen, Wissensentdeckung und Adaptivität) conference proceedings, pp. 324–229. Martin-Luther-University Halle-Wittenberg (2007)

[Koh06]   Kohlhase, M.: OMDoc – An Open Markup Format for Mathematical Documents [version 1.2]. LNCS (LNAI), vol. 4180. Springer, Heidelberg (2006)

[KP06]   Kerber, M., Pollet, M.: A tough nut for mathematical knowledge management. In: Kohlhase, M. (ed.) MKM 2005. LNCS (LNAI), vol. 3863, pp. 81–95. Springer, Heidelberg (2006)

[KWZ08]   Kamareddine, F., Wells, J.B., Zenglere, C.: Computerizing mathematical text with mathlang. Electron. Notes Theor. Comput. Sci. 205, 5–30 (2008)

[MAH01]   Mossakowski, T., Autexier, S., Hutter, D.: Extending development graphs with hiding. In: Hussmann, H. (ed.) FASE 2001. LNCS, vol. 2029, pp. 269–284. Springer, Heidelberg (2001)

[RK08]   Rabe, F., Kohlhase, M.: An exchange format for modular knowledge. In: Rudnicki, P., Sutcliffe, G. (eds.) Knowledge Exchange: Automated Provers and Proof Assistants (KEAPPA) (November 2008)

[RK09]   Rabe, F., Kohlhase, M.: A web-scalable module system for mathematical theories. The Journal of Symbolic Computation (manuscript, to be submitted) (2009)

[Sch09]   Charles, F.: Schmidt. Productive thinking...the gestalt emphasis (2009), http://www.rci.rutgers.edu/~cfs/305_html/Gestalt/gestalt.html

[SRWB86]   Snow, D.A., Rochford, E.B., Worden, S.K., Benford, R.D.: Frame alignment processes, micromobilization, and movement participation. American Sociological Review 51(4), 464–481 (1986)

[Win01]   Windsteiger, W.: On a solution of the mutilated checkerboard problem using the theorema set theory prover. In: Linton, S., Sebastiani, R. (eds.) Proceedings of the Calculemus 2001 Symposium, Siena, Italy, pp. 28–47 (2001)

[Win06]   Winograd, T.: The spreadsheet. In: Winograd, T., Bennett, J., de Young, L., Hartfield, B. (eds.) Bringing Design to Software, 1996, pp. 228–231. Addison-Wesley, Reading (2006)

Compensating the Computational Bias of Spreadsheets with MKM Techniques

Andrea Kohlhase and Michael Kohlhase

German Center for Artificial Intelligence (DFKI), Bremen, Germany
{Andrea,Michael}.Kohlhase@dfki.de

Abstract. Spreadsheets are mathematical documents that are heavily employed in administration, financial forecasting, education, and science because of their intuitive, flexible, and direct approach to computation. In this paper we show that spreadsheets are interesting applications for MKM techniques which can alleviate usability and maintenance problems as spreadsheet-based applications grow evermore complex and long-lived. We present the software and information architecture of a semantic enhancement of MS Excel spreadsheets that aims at compensating the computational bias in spreadsheets.

1 Introduction

Spreadsheets programs are mathematical software systems: they contain mathematical formulae and are used in financial forecasting, education, and science because of their intuitive, flexible, and direct approach to mathematical computation. It has been estimated that each year tens of millions professionals and managers create hundreds of millions of spreadsheets [Pan00]. This probably makes spreadsheets the most heavily used mathematical software systems at this point of time and should therefore be an interesting testbed for MKM applications. But it seems that the MKM community has not risen to this opportunity; possibly since the mathematical aspects are geared almost exclusively to computational concerns and the declarative aspects of mathematical knowledge that are the concern of the MKM community seem to play a subordinate role in spreadsheets at first glance.

In this paper we show that semantic knowledge management techniques can be used to enhance the interaction with spreadsheets and alleviate usability problems appearing with spreadsheet complexity: in many spreadsheet-based applications even longtime users cannot interpret all data and are not certain about their origins (see [AE06] and its references for a discussion), which often results in errors on the data level and misinterpretation or misapprehension of the underlying model.

In the next section we will start out with an analysis of the semantic layers of spreadsheets. To compensate the diagnosed computational bias we propose to augment the two existing semantic layers of a spreadsheet — the surface structure and the formulae — by a third that makes the intention of the spreadsheet

J. Carette et al. (Eds.): Calculemus/MKM 2009, LNAI 5625, pp. 357–372, 2009.
© Springer-Verlag Berlin Heidelberg 2009

author explicit. In the SACHS project we encode the intention as an accompanying OMDoc [Koh06] document and can thereby provide multi-layered, semantic help services.

2   Semantic Layers in Spreadsheets

Instead of developing the general theory, we will expose the salient parts of our approach using WINOGRAD's example spreadsheet from [Win06] (re-represented in MS Excel in Figure 1 on page 359) as a running example. We will differentiate the three semantic layers in turn and draw conclusions viewing this spreadsheet as both an active and a semantic document.

2.1   Active and Semantic Documents

We call a document **semantic**, iff it contains an infrastructure that distinguishes between content and form. Note that such an infrastructure can range from superficial styling information in PowerPoint slide masters or LaTeX document classes, over RDFa [W3C08] annotations in web pages to formal logic specifications of program behaviors. The idea is that this infrastructure makes relations between the objects described in the document explicit, so that they can be acted upon by machines. In particular, semantic documents can be interpreted by "presentation engines" that operationalize the semantic relations by allowing the reader to interact with various aspects of the semantic properties. We call the combination of a semantic document with a presentation engine that can adapt the surface structure of the document to the environment and user input an **active document**. Our definition is between the concept of embedding semantic networks into hyper-documents employed by GAINES and SHAW in [GS99] and the rather visionary notion of documents that can answer arbitrary questions about their content proposed by HEINRICH and MAURER [HM00]. Crucially both presuppose some kind of content representation in or near the document and a suitable "presentation engine".

For the purposes of this paper, we will neglect the fact that most presentation engines also incorporate editing facilities and concentrate on the interaction with active documents for reading and exploration. This view is similar to what [UCI+06] call *"intelligent documents"*.

A paradigmatic example of an active document is a MATHEMATICA notebook [Wol02], where equations and mathematical objects can be inspected, visualized, and manipulated by the user. Here, the semantic document is written in the MATHEMATICA markup language which includes a content markup scheme for mathematical formulae and a high-level programming language. The presentation engine is the MATHEMATICA front-end which presents interactive documents to the user and calls the MATHEMATICA kernel for evaluation of program fragments and computation of mathematical properties.

Spreadsheets are another paradigmatic class of active documents. Here the semantic document contains representations of the cell values or formulae together with display information such as cell color, font information, and current viewport. The presentation engine is a spreadsheet program like MS Excel, which presents the semantic document to the user by giving it a grid layout and recalculates values from formulae after each update. But what is the underlying semantic model, i.e. what is the "activeness" of spreadsheets based on?

2.2   The Surface/Data Layer

If we look at the example in Figure 1, we see that the grid of cells can be roughly divided into three areas. The darker, ochre area in the center contains values of actual and past expenses and revenues; the lighter, yellow box on the right contains values projected from these. The white region that surrounds both boxes supplies explanatory text or header information that helps users to interpret these numbers. Generally, non-empty cells that do not contain input or computed values usually contain text strings that give auxiliary information on the cells that do; we call these cells collectively the **legend** of the spreadsheet, since they serve the same purpose as the legend of a map.

Observe that row 17 displays the central values of the spreadsheet: the profit/loss situation over time (i.e., in the years 1984-1988 as indicated by the values in row 4). Moreover note that the meaning of the values in row 17 is that they represent profits and losses as a *function* π of time: recall that a function is a right-unique relation — i.e., a set of pairs of input values and output values. In our example the pair $\langle 1984, 1.662 \rangle$ of values of the cells [B4] and [B17] is one of the pairs of π. We will call

	A	B	C	D	E	F
1	**Profit and Loss Statement**					
2						
3	(in Millions)		Actual		Projected	
4		1984	1985	1986	1987	1988
5						
6	Revenues	3,865	4,992	5,803	6,022	6,481
7						
8	Expenses					
9	Salaries	0,285	0,337	0,506	0,617	0,705
10	Utilities	0,178	0,303	0,384	0,419	0,551
11	Materials	1,004	1,782	2,046	2,273	2,119
12	Administration	0,281	0,288	0,315	0,368	0,415
13	Other	0,455	0,541	0,674	0,772	0,783
14						
15	Total Expenses	2,203	3,251	3,925	4,449	4,573
16						
17	Profit (Loss)	1,662	1,741	1,878	1,573	1,908
18						

Fig. 1. A Simple Spreadsheet after [Win06]

such a grid region a **functional block**, and the function it corresponds to its **intended function**. Empirically, all non-legend, semantically relevant cells of a spreadsheet can be assigned to a functional block, so we will speak of *the* functional block and *the* intended function of a cell.

Often a functional block consists of multiple rows and columns and represents a binary function whose values depend on two (main) parameters which are usually in the cells of the column on the left and the row on top of the block. In our example the block with cells [B9:D13] represents a binary function that ranges over expense categories (given in [A9:A13]) and years (given in [B4:D4]). In the general case, the intended function of a functional block can have any arity; its arguments again correspond to functional blocks, which we call **input blocks**.

Our notion of a functional block is related to, but different from ABRAHAM and ERWIG's of a *"table"* [AE04, AE06], which also contains row and column header and footer blocks. Our functional blocks roughly correspond to their *"table core"*, whereas we would consider their header blocks as input blocks or legend cells (but there may be non-header input blocks in our model) and their footer blocks which contain aggregation cells as separate functional blocks.

2.3   The Formula Layer

A spreadsheet cell c may not only be associated with a simple data item (or **value**) $[\![c]\!]$, it may also be connected with a **formula** $[\![\![c]\!]\!]$, which evaluates to the cell value. A formula is an expression built up from constants, an extended set of numeric and logic operators, and references to other cells.

In our example, the value of π can be computed from the yearly revenues in $\mathcal{R}:= [\text{B6:F6}]$ and the total expenses in $\mathcal{E}:= [\text{B15:F15}]$ by a simple subtraction, the total expenses can in turn be computed by summing up the various particular expense categories listed in cells $[\text{A9:A13}]$.

Note that the formulae of cells in a functional block have to be *"cp-similar"* [BSRR02], i.e., they can be transformed/copied into each other by adjusting the respective rows and columns. We call a functional block **computed** if all of its formulae are cp-similar. In our example, the functional block $\mathcal{P}:= [\text{B17:F17}]$ is computed: let γ range over the columns B to F in $\mathcal{P}$. Note that the formulae $[\![\![\gamma 17]\!]\!] = \gamma 6 - \gamma 15$ in cells $[\gamma 17]$ are indeed cp-similar. Together, they make up the function

$$\mathcal{F}(\mathcal{P}):= \{\langle [\![\gamma 6]\!], [\![\gamma 15]\!], [\![\gamma 17]\!]\rangle | \gamma \in \{\text{B}, \ldots, \text{F}\}\}$$

We call $\mathcal{F}(\mathcal{P})$ the function **induced by** the (formulae in) block $\mathcal{P}$. But we also observe that not all functional blocks in a spreadsheet are computed, for instance the formulae in the block $[\text{B9:D13}]$ are all different constants representing the measured values, so they cannot be cp-similar. We call such blocks **data blocks** and note that the property of being a functional block only depends on a functional correspondence (a conceptual aspect of the data) and not on the existence of formulae (a property of the spreadsheet).

With spreadsheet formulae, users can express data dependencies on a generic level, so that the spreadsheet program can do much computational work in the background. By this virtualization of the traditional ledger sheet (see above), the user's role is lifted to a layman programmer and offers according potential. But ABRAHAM and ERWIG report an error rate of up to 90% (!) in spreadsheets [AE06], which shows that this potential comes with a substantial risk. They analyze the source of many of these errors to be in a mismatch between what the spreadsheet author wants to express and the formulae he writes. They try to address this situation by static analysis techniques (type checking) of the formulae and supplying the author with *"spreadsheet templates"*. To understand this mismatch better, let us now turn to the model the author intends to convey.

2.4   The Intention Layer

Note that even though $\mathcal{F}(\mathcal{P})$ and π compute the same values, they are completely different functions. π is *defined* via the actual or projected profits or losses of an organization, while $\mathcal{F}(\mathcal{P})$ is a finite, partial binary arithmetic function. Even when we compose $\mathcal{F}(\mathcal{P})$ with $\mathcal{F}(\mathcal{R})$ and $\mathcal{F}(\mathcal{E})$ and restrict them to the years 1984-86 yielding $\mathcal{F} := \mathcal{F}(\mathcal{P}) \circ \langle \mathcal{F}(\mathcal{R}), \mathcal{F}(\mathcal{E}) \rangle |_{\text{[B4:D4]}}$, the functions $\mathcal{F}$ and π are only *extensionally* equal (they are equal as input/output relations) and still differ *intensionally*.

Surprisingly, only $\mathcal{F}$ is explicitly represented in the spreadsheet of Figure 1, moreover, this explicit representation is invisible to the user if she doesn't look at the formula boxes — thus, leaving the user to figure out the 'intention' (the function π) from the implicit information given in the white part by herself. This is why we speak of a **computational bias** of spreadsheet programs, as some layers of the semantics are explicated but others are not.

Generally, we can assume that spreadsheet program authors use spreadsheets to express and compute (measurable) properties of situations; if we look a little closer then we see that these are not properties of the world as such, but of a high-level, abstract, or mental model of the world, which we subsume under the term **intention of the spreadsheet**. In our example, the function π could be seen as a concept from the intention, whereas the function $\mathcal{F}$ can be seen as its implementation. In our simple example the intention is easy to deduce from the text in the legend and basic financial accounting knowledge.

But even here, some parts of the spreadsheet's intention remain unclear: e.g. for what company or department are the profits and losses computed or what are the methods of projection for the years 1987/8. Let us now take stock of what the cells in the spreadsheet mean and what information we need to infer from this. As we already remarked above, the values of cells [B17:D17] are (the scalar parts of) the actual profits/losses in the years 1984-1986. We need information from cell [A3] for the unit of measurement, from cells [B3:D3] that they are actual, and from [A17] for the interpretation as a 'profit/loss'. To understand the full meaning of these cells, we also need to know about profits and losses of companies — e.g. that high profits of a company I am employed by or that I own stock in are good for me, the fact that the company is based in the Europe and therefore calculates finances in €, and that values that are actual are computed from measured values. Finally, we need to know that the profit/loss of an organization over a time interval is *defined* as the difference between its revenues and expenses over this interval. This knowledge allows to compute the values of cells in $\mathcal{P}$ with the respective formulae from the values of cells in $\mathcal{R} \cup \mathcal{E}$ (i.e., using the function $\mathcal{F}$). The values of the cells in $\mathcal{E}$ can be similarly computed from the values of the cells [B9:D13]. Note that while the definition of profits and losses above is general accounting knowledge, this definition is particular to the respective company, as the applicable expenses vary with the organization.

A similar account can be given for the projected profits/losses in cells [E17:F17], only that the interpretation of the cells wrt. the intention is even more difficult — even though the situation is simple if taken at face value. Cell

[[E17]] is the projected profit in the year 1987, which is computed from the revenue and expenses in column E. But in contrast to the values in the actual block [B6:D6] ∪ [B9:D13], the values in the projected block [E6:F6] ∪ [E9:F13] are not measured, but projected from the actual values by some financial forecasting method that is reflected in the respective formulae. Note that the correspondence of the formula need not be as direct as in the case of the total expenses above. It might be that the forecasting method is defined abstractly, and the concrete formula is derived from it making some simplifying assumptions. Furthermore, to fully understand the values we need to know what assumptions the forecasting method makes, what parameter values are employed and why, how reliable it is, etc. All of these concerns are not addressed at all in the spreadsheet as an active document. ABRAHAM and ERWIG describe this situation as follows:

> There is a high level of ambiguity associated with spreadsheet template inference since spreadsheets are the result of a mapping of higher-level abstract models in the user's mind to a simple two-dimensional grid structure. Moreover, spreadsheets do not impose any restrictions on how the users map their mental models to the two-dimensional grid (flexibility is one of the main reasons for the popularity of spreadsheets). Therefore the relationship between the model and the spreadsheet is essentially many-to-many [...].   [AE06, p. 5]

3   Compensating the Computational Bias

Our analysis of the example above has shown us that large parts of the intention of a spreadsheet is left implicit, even though it is crucial for a user's comprehension. In particular, a user needs to know the following for a spreadsheet:

- The **ontology**, i.e., background information about relations between concepts and objects of the intention. The objects in the intention include the functions represented in the spreadsheets e.g. π, their properties, e.g. the *units* of their arguments and values, and thus of the values in the cells.
- The **provenance** of data in a cell, i.e., how the value of this data point was obtained, e.g. by direct measurement, by computation from other values via a spreadsheet formula, or by import from another source; see [MGM+08] for a general discussion of provenance.
- The **interpretation**, i.e., a correspondence between functional blocks and concepts or objects of the intention. We distinguish three parts here
  - The **functional interpretation**, that specifies the intended function of the functional block.
  - The **value interpretation**, i.e. a bijective function that specifies how to interpret the values of the block cells as ontology objects.
  - The **formula interpretation** that links the formulae of a block to an object in the ontology. This mapping must be a *refinement* in the sense that the interpretations of proper formulae compute the same value as the formulae itself and the pseudo-formulae input is mapped to a provenance object.

In some spreadsheets that are the digital equivalent to "back-of-the-envelope calculations", the interpretation, provenance, and ontology information is simple to infer, so that the un-documented situation is quite tolerable. Indeed this shows the cognitive strength of the table metaphor, in our example it is no problem for the human reader to interpret the legend item "*(in Millions)*" as a specification of the value interpretation of the cells [B6:F17] (but not of the years $\mathcal{Y}$:= [B4:D4]).

In many cases spreadsheets have developed into mission-critical tools that are shared amongst whole departments, because they encapsulate important, non-trivial, institutional knowledge. The intention of such spreadsheets is much harder to infer, a fact that is witnessed by the fact that companies spend considerable energy to train employees in the usage (and intention) of such spreadsheets.

In this situation, it would be natural to make spreadsheets even more active to support the user's comprehension of the spreadsheet intention. Thus, in light of the discussion of Section 2.1, we suggest to compensate for the computational bias diagnosed above by extending the underlying semantic document of a spreadsheet. Concretely, we propose to represent the intention (as the provenance and ontology) and to tie the cells in the spreadsheets to concepts in the intention (via an interpretation function).

3.1 Fixing the Ontology

We have (at least) two possibilities to extend spreadsheets with an ontology and provenance component: we can either extend spreadsheets by ontology and provenance facilities or we can extend them by interpretation facilities that reference external representations of the intention. As we have seen above, the intention contains quite a lot of information, and making it explicit in a software framework means a large investment. Therefore we contend that an external representation of the intention is more sensible, since it can leverage pre-existing tools and profit from interoperability.

We use the OMDoc format (Open Mathematical Documents, see [Koh06]) to represent the intention model. OMDoc is an XML-based format for representing semi-formal, semantically structured document collections. It allows to factorize knowledge into "content dictionaries" that serve as constitutive contexts for knowledge elements. OMDoc provides a mathematically inspired infrastructure for knowledge representation: document fragments are classified by their epistemic role as e.g. axioms, definitions, assertions, and proofs and mathematical formulae are represented in the OPENMATH [BCC+04] or MATHML [ABC+03] formats. Furthermore, OMDoc provides a strong, logically sound module system based on structured "theories" (content dictionaries (CD) extended by concept inheritance and views) [RK09]. Finally, the language has been extended to deal with units and measurable quantities [HKS06] as a prerequisite for interacting with the physical world. We make use of all of these features for modeling the intentions of spreadsheets. In contrast to other ontology modeling languages like OWL [MvH04], the OMDoc format does not commit to a formal logical language, and therefore lacks a native concept of inference but also does not force the author to fully formalize the spreadsheet intention and to work around the expressivity limitations of the underlying logical system. Instead, OMDoc allows to locally formalize elements — and thus provide partial inference — with whatever formal system is most suitable; in our application, we mainly use an formal system for arithmetics as a counterpart for spreadsheet formulae.

For the intention model in our example we divide the background knowledge into theories that inherit functionalities from each other via the `imports` relation. At the very basis we would have a CD Revenues that defines the concept of the revenue of an organization over a time interval. This theory defines the concept of a binary revenue function ρ, such that given an organization o and a natural number n the value $\rho(o, n)$ is the revenue (as a monetary quantity) of o over the span of the year n (AD) in the Gregorian calendar. Note that we use this very naive notion of revenues for didactic purposes only. For instance ρ was chosen as a binary function to highlight that there is no automatic agreement between functional correspondences in the table and objects of the intention. We would be perfectly free to analyze the concepts more thoroughly, embarking into representing monetary systems, theories of time, etc. For the purposes of this paper, we assume that we can either appeal to the intuition of the user or inherit these representations from a suitable foundational ontology.

In the same way we proceed with a CD Expenses, which imports from a CD Salaries. Finally, we build a CD Profits that imports from both. In the OMDoc document pictured in Figure 2 we have summarized some of the relevant CDs and the concepts they introduce.

3.2   Fixing the Provenance

We enrich our ontology with provenance information: As we have required the formula interpretation to be a refinement, we need to represent an abstract notion of spreadsheet computation in the ontology. This can be readily done making use of e.g. the CASL libraries [CoF04]. For modeling the provenance of user inputs in spreadsheets, we can be extremely minimalistic, just establishing a stub content dictionary that lifts the concept of "user input" to the ontology level. But in our example we can already see what a more elaborate provenance model could give us: We could specify that the salary values in [B9:F9] are not only user inputs, but really manually copied over from another spreadsheet — the top spreadsheet "Salaries" in Figure 2. To take advantage of this (see details in the next section) we have to develop CDs for provenance, adapting and extending first formalizations reported on in [MGM⁺08]. As this goes beyond the scope of this paper, we leave this to future work.

3.3   Fixing the Interpretation

To interpret the cells in $\mathcal{P}$ for example, we need to

- *fix the functional interpretation*: Identify that $\mathcal{P}$ and $\mathcal{R}$ form functional blocks with input row $\mathcal{Y}$.

In Section 2.2 we have already discussed the functional interpretation of $\mathcal{P}$: it is just the intended function π. Similarly, we link the revenues block $\mathcal{R}$ with a unary function $\rho(c, \cdot)$, where c is the representation of our example company. Then we have to express the semantic counterpart of the spreadsheet formulae. In our OMDoc format we can simply represent this function as $\lambda y.\rho(c, y)$ in OPENMATH or MATHML.

– *fix the value interpretation*: In our example we observe that the values in $\mathcal{P}$ are actually only the scalar parts of measurable quantities, in this case measured "in millions" (of € presumably) according to the spreadsheet legend. Similarly, the (string) values Salaries, Utilities, ... in [A9:A13] have to be interpreted as objects in the ontology. Thus in our example, we choose $i := \{y \mapsto y(AD)\}$ as the value interpretation of block $\mathcal{Y}$ and $j := \{x \mapsto 10^6 x \text{€}\}$ for block $\mathcal{P}$; obviously both are bijective.

– *fix the formula interpretation*: our example the formulae $\gamma 6 - \gamma 15$ in the functional block $\mathcal{P}$ would be linked to the formula $\pi(year) = \rho(year) - \epsilon(year)$ in the Profit/Year definition. Furthermore, we would link $\mathcal{R}$ to a semantic provenance object "imported from Salaries.xsl".

In Figure 2 we show the functional interpretation mappings as red, dotted arrows and the formula interpretation as purple, dot-dash-arrows. The totality of cell interpretations in a spreadsheet induces an associated set of CDs we call the **intention model**. Note that this is indeed a representation of the intention of this spreadsheet.

We can think of the value interpretations as parser/generator pairs that mediate between the scalar function represented by the formulae in the spreadsheet and the intended function in the intention — which is usually a function between measurable quantities. In particular the functions $\mathcal{F}$ and π are related via the commutative diagram on the right , where the function $\mathcal{F}$ is induced by the spreadsheet formulae as discussed in Section 2.4 above. We see that the three components of the interpretation fully specify the correspondence between functional blocks in the spreadsheet and objects induced by the intention model. To see the strength of this construction let us return to our example and look at the import of salaries from Salaries.xsl. There we have a different value interpretation for the functional block [F6:F6]: this spreadsheet does not calculate in millions, so we chose $k := \{x \mapsto x\}$ and get the import functions $k \circ j^{-1} = \{x\text{€} \mapsto 10^{-6} x\text{€}\}$ in the intention and $j^{-1} \circ k = \{x \mapsto 10^{-6} x\}$ in the spreadsheet.

In conclusion we observe that setting the ontology, provenance, and the interpretation of a functional block gives us a full and explicit account of its intention, and we can relegate all further semantic services to the intention model. For instance we can verify (using inference in the intention model) this part of the spreadsheet by establishing the equality of $j^{-1} \circ \mathcal{F} \circ j$ and $\rho(c)$.

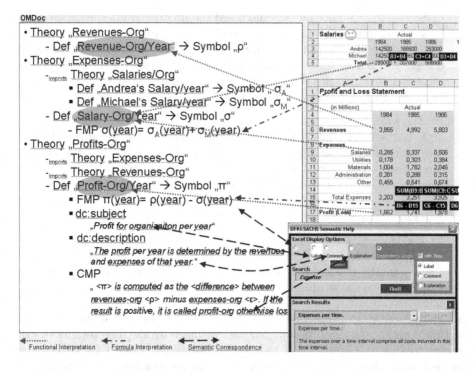

Fig. 2. The SACHS Information Architecture and Control/Search Panel

4   A Semantic Help System for Spreadsheets with Intentions

The SACHS system is a work-in-progress add-in for MS Excel (written in Visual Basic for Applications) that aims at providing semantic help facilities for spreadsheets. Though it has been developed for the DFKI controlling system, it works for any spreadsheet whose intention has been encoded as an OMDoc document, e.g. our running example, and mash-up information has been provided[1]. We have designed SACHS as *invasive technology* that extends well-used (and therefore well-known) software systems from within to overcome usage hurdles — see [Koh05b] for a discussion. We designed the semantic services in SACHS to allow the user to explore the intention of a spreadsheet from her micro-perspective, i.e., from her here-and-now. In particular, all semantic help facilities start from the cell level. Moreover, we tried to implement a process-based interaction design, i.e., a design where the semantic help evolves in a user-steered process.

For services that visualize the intention, the cells in the spreadsheet must be interpreted, i.e., linked to elements of the accompanying OMDoc document as e.g. shown in Figure 2. Generally, all cells in a functional block are linked to

---

[1] We will spare the reader the technical details of this mash-up for lack of space.

an OMDoc definition — the definition of its intended function, while OMDoc assertions justify their formulae. This assignment is internally represented by an extra worksheet within the spreadsheet we call the "SACHS interpretation". This is manually maintained by the spreadsheet author. Once the interpretation is established we can directly make use of the various elements of the OM-Doc information for the respective objects (see the dashed arrows in Figure 2). Concretely, for instance, the Dublin Core metadata element dc:subject of an OMDoc definition can be used as a *SACHS label* for the cell it interprets. MS Excel's comment functionality is hijacked to create *SACHS comments* that draw on the respective dc:description element, which contains a concise description of the object in question. In contrast, the CMP element of the OMDoc definition contains a detailed explanation using semantically marked up representations of objects from the intention model. These can be mathematical formulae encoded as OPENMATH objects like the revenue function ρ or technical terms like "difference" which we have decorated in angle brackets in Figure 2. The added value of semantic annotation here is that the meaning of both can be further explored: The front end item *"SACHS explanations"* allows this by providing "jump points" from within the text to those cells that are assigned to the definitions of those symbols via the SACHS interpretation sheet. Once jumped the user can look up the available semantic information of that particular cell and so on.

A *formula* underlying a cell is mirrored in the formula element FMP of the respective definition in the semantic document (see Figure 2) in the OPENMATH format, this allows us to present it to the user in math notation: $\sum_{i=1}^{5} \epsilon_i(1985)$ is more readable than "=SUM(C9:C13)".

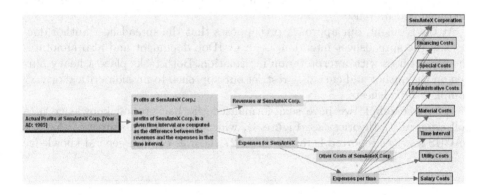

Fig. 3. Navigating the Spreadsheet Intention

In Figure 2 we have already shown the main control panel of the SACHS system in the right bottom hand corner. This allows the user to enable various semantic enhancements in the spreadsheet and also search the intention semantically. From a semantic point of view, contextual views of the spreadsheet intention as the one in Figure 3 are probably most interesting and helpful.

Such views allow a user to understand the connections between spreadsheet cells and background information. This view aggregates information about

- the intention of a cell in terms of the intended function of its functional block, and the intended arguments.
- how the cell value is computed from the values of its arguments.
- and the intentions of the arguments.

All information points in the pop-up are hyperlinked to their respective sources in the OMDoc document or in the spreadsheet so that the graph view can also serve as a meaning-driven navigation tool. In a future version of SACHS, these hyperlinks could pre-instantiate the intention model with the argument values and allow an exploration from the view of the current cell — in our example in Figure 3 the intention specialized to the year 1985. Note that our example here only shows the situation for a formula-computed cell. For other provenances, the pop-up would visualize the provenance object given by the formula interpretation. For instance, for cell [B9] the provenance object is "imported from Salaries.xls[B5]", so we can visualize the data from that cell using the existing SACHS pop-up.

As OMDoc is organized by theories, the provision of multi-level theory graphs as in the CPOINT system [Koh05a] are nice-to-have services one can think of.

4.1   Evaluation: Estimating the Semantic Overhead in SACHS

As always the interesting question at the end of the implementation of exciting new ideas is whether it was worth the effort. Even though we can't answer that yet objectively, we can at least give an estimate for the costs of the semantic overhead and its reuse.

At the moment, our approach presupposes that the spreadsheet author documents the spreadsheet intention as an OMDoc document and also annotates the spreadsheet with interpretation information. Both tasks place a heavy burden on the author and currently restrict our approach to mission-critical or very complex spreadsheets.

For our example we have semi-formalized the background knowledge in a collection of 47 theories (see Figure 4), which serve as the source of the new SACHS services reported here. Of these, 27 theories contain general knowledge

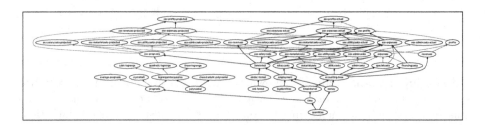

Fig. 4. The Theory Graph of Background Knowledge for our Example Spreadsheet

about quantities, units, basic cost accounting, and prognosis. Only 20 were developed specially for our example: a theory about the SemAnteX Corp. itself and union theories specializing accounting and prognosis to the SemAnteX Corp. peculiarities. A dozen general theories about real arithmetic were not taken into account here, since they were available externally.

By size, the specific theories only amounted to ca. 25% of the semi-formalization. We are currently developing a background knowledge corpus for a central part of the DFKI controlling system, which has three orders of magnitude more cells than the expository example in this paper. But the semi-formalization of this is only double in size, conceivably since the controlling system is relatively regular, for instance it is organized in about 20 analogous sheets for the respective DFKI departments.

In general, we expect the burden of specifying the ontology to decrease as more and more OMDoc content dictionaries for common models (e.g. standard accounting techniques) appear. Our case study has shown that for an efficient development of background ontologies we need to integrate editing facilities into the graph view in Figure 3. For the interpretations, we plan to adapt techniques of header, unit, and template inference [AE04, AE06] to partially automate the annotation process via suggestions.

For evaluating the usefulness of the SACHS system we are currently undertaking a formal user study on the DFKI controlling system — our expository example in Figure 2 is too simple to require a help system really. First informal feedback was encouraging, for instance, users highly appreciated being made aware of the differing reference periods data that were implicit in the concrete spreadsheet layout.

5   Conclusion and Outlook

We have analyzed the reasons for users' difficulties in understanding and appropriating complex spreadsheets, as they are found e.g. in financial controlling systems. We claim that the ultimate cause is that spreadsheets are weak as active documents, because their underlying semantic documents are biased to computational aspects and fail to model the provenance, interpretation, and ontological relations of the objects and concepts operationalized by the system. To remedy the situation we propose to explicitly model the intention of a spreadsheet as an intention model in a collection of OMDoc content dictionaries which serve as an explicit knowledge base for the spreadsheet. Finally, we present the work-in-progress SACHS system that draws on such intention models to offer various semantic services that aid the user in understanding and interacting with the spreadsheets.

In essence, our approach makes double use of the following duality identified by FENSEL in [Fen08]

  – *Ontologies define* formal *semantics for information, consequently allowing information processing by a computer.*

- *Ontologies define* real-world *semantics, which makes it possible to link machine processable content with meaning for humans based on* consensual *terminologies.*

In the analysis we look at the formal semantics of spreadsheets and find a computational bias that hampers understanding since it fails to model consensual terminologies and therefore leads to real-world usability problems. In our proposed solution we extend the formal semantics of spreadsheets to draw on explicitly represented consensual terminologies.

While a semantic help system for a spreadsheet-based controlling system was the original motivation for our analysis, we feel that an explicit representation of the intention model of a spreadsheet has many more applications: it can be used for verification of the formulae, for change management along the lines of [MW07], and automated generation of user-adapted spreadsheets.

Our approach seems to be related to *"Class Sheets"* [EE05] introduced by ENGELS and ERWIG. Their class descriptions can also be counted as an external, structured intention model, which is referenced by an interpretation mapping. We will have to study the relation to our work more closely, but it seems that their work suffers the same computational bias as the spreadsheets themselves. But classes with extensive documentation or UML diagrams might go some ways towards generating a help system like SACHS.

In the future we plan to extend the repertoire of semantic services of the SACHS system. For instance, we envision a dual service to the one in Figure 3 which could visualize where the value of a cell is used to get an intuition for the relevance of a cell value.

As a theoretical extension, we can see the interpretation mappings as logic morphisms [RK09], and logic morphisms in the domain knowledge as mathematical framings (ways to view mathematical objects in terms of well-understood ones). This allow us to re-use much of the higher-level MKM techniques and OMDoc functionalities based on theory graphs: we have already extended the user interaction of the SACHS to be framing-aware [KK09], which allows the user to customize the interaction to her subjective background and goals.

But future extensions are not limited to spreadsheets. Note that the semantic analysis in Section 2 is largely independent of the computational information that is at the heart of spreadsheets. The semantic information we are after only pertains to the use of data grids as a user interface to complex functions. In the future we plan to generalize the intention model architecture presented in Section 3 to the case of data grids — e.g. tables of experimental results or raster data from satellite images. Moreover, we want to develop a native infrastructure for representing "data grids" as a *user interaction feature* in OMDoc: In accordance with the format, the provenance and interpretation functionality would allow to link grid presentations to their intention without leaving the language. OMDoc-based systems could then pick up the semantics and offer complex interactions based on them. This would lead to much more general active documents. We could even turn the approach presented in this paper around and generate SACHS spreadsheets from OMDoc documents as active documents.

Acknowledgements. This paper contains our view on the foundations of the SACHS project which aims to extend semantic document modeling to spreadsheets. Bernd Krieg-Brückner has been proposing for years to do this. He always maintained that spreadsheets would be ideal targets for semantic preloading. Indeed our research shows that spreadsheets are semantically much more interesting than anticipated. We also thank the other SACHS project members: Dieter Hutter, Achim Mahnke, as well as Klaus Hofmann for valuable discussions.

References

[ABC+03]   Ausbrooks, R., Buswell, S., Carlisle, D., Dalmas, S., Devitt, S., Diaz, A., Froumentin, M., Hunter, R., Ion, P., Kohlhase, M., Miner, R., Poppelier, N., Smith, B., Soiffer, N., Sutor, R., Watt, S.: Mathematical Markup Language (MathML) version 2.0 (second edition). W3C recommendation, World Wide Web Consortium (2003)

[AE04]   Abraham, R., Erwig, M.: Header and unit inference for spreadsheets through spatial analysis. In: IEEE International Symposium on Visual Languages and Human-Centric Computing, pp. 165–172 (2004)

[AE06]   Abraham, R., Erwig, M.: Inferring templates from spreadsheets. In: ICSE 2006: Proceedings of the 28th international conference on Software engineering, pp. 182–191. ACM, New York (2006)

[BCC+04]   Buswell, S., Caprotti, O., Carlisle, D.P., Dewar, M.C., Gaetano, M., Kohlhase, M.: The Open Math standard, version 2.0. Technical report, The Open Math Society (2004)

[BSRR02]   Burnett, M.M., Sheretov, A., Ren, B., Rothermel, G.: Testing homogenous spreadsheet grids with the "what you see is what you test" methodology. IEEE Transactions on Software Engineering 29(6), 576–594 (2002)

[CoF04]   CoFI (The Common Framework Initiative). In: Mosses, P.D. (ed.) CASL Reference Manual. LNCS (IFIP Series), vol. 2960. Springer, Heidelberg (2004)

[EE05]   Engels, G., Erwig, M.: ClassSheets: Automatic generation of spreadsheet applications from object oriented specifications. In: 20th IEEE/ACM International Conference on Automated Seofware Engineering, pp. 124–155. IEEE Computer Society, Los Alamitos (2005)

[Fen08]   Fensel, D.: Foreword. In: Hepp, M., De Leenheer, P., de Moor, A., Sure, Y. (eds.) Ontology Management: Semantic Web, Semantic Web Services, and Business Applications, Semantic Web and beyond: Computing for Human Experience, pp. 9–11. Springer, Heidelberg (2008)

[GS99]   Gaines, B.R., Shaw, M.L.G.: Enbedding formal knowledge models in active documents; creating problem-solving documents. Communications of the ACM 42(1), 57–63 (1999)

[HKS06]   Hilf, E., Kohlhase, M., Stamerjohanns, H.: Capturing the content of physics: Systems, observables, and experiments. In: Borwein, J.M., Farmer, W.M. (eds.) MKM 2006. LNCS (LNAI), vol. 4108, pp. 165–178. Springer, Heidelberg (2006)

[HM00]   Heinrich, E., Maurer, H.: Active documents: Concept, implementation, and applications. Journal of Universal Computer Science 6(12), 1197–1202 (2000)

[KK09]   Kohlhase, A., Kohlhase, M.: Spreadsheet interaction with frames: Explor-
            ing a mathematical practice. In: Carette, J., Dixon, L., Sacerdoti Coen, C.,
            Watt, S.M. (eds.) Calculemus/MKM 2009. LNCS (LNAI), vol. 5625,
            pp. 341–356. Springer, Heidelberg (2009)
[Koh05a]   Andrea Kohlhase. Cpoint (2005), http://kwarc.info/projects/CPoint/
[Koh05b]   Kohlhase, A.: Overcoming Proprietary Hurdles: CPoint as Invasive Edi-
            tor. In: de Vries, F., Attwell, G., Elferink, R., Tödt, A. (eds.) Proceedings
            of Open Source for Education in Europe: Research and Practise, Heerlen,
            The Netherlands, November 2005, pp. 51–56. Open Universiteit Neder-
            land, http://hdl.handle.net/1820/483
[Koh06]   Kohlhase, M.: OMDoc – An Open Markup Format for Mathematical
            Documents [version 1.2]. LNCS, vol. 4180. Springer, Heidelberg (2006)
[MGM+08]   Moreau, L., Groth, P., Miles, S., Vazquez, J., Ibbotson, J., Jiang, S.,
            Munroe, S., Rana, O., Schreiber, A., Tan, V., Varga, L.: The provenance
            of electronic data. Communications of the ACM 51(4), 52–58 (2008)
[MvH04]   McGuinness, D.L., van Harmelen, F.: OWL web ontology language
            overview. W3C recommendation, W3C (February 2004)
[MW07]   Müller, N., Wagner, M.: Towards Improving Interactive Mathematical
            Authoring by Ontology-driven Management of Change. In: Hinneburg,
            A. (ed.) Wissens- und Erfahrungsmanagement LWA (Lernen, Wissensent-
            deckung und Adaptivität) conference proceedings, pp. 289–295. Martin-
            Luther-University Halle-Wittenberg (2007)
[Pan00]   Panko, R.R.: Spreadsheet errors: What we know. what we think we can
            do. In: Symp. of the European Spreadsheet Risks Interest Group (Eu-
            SpRIG) (2000)
[RK09]   Rabe, F., Kohlhase, M.: A web-scalable module system for mathematical
            theories. The Journal of Symbolic Computation (2009) (manuscript to be
            submitted)
[UCI+06]   Uren, V., Cimiano, P., Iria, J., Handschuh, S., Vargas-Vera, M., Motta,
            E., Ciravegna, F.: Semantic annotation for knowledge management: Re-
            quirements and a state of the art. Web Semantics: Science, Services, and
            Agents on the World Wide Web 4(1), 14–28 (2006)
[W3C08]   RDFa Primer (2008), http://www.w3.org/TR/xhtml-rdfa-primer/
[Win06]   Winograd, T.: The spreadsheet. In: Winograd, T., Bennett, J., de
            Young, L., Hartfield, B. (eds.) Bringing Design to Software, pp. 228–231.
            Addison-Wesley, Reading (2006)
[Wol02]   Wolfram, S.: The Mathematica Book. Cambridge University Press,
            Cambridge (2002)

MathLang Translation to Isabelle Syntax

Robert Lamar, Fairouz Kamareddine, and J.B. Wells

ULTRA Group, Heriot-Watt University
http://www.macs.hw.ac.uk/ultra/

Abstract. Converting mathematical documents from a human-friendly natural language to a form that can be readily processed by computers is often a tedious, manual task. Translating between varied computerised forms is also a difficult problem. MathLang, a system of methods and representations for computerising mathematics, tries to make these tasks more tractable by breaking the translation down into manageable portions. This paper presents a method for creating rules to translate documents from MathLang's internal representation of mathematics to documents in the language of the Isabelle proof assistant. It includes a set of example rules applicable for a particular document. The resulting documents are not completely verifiable by Isabelle, but they represent a point to which a mathematician may take a document without the involvement of an Isabelle expert.

1 Introduction

When a mathematician describes a piece of mathematics in written form, it may be of interest to use computers to process this text in a variety of ways. They may use editing software to write the text as they are developing the ideas. Similarly, it is very common for such a text to be typeset so it is pleasing to the eye. It may also be useful to identify the semantics, for presentation to those with disabilities, or there may exist in the document partial calculations which the computer should process for the benefit of the original author or other readers. These are a few ways in which a computer may be used to directly benefit a human who wants to understand the mathematics.

In addition to providing output for humans, a computer may be useful for verifying the correctness of a document, in a variety of ways. These include checking of spelling and grammar in the natural language; syntax checking for mathematical notation; checking soundness of the interrelations between definitions, theorems, and proofs; and formally verifying the logical structure of the document, at different levels of rigour. These are ways in which the computer may provide evaluation and feedback on the document.

For mathematicians, systems falling at the formal end of that spectrum are currently of limited use, as they require considerable investment to gain proficiency. It is possible for a mathematician to author a paper and then pass it on to an expert in some formal system, but this requires the expert to completely comprehend each natural-language document received so as to ensure a faithful

J. Carette et al. (Eds.): Calculemus/MKM 2009, LNAI 5625, pp. 373–388, 2009.

translation. The current methods of translation are almost entirely manual. This leaves a vast chasm between original and formalised documents, in which may be introduced semantic discrepancies. Furthermore, if the original document is changed, there is some risk that necessary changes in the formalised counterpart will be overlooked when the time comes for revision.

1.1 Contributions

This paper describes developments in MathLang (a system for computerising mathematics, described in Section 2) which may help close the gap between natural language and formalised documents. MathLang is a system which tries to give as much flexibility to the user as possible, trying to process any style that a person may use to express their mathematics. The existing parts of MathLang, which are assumed to be the starting point for the developments in this paper, are overviewed in Section 2.1.

This paper offers a *process for arriving at rules for translation*. Our goal is to produce, from the already-computerised MathLang document, a text in the language of a formal system. Our chosen target language is Isabelle. To do this, rules are created which operate recursively on a document. The nature of these rules is described in Section 3.1, and is illustrated by a detailed example.

1.2 Related Work

In the context of the MathLang project, this work provides tools which provide output which is closer to existing formal systems than ever before. Specifically, it uses the facilities provided by [1,2] to translate the bulk of a text to the syntax of the proof assistant Isabelle. This is in parallel with [3], which was translating the same kind of document to Mizar [4]. However, in that work the focus was on identifying relevant theories from the Mizar Mathematical Library to include in the environment of a new Mizar document, moreso than translating the main text of a MathLang document to Mizar.

In the larger field, there are a number of projects which are attempting to bridge the gap between human-friendly mathematical texts and easily-processed and -verified computer documents. Most focus primarily on one side or another. On the formal end of the spectrum are projects like Mizar [4], Theorema [5] and Isar [6,7]. These three computer proof systems are designed with syntax that is constructed to be similar to the way that mathematicians write in natural English. A similar approach is being taken in the work of Muhammad Humayoun on MathNat [8], in which he tries to express both mathematical proofs and natural language. In the cases of Mizar and Isar, the language is *like* natural English, but does not provide the author with much flexibility. MathNat and Theorema allow much greater flexibility – in MathNat's case due to its incorporation of GF [9] – but still force the user to employ a controlled language, which may often be a subset of what an author would normally employ. MathLang endeavours to accommodate any writing style through the use of flexible annotation, accommodating documents that were never intended to be computerised, such as Euclid's *Elements* [10] and Landau's *Grundlagen der Analysis* [11].

On the other side of the natural–formal gap, Aarne Ranta's system GF [9] has a flexible system for defining grammars, and provides an API for interfacing with other programs, but is not, itself, designed to process documents to further formal states. We wish to process such documents in other interesting ways.

Finally, there are systems such as Isabelle [6] and Coq [12], which are systems for computer proof, along with Logiweb [13], which is a system for document processing that interfaces with arbitrary systems. Each of these allow natural language text to be interleaved with formal expressions in a kind of literate proof document in the manner of CWEB [14]. However, care during revision is necessary to ensure that natural language and formalism remain consistent.

1.3   Conclusion

The rules presented in this document are very limited, but they represent a pattern which may be used to develop a library of rules which may be useful in translating a variety of documents to Isabelle syntax. The documents that result from translation are very incomplete, but they may be a useful middle ground between a mathematician who has little knowledge of proof systems, and an Isabelle expert who is trying to formalise the mathematics that the original author has written. The system needs to be extended and tested extensively, but the current work shows a valuable proof-of-concept which merits further investigation.

2   Relevant Systems

This section describes systems and theories that the current work relies upon, but which the reader may not be familiar with. First it describes the system called MathLang. It first provides an overview, then a detailed description of the relevant portions of operational theory. A second section gives an example Isabelle document, drawing the reader's attention to relevant features of the language.

2.1   System Aspects

The current MathLang system [16] is designed to computerise mathematical texts like that seen in Fig. 1. By *computerising*, we mean operating on documents which are easily accessed and modified by computers. We also mean processing documents so they may be easily accessed and modified. Currently MathLang provides several ways to achieve this. They are classified into domains called *aspects* of MathLang. The current aspects are the Text and Symbol aspect [1,17], the Core Grammatical aspect [2,17], and the Document Rhetorical aspect [3]. For the current paper we restrict our focus to the first two.

Rings

Definition 1. *A ring R is a nonempty set with two binary operations, addition (denoted by $a+b$) and multiplication (denoted by ab), such that for all a,b,c in R:*

1. $a+b=b+a$.

2. $(a+b)+c=a+(b+c)$.

3. *There is an additive identity 0. That is, there is an element 0 in R such that $a+0=a$ for all a in R.*

4. *There is an element $-a$ in R such that $a+(-a)=0$.*

5. $a(bc)=(ab)c$.

6. $a(b+c)=ab+ac$ and $(b+c)a=ba+ca$.

Theorem 2.

1. $a0=0a=0$.

2. $a(-b)=(-a)b=-ab$.

Proof.

Consider rule 1.

Clearly,

$$0+a0=a0=a(0+0)=a0+a0.\tag{1}$$

So, by cancellation, $0=a0$. Similarly, $0a=0$.

To prove rule 2, we observe that $a(-b)+ab=a(-b+b)=a0=0$.

□

Adding $-(ab)$ to both sides yields $a(-b)=-(ab)$. The remainder of rule 2 is done analogously.

Fig. 1. Ring theory text as taken from *Contemporary Abstract Algebra* [15]

The Text and Symbol aspect. The Text and Symbol aspect (TSa) is the aspect of MathLang which is directly concerned with the ways in which documents present mathematics to a reader or author. A particular focus of the aspect is the way in which mathematical concepts are abbreviated, such as combining a pair of equations $a=b$ and $b=c$ into the single string $a=b=c$.

It also provides the details for sensible presentation of the fruits of other aspects of MathLang: these others augment the text with information, but TSa governs how this extra information is *symbolised* in relation to the original *text*. The document in Fig. 1, for instance, has been enhanced in Fig. 2 to show the extra information provided with the Core Grammatical aspect, below.

The Core Grammatical aspect. The Core Grammatical aspect (CGa) provides analysis for the sentence level of the document. It provides a type system for objects, definitions, and assertions and a means for checking the document for type correctness. The types of CGa are summarised in Table 1.

For instance, the variable x and number 1 could be **decl**ared as having the type **term**. The operation $+$ might have type **term** $\rightarrow$ **term** $\rightarrow$ **term**. Thus, $x+1$ would also be considered a **term**. However, this expression can also perhaps be considered as an instance of the **noun** polynomial. Furthermore, the **adjective**

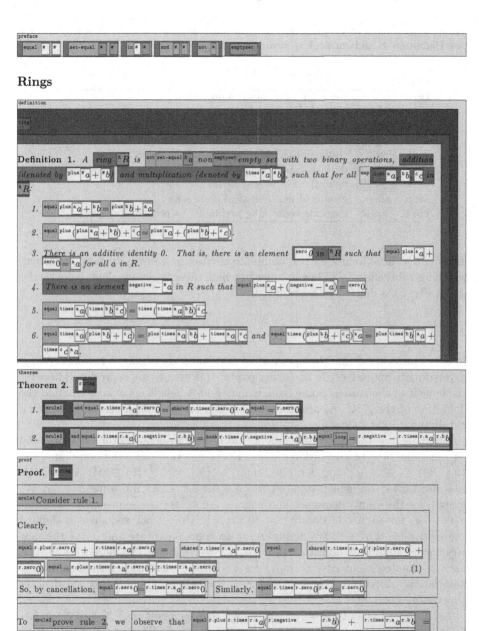

Fig. 2. Ring theory text (Fig. 1) enhanced with CGa information (See Sec. 2.1)

Table 1. Elements of $\mathcal{C}$ (primitive grammatical categories), with associated colour (see Definition 4) and ontological meaning

Cat.	Colour	Description
term	blue	common mathematical objects
set	red	collections of **terms**
noun	orange	families of **terms**
adj	yellow	adjectives used to construct new **nouns** from old
stat	green	statements which have some truth value
decl	dk. gray	declarations of new **terms**, **sets**, **nouns**, **adjs**, or **stats**
defn	lt. gray	definitions of new symbols
step	salmon	groups of mathematical assertions
cont	purple	contexts containing preliminaries prior to a step

linear can be used to modify polynomial to create a new **noun**, linear polynomial, which also classifies $x+1$. The manner in which these types are shown to a reader (as in Fig. 2) is defined in Section 2.2.

2.2   Operational System

In this section we define the operational system of MathLang. This definition is covered more extensively in an earlier paper [1]. Portions are reproduced here for the benefit of the reader. Some readers may find it beneficial to keep in mind the definition of the XML XPath data model [18], as there exist strong conceptual parallels with the following definition.

Let $\mathbb{N}$ denote the natural numbers, use $(-;-)$ to denote ordered pairs, and let functions be sets φ of ordered pairs. Every function has a domain $\mathrm{dom}(\varphi) = \{a \mid \exists b \ni (a;b) \in \varphi\}$ and a range $\mathrm{ran}(\varphi) = \{b \mid \exists a \ni (a;b) \in \varphi\}$. A sequence is a function σ for which $\mathrm{dom}(\sigma) = \{n \mid 0 \leq n < k\}$ for some $k \in \mathbb{N}$. We write $[]$ for the empty sequence and $[x_0, x_1, \ldots, x_n]$ for the sequence σ such that $\sigma(i) = x_i$ for each $i \in \mathrm{dom}(\sigma) = \{0, \ldots, n\}$. Upon that sequence is defined the metric $|\sigma| = n + 1$. We define σ_1, σ_2 to concatenate σ_1 and σ_2 as the new sequence σ such that $\mathrm{dom}(\sigma_1, \sigma_2) = \{0, \ldots, |\sigma_1| + |\sigma_2| - 1\}$ where $\sigma(i) = \sigma_1(i)$ for $i \in \mathrm{dom}(\sigma_1)$ and $\sigma(i) = \sigma_2(i)$ for $i - |\sigma_1| \in \mathrm{dom}(\sigma_2)$. Concatenation is associative. Moreover, $[], \sigma = \sigma$ and $\sigma, [] = \sigma$. For any set S, say $[S]$ denotes $\{\sigma \mid \mathrm{ran}(\sigma) \subseteq S\}$.

Let $\mathcal{L} = \mathcal{F} \cup \mathcal{G} \cup \mathcal{S}$ be a set of labels such that elements of $\mathcal{F}$, $\mathcal{G}$ and $\mathcal{S}$ are formatting, grammatical, and souring labels, respectively. The set $\mathcal{F}$, of *formatting instructions*, varies according to rendering system. We define $\mathcal{G} = \mathcal{C} \times \mathcal{I}$, where $\mathcal{C} = \{\mathbf{term}, \mathbf{set}, \mathbf{noun}, \mathbf{adj}, \mathbf{stat}, \mathbf{decl}, \mathbf{defn}, \mathbf{step}, \mathbf{cont}\}$, and contains identifiers for the primitive grammatical categories of Table 1. The set $\mathcal{I}$ consists of strings used for identifying abstract interpretations. We let ℓ, f, g, c and i range over $\mathcal{L}$, $\mathcal{F}$, $\mathcal{G}$, $\mathcal{C}$ and $\mathcal{I}$, respectively.

We let s range over $\mathcal{S} = \mathcal{S}_u \cup \mathcal{S}_i$ where $\mathcal{S}_u$ contains *souring identifiers* to be employed directly by the user while $\mathcal{S}_i$ holds several identifiers used internally for rewriting. $\mathcal{S}_u$ and $\mathcal{S}_i$ are disjoint, defined as follows:

$$\mathcal{S}_u = \{\texttt{fold-left},\texttt{fold-right},\texttt{map},\texttt{base},\texttt{list},\texttt{hook},\texttt{loop},\texttt{shared}\} \cup (\{\texttt{position}\} \times \mathbb{N})$$

$$\mathcal{S}_i = \{\texttt{hook-travel},\texttt{head},\texttt{tail},\texttt{daeh},\texttt{liat},\texttt{right-travel},\texttt{left-travel}\} \cup (\{\texttt{cursor}\} \times \mathbb{N})$$

Definition 1 (Document). *Let $\mathcal{D}$ be the smallest set such that:*

1. *$[] \in \mathcal{D}$,*
2. *if $d \in \mathcal{D}$ and $\ell \in \mathcal{L}$ then $[(\ell; d)] \in \mathcal{D}$, and*
3. *if both d_1 and d_2 are elements of $\mathcal{D}$ then $(d_1, d_2) \in \mathcal{D}$.*

A MathLang document is an element of the set $\mathcal{D}$. In addition, we denote by $\mathcal{D}_\mathcal{F}$, $\mathcal{D}_\mathcal{G}$, $\mathcal{D}_{\mathcal{F} \cup \mathcal{G}}$, $\mathcal{D}_{\mathcal{G} \cup \mathcal{S}}$ and $\mathcal{D}_{\mathcal{F} \cup \mathcal{G} \cup \mathcal{S}}$ the sets of documents whose labels are restricted to the respective subscripted set. The variables d, d_n (where $n \in \mathbb{N}$) denote members of $\mathcal{D}$, unless otherwise noted.

Remark 2 (Notational convention). We use $\ell \langle d \rangle$ to denote $[(\ell; d)]$. When not ambiguous, ℓ denotes $\ell \langle [] \rangle$. A box with black border and **coloured background**, $\boxed{{}^i d}$, is used to represent $(c; i) \langle d \rangle$ (a document with grammatical label), where the background colour of the box corresponds to c as shown in Table 1 (See Example 6, below). Similarly, a box with thick pink border and white background, $\boxed{{}^s d}$, is used to represent $s \langle d \rangle$, documents with souring labels.

It is worth noting that these notations have been developed for ease of reading, and particularly interactive annotation of texts. One of MathLang's biggest motivations is for humans to be able to type a mathematical text in a natural way on the computer, and then add the grammatical and souring information with ease. The prototype based on TeXmacs is described in detail in [17].

Formatting systems are treated as a set of formatting instructions $\mathcal{F}$, a blank formatting instruction ε, a concatenation operator $\bullet$, and a hole-filling function fill : $\mathcal{F} \times [\mathcal{F}] \rightarrow \mathcal{F}$, which takes two arguments, a formatting instruction f and a sequence of instructions σ. Instruction f may have holes, denoted $\boxed{n}$, where $0 \leq n < |\sigma|$. The instruction f is rewritten so that each $\boxed{n}$ is replaced by $\sigma(n)$.

Definition 3 (Souring). Souring *is a rewriting process that was described in [1]. The particulars of the procedure are not important to this paper. We may regard the souring function as a black box function* sour : $\mathcal{D} \rightarrow \mathcal{D}_{\mathcal{F} \cup \mathcal{G}}$.

The motivation for souring is as follows: as syntactic sugar is added to a formal document to make it easier to read for humans, syntax souring is added to natural-language documents to make them easier for a computer to process. Before processing a document with the rules in Section 3.1, we typically *sour* the document by applying this function to the document, then further processing the result. An example of a typical souring operation would be to convert $a = b = c$ to $a = b,\ b = c$.

Definition 4 (Rendering functions). *Let* $\mathrm{r} : \mathcal{D} \to \mathcal{F}$ *be defined as*

$$\mathrm{r}([]) = \varepsilon \tag{REN1}$$

$$\mathrm{r}(f\langle d\rangle) = \mathrm{fill}(f, [\mathrm{r}(d(0)), \ldots, \mathrm{r}(d(|d|-1))]) \tag{REN2}$$

$$\mathrm{r}((c; i)\langle d\rangle) = \boxed{^{i}\,\mathrm{r}(d)} \tag{REN3}$$

$$\mathrm{r}(s\langle d\rangle) = \boxed{^{s}\,\mathrm{r}(d)} \tag{REN4}$$

$$\mathrm{r}(d_1, d_2) = \mathrm{r}(d_1) \bullet \mathrm{r}(d_2) \tag{REN5}$$

where the background colour of the box given by (REN3) is the colour from Table 1 (i.e., $\mathrm{r}((\mathbf{term}; i)\langle d\rangle) = \boxed{^{i}\,\mathrm{r}(d)}$, $\mathrm{r}((\mathbf{set}; i)\langle d\rangle) = \boxed{^{i}\,\mathrm{r}(d)}$, *etc.)*

Definition 5 (Extract original document). *Usually, some* $d \in \mathcal{D}$ *consists of a "typical" mathematical text plus some information which is stored in the labels from* $\mathcal{G} \cup \mathcal{S}$*. For any document which has this property, it may be useful to filter d with the function* $\mathrm{od} : \mathcal{D} \to \mathcal{D}_{\mathcal{F}}$*, defined as*

$$\mathrm{od}([]) = [] \tag{OD1}$$

$$\mathrm{od}(\ell\langle d\rangle) = \begin{cases} \ell\langle\mathrm{od}(d)\rangle & \text{if } \ell \in \mathcal{F} \\ d & \text{otherwise} \end{cases} \tag{OD2}$$

$$\mathrm{od}(d_1, d_2) = \mathrm{od}(d_1), \mathrm{od}(d_2). \tag{OD3}$$

It is then possible to obtain the mathematician's original text as $\mathrm{r}(\mathrm{od}(d))$.

Example 6. In this example, formatting instructions are taken to be from the LATEX typesetting system. Consider the document d given as

␣\$$\boxed{0}$$\$$^{\neg}$$⟨(**stat**; **equal**)⟨␣$\boxed{0}$⊨$\boxed{1}$$^{\neg}$⟨[(**term**; **times**)⟨

[(**term**; **zero**)⟨0⟩, (**term**; a)⟨a⟩]⟩, (**term**; a)⟨a⟩]⟩⟩⟩.

The document will be rendered, $\mathrm{r}(d)$, as $\boxed{\text{equal}\,\boxed{\text{times}\,\boxed{\text{zero}\,\boxed{0}\,\boxed{a}\,a}} = \boxed{a}\,a}$, while the filtered document $\mathrm{od}(d)$, ␣\$$\boxed{0}$$\$$^{\neg}$$⟨␣$\boxed{0}$⊨$\boxed{1}$$^{\neg}$⟨[␣$\boxed{0}$⊨$\boxed{1}$$^{\neg}$⟨0, a, a]⟩⟩⟩, would be rendered as $0a = a$.

2.3   Overview of Isabelle

For the benefit of the reader, the following provides a brief overview of pertinent parts of the proof assistant Isabelle. Isabelle/HOL was chosen as a target for translation from MathLang because it is a popular, mature system with extensive

```
 7  theory Groebner_Basis
 8  imports NatBin
    ...
14  begin
    ...
259  locale gb_field = gb_ring +
260    fixes divide :: "'a \<Rightarrow> 'a \<Rightarrow> 'a"
261      and inverse:: "'a \<Rightarrow> 'a"
262    assumes divide: "divide x y = mul x (inverse y)"
263      and inverse: "inverse x = divide r1 x"
    ...
338  lemma no_zero_divirors_neq0:
339    assumes az: "(a::'a::no_zero_divisors) \<noteq> 0"
340      and ab: "a*b = 0" shows "b = 0"
341  proof -
342    { assume bz: "b \<noteq> 0"
343      from no_zero_divisors [OF az bz] ab have False by blast }
344    thus "b = 0" by blast
345  qed
    ...
440  end
```

Fig. 3. Excerpts of code [19] from Isabelle/HOL distribution

documentation. The authors of the current paper do not consider themselves to be Isabelle experts, but it was straightforward to learn the basics of the system in order to begin making basic proof documents. Isabelle [6] allows a user to express and record formulae and reasoning steps. It is designed to work with a variety of logical foundations, the most popular being HOL. Isar [7] enhances the language of Isabelle for a more declarative proof style.

What follows is a summary of certain Isabelle features which may be useful in understanding the remainder of this document, referencing Figure 3.

In Isabelle, formal developments are organised into *theories*, which are given unique names and stored in separate text files. The theory in Figure 3, for instance, is stored in **Groebner_Basis.thy** and begins with the indicated line 7 (The previous lines in the file are all comments). Line 8 of this example shows that the current theory may need to use results formalised in the Isabelle theory NatBin. This **imports** directive allows access to definitions and results of the other theory. The rest of the file, which will develop new formalisations, is enclosed with the keywords **begin** (line 14) and **end** (line 440).

A **locale** is an Isabelle construct which defines a local scope in which assumptions and symbols are declared. Theorems may then depend on locales for the premises on which they are proved. Line 259 starts the declaration of a locale, which in this case has the name **gb_field** and inherits the properties of locale **gb_ring**. It starts by declaring (**fixing**) a pair of constants with type signatures and stating two axioms for the locale. This is followed by a lemma and proof.

3   Rules for Translating Documents

Section 2.2 described the operational representation of MathLang documents. The documents are stored as an assembly of labels, each of which has a particular role (formatting, grammatical, or souring). It may be of interest to reformat the document into another form. In this section, we give a set of translation rules which could be recursively applied to a MathLang document, and easily extended to be applied to other documents. This set of rules converts some of the information of the document to Isabelle syntax.

3.1   Example Rules for Translating to Isabelle

In this section, we describe a set of rules which are sufficient to translate the document in Figure 2 to the language of Isabelle. These are given to show how rules can be created to cover different cases in MathLang documents. Suppose that d is a document which has been soured (see Definition 3). Then we apply mutually recursive translation rules $\mathcal{T} : \mathcal{D}_\mathcal{G} \to \mathcal{D}_\mathcal{F}$ as partial translations of d into the syntax of Isabelle. These are defined from the top down: each rule may rely on other rules which are defined later in the section. Figure 4, in Section 3.2, shows the translation given by the rules.

In the first rule, (*name*) is an Isabelle comment which should be replaced with a name for the theory. Similarly, (*theories*) is a list of other theories which contain required prior knowledge. Constructing this list of theories is outside the scope of this paper. For the current work, we leave this task to an Isabelle expert, to fill in the blanks. The root document tree may be translated by the following rule.

$$\mathcal{T}_{\text{root}}(d) = \text{fill}(\llcorner\texttt{theory (*name*) imports (*theories*)}$$
$$\texttt{begin } \boxed{1} \texttt{ end}\urcorner, [\mathcal{T}_{\text{main}}(d)]) \quad (\text{ROOT1})$$

This inserts the main frame for the theory and then invokes $\mathcal{T}_{\text{main}}$, as defined below. Note that the aforementioned (*theories*) list would likely depend on the contents of any preface, but that is outside of the scope of this paper, so (MAIN1) returns an empty string, ignoring its contents.

$$\mathcal{T}_{\text{main}}\left(\boxed{\texttt{preface } d}\right) = \llcorner\urcorner \qquad\qquad (\text{MAIN1})$$

$$\mathcal{T}_{\text{main}}\left(\boxed{\texttt{definition } d}\right) = \mathcal{T}_{\text{def}}(d) \qquad\qquad (\text{MAIN2})$$

$$\mathcal{T}_{\text{main}}\left(\boxed{\texttt{theorem } \boxed{i \; \boxed{\;} \; i' \; \boxed{\;}}, d}\right) = \mathcal{T}_{\text{thm}}(i', d) \qquad (\text{MAIN3})$$

$$\mathcal{T}_{\text{main}}\left(\boxed{\texttt{proof } d}\right) = \mathcal{T}_{\text{pf}}(d) \qquad\qquad (\text{MAIN4})$$

$$\mathcal{T}_{\text{main}}(d_1, d_2) = \mathcal{T}_{\text{main}}(d_1) \bullet \mathcal{T}_{\text{main}}(d_2) \qquad (\text{MAIN5})$$

When the main text contains a **definition** annotation surrounding **nouns**, this kind of annotation may be translated with the following rule.

$$\mathcal{T}_{\text{def}}\left(\boxed{\ \boxed{\ }\ \boxed{^{i'}\text{props}\ d}\ } \right) = \text{fill}\left(\llcorner\texttt{locale}\ \boxed{0}\texttt{=}\boxed{1}\lrcorner, [i, \mathcal{T}_{\text{def}}(d)]\right) \tag{D1}$$

$$\mathcal{T}_{\text{def}}\left(\boxed{^{i}\boxed{\ }\ ^{i'}\boxed{\ }} \right) = \text{fill}\left(\llcorner\texttt{fixes}\ \boxed{0}\texttt{::"'r"}\ \texttt{assumes}\ \texttt{"}\boxed{0}\texttt{:}\boxed{1}\texttt{"}\lrcorner, [i, i']\right) \tag{D2}$$

$$\mathcal{T}_{\text{def}}\left(\boxed{^{i}\ d} \right) = \text{fill}\left(\llcorner\texttt{fixes}\ \boxed{0}\texttt{::"}\boxed{1}\texttt{"}\lrcorner, \left[i, \mathcal{T}_{\text{ty}}\left(\boxed{^{i}\ d}\right)\right]\right) \tag{D3}$$

$$\mathcal{T}_{\text{def}}\left(\boxed{^{i}\boxed{\ }} \right) = \text{fill}\left(\llcorner\texttt{fixes}\ \boxed{0}\texttt{::"'r set"}\lrcorner, [i]\right) \tag{D4}$$

$$\mathcal{T}_{\text{def}}\left(\boxed{^{i}\ d} \right) = \text{fill}\left(\llcorner\texttt{assumes}\ \texttt{"}\boxed{1}\texttt{"}\lrcorner, \left[\mathcal{T}_{\text{pfx}}\left(\boxed{^{i}\ d}\right)\right]\right) \tag{D5}$$

$$\mathcal{T}_{\text{def}}(d_1, d_2) = \mathcal{T}_{\text{def}}(d_1) \bullet \mathcal{T}_{\text{def}}(d_2) \tag{D6}$$

For $\mathcal{T}_{\text{ty}}$, for $i \in \mathcal{I}, d \in \mathcal{D}$ we have $\boxed{^{i}\ d} \in \left\{ \boxed{^{i}\ d}, \boxed{^{i}\ d}, \boxed{^{i}\ d} \right\}$ (term, set, or statement). This rule extracts the type signature for the given expression.

$$\mathcal{T}_{\text{ty}}\left(\boxed{^{i}\ d} \right) = \text{fill}\left(\llcorner\boxed{0}\texttt{=>}\boxed{1}\lrcorner, [\mathcal{T}_{\text{ty}}(d), i]\right) \tag{TY2}$$

$$\mathcal{T}_{\text{ty}}\left(\boxed{^{i}\ d}, d' \right) = \text{fill}\left(\llcorner\boxed{0}\texttt{=>}\boxed{1}\lrcorner, \left[\mathcal{T}_{\text{ty}}(d'), \mathcal{T}_{\text{ty}}\left(\boxed{^{i}\ d}\right)\right]\right) \tag{TY3}$$

$$\mathcal{T}_{\text{ty}}\left(\boxed{^{i}\boxed{\ }} \right) = \llcorner\texttt{'r}\lrcorner \qquad \mathcal{T}_{\text{ty}}\left(\boxed{^{i}\boxed{\ }} \right) = \llcorner\texttt{'r set}\lrcorner \qquad \mathcal{T}_{\text{ty}}\left(\boxed{^{i}\boxed{\ }} \right) = \llcorner\texttt{bool}\lrcorner \tag{TY1}$$

Example 7. When (D3) is applied to the annotated expression

addition (denoted by $\boxed{^{\text{plus}}\ \boxed{^{\#}\text{a}}\texttt{+}\boxed{^{\#}\text{b}}}$)

the result of the translation is

```
8    fixes    plus    ::  "'r => 'r => 'r"
```

where all three of the symbols in the type signature are `'r` because the three inner boxes were all **terms**.

If, on the other hand, we want to convert several boxes – again, for $i \in \mathcal{I}, d \in \mathcal{D}$ we have $\boxed{^{i}\ d} \in \left\{ \boxed{^{i}\ d}, \boxed{^{i}\ d}, \boxed{^{i}\ d} \right\}$ (term, set, or statement) – the following rules turn the boxes into a prefix notation that is Isabelle-friendly, although it is not perfect (See Note 9, below).

$$\mathcal{T}_{\text{pfx}}\left(\boxed{^{i}\boxed{\ }} \right) = i \tag{PFX1}$$

$$\mathcal{T}_{\text{pfx}}\left(\boxed{^{i}\ d} \right) = \text{fill}\left(\llcorner\boxed{0}\ \boxed{1}\lrcorner, [i, \mathcal{T}_{\text{pfx-inner}}(d)]\right) \tag{PFX2}$$

$$\mathcal{T}_{\text{pfx-inner}}\left(\boxed{^{i}\boxed{\ }}, d' \right) = \text{fill}\left(\llcorner\boxed{0}\ \boxed{1}\lrcorner, [i, \mathcal{T}_{\text{pfx-inner}}(d)]\right) \tag{PFX3}$$

$$\mathcal{T}_{\text{pfx-inner}}\left(\boxed{^{i}\ d}, d' \right) = \text{fill}\left(\llcorner(\boxed{0}\ \boxed{1})\ \boxed{2}\lrcorner, [i, \mathcal{T}_{\text{pfx-inner}}(d), \mathcal{T}_{\text{pfx-inner}}(d')]\right) \tag{PFX4}$$

Example 8. We see that the annotated expression

which may then be manipulated to

```
7    assumes "not (set-equal R emptyset)"
```

Note 9. It is possible, on a case-by-case basis, to translate expressions such as `emptyset` to the more Isabelle-friendly `{}`, or even `equals zero (times a zero)` to `zero = a * zero`, but this kind of automated translation may not be useful or even desirable for the user. We leave it, for the moment, to future work.

Example 10. To illustrate the way that $\mathcal{T}_{\text{def}}$, $\mathcal{T}_{\text{ty}}$, and $\mathcal{T}_{\text{pfx}}$ work together, note

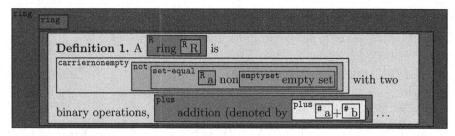

would be translated into

```
5    locale ring =
6       fixes    R :: "'r set"
7       assumes "not (set-equal R emptyset)"
8       fixes    plus  :: "'r => 'r => 'r"
```

$$\mathcal{T}_{\text{thm}}\left(p, \boxed{{}^{i}\;[]\;{}^{i}d}\right) =$$

$$\text{fill}(\llcorner\texttt{theorem (in } \boxed{0}\texttt{) } \boxed{1}\texttt{: "}\boxed{2}\texttt{"}\lrcorner, [p, i, \mathcal{T}_{\text{pfx}}\left(\boxed{{}^{i}d}\right)]) \quad \text{(THM1)}$$

$$\mathcal{T}_{\text{thm}}(p, (d_1, d_2)) = \mathcal{T}_{\text{thm}}(p, d_1) \bullet \mathcal{T}_{\text{thm}}(p, d_2) \quad \text{(THM2)}$$

Example 11. If the above rule is applied to the theorem in Figure 2,

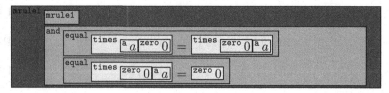

it would result in the output

```
28  theorem (in ring) mrule1:
29  shows "and (equal (times a zero) (times zero a))
30              (equal (times zero a) zero)"
```

The final rules are filled in as follows. We note that in Isabelle, theorems pass their locale information on to their associated proof. Thus, although we see the declaration of a ring as context for both theorems and definitions (as denoted $\boxed{\text{r}}\,\boxed{\text{ring}}$ in Figure 2), and this is necessary for MathLang's internal type checking, we do not need this information in the translation. Thus, (PF1) returns an empty string.

$$\mathcal{T}_{\text{pf}}\left(\boxed{d}\right) = {}_{\llcorner}{}^{\urcorner} \tag{PF1}$$

$$\mathcal{T}_{\text{pf}}\left(\boxed{^{i}\,d}\right) = \text{fill}\left({}_{\llcorner}\text{have } "\boxed{0}"{}^{\urcorner}, \left[\mathcal{T}_{\text{pfx}}\left(\boxed{^{i}\,d}\right)\right]\right) \tag{PF2}$$

Example 12. This will translate $\boxed{\begin{array}{c}\text{equal}\\ \boxed{\text{zero } 0} = \boxed{\text{times}\,\boxed{a}\,\boxed{\text{zero } 0}}\end{array}}$ to the code

```
40  have "equal zero (times a zero)"
```

3.2 Resulting Code

With the aid of the rules from Section 3, the Isabelle code in Figure 4 may be constructed (again based on the annotations of the small ring theory in Figure 2). The rules described in this section are sufficient to translate the document given in Figure 2 to Isabelle syntax, and even to get the user very close to a formal proof sketch, but the rules as defined are only sufficient for an extremely small subset of examples. It is not difficult to find a new document for which the translation rules give us an Isabelle-like text which is an insufficient representation of the original mathematics.

The translation shown in Figure 4 shows several specific drawbacks: First, the document does not successfully pass through the Isabelle system for several reasons. There are some trivial things, like the theory name on Line 1, which are simple to add but are not easily provided by an intelligent system. Providing the list of imported theories, also, is difficult for a person who does not know the existing libraries nor how to search them for relevant information. The main failure of the resulting locale definition is the form of expressions such as equality. On a case-by-case basis, such things could be converted (in the case of **equal** and **set-equal**, an infix '=' would satisfy Isabelle nicely), but it is hard to say that such transformations would be generally useful without being highly context-sensitive.

In addition to these shortcomings, the relationship between theorems and proofs is not ordered well. In the original text, it makes perfect sense for the author to write what are essentially two theorems, then prove them in the same order. However, Isabelle prefers proofs to directly follow their assertions, and the fact that lines 36–41 should be moved just before line 35 is not addressed well. It may not even immediately evident to the human eye that it is these lines,

```
     theory (* name *)
     imports (* theories *)
     begin

 5   locale ring =
         fixes   R :: "'r set"
         assumes "not (set-equal R emptyset)"
         fixes   plus  :: "'r => 'r => 'r"
         fixes   times :: "'r => 'r => 'r"
10       fixes   a :: "'r"
         assumes "a : R"
         fixes   b :: "'r"
         assumes "b : R"
         fixes   c :: "'r"
15       assumes "c : R"
         assumes "equal (plus a b) (plus b a)"
         assumes "equal (plus (plus a b) c) (plus a (plus b c))"
         fixes   zero :: "'r"
         assumes "zero : R"
20       assumes "equal (plus a zero) a"
         fixes   negative :: "'r => 'r"
         assumes "equal (plus a (negative a)) zero"
         assumes "equal (times a (times b c)) (times (times a b) c)"
         assumes "equal (times a (plus b c)) (plus (times a b) (times a c))"
25       assumes "(times (plus b c) a) (plus (times b a) (times c a))"


     theorem (in ring) mrule1:
     shows "and (equal (times a zero) (times zero a))
30                (equal (times zero a) zero)"

     theorem (in ring) mrule2:
     shows "and (equal (times a (negative b)) (times (negative a) b))
                (equal (times (negative a) b) (negative (times a b))"
35
     have "equal (plus zero (times a zero)) (times a zero)"
     have "equal (times a zero) (times a (plus zero zero))"
     have "equal (times a (plus zero zero))
                 (plus (times a zero) (times a zero))"
40   have "equal zero (times a zero)"
     have "equal (times zero a) zero"

     have "equal (plus (times a (negative b)) (times a b))
                 (times a (plus (negative b) b))"
45   have "equal (times a (plus (negative b) b)) (times a zero)"
     have "equal (times a zero) zero"
     have "equal (times a (negative b)) (negative (times a b))"

     end
```

Fig. 4. Isabelle code created using rules from Section 3 on annotations in Figure 2

exactly, which should be associated with theorem `mrule1`. There is the smaller matter that these proofs should be surrounded with `proof ... qed` pairs, but this issue goes hand-in-hand with the aforementioned problem of discerning which proof lines ago with which theorem.

The major hurdle, however, is that for Isabelle to find this theory correct, it requires much more information. None of the proof claims (`have "..."`) are justified, and there are significant holes in the reasoning. This is largely due to the fact that the original author simply left many holes which would be evident to a human reader, considering them unnecessary. When this theory file is developed to a point at which Isabelle is completely satisfied, it is approximately 10 times longer.

While these problems are significant, we believe that the current end-result has merit. One of the major benefits is that this can be performed by a mathematician who knows little-to-nothing about Isabelle. The (very incomplete) theory in Figure 4 can then be given to an Isabelle expert for development into a robust theory. This way, they have a starting point in Isabelle syntax, which may save them time in understanding the intent of the document.

References

1. Kamareddine, F., Lamar, R., Maarek, M., Wells, J.B.: Restoring natural language as a computerised mathematics input method. In: MKM 2007 [20], pp. 280–295 (2007)
2. Kamareddine, F., Maarek, M., Wells, J.B.: Toward an object-oriented structure for mathematical text. In: Kohlhase, M. (ed.) MKM 2005. LNCS, vol. 3863, pp. 217–233. Springer, Heidelberg (2006)
3. Kamareddine, F., Maarek, M., Retel, K., Wells, J.B.: Narrative structure of mathematical texts. In: MKM 2007 [20], pp. 296–311 (2007)
4. Rudnicki, P.: An overview of the Mizar project. In: Proceedings of the 1992 Workshop on Types for Proofs and Programs (1992)
5. Buchberger, B., Crǎciun, A., Jebelean, T., Kovács, L., Kutsia, T., Nakagawa, K., Piroi, F., Popov, N., Robu, J., Rosenkranz, M., Windsteiger, W.: Theorema: Towards Computer-Aided Mathematical Theory Exploration. Journal of Applied Logic, 470–504 (2006)
6. Nipkow, T., Paulson, L.C., Wenzel, M.T.: Isabelle/HOL. LNCS, vol. 2283. Springer, Heidelberg (2002)
7. Wenzel, M.T.: Isar – a generic interpretative approach to readable formal proof documents. In: Bertot, Y., Dowek, G., Hirschowitz, A., Paulin, C., Théry, L. (eds.) TPHOLs 1999. LNCS, vol. 1690, pp. 167–184. Springer, Heidelberg (1999)
8. Humayoun, M.: Software specifications and mathematical proofs in natural languages. Poster in Journées scientifiques du cluster, ISLE Rhône-Alpes, Domaine universitaire, Grenoble, France (2008)
9. Ranta, A.: Grammatical framework: A type-theoretical grammar formalism. Journal of Functional Programming 14(2), 145–189 (2004)
10. Heath, T.L.: The 13 Books of Euclid's Elements. Dover, New York (1956); in 3 volumes. Sir Thomas Heath originally published this in 1908
11. Landau, E.: Grundlagen der Analysis. Chelsea (1930)
12. LogiCal Project, INRIA, Rocquencourt, France. The Coq Proof Assistant Reference Manual – Version 8.0 (June 2004), `ftp://ftp.inria.fr/INRIA/coq/V8.0/doc/`
13. Grue, K.: The layers of Logiweb. In: Towards Mechanized Mathematical Assistants (Calculemus 2007 and MKM 2007 Joint Proceedings), pp. 250–264 (2007)
14. Knuth, D.E., Levy, S.: The CWEB System of Structured Documentation: Version 3.0. Addison-Wesley/ Longman Publishing Co., Inc., Boston (1994)
15. Gallian, J.A.: Contemporary Abstract Algebra, 5th edn. Houghton Mifflin Company (2002)

16. Kamareddine, F., Wells, J.B.: Computerizing mathematical text with MathLang. In: Ayala-Rincon, M., Heusler (eds.) Proc. Second Workshop on Logical and Semantic Frameworks, with Applications, Ouro Preto, Minas Gerais, Brazil, pp. 5–30. Elsevier, Amsterdam (2008); The LSFA 2007 (post-event) proceedings is published as vol. 205 (2008-04-06) Elec. Notes in Theoret. Comp. Sci.
17. Maarek, M.: Mathematical Documents Faithfully Computerised: the Grammatical and Text & Symbol Aspects of the MathLang Framework. PhD thesis, Heriot-Watt University, Edinburgh, Scotland (June 2007)
18. Clark, J., DeRose, S.: XML Path Language (XPath) Version 1.0. W3C (World Wide Web Consortium) (1999), http://www.w3.org/TR/xpath
19. Chaieb, A.: Semiring normalization and Groebner bases. File Groebner_Basis.thy, in Isabelle2007 distribution (October 2007)
20. Kauers, M., Kerber, M., Miner, R., Windsteiger, W. (eds.): MKM/Calculemus 2007. LNCS (LNAI), vol. 4573. Springer, Heidelberg (2007)

A Mathematical Approach to Ontology Authoring and Documentation

Christoph Lange and Michael Kohlhase

Computer Science, Jacobs University, Bremen, Germany
{ch.lange,m.kohlhase}@jacobs-university.de

Abstract. The semantic web ontology languages RDFS and OWL are widely used but limited in both their expressivity and their support for modularity and integrated documentation. Expressivity, modularity, and documentation of formal knowledge have always been important issues in the MKM community. Therefore, we try to improve these ontology languages by well-tried MKM techniques.

Concretely, we propose embedding the language concepts into OMDoc to make use of its modularity and documentation infrastructure. We show how OMDoc can be made compatible with semantic web ontology languages, focusing on knowledge representation, modular design, documentation, and metadata. We evaluate our technology by re-implementing the Friend-of-a-friend (FOAF) ontology and applying it in a novel metadata framework for technical documents (including ontologies).

1 Introduction

The concept of an "ontology" as a formalization of a shared conceptualization is at the heart of the semantic web – the web of data and intelligent agents. RDFS (RDF Schema/Vocabulary Description Language [BG04]) and OWL (Web Ontology Language [MvH04]), the major semantic web ontology languages, have a limited expressivity: The common OWL sublanguages OWL-Lite and OWL-DL implement two different description logics – decidable subsets of first-order logic [BCM+07]. This was a deliberate design goal as decidability is a prerequisite for web scalability. A common experience in ontology design is, however, that certain axioms in the domains to be modeled exceed the expressivity of the languages chosen for implementation. Sometimes, dumbing down the model to less expressive special cases[1] is sufficient, whereas in other cases, a prose description of the actual axiom is added to the documentation of the ontology.

An example for the latter can be seen in the Friend-of-a-Friend (FOAF) ontology [BM07] for modeling user profiles and simple social relationships: Usually, a *foaf:Group* has members of type *foaf:Agent*, where an agent can be a group, a person, or an organization. The *foaf:membershipClass* property can be used to be more specific about the type of the members of a group by linking an instance

---

[1] as has, e. g., been done for the DOLCE ontology, a simplified version of which has been formalized in OWL-DL; cf. http://www.loa-cnr.it/DOLCE.html

J. Carette et al. (Eds.): Calculemus/MKM 2009, LNAI 5625, pp. 389–404, 2009.

of *foaf:Group* to an RDFS or OWL *Class*. We can, e. g., require that all members of the KWARC research group in Bremen be computer scientists. Then, if we state that Michael is a member of KWARC, we would like a reasoner to infer that he is a computer scientist, or, vice versa, to complain, if he is classified as a type of person that is not consistent with being one. This combination of ABox and TBox (instance- and terminology-level) reasoning is not supported by OWL reasoners, though. Therefore, *foaf:membershipClass* is not formally described in the OWL-DL implementation of FOAF, but an informal text in the specification explains how application developers can implement hand-crafted support for the missing inference step[2]. Such informal descriptions are often ambiguous[3] and have to be turned into algorithms manually.

In the MKM domain, tensions between high expressivity desired by authors and decidability or even tractability required for web-scalable automated inference are well-known. Earlier, we have discussed the problem of representing expressive mathematical knowledge, such as the theorem that all differentiable functions are continuous and its proof, which involves higher order logic, in semantic web systems [LK08]. This paper proposes a solution by applying well-tried techniques from mathematical knowledge representation to semantic web ontology engineering. We show how the expressive mathematical markup language OMDOC can be used to express and document semantic web ontologies in a way that complies with existing semantic web tools. We also discuss the particular requirement of extensible metadata vocabularies for ontology documentation, which we address by applying our technologies. We evaluate our approach by applying it to FOAF and conclude with a survey of related work and a summary and outlook. This paper is based on [LK09], which provides additional details.

2   Mathematical Semantic Markup with OMDoc

OMDOC [Koh06] is a three-layered semantic markup language for mathematical knowledge. On every layer, the author is free to choose the degree of formality; anything from informal text to shallow annotations to a full formalization (as needed for symbolic computation or automated deduction) is possible. **Objects** can be complex numbers, derivatives, etc. They are usually composed of *symbols* and represented in content markup, using OpenMath [BCC+04] or MathML [W3Ca]. **Statements** are made about objects and model knowledge about our environment in the respective domain. Statement types include model assumptions, their consequences, hypotheses. They have in common that they state relationships between objects and have to be verified or falsified in theories or experiments. A model is fully determined by its assumptions

---

[2] Note that a way has been found to replace *foaf:membershipClass* by a semantically equivalent OWL-DL-compatible construct using property restrictions [Alf07]. Nevertheless, we keep this as an example as it is easy to understand, the proposed solution has not yet been officially implemented, and is less intuitive for non-experts.

[3] as can be seen in the mail thread following [Alf07].

(also called *axioms*); all consequences are deductively derived from them (via *theorems* and *proofs*); hence, their experimental falsification uncovers false assumptions of the model. **Theories** put symbols and statements into a context. Even the meaning of a single symbol is determined by its context – e. g., the identifier h can stand for the height of a triangle or Planck's quantum of action, and – depending on the current assumptions – a statement can be true or false. While mathematicians fix and describe the context of a statement, these structures have to be modeled explicitly for computer-supported management. For instance, in logic, a theory is the deductive closure of a set of axioms, that is, the (often infinite) set of logical consequences of the model assumptions. Even though, in principle, this fully explains the phenomenon of context, important aspects like the reuse of theories, knowledge inheritance, and the management of theory changes are disregarded completely. Finally, **documents** consist of narrative and content layers. Content layers contain statements or theories, whereas narrative layers sequentially order snippets from content layers. This facilitates the reuse of content from a shared knowledge base (also called "content commons") in documents that are actually consumed by humans: scientific articles, books, or slide shows.

One can easily identify the following correspondences between the semantic web ontology languages RDFS/OWL and OMDoc: **Classes**, **Properties**, and **Individuals** correspond to objects or symbols. **Axioms** and **Rules** correspond to statements, as they state properties of resources. However, a distinction between proper axioms and facts derived from them is not usually made in ontologies. OMDoc, following the "little theories" approach [FGT92], allows for modeling this distinction and thus reducing theories to their core, while still enabling authors to document selected logical consequences of this core within the same theory. **Ontologies** correspond to theories. Both are often designed modularly and import other ontologies or theories. Both entities of an ontology and symbols of an OMDoc theory are identified by URIs (Uniform Resource Identifiers [BLFM05]) within the namespace of the whole theory/ontology.

We claim that OMDoc particularly performs better in *integrated documentation* and *modularity*. It supports mixing formal, semiformal, and informal knowledge in a literate-programming style, and integrating this into documents that can then be adapted to human audiences (cf. Sect. 3.4). As RDFS/OWL axioms could be *reified*, i. e. treated as resources of their own, by giving them a URI, one could in principle attach documentation to all parts of an ontology. In practice, this is supported less well. RDFa as a way of embedding ontologies into XHTML documents [ABMP08] and certain semantic wikis supporting ontology authoring (e. g. IkeWiki [Sch06]) are notable, but to date not yet completely adequate exceptions (cf. Sect. 6 for a discussion of RDFa for ontology authoring). Modularity in semantic web ontologies is optional at best: In RDFS, entities from external ontologies can be reused without restrictions, just by writing down their URIs. This does not make dependencies explicit at all and can easily lead authors into creating inconsistency. If possible at all, one would have to collect all URI references mentioned in an RDFS ontology and then apply some heuristics to these URIs in order to get hold of the actual ontologies depended upon. OWL improves on this by allowing explicit imports of ontologies via the *owl:imports* declarative – which only permits literal reuse of imported symbols, though. OM-DOC greatly enhances modularity by supporting imports via theory morphisms (symbol or formula mappings) and allows for parametric theories. Even literal imports are not yet widely used in web ontologies, and tools usually do not enforce their usage; improvements are to be expected with a more widespread adoption of OWL 2 [CGHM+08]. OMDOC applications rely on proper imports and can already check their consistency; see [RK08] for details.

3   OMDoc as a Semantic Web Ontology Language

OMDOC is XML-based and thus complies with basic web standards like URIs, and any desired logical foundation can be formalized in OMDOC. We can thus make use of the similarities to semantic web ontology languages pointed out above and use OMDOC for modeling ontologies – provided that we overcome certain obstacles, which are addressed in the following subsections: 1. Since OM-DOC is uncommitted to a particular logical foundation, it does not have a native understanding of the RDF[4], RDFS, and OWL(-DL) syntax and semantics. Therefore, these foundations have to be modeled as OMDOC meta-theories first. 2. OMDOC theories can import other theories for a modular design, but they cannot directly reference existing semantic web ontologies in order to enhance them. Therefore, we have to specify an import syntax and semantics. 3. OMDOC itself is not supported by any description logic reasoner[5]. Therefore, we need to provide a way to extract semantic web ontologies from theories.

3.1   Knowledge Representation

As a foundation for expressing semantic web ontologies in OMDOC, we wrote theories for RDF, RDFS, and OWL, which declare as symbols all classes, properties, and individuals of these languages. An ontology is then written as follows: Classes, properties, and individuals are declared as *symbols* with a *type*[6]. The type of an object property is, e.g., *owl#ObjectProperty*, i.e. the symbol *ObjectProperty* from our *owl* theory. Class definitions like "Student = Person ⊓ ≥ 1 enrolledIn" ("A student is a person, and is enrolled at least once") are given

---

[4] RDF (Resource Description Framework [RDF04]) is the foundation of knowledge representation on the semantic web. It represents knowledge as a graph, where nodes are instances of *classes* defined in ontologies, edges are instances of *properties*. An edge is usually read as a "subject predicate object" triple.

[5] There are converters from and to the native languages of several common first-order or higher-order theorem provers, though, which demonstrate OMDOC's utility as a mathematical exchange format.

[6] OMDOC has a foundationally unconstrained infrastructure for type systems: objects can be associated with types that are objects themselves. The particular choice of types is only governed by the available theories. Here we define types as part of the RDF, RDFS, and OWL theories.

as OMDOC *definitions* (cf. Listing 1.1[7]). This is a machine-oriented representation that a user would not usually see, but which would render as three lines in Figure 2 and be edited by a tool like the OMDOC-based semantic wiki SWIM [Lan08b, LGP08] using a dedicated formula editor (cf. Fig. 1).

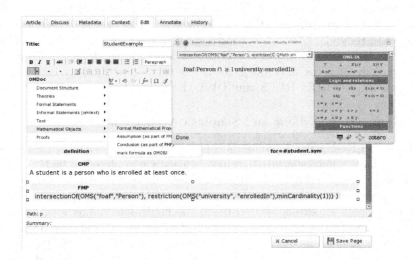

Fig. 1. The definition in SWIM, using the Sentido formula editor. The formula can be edited in OWL abstract syntax [MvH04], or using the tool palette.

Listing 1.1. An OWL ontology in OMDOC: class definition and documentation

```
<theory name="university">
  <imports from="owl.omdoc#owl"/> <!-- The OWL meta-theory -->
  <imports from="foaf.omdoc#foaf"/> <!-- OMDoc wrapper for FOAF -->
  <omtext type="introduction"><CMP>For our "university" ontology, we first import
    FOAF and then introduce the concept of a student. ...</CMP></omtext>
  <symbol name="Student" xml:id="student.sym">
    <metadata>
      <meta property="dc:description">A student</meta>
    </metadata>
    <type system="owl">
      <OMOBJ xmlns="http://www.openmath.org/OpenMath">
        <OMS cd="owl" name="Class"/></OMOBJ></type>
  </symbol>
  <!-- left out a similar declaration of enrolledIn -->
  <definition for="#student.sym" type="simple">
    <CMP>A student is a person who is enrolled at least once.</CMP>
    <OMOBJ xmlns="http://www.openmath.org/OpenMath">
      <OMA>
        <OMS cd="owl" name="intersectionOf"/>
        <OMS cd="foaf" name="Person"/>
        <OMA>
          <OMS cd="owl" name="Restriction"/>
          <OMS cd="university" name="enrolledIn"/>
          <OMA>
```

---

[7] OMS is the OpenMath syntax for a symbol. OMA applies a symbol (usually a function or an operator) to some arguments. OMI is an integer.

```
      <OMS cd="owl" name="minCardinality"/>
      <OMI>1</OMI></OMA></OMA></OMA></OMOBJ>
  </definition></theory>
```

All other statements can be expressed as OMDoc *axioms* in such a way that a property is applied to two arguments: a subject and an object. This is the most direct way of representing RDF in OMDoc but does not take advantage of the higher expressivity of OMDoc. However, the author has the possibility to annotate redundant axioms (as introduced in Sect. 2) as *theorems* instead, which can then be proven on the OMDoc level, using other axioms of the same ontology plus the inference rules of the respective ontology language, as represented in the RDF, RDFS, and OWL theories.

3.2   Connecting OMDoc and Semantic Web URIs

OMDoc and RDF have different ways of giving URIs to symbols. RDF-based ontologies have a namespace URI, which is usually considered to be the URI of the ontology, and all entities within the ontologies have local names. An absolute URI is formed by concatenating the namespace URI and a local name.

OMDoc uses an extended URI-based mechanism for addressing semantic objects. Following the addressing schemes of OPENMATH and MATHML3, we can address objects by their local name n in their home theory θ, which in turn is referenced by an import path in an OMDoc document identified by a URI g. Thus the URI of a semantic object is of the form $g?\theta?n$; see [RK08] for details. OMDoc allows theory inheritance via renamings – a crucial feature for modularity and ontology interoperability. As a consequence the semantic URIs of OMDoc go beyond traditional URIs and allow to reference objects that are only virtually represented by inheritance.

This difference is largely conventional and does not hinder the integration of OMDoc with RDF-based semantic web ontologies. The only situation where the difference needs to be overcome is where an existing semantic web ontology is rewritten in OMDoc, e.g. for the purpose of documenting it or making its modular structure more explicit, and whenever an OMDoc ontology imports a semantic web ontology. In order to have OMDoc ontologies generate RDF-style URIs, we allow for attaching the namespace URI of the original ontology to a theory via the special metadata field `odo:semWebBase`, which is recognized by our OMDoc→OWL translation presented in the following section. Here is how this would be done for FOAF:

```
<theory name="foaf">
  <metadata>
    <link rel="odo:semWebBase" href="http://xmlns.com/foaf/0.1/"/>
    <meta property="dc:title">Friend of a Friend (FOAF) vocabulary</meta>
  </metadata>
  <!-- imported theories and ontologies left out -->
  <symbol name="Agent"><!-- declaration omitted --></symbol>
  <!-- ... --></theory>
```

This makes sure that the OMDoc→OWL translation gives the *Agent* class its correct URI, i.e. `http://xmlns.com/foaf/0.1/Agent`. We can create an OMDoc theory from a semantic web ontology by simply providing a suitable `odo:semWebBase` metadata field, only adding symbol declarations, definitions, axioms, etc., later. This is a low-cost way for starting OMDoc-based ontologies which, does not preclude making use of OMDoc's possibilities for documentation and expressive knowledge representation later. Thus we have a suitable migration path from web ontologies to OMDoc.

3.3   Reasoning

Our intention with promoting OMDoc as a more expressive semantic web ontology language is not to replace well-tried technologies for semantic web *reasoning*. While OMDoc does, in principle, allow for alternative approaches to reasoning, being an exchange format for automated theorem provers, this is not the focus of this paper. So in order to allow for writing expressive ontologies in OMDoc while still being able to use optimized reasoners on their tractable/decidable fragments, we defined and implemented a translation from OMDoc to OWL as a module within our Krextor XML→RDF extraction framework [Lan08a]. While the implementation is hard-coded, we aim at giving an exact specification by OMDoc axioms: There is, for example, a set of direct subject–predicate–object axioms (cf. Sect. 3.1) in the OWL theory that state that any application of the *owl#Restriction* symbol to suitable arguments translates to an anonymous RDF resource of type *owl:Restriction* that has certain RDF properties. Extracting RDF triples from OMDoc symbol declarations and axioms is mostly straightforward, but the generation of correct URIs for entities of semantic web ontologies is more involved. We traverse the graph of theory imports and collect the namespace URIs of all theories that carry an `odo:semWebBase` metadatum. Whenever we encounter a reference to a symbol *onto#sym* for an ontology that is implemented as an OMDoc theory *onto*, we generate the semantic web compliant URI as the concatenation of the namespace URI of the theory and the name of the symbol. Here is the RDF generated from the example introduced in Listing 1.1 above[8]:

```
<.../uni.omdoc?university>      rdf:type    owl:Ontology ;
                     owl:imports foaf: .
<.../uni.omdoc?university?Student> rdf:type    owl:Class ;
                     owl:equivalentClass _:d24e43 .
_:d24e43             owl:intersectionOf  _:collection-d24e44 .
_:collection-d24e44         rdf:first   foaf:Person ;
                     rdf:rest    _:collection-d24e44-1 .
_:collection-d24e44-1       rdf:first   _:d24e47 ;
                     rdf:rest    rdf:nil .
_:d24e47             rdf:type    owl:Restriction ;
                     owl:onProperty
                     <.../uni.omdoc?university?enrolledIn> ;
                     owl:minCardinality "1"^^xsd:nonNegativeInteger .
```

The result looks is somewhat illegible (compared e. g. to Fig. 2); in fact there are less technical representations of OWL [HPS08], but in practice it does not

---

[8] This is Turtle, a text-oriented serialization for the RDF data model. Identifiers prefixed with _ denote anonymous ("blank") nodes that are only accessible within the current RDF graph. The class, which a student is defined to be equivalent to, is represented as a union class of a set of classes, represented as a linked list.

make a difference, as all OWL tools are required to support the RDF representation. Most of the statement- and theory-level structure of OMDoc, such as the distinction between defined and inferred statements and theory morphisms, is lost and uniformly translated to less expressive OWL axioms. Thus, our translation works like a compiler and linker that creates (OWL/RDF) object code from a higher-level OMDoc source code.

3.4   Documentation and Presentation

OMDoc comes with an elaborate, adaptive presentation framework for creating human-readable documents from semantic markup [KMR08]. Mathematical formulae are rendered as Presentation MathML; structures on the statement and theory levels, and complete documents, are rendered as XHTML. For every mathematical symbol, one or more *notation*s can be defined – compare, e.g., our initial OWL example in the German DL notation (Student = Person $\sqcap \geq$ 1 enrolledIn) vs. the Manchester syntax [HPS08]:

```
Class: Student
EquivalentTo: Person that enrolledIn min 1
```

A default notation is usually provided by the author of a theory, but users can also author their own ones to customize the presentation to their preferences. Initially, the renderer collects all available notation definitions from all imported theories. For every symbol in a content formula as the one in Listing 1.1, the renderer selects from those notation definitions that match the symbol the most appropriate one for the current presentation context, which is made up of, e. g., the language of the enclosing document, the domain of application, or user preferences. The output is parallel markup [W3Ca, section 5.4], which allows for implementing additional services that facilitate browsing and reading – for example linking rendered symbols to the place where they are introduced. A reader who does not know, e.g., the symbol $\sqcap$ in our sample formula, can click on it and thus navigate to the section of the document rendered from the *owl* OMDoc theory that declares (and documents!) the symbol *owl:intersectionOf*. We have implemented this in SWiM using XLinks; the JOBAD active document framework even displays definitions as tooltips without forcing the user to leave the document [GLR09]. Documentation can be given in metadata blocks (cf. Sect. 4), which can be attached to any element on the statement and theory level (cf. Listing 1.1). Textbook or literate-programming style is also possible: A theory can not only contain formal statements but also informal text sections, and *definitions*, *axioms*, and *theorems* can have both formal and informal content (*CMP* and *FMP*; cf. Listing 1.1).

4   Scalable Metadata for Technical Specifications

In the previous sections, we have already used metadata for documenting ontologies. Simple metadata vocabularies like Dublin Core (DC [Dub08]) or Creative Commons licensing information (CC [AALY08]) are suitable for retrieval, e. g.

Friend of a Friend (FOAF) vocabulary

imports from: wordnet, dc, simpletypes, owl, quant1, logic1

AXIOM:

The foaf:Person class is a sub-class of the foaf:Agent class, since all people are considered 'agents' in FOAF.

Person ⊑ Agent

AXIOM: Person ⊓ Organization = ⊥

CONCEPT: **made**

The foaf:made property relates a foaf:Agent to something foaf:made by it.

TYPE: **(owl)**

ObjectProperty

TYPE: **(simpletypes)**

(Agent, Thing)

AXIOM: made = maker⁻

LEMMA: maker = made⁻

PROOF: 1. We know that made = maker⁻ .

2. Interpreted using the model-theoretic semantics, this means that

$made^I = (maker^-)^I = (maker^I)^-$.

3. Now we apply the inverse on both sides, eliminate double inverses, and obtain

$(made^I)^- = ((maker^I)^-)^- = maker^I$

4. This is just the interpretation of maker = made⁻ , which we had to prove.

CONCEPT: **membershipClass**

The foaf:membershipClass property relates a foaf:Group to an RDF class representing a sub-class of foaf:Agent whose instances are all the agents that are a foaf:member of the foaf:Group. See foaf:Group for details and examples.

AXIOM: $\forall m, g, C.(g \ni_{member} m \wedge membershipClass(g, C) \Rightarrow m :_{type} C)$

Fig. 2. FOAF in OMDOC, rendered (slightly shortened cf. Sect. 5). We defined some custom notations, e. g. rendering *foaf:member* like set membership, and we combined domain and range of a property into a "relation type".

using a search engine. Specialized document and knowledge management tasks require more complex metadata. In practical application scenarios where OM-DOC is used to author formal specifications of safe and secure technical devices, we have particularly experienced a need for documenting the change history of a formal document within that document. Note that a revision log within a document is not intended to replace a versioned repository on the server side – which we also use for OMDOC –, but as an extension for certain use cases. Sometimes, for example, a persistent revision log is required for legal reasons.

4.1 Metadata in OMDOC 1.2

OMDOC allows for attaching a metadata record to any element on the document, theory, and statement level [Koh06, chapter 12]. The current version 1.2 provides XML syntax for all DC and CC properties, plus a few extensions[9], most notably a simple vocabulary for recording revision histories, which has been added to the `dc:date` XML element: The additional `who` attribute refers to the URI of a `dc:creator` or `dc:contributor` element in the same metadata record, and the `action` attribute can have values like "updated", "created", or "imported".

This way of representing metadata has various drawbacks: The vocabulary is hard-coded and not extensible. There is no easy way of adding other vocabularies to OMDOC. Secondly, OMDOC is not aware of the formal semantics of these vocabularies. They have been integrated into the *syntax* of OMDOC, but their

---

[9] Here, we only give a short summary. Please see the OMDOC 1.2 specification [Koh06] and the extended version of this paper [LK09] for details.

semantics is only available informally as a part of the natural-language specification of OMDOC [Koh06, chap. 12]. More formal semantics for DC and CC
would be available as RDFS ontologies, but those have not been incorporated
into OMDOC. Even worse, OMDOC's DC extension for revision histories does
not have any formal semantics at all. This lack of formal semantics has restricted
the attractivity of OMDOC's metadata for application developers. So far, support for them has not been implemented by any OMDOC-aware application,
except our own semantic wiki SWiM [Lan08b] and the e-learning environment
ActiveMath [GUM+04]. ActiveMath makes use of additional vocabularies for
educational metadata, but they are hard-coded into the XML schema in an even
less extensible way than in OMDOC 1.2, as they are not distinguished by different
namespaces [GUM+04] (ActiveMath's document format forked off OMDOC1.1.).

4.2   The New Metadata Framework

Requirements for a new metadata framework for OMDOC were as follows:

1. Stay backwards-compatible with OMDOC 1.2 concerning expressivity. That
   is, continue supporting DC and CC, and the custom extensions.
2. Make the formal semantics of vocabularies available to OMDOC applications.
3. Incorporate vocabularies for versioning (for technical documents in particular) and people (for bibliographical data).
4. Don't hard-code a fixed set of vocabularies into the language but stay flexible
   and extensible for many applications, even future and unknown ones.

Given the fact that many existing metadata vocabularies, including DC and
CC, have an RDF semantics, and that with RDFa [ABMP08] a standard for
flexibly embedding metadata into X(HT)ML documents had recently stabilized, we chose to incorporate a subset of RDFa into OMDOC, and to look
for RDF-compatible metadata vocabularies to satisfy our further requirements.
So far, RDFa has only been specified for the "host languages" XHTML and SVG
(cf. [W3Cb]), but the specification foresees the integration into other XML-
based languages. The new metadata framework introduces the elements `meta`
and `link` with the same semantics as their XHTML counterparts as children of
any `metadata` block. Resources with document-local identifiers only, i.e. *blank
nodes*, can be created using the `resource` element:

Element	Attributes	Children
meta	property, content, datatype	literal text or XML (optional)
link	rel, rev, href	(resource|meta|link)*
resource	about, typeof	(meta|link)*

Due to the inherent flexibility of RDFa, any metadata vocabulary can be
used. However, we give particular recommendations for metadata in the above-
mentioned domains of special interest. Using DC and CC metadata with the
new RDFa syntax for OMDOC is trivial. Our previous DC extensions for revision logs were not immediately RDF-compatible, as they were given as additional

annotations to triples, and no formal semantics was defined for them. Therefore, we replaced them by a completely re-engineered versioning ontology. This
ontology reuses the core of the ModelDriven.org versioning ontology [Mod08],
with classes *DataAsset* (of which anything on the statement, theory, or document level of OMDOC is a subclass), *Revision*, and *Change*, where an *DataAsset*
has *Revisions*, and a *Change* represents a transition from one *Revision* to the
following one. As we made *Change* a subclass of the *Event* class from the event
ontology [RA07], a change can have a date and an agent. Instead of a generic
Change, a more specific subclass can be chosen. In future, we plan to introduce
specific change types (e. g. for adding a type declaration to a symbol), in a similar way as the OMV Ontology Metadata Vocabulary does for semantic web
ontologies [HPHGP07].

Here is a part of the metadata block of a digital library edition of Fermat's last
theorem that documents the revision history. The resource has two revisions; for
each, the act of creation has an author and a date given as additional metadata:

```
<link rel="rev:created_by_act" href="[_:creation]"/>
<link rel="rev:current_version" href="[_:current]"/>
<link rel="rev:has_version">
  <resource about="[_:v1]" typeof="rev:Revision">
    <link rel="rev:content" href="fermats−last−theorem?rev=1"/>
    <link rel="rev:created_by_act">
      <resource about="[_:creation]" typeof="chg:Creation">
        <link rel="event:agent" href=".../Pierre_de_Fermat"/>
        <meta property="dc:date">1637−06−13T00:00:00</meta>
      </resource></link></resource></link>
<!−− revision 2 (proof by Wiles) omitted to save space −−>
<link rel="rev:has_version">
  <resource about="[_:current]" typeof="rev:Revision">
    <link rel="rev:content" href="fermats−last−theorem?rev=3"/>
    <link rel="rev:created_by_act">
      <resource typeof="chg:Import">
        <link rel="event:agent" href="http://.../kohlhase"/>
        <meta property="dc:date">2006−08−28T00:00:00</meta>
        <link rel="rev:prior_version" href="[_:v2]"/>
      </resource></link></resource></link>
```

As we modeled our metadata ontologies in OMDOC, we are now able to extend
it by a formal specification of certain rules that had only informally been stated
in the OMDOC 1.2 specification: for example, that most DC metadata propagate from document sections down into subsections unless subsections specify
different values, or that any *dc:creator* of a subsection of a document becomes
a *dc:contributor* to the whole document.

4.3   Extracting Metadata to RDF

Similarly to the extraction of RDF representations of OWL ontologies written in
OMDOC (cf. Sect. 3.3), we implemented a Krextor extraction module for RDFa.
We then divided the RDFa extraction rules into XHTML-specific ones and into
generic ones, the latter of which we combined with support for our OMDOC-
specific metadata syntax. The extraction of RDFa from OMDOC is performed
both in the extraction of OWL from OMDOC, where it enriches the extracted
ontologies with metadata, and in the extraction of RDF outlines from OMDOC
in terms of the OMDOC's own document ontology. The latter is a foundation

for semantic web applications having OMDOC (and not OWL) as their native language, such as the semantic wiki SWiM [Lan08b].

4.4 Annotation

As the listing in Sect. 4.2 shows, the new RDFa-based metadata syntax is much more verbose than the old one of OMDOC 1.2. Therefore, we suggest two ways of facilitating the annotation: For manual authoring, we keep the old, "pragmatic" OMDOC 1.2 syntax and specify a transformation of such annotations to the new, "strict" RDFa syntax – implementable, e. g., in XSLT. Having a rich pragmatic syntax that is convenient to author and a strict syntax that is more suited for automated processing and validation is actually a general strategy that we first introduced in MathML 3 and also employ for other aspects of OMDOC. In certain application settings, we can generate part of the metadata automatically. In the SWiM wiki [Lan08b], for example, the names of the author and the contributors of a document are known from the user profiles of these persons and only inserted into the metadata record of a document when it is exported from the wiki to a file. The same holds for the revision history.

5   Evaluation and Discussion

We evaluated our approach on a reimplementation of FOAF in OMDOC. From studying the OWL implementation and the specification of FOAF, we noticed the following problems, which we were able to solve using OMDOC:

1. FOAF references entities from other ontologies (DC, WordNet, Geo Positioning, etc.), but it does not import them. With OMDOC tools (as described in [RK08]), we can identify imports missing in an OMDOC ontology, and our OMDOC→OWL translation (Sect. 3.3) adds them to the OWL ontology resulting from the translation.
2. The source code contains notes for developers as XML comments. In the OMDOC version of FOAF, we were instead able to create informal text sections for them. Other XML comments divide the ontology into sections like "naming properties". In OMDOC, we were able to model document sections without disrupting the logical structure of the ontology.
3. Some of these comments were attached to individual triples, e. g. *foaf:mbox_ sha1sum rdf:type owl:DatatypeProperty*. Thanks to literate programming in OM-DOC, we could precisely add them as informal comments (*CMPs*) to the respective OMDOC statements.
4. The following properties are inverses of each other: $foaf{:}maker = foaf{:}made^-$, $foaf{:}depiction = foaf{:}depiction^-$, $foaf{:}topic = foaf{:}page^-$, and $foaf{:}primaryTopic = foaf{:}isPrimaryTopicOf^-$. While for each $p = q^-$, an OWL reasoner can infer $q = p^-$, using its built-in axioms for DL reasoning, FOAF redundantly declares each inverse relationship for both participating properties for the purpose of documentation. OMDOC allows for making the difference explicit: For any of the above p, q property pairs, we picked one p and stated $p = q^-$ as an axiom, but $q = p^-$ as an *assertion* that can (provably) be derived from the axiom and the semantics of *owl:inverseOf*, as shown in Fig. 2. Domain and range of inverse properties can be handled similarly.

5. We were able to express the non-OWL semantics of *foaf:membershipClass* (cf. Sect. 1). We chose the first-order-logic representation shown in Fig. 2.
6. The correspondence of *foaf:maker* to *dc:creator* is only defined in prose. The specification suggests using *foaf:maker* whenever the agent who created something is known by URI, and to use the less semantic *dc:creator*, which neither has range nor domain declared, when the creator is only known by a string. Then, it also informally states a rule that the *foaf:name* or *rdfs:label* of the *foaf:maker* of something is the same as the *dc:creator* of that thing. The rule can be captured by a first-order-logic expression in OMDOC, or alternatively by an OWL 2 property chain inclusion [CGHM$^+$08]. The notion that *foaf:maker* is similar to *dc:creator* but has a stronger semantics can be captured by having the FOAF theory import the DC theory and defining a *view* on DC, namely a morphism that maps *dc:creator* to *foaf:maker*. Views frequently occur in mathematics. We can, for example, model the theory of integers by a view $\{\circ \mapsto +, e \mapsto 0\}$ on the theory of monoids, where $\circ$ is the binary operation and e the unit element of the monoid.
7. Finally, we were able to include the informal sections and descriptions of the FOAF specification [BM07] right into the ontology document. This allows for a unified management of the formal specification and its informal explanation, including the introductory chapters and the change log, in a single, coherent document, of which both OWL and XHTML can be generated. The original FOAF specification is generated from the OWL ontology and a set of HTML snippet files with detailed informal descriptions as input using a script, a FOAF-independent version of which is also available [Boj].

This enhanced expressivity of the OMDOC implementation comes at the expense of much more verbosity. While in RDF one can easily attach another axiom to a class (stating, e.g., a subclass relationship or disjointness), most of these triples have to be represented as a individual axiom in OMDOC, unless there is an intuitive way of capturing their semantics as types. While better annotation tools could help (cf. Sect. 4.4), there is also a mathematical approach to improving this: One could add additional axioms to the OMDOC theory for OWL, which introduce operators for shorthand notations (such as pairwise disjointness of a whole set of classes) that imply multiple atomic statements– but then all these axioms would have to be *applied* before generating OWL from OMDOC. This can be done by supporting λ-calculus at the meta level and β-reducing all OMDOC axioms before generating OWL.

As the new metadata framework has not yet been deployed to the OMDOC users, our evaluation focused on the coverage of the RDFa extraction. We first implemented XHTML+RDFa support and then generalized that, so we could evaluate our implementation against the W3C RDFa test suite [HY07], of which it currently passes 90 out of 100 test cases.

6   Related Work

Concerning **expressive ontologies**, the Common Algebraic Specification Language (CASL) and its extensions for various logics are related to OMDOC and its module system. For the CASL-based Heterogeneous Tool Set (HETS), it has been investigated how to integrate OWL-DL and more expressive logics within a logical framework [KLMN08]. However, CASL is a purely formal language and does not allow for integrating documentation. Concerning **integrated ontology documentation**, RDFa in combination with RDF-based ontologies is similar to our approach. RDFa has mainly been used for ABox knowledge so far; we are only aware of one application of RDFa for TBox knowledge: Ontology Online is a web site for browsing and querying OWL and RDFS ontologies. Every page visualizes one entity of an ontology – as XHTML with the original OWL or RDFS embedded as RDFa annotations [Dec07]. **Metadata for technical specifications** are supported by DocBook [Wal08], a semantic markup language that had originally been conceived for software documentation. DocBook has hard-coded, non-extensible markup for metadata, covering general DC-like metadata, revision histories, and more. None of this has an RDF semantics. There is, however, a workaround for adding RDF-compatible annotations to DocBook: Any DocBook element can carry XLink attributes, from which RDF can be harvested [Dan00].

7   Conclusion and Further Work

By connecting the semantic markup language OMDOC and techniques from MKM to the semantic web standards OWL and RDFa, we contributed a language for ontologies and technical specifications that supports different levels of expressivity and formality but still remains compatible with the existing semantic web infrastructure. As an anonymous reviewer pointed out, we use (MK)M – i. e. techniques from the management of mathematical knowledge – for M(KM) – i. e. for managing knowledge in a mathematical way. We consider our scalable metadata framework applicable to other MKM languages, such as OpenMath CDs, as well. For example, it has been proposed to add a metadata field for the author to OpenMath 3 content dictionaries[10]. Simply employing RDFa with an appropriate ontology would facilitate such decisions.

Our next planned step is identifying further possibilities to modularize the ontologies that we have implemented in OMDOC so far – including the OMDOC formalizations of RDF, RDFS, and OWL –, or making existing modularity more explicit, and then integrating our OMDOC→OWL translation with HETS to enable heterogeneous reasoning [KLMN08][11]. Finally, we want to apply existing OMDOC applications to ontologies written in OMDOC, enhance the SWiM wiki by user interface elements for more conveniently editing and browsing such ontologies, and evaluate its usability in a case study involving ontology engineers. Our group is working on a distributed database for mathematical documents (TNTBase [Zho09]). This database will also employ our document renderer and then follow the practice of content negotiation that is well established on the semantic web [SC08]: OMDOC-aware clients will get OMDOC, semantic web clients will get extracted RDF, and web browsers will get XHTML+MathML – a foundation for a *mathematical semantic web*.

---

[10] See `http://trac.mathweb.org/OM3/ticket/12`

[11] See `http://trac.kwarc.info/krextor/roadmap` for work in progress.

Acknowledgments. We would like to thank Florian Rabe for help with modeling RDF, RDFS, and OWL as OMDOC theories, Siarhei Kuryla for his contributions to the implementation, and Richard Cyganiak for sharing insights about RDFa. This work was supported by JEM-Thematic-Network ECP-038208.

References

[AALY08]   Abelson, H., Adida, B., Linksvayer, M., Yergler, N.: ccREL: The Cre-
           ative Commons Rights Expression Language. Technical report, Creative
           Commons (2008),
           `http://wiki.creativecommons.org/Image:Ccrel-1.0.pdf`
[ABMP08]   Adida, B., Birbeck, M., McCarron, S., Pemberton, S.: RDFa in XHTML:
           Syntax and processing. Recommendation, W3C (2008)
[Alf07]   Alford, R.: Proposal: Deprecate membershipClass, add memberOf.
           E-mail (2007),
           `http://lists.foaf-project.org/pipermail/foaf-dev/2007-May/`
           `008551.html`
[BCC+04]   Buswell, St., Caprotti, O., Carlisle, D.P., Dewar, M.C., Gaetano, M.,
           Kohlhase, M.: The Open Math standard, version 2.0. Technical report,
           Open Math Society (2004)
[BCM+07]   Baader, F., Calvanese, D., McGuinness, D.L., Nardi, D., Patel-Schneider,
           P.F. (eds.): The Description Logic Handbook: Theory, Implementation,
           and Applications, 2nd edn. Cambridge University Press, Cambridge (2007)
[BG04]   Brickley, D., Guha, R.V.: RDF vocabulary description language 1.0: RDF
           Schema. Recommendation, W3C (2004)
[BLFM05]   Berners-Lee, T., Fielding, R., Masinter, L.: Uniform resource identifier
           (URI): Generic syntax. RFC 3986, IETF (2005)
[BM07]   Brickley, D., Miller, L.: FOAF vocabulary specification 0.91. Technical
           report, ILRT (2007)
[Boj]   Bojārs, U.: SpecGen 4 – ontology specification generator for RDFS and
           OWL, `http://code.google.com/p/specgen`
[CGHM+08] Cuenca Grau, B., Horrocks, I., Motik, B., Parsia, B., Patel-Schneider, P.,
           Sattler, U.: OWL 2: The next step for OWL. Web Semantics: Science,
           Services and Agents on the World Wide Web 6(4) (2008)
[Dan00]   Daniel Jr., R.: Harvesting RDF statements from XLinks. Note, W3C
           (2000)
[Dec07]   Decraene, D.: Online ontology visualisation: Embedding OWL-RDFS
           syntax in XHTML with RDFa (2007),
           `http://ontologyonline.blogspot.com/2007/11/`
           `embedding-owl-rdfs-syntax-in-xhtml-with.html`
[Dub08]   Dublin Core metadata element set, version 1.1. DCMI (2008)
[FGT92]   Farmer, W., Guttman, J., Thayer, X.: Little theories. In: Kapur, D. (ed.)
           CADE 1992. LNCS, vol. 607. Springer, Heidelberg (1992)
[GLR09]   Giceva, J., Lange, C., Rabe, F.: Integrating web services into active math-
           ematical documents. In: MKM/Calculemus 2009 Proceedings. LNCS
           (LNAI). Springer, Heidelberg (in press, 2009)
[GUM+04]   Goguadze, G., Ullrich, C., Melis, E., Siekmann, J., Gross, C., Morales,
           R.: LeActiveMath Structure and Metadata Model. Deliverable D6 (2004)
[HPHGP07] Hartmann, J., Palma, R., Haase, P., Gómez-Pérez, A.: Ontology Meta-
           data Vocabulary – OMV (2007), `http://omv.ontoware.org`

[HPS08]   Horridge, M., Patel-Schneider, P.F.: OWL 2 web ontology language: Manchester syntax. Working draft, W3C (2008)

[HY07]   Hausenblas, M., Yung, W.C.: RDFa test suite. Editor's Draft, W3C (2007), http://www.w3.org/2006/07/SWD/RDFa/testsuite/

[KLMN08]   Kutz, O., Lücke, D., Mossakowski, T., Normann, I.: The OWL in the CASL – designing ontologies across logics. In: Sattler, U., Dolbear, C., Ruttenberg, A. (eds.) OWL: Experiences and Directions (2008)

[KMR08]   Kohlhase, M., Müller, C., Rabe, F.: Notations for living mathematical documents. In: Autexier, S., Campbell, J., Rubio, J., Sorge, V., Suzuki, M., Wiedijk, F. (eds.) AISC 2008, Calculemus 2008, and MKM 2008. LNCS (LNAI), vol. 5144, pp. 504–519. Springer, Heidelberg (2008)

[Koh06]   Kohlhase, M.: OMDoc – An Open Markup Format for Mathematical Documents [version 1.2]. LNCS (LNAI), vol. 4180. Springer, Heidelberg (2006)

[Lan08a]   Lange, C.: Krextor – the KWARC RDF extractor (2008), http://kwarc.info/projects/krextor/

[Lan08b]   Lange, C.: SWiM – a semantic wiki for mathematical knowledge management. In: Bechhofer, S., Hauswirth, M., Hoffmann, J., Koubarakis, M. (eds.) ESWC 2008. LNCS, vol. 5021, pp. 832–837. Springer, Heidelberg (2008)

[LGP08]   Lange, C., González Palomo, A.: Easily editing and browsing complex OpenMath markup with SWiM. In: Libbrecht, P. (ed.) Mathematical User Interfaces Workshop (2008)

[LK08]   Lange, C., Kohlhase, M.: A Semantic Wiki for Mathematical Knowledge Management. In: Rech, J., Decker, B., Ras, E. (eds.) Emerging Technologies for Semantic Work Environments. IGI Global (2008)

[LK09]   Lange, C., Kohlhase, M.: A mathematical approach to ontology authoring and documentation (2009), https://svn.omdoc.org/repos/omdoc/trunk/doc/blue/foaf/note.pdf

[Mod08]   ModelDriven.org versioning ontology (2008), http://modeldriven.org/2008/ArchitectureOntology/doc/Versioning.html

[MvH04]   McGuinness, D.L., van Harmelen, F.: OWL web ontology language overview. Recommendation, W3C (2004)

[RA07]   Raimond, Y., Abdallah, S.: The event ontology. Technical report (2007), http://motools.sourceforge.net/event/

[RDF04]   Resource description framework (RDF) (2004), http://www.w3.org/RDF/

[RK08]   Rabe, F., Kohlhase, M.: An exchange format for modular knowledge. In: Rudnicki, P., Sutcliffe, G. (eds.) Knowledge Exchange: Automated Provers and Proof Assistants (KEAPPA) (2008)

[SC08]   Sauermann, L., Cyganiak, R.: Cool URIs for the semantic web. Working Draft, W3C (2008)

[Sch06]   Schaffert, S.: IkeWiki: A semantic wiki for collaborative knowledge management. In: 1st Workshop on Semantic Technologies in Collaborative Applications (STICA)(2006)

[W3Ca]   W3C. Mathematical Markup Language (MathML) 3.0, 3rd edn.

[W3Cb]   W3C. Scalable Vector Graphics (SVG) Tiny 1.2

[Wal08]   Walsh, N.: DocBook 5.0: The Definitive Guide. O'Reilly, Sebastopol (2008)

[Zho09]   Zholudev, V.: TNTBase (2009), https://trac.mathweb.org/tntbase/

A Logically Saturated Extension of $\bar{\lambda}\mu\tilde{\mu}$

Lionel Elie Mamane, Herman Geuvers, and James McKinna

Institute for Computing and Information Sciences
Radboud University Nijmegen
lionel@mamane.lu, herman@cs.ru.nl, james@cs.ru.nl

Abstract. This paper presents a proof language based on the work of Sacerdoti Coen [1,2], Kirchner [3] and Autexier [4] on $\bar{\lambda}\mu\tilde{\mu}$, a calculus introduced by Curien and Herbelin [5,6]. Just as $\bar{\lambda}\mu\tilde{\mu}$ preserves several proof structures that are identified by the λ-calculus, the proof language presented here aims to preserve as much proof structure as reasonable; we call that property being *logically saturated*. This leads to several advantages when the language is used as a generic exchange language for proofs, as well as for other uses.

We equip the calculus with a simple rendering in pseudo-natural language that aims to give people tools to read, understand and exchange terms of the language. We show how this rendering can, at the cost of some more complexity, be made to produce text that is more natural and idiomatic, or in the style of a declarative proof language like Isar or Mizar.

1 Introduction

Effective Mathematical Knowledge Management requires languages for the representation of proofs at a level that is aware of the logical reasoning without committing to the technical details of a proof representation in e.g. a proof assistant or a derivation system. We aim at a proof language that is general enough to capturase different notions of proofs and that captures the logical structure of a proof in detail; a language that differentiates distinct proofs, but identifies two texts that represent the same proof. Such a proof language can be used as a common ground for interchange between different systems, as a language to speak about proofs and transformations thereof (e.g. automatic proof enhancement, rendering into natural language, ...).

Another requirement of such a proof language would be to have a nice natural language-style pretty-printing; the latter transformation ideally being simple enough to be done in one's head, so that a term in that language be readable by itself for someone that knows the language. In this respect, the natural language transformation of $\bar{\lambda}\mu\tilde{\mu}$ in [1] is very attractive: the transformation is purely structural, and the term is read strictly from left to right. However, it does not satisfyingly treat the whole calculus and $\bar{\lambda}\mu\tilde{\mu}$ (as extended to predicate logic in Fellowship [3]) still identifies proofs we'd like to differentiate.

This paper presents a more discerning extension of $\bar{\lambda}\mu\tilde{\mu}$ and a basic rendering of it in pseudo-natural language. We show how the rendering can be enhanced to

J. Carette et al. (Eds.): Calculemus/MKM 2009, LNAI 5625, pp. 405–421, 2009.
© Springer-Verlag Berlin Heidelberg 2009

produce text that is more pleasing to read, and sketch how $\bar{\lambda}\mu\tilde{\mu}$ can be translated into the input language of proof assistants. The language presented here covers only implicational logic and disjunction (with an extension to full propositional logic given in the appendix and in [7]) and only proofs where every step taken is an atomic step of reasoning. So, seen from the viewpoint of proof assistants, we only deal with proofs where no automation is used. Naturally, the language will be extended in future work to address these limitations.

1.1   What Is a Proof?

From a logician's point of view, a proof is a derivation in a formal system of rules. From a more general mathematician point of view, it is a text that convinces his peers, for example a text that convinces that if they would spend enough time, they would be able to produce a fully formal derivation.

With our view centred on proof assistants, we consider the user input to the proof assistant to be a good candidate for the right notion of proof. A good test for the suitability of a candidate proof format is thus how well it captures these "proof assistant proofs".

This notion of proof is both coarser and finer than logician's proofs:

– It is coarser, because it glosses over automation done by the proof assistant; if the automation procedure changes, and finds a different logician's proof of a step done by automation, we still consider it the same proof.
– It is finer, because it separates cases where the same logician's proof (e.g. a natural deduction derivation) is produced in different ways, e.g. by a top-down proof or by a bottom-up proof, or by a proof that is partially top-down and partially bottom-up.

1.2   Design

Differentiating Power. We have already mentioned that we want our proof language to distinguish texts that code for different proofs, but identify texts that code for the same proof. This naturally begs the question: when do two texts represent different proofs and when do they represent the same proof?

We want to preserve the *intentional* content of a proof, the story that is being told. For example, a proof that first establishes A, then B and from these two concludes C is not the same as a proof that first established B, then A and then concludes. So we want our language to distinguish the order in which things happen, and to distinguish a forward-style (bottom-up) proof from a backward-style (top-down) proof, and to distinguish these from proofs that are done partially forwards and partially backwards.

As we focus on the logical content of proofs, it seems natural that we identify texts that vary only by purely linguistic differences. For example, the proofs at the right differ only linguistically from the corresponding proof at the left:

case 1: A holds ...	either A ...
case 2: $\neg A$ holds ...	or $\neg A$ ...

H: A by B	H: A by B
hence C	thus C by H

But if the difference comes from application of a different reasoning step, a different deduction rule, then it is not the same proof and the language should distinguish them. For example:

we have $A \rightarrow C \wedge B$	we have $A \rightarrow C \wedge B$
in particular we have $A \rightarrow C$	that is, we have $A \rightarrow C$ $\qquad (x)$
we already established A in lemma 5	and we have B $\qquad (y)$
thus C	by lemma 5 and x, we conclude C

The left proof uses a projection (from $A \wedge B$, we deduce *only* A), while the right proof uses a full decomposition (from $A \wedge B$, we deduce *both* A and B), it is not the same deduction rule.

Saturated System. In the design of a proof language, one usually tries to make it *minimal* at the logical level: A deduction rule that can be derived from others is considered superfluous and is therefore removed. We however, aim for a language that is *saturated*: Any step that an author can reasonably see as an atomic step, as a rule of reasoning that his reader will not doubt, should be a rule of the language. Any deduction rule that a proof assistant, or a logic, can reasonably choose as part of its "minimal set" should be a rule of the language. We don't claim that our system is the final answer to the quest for a saturated system, but we think we have come a long way. It should be tested on concrete proof examples to see whether anything is missing. We now give a pointwise discussion of the use of a saturated language for proofs.

• When the language is used as a proof *interchange* language, it allows the language to be neutral towards the choices of primitives made by different proof assistants; the language then is not closer to any one specific family of proof assistants than to another. By its saturation, it is close to all of them. A prime example of this is the implementation of classical logic: Some proof assistants implement intuitionistic logic augmented with an axiom, e.g. excluded middle. Others, such as PVS, use the reasoning with multiple goals of sequent calculus, where proving any one of the goals finishes the proof. A proof in one style can be transformed into the other style mechanically, but this produces a *different* proof of the same proposition. A language that provides classical logic solely through the double negation rule cannot faithfully represent PVS proofs.

• It makes the language particularly well suited to talk about proof manipulations, e.g. an algorithm that transforms proofs from one system into another. Because the saturated language has the concepts and rules from both systems, the transformation can be expressed as a transformation of terms of that language.

• It gives a tool to study and characterise what kinds of proofs a system can handle, because these can be expressed as sublanguage of the saturated language.

• When the language is used as a proof *authoring* language, it has the advantage to present all choices in a uniform way, without arbitrary distinction between which (in the user's intuition) atomic step is atomic for the system and which step is a lemma application. There is no reason the user should have to care about that distinction.

Taking all this together, imagine a user that wants to work in classical logic, but is used only to its expression as intuitionistic logic plus double negation law. He ploughs on with his proof, but his proof assistant keeps track of the alternative goals he could be proving instead of the goal he is thinking of, and informing him of that list in a side-window. The user keeps an eye on it, and notices that this other goal seems easier to prove at this point. He does so. He doesn't understand the proof he has written, but he asks the system to transform it into an intuitionistic logic plus excluded middle proof, and he has a proof he can read and understand. By being based on a saturated logical system, the proof assistant has made its user's life easier.

Naturally, the kernel of the proof assistant can still happen in a minimal system, interpreting the other rules as lemma applications. A next version of the proof assistant may actually use a different minimal set of rules, and no one will notice. This is a kind of "abstract datatype" approach to logic: One does not need to look into the choices the proof assistant has made; one is free to do the things the logic one works in allows, the system implementing the abstract signature maps some steps to atomic steps and some others to lemmas, but this is none of our concern.

Human Factor. The proof language should also cater for human use, which amounts to the following two criteria.

• **Understandability.** Expressions of the language should *mean* something to a reader, be understandable. This is ensured if the user knows a transformation to natural language that is simple enough that he can do it in his head, if every construct of the language has a clear semantics and maps to a concept or a rule that the reader recognises.

• **Flexibility.** The language should capture different notions of a human's natural language view of a rigorous proof.

2 $\bar{\lambda}\mu\tilde{\mu}$

The $\bar{\lambda}\mu\tilde{\mu}$ calculus, which covers implication logic, is made up of three interdependent syntactical categories, namely *terms*, *environments* and *commands*:

Syntax	Typing judgement
$v ::= x \mid \lambda x : T.v \mid \mu\alpha : T.c$	$\Gamma \vdash v : T \mid \Delta$
$e ::= \alpha \mid v \circ e \mid \tilde{\mu}x : T.c$	$\Gamma \mid e : T \vdash \Delta$
$c ::= \langle v \| e \rangle$	$c : (\Gamma \vdash \Delta)$

Its typing makes use of a *hypothesis context* (Γ, which is a set of declarations $\{x_1 : T_1, \ldots, x_n : T_n\}$ where the T_i are simple types and all x_i are different) and a *goal context* (Δ, which is a set of declarations $\{\alpha_1 : T_1, \ldots, \alpha_n : T_n\}$ where the

T_i are simple types and all α_i are different). The part between $\vdash$ and $|$ is the *stoup* and contains a distinguished nameless formula.

$$\Gamma, x : T \vdash x : T \mid \Delta \qquad \Gamma \mid \alpha : T \vdash \alpha : T, \Delta$$

$$\frac{c : (\Gamma \vdash \alpha : T, \Delta)}{\Gamma \vdash (\mu\alpha : T.c) : T \mid \Delta} \quad \frac{c : (\Gamma, x : T \vdash \Delta)}{\Gamma \mid (\tilde{\mu}x : T.c) : T \vdash \Delta} \quad \frac{\Gamma \vdash v : T \mid \Delta \qquad \Gamma \mid e : T \vdash \Delta}{\langle v \| e \rangle : (\Gamma \vdash \Delta)}$$

$$\frac{\Gamma \vdash v : T \mid \Delta \qquad \Gamma \mid e : T' \vdash \Delta}{\Gamma \mid v \circ e : T \to T' \vdash \Delta} \qquad \frac{\Gamma, x : T \vdash v : T' \mid \Delta}{\Gamma \vdash (\lambda x : T.v) : T \to T' \mid \Delta}$$

Intuitionistic Fragment. Intuitionistic logic is obtained by restricting the use of environment variables to only the most recently (innermost) bound one.

Definition 1. α *is said to be used intuitionistically in c iff it occurs only in positions where it is the most recently bound environment variable. Equivalently, no path from the μ that binds α to an occurrence of α traverses a μ.*

The three syntactical categories are to be understood as:

term v proves the sequent $\Gamma \vdash T, \Delta$. T is singled out as the thesis one is currently working on; switching is allowed, but is an explicit step: a μ captures the current thesis, gives it a name so that it can be referred back to later and goes to a "neutral" state where no formula is distinguished.

The natural language rendering of a term v thus naturally is some text that is a proof of the sequent that types v, with the type of v as focus (current thesis) at the beginning of the text.

environment e expects (consumes) a proof of T (a term v typed by $\Gamma \vdash v : T \mid \Delta$) and continues further with the proof of sequent $\Gamma, T \vdash \Delta$, using the v it has consumed. A $\tilde{\mu}$ is the dual of μ; it captures the consumed proof and gives it a name.

The natural language rendering of an environment thus naturally is a context; that is some text containing a placeholder, a hole, such that if the placeholder is filled in with a proof of $\Gamma \vdash T, \Delta$, then the result is a proof of $\Gamma \vdash \Delta$. Furthermore, in this paper, the placeholder will, in the spirit of [1], always be at the very beginning of the rendering.

command combination of a term and an environment (a provider and a consumer) typed by the same Γ, T and Δ into a "closed" whole proving the sequent $\Gamma \vdash \Delta$, which is the type of c. The type of commands does not have a stoup (no singled out formula).

The natural language rendering of the command $\langle v \| e \rangle$ thus naturally is the rendering of e with the placeholder filled in with the rendering of v. In this paper, this amounts to the concatenation of the rendering of v and the rendering of e.

Definition 2. *A command whose environment ends in α is said to* conclude α. *It* ultimately concludes α *if it concludes α or ends in a binder (i.e. $\tilde{\mu}x : T.c$) whose binding domain (i.e. c) ultimately concludes α.*

For example, $\langle v \| v' \circ \alpha \rangle$ concludes α, but $\langle v \| v' \circ \tilde{\mu}y : T.\langle v_0 \| v_1 \circ \alpha \rangle \rangle$ does not. The latter ultimately concludes α.

3   Basic Pseudo-natural Language Rendering

The purpose of this rendering is to be a purely depth-0 structural, left-to-right reading of $\bar{\lambda}\mu\tilde{\mu}$ expressions, that is faithful to the proof the expression codes for. It is the rendering of [1], extended to handle the whole calculus and not only the intuitionistic fragment. $\boxed{\hookrightarrow}$ is an increase in indentation level and $\boxed{\hookleftarrow}$ a decrease.

$$[\![x]\!] := \text{by } x \qquad\qquad [\![\alpha]\!] := \boxed{\hookleftarrow} \text{ done proving } \alpha$$

$$[\![\lambda x : T.v]\!] := \text{assume } T \ (x) \ [\![v]\!] \qquad [\![v \circ e]\!] := \text{and } [\![v]\!] \ [\![e]\!]$$

$$[\![\mu\alpha : T.c]\!] := \text{thesis } T \ (\alpha) \qquad\qquad [\![\tilde{\mu}x : T.c]\!] := \text{we have proven } T \ (x)$$

$$\boxed{\hookrightarrow} [\![c]\!] \qquad\qquad\qquad\qquad [\![c]\!]$$

$$[\![\langle v \| e \rangle]\!] := [\![v]\!] \ [\![e]\!]$$

Example 1. This term

$$\lambda x_R : P \to R.\lambda x_P : Q \to S \to P.\lambda y_S : S.\lambda y_Q : Q.\mu\alpha : R.$$
$$\langle x_P \| y_Q \circ y_S \circ \tilde{\mu}y_P : P. \langle x_R \| y_P \circ \alpha \rangle \rangle$$

of type $(P \to R) \to (Q \to S \to P) \to S \to Q \to R$ renders as

assume $P \to R$	(x_R)
assume $Q \to S \to P$	(x_P)
assume S	(y_S)
assume Q	(y_Q)
thesis R	(α)
by x_P and by y_Q and by y_S	
we have proven P	(y_P)
by x_R and by y_P	
done proving α	

The following term of the same type:

$$\lambda x_R : P \to R.\lambda x_P : Q \to S \to P.\lambda y_S : S.\lambda y_Q : Q.\mu\alpha : R.$$
$$\langle x_R \| (\mu\beta : P. \langle x_P \| y_Q \circ y_S \circ \beta \rangle) \circ \alpha \rangle$$

renders as

assume $P \to R$	(x_R)
assume $Q \to S \to P$	(x_P)
assume S	(y_S)
assume Q	(y_Q)
thesis R	(α)
by x_R and thesis P	(β)
by x_P and by y_Q and by y_S	
done proving β	
done proving α	

The previous term was a forward (bottom-up) proof, this is backward (top-down) proof. In this manner, $\bar{\lambda}\mu\tilde{\mu}$ allows to choose at every step whether it is done backwards or forwards.

Classical Logic. Sequent calculus handles classical logic by allowing a set of formulas on the right hand side of the ⊢. In terms of ordinary logical arguments, this means that one maintains a *set* of goals throughout the reasoning; concluding any one of these goals concludes the whole proof. In a $\bar{\lambda}\mu\tilde{\mu}$ term, there is one goal "in focus" (the one before the stoup) and the other ones are named by environment variables; we can switch to another goal by using its name. To show how our extension of the [1] rendering to classical logic works we give as an example a proof of Peirce's law, $((P \to Q) \to P) \to P$. This uses the "goal switching" facility.

$$\lambda x : (P \to Q) \to P.\mu\alpha : P.\langle x \| (\mu\beta : P \to Q.\langle \lambda y : P.\mu\gamma : Q.\langle y \| \alpha \rangle \| \beta \rangle) \circ \alpha \rangle$$

which renders as

assume $(P \to Q) \to P$	(x)
thesis: P	(α)
by x and thesis $P \to Q$	(β)
assume P	(y)
thesis: Q	(γ)
by y	
done proving α	
done proving β	
done proving α	

The classical logic step is the first "done proving α". Intuitionistically, one would have to prove γ at this point, but we conclude α instead, which concludes the whole proof.

4   Enhanced Pseudo-natural Language

We present several enhancements to the basic transformation, that do not break faithfulness to the proof the expression embodies. In order to keep the presentation simple, we will not discuss the interaction between the enhancements explicitly, unless there is an interesting or problematic point.

Backwards Proofs. This enhancement, namely replacing "and thesis" by "the thesis is reduced to", was already proposed in [1].

$$[\![(\mu\alpha : T.c) \circ e]\!] := \text{the thesis is reduced to } T \qquad (\alpha)$$
$$\boxed{\hookrightarrow} \, [\![c]\!] \, [\![e]\!]$$

It makes the intent of backwards proofs much more clear. Read the previous example again while mentally doing the replacement.

Intuitionistic Logic. Here, we recognise when single-goal logic (*i.e.* intuitionistic logic, plus eventually a classical logic axiom) is used and adapt the rendering. This consists in omitting the "(α)" when rendering a $\mu\alpha : T.c$ when α is used intuitionistically in c, combined with these rules; ⌐ is a line break.

$$[\![\alpha]\!] \text{ when the innermost parent } \mu \text{ binds } \alpha$$
$$:= \boxed{\hookleftarrow} \text{ done}$$
$$[\![\mu\alpha : T.c]\!] \text{ when } c \text{ ultimately concludes } \alpha$$
$$:= \text{we have to prove } T\ (\alpha) \lrcorner \boxed{\hookrightarrow} [\![c]\!]$$

With this improvement, our natural language translation gives the same result as the one in [1] on single-goal (sub)proofs, and also handles multiple-goal proofs.

Announcing Thesis Changes. The basic rendering informs the reader that what has been proven was not what the reader thinks of as "the current thesis" only *at the end* of a subproof. We see that e.g. in "done proving β" or "we have proven $T\ (x)$". That is essentially inherited from a prefix depth-first left-right reading of $\bar{\lambda}\mu\tilde{\mu}$ terms. It enhances the readability of the proof if such changes are announced at the start of the corresponding subproof, rather than at the end. This is typically also required in the proof input language of proof assistants. There are essentially three situations where such a thesis change happens:

Switching to another goal in Δ, a thing that is implicit in the standard sequent calculus, but is made explicit by the stoup structure of $\bar{\lambda}\mu\tilde{\mu}$. This corresponds to the pattern $\mu\alpha : T.c$ where c does not ultimately conclude α (but, say $\beta : T'$). We want to announce the thesis $T'(\beta)$, however, in a pattern like $\mu\alpha : T. \langle v \| \tilde{\mu}x : T''. \langle v' \| c \rangle \rangle$ we want to delay the announcement until we are under the $\tilde{\mu}$ binder. As a solution, we use a subscript in the transformation to keep track of the thesis currently active in the natural language text.

$$[\![\langle v \| e \rangle]\!]_\beta \text{ when } e \text{ does not conclude } \beta \text{ and concludes } \alpha$$
$$:= \text{we now consider thesis } \alpha$$
$$[\![v]\!]_\alpha [\![e]\!]_\alpha$$
$$[\![\mu\beta : T.c]\!]_\alpha := \text{thesis } T\ (\beta) \lrcorner \boxed{\hookrightarrow} [\![c]\!]_\beta$$

This rendering keeps implicit in the natural language text that the active thesis is the most recently introduced one.

A cut. If the root of the term of a command is a μ, then the basic rendering already makes the announcement; we just tweak the text a bit:

$$[\![\langle \mu\alpha : T.c \| e \rangle]\!] := \text{we now prove } T\ (\alpha)$$
$$\boxed{\hookrightarrow} [\![c]\!]$$
$$[\![e]\!]$$

As to the pattern $\langle \lambda x : T.v \| e \rangle$, it is dismissed as "bad style": It is a proof that does a thesis change, but refuses to announce it; fixing it crosses the line of showing a *better* proof than the one written, not the proof written. It is suggested to η-expand this term to $\langle \mu\alpha : T. \langle v \| \alpha \rangle \| e \rangle$, which can be done programmatically as part of a "proof enhancement" transformation. A similar thing happens with the pattern $v \circ e$, where the root of v is a recursive constructor other than μ (e.g. λ), with the same solution.

The pattern $\langle v \| v_1 \circ \ldots \circ v_n \circ \tilde{\mu}x : T.c \rangle$. We can describe the environment part more succinctly by writing $e(\tilde{\mu}x : T.c)$: it is an environment that finishes with $\tilde{\mu}x : T.c$ and $e(\cdot)$ is that environment with the $\tilde{\mu}$ removed and replaced by a placeholder $\cdot$. A rendering would be

$$[\![\cdot]\!] := \boxed{\hookleftarrow} \text{ done}$$
$$[\![\langle v \| e(\tilde{\mu}x : T.c)\rangle]\!] := \text{we now prove } T\ (x)$$
$$\boxed{\hookrightarrow} [\![v]\!]\ [\![e(\cdot)]\!]$$
$$[\![c]\!]$$

It forms a critical pair with the "detect a cut" rule, resolved with

$$[\![\langle \mu\alpha : T.c \| \tilde{\mu}x : T.c' \rangle]\!] := \text{we now prove } T\ (x, \alpha)$$
$$\boxed{\hookrightarrow} [\![c]\!]$$
$$[\![c']\!]$$
$$[\![\langle \mu\alpha : T.c \| e(\tilde{\mu}x : T'.c') \rangle]\!] := \text{we now prove } T'\ (x)$$
$$\boxed{\hookrightarrow} \text{we now prove } T\ (\alpha)$$
$$\boxed{\hookrightarrow} [\![c]\!]$$
$$[\![e(\cdot)]\!]$$
$$[\![c']\!]$$

These rules catch occurrences of μ when its type is not the current thesis in the text; this allows to enhance the rendering of the other occurrences (those that capture the current thesis and give it a name):

$$[\![\mu\alpha : T.c]\!] := \text{left to prove: } T\ (\alpha)$$
$$\boxed{\hookrightarrow} [\![c]\!]$$

Remark 1. The rules introduced here have the big disadvantage that their structural depth (the depth at which they have to look into an expression before deciding how to render its root) is unbounded. This can be fixed by changing the syntax a bit, so that in a command the terminal constructors of the environment are available at depth 1:

$$E ::= \cdot \mid v \circ E \qquad \text{all non-terminal environment constructors}$$
$$e ::= \alpha \mid \tilde{\mu}x : T.c \qquad \text{all terminal environment constructors}$$
$$c ::= \langle v | E | e \rangle$$

The meaning of the new syntax command $\langle v | v_1 \circ \ldots \circ v_n \circ \cdot | e \rangle$ is just $\langle v \| v_1 \circ \ldots \circ v_n \circ e \rangle$ The typing rules can be adapted accordingly.

From Binary to n-ary. $\circ$ is a binary constructor, but one can recognise sequences of it and treat it as an n-ary constructor; this is here combined with controlling its interaction with term variables and μ more closely:

$$[\![x_0 \circ x_1 \circ \cdots \circ x_n \circ e]\!] := \text{by } x_0, x_1, \ldots, x_{n-1} \text{ and } x_n \; [\![e]\!]$$
$$[\![\langle x \| x_0 \circ x_1 \circ \cdots \circ x_n \circ e \rangle]\!] := \text{by } x, x_0, x_1, \ldots, x_{n-1} \text{ and } x_n \; [\![e]\!]$$
$$[\![(\mu\alpha_0 : T_0.c_0) \circ \cdots \circ x_j \circ \ldots$$
$$\cdots \circ (\mu\alpha_i : T_i.c_i) \circ \cdots \circ e]\!] := \text{the thesis is reduced to:}$$

- $T_0 \; (\alpha_0)$
  $\boxed{\hookrightarrow} [\![c_0]\!]$

  $\ldots$

- assumption x_j

  $\ldots$

- $T_i \; (\alpha_i)$
  $\boxed{\hookrightarrow} [\![c_i]\!]$

  $\ldots$

  $[\![e]\!]$

5   Saturation

There is a variety of alternative ways to handle implication in the calculus in [6]. In our quest for a saturated calculus, we examine them all and give them a natural language rendering.

ι_2 The first, and the only one we decide to keep as is, is called ι_2 in [6], but this conflicts with another constructor with the same notation; we thus rename it to $\lambda\_ : A$, in line with the convention that $\_$ is a special name for "do not bind":

$$v ::= \ldots \mid \lambda\_ : A.v \qquad\qquad \frac{\Gamma \vdash v : T' \mid \Delta}{\Gamma \vdash (\lambda\_ : T.v) : T \to T' \mid \Delta}$$

$$[\![\lambda\_ : T.v]\!] := \text{assumption } T \text{ in thesis is not necessary. } [\![v]\!]$$

If the transformation has access to typing information (e.g. because every expression is annotated with its type), then one can use:

$$[\![\lambda\_ : T.v]\!] := \text{it suffices to prove } t(v). \; [\![v]\!]$$

where $t(v)$ is the type of v. This introduces some redundancy if a μ follows immediately; this redundancy can be avoided with

$$[\![\lambda\_ : A.\mu\alpha : B.c]\!] := \text{it suffices to prove } B \; (\alpha)\lrcorner\boxed{\hookrightarrow}[\![c]\!]$$

ι_1 As to its companion ι_1,

$$v ::= \ldots \mid \iota_1(e) \qquad \qquad \frac{\Gamma \mid e : A \vdash \Delta}{\Gamma \vdash \iota_1(e) : A \to B \mid \Delta}$$

while the logical step performed is naturally and immediately accepted as admissible, it is not intuitively seen as one atomic step; it is more natural to decompose it into two steps, namely assuming A and erasing goal B. We thus introduce a "no binding" version of μ:

$$v ::= \ldots \mid \mu_{\_} : T.c \qquad \qquad \frac{c : (\Gamma \vdash \Delta)}{\Gamma \vdash (\mu_{\_} : T.c) : T \mid \Delta}$$

$$[\![\mu_{\_} : T.c]\!] := \text{we give up on the current thesis} \lrcorner \boxed{\hookrightarrow} [\![c]\!]$$

This $\mu_{\_} : T.c$ has a natural dual in a putative $\tilde{\mu}_{\_} : \boldsymbol{T.c}$:

$$e ::= \ldots \mid \tilde{\mu}_{\_} : T.c \qquad \qquad \frac{c : (\Gamma \vdash \Delta)}{\Gamma \mid (\tilde{\mu}_{\_} : T.c) : T \vdash \Delta}$$

but while $\mu_{\_} : T.c$ fulfils a real role (e.g. in $\lambda x : T_x.\mu_{\_} : \bot.\langle x \| \beta \rangle$, the current goal $\bot$ really is not useful in the rest of the proof; it is thus not useful to bother to give it a fresh name α to never refer to it later), any instance of $\tilde{\mu}_{\_} : T.c$ shows that the proof contains a completely non useful part: it takes the effort to prove T, but then just throws that result away. $\langle x \| y \circ \tilde{\mu}_{\_} : T. \ldots \rangle$ would correspond to something like "by x and y, we have proven T, but don't use that fact in the rest of the proof ...".

Furthermore, ι_1 is an instance of a bigger problem in the context of the rest of the natural language translation: term constructors that syntactically recurse into the environment category do not fit well. We have not found a nice phrase that turns what follows (which, being the translation of an environment, consumes a proof) into something which provides a proof. The best we could do was a rather weak and unnatural "the other goals follow from $A \lrcorner [\![e]\!]$", which had to be combined with changing the translation for $\tilde{\mu} x : T.c$ to "we can now assume $T \lrcorner [\![c]\!]$", because e.g. in the expression $\iota_1(\tilde{\mu} x : A.c)$ of type $A \to B$, A has not been proven, but assumed, so "we have proven" does not fit anymore. It makes the translation of $\tilde{\mu}$ in other situations weaker, but not wrong:

	thesis T	(α)
	$\ldots$	
$\langle \mu\alpha : T. \ldots \| \tilde{\mu} x : T. \ldots \rangle$	done proving α	
	we can now assume T	(x)
	$\ldots$	

A construct that has no good natural language rendering cannot be a natural proof step for vernacular proofs and is thus not necessary to form a saturated system.

$\lambda(x : A, \alpha : B).c$ naturally feels like two steps and is advantageously replaced by $\lambda x : A.\mu\alpha : B.c$.

$$v ::= \ldots \mid \lambda(x : A, \alpha : B).e \qquad \frac{c : (\Gamma, x : A \vdash \alpha : B, \Delta)}{\Gamma \vdash (\lambda(x : A, \alpha : B).c) : A \to B \mid \Delta}$$

$\lambda\alpha : B$ takes an environment as argument, which raises the problems already discussed.

$$v ::= \ldots \mid \lambda\alpha : B.e \qquad \frac{\Gamma \mid e : A \vdash \alpha : B, \Delta}{\Gamma \vdash (\lambda\alpha : B.e) : A \to B \mid \Delta}$$

$[\![\lambda\alpha : B.e]\!] :=$ we now prove B, which follows from A

6   Link to Proof Assistants

We here succinctly treat the transformation of intuitionistic-logic $\bar{\lambda}\mu\tilde{\mu}$ terms (not the extended calculus) to Isar proofs by way of an example. A transformation into Mizar is similar (except that Mizar can *only* do forward steps, no backward step). Also a transformation into PVS has been designed, but it is less direct. Both are not included here for lack of space. All these translations have been tested by evaluating them manually on a few examples and having the corresponding prover accept the result. It should be noted that not all $\bar{\lambda}\mu\tilde{\mu}$ terms can be mapped faithfully to an Isar proof: there are proof constructs in our saturated system that Isar cannot capture. We could syntactically single out the $\bar{\lambda}\mu\tilde{\mu}$ terms that map to an Isar proof, but we take a different approach by describing a transformation of $\bar{\lambda}\mu\tilde{\mu}$ terms. The terms that are invariant under this transformation are the ones corresponding to Isar proofs.

Replace any pattern in the left column by the one in the right column:

$\langle v \| \cdots \circ (\lambda x : T.v') \circ \ldots \rangle$	$\langle v \| \cdots \circ (\mu\alpha : T_\alpha. \langle \lambda x : T.v' \| \alpha \rangle) \circ \ldots \rangle$
$\langle \lambda x : T.v \| v_0 \circ \ldots \rangle$	$\langle \lambda x : T.v \| \tilde{\mu} y : T'. \langle x \| v_0 \circ \ldots \rangle \rangle$
$\langle v \| \cdots \circ (\mu\alpha : T_\alpha.c) \circ x \circ \ldots \rangle$	$\langle v \| \cdots \circ (\mu\alpha : T_\alpha.c) \circ (\mu\beta : T_\beta. \langle x \| \beta \rangle) \circ \ldots \rangle$
$\langle \mu\alpha : T_\alpha.c \| e \rangle$	$c\{\alpha := e\}$
$\lambda x : T.x'$	$\lambda x : T.\mu\alpha : T'. \langle x \| \alpha \rangle$

Most of these transformation rules are interesting by themselves and fall under "the pattern on the left is bad proof style", but others have their origins in idiosyncrasies of Isar.

We do not deal here with mapping Isar proofs to $\bar{\lambda}\mu\tilde{\mu}$, but we claim that any Isar proof that does not use automation can be mapped faithfully to $\bar{\lambda}\mu\tilde{\mu}$. The Isar commands not used by the transformation below are mostly either syntactic sugar or logically equivalent to a basic command that we do use, or are concerned with automation.

The transformation follows. The purpose of cl is to count the lambdas in a term; we don't spell out its definition. We use the alternate syntax of page 413 for clarity.

$$[\![\mu\alpha : T_\alpha. \langle v|E|\tilde{\mu}x : T_x.c\rangle]\!] := \mathbf{have}\ x : T\ [\![\langle v|E|\alpha\rangle]\!]$$
$$[\![\mu\alpha : T_\alpha.c]\!]$$
$$[\![\mu\alpha : T.c]\!] := \mathbf{show}\ T\ [\![c]\!]$$
$$[\![\lambda x : T.v]\!] := \mathbf{assume}\ x : T\ [\![v]\!]$$
$$[\![\langle x|y_0 \circ \cdots \circ y_{n-1} \circ \cdot|\alpha\rangle]\!] := \mathbf{by}\ (\mathbf{rule\ mp},\ \ldots(n\ \text{times})\ldots,\ \mathbf{rule\ mp},$$
$$\mathbf{fact}\ x,\ \mathbf{fact}\ y_0,\ \ldots,\ \mathbf{fact}\ y_{n-1})$$

$$[\![\langle x|y_0 \circ \cdots \circ y_{n-1} \circ v_0 \circ \cdots \circ v_{p-1} \circ \cdot|\alpha\rangle]\!] :=$$
$$\mathbf{proof}\ (\mathbf{rule\ mp},\ \ldots(n+p\ \text{times})\ldots,\ \mathbf{rule\ mp},$$
$$\mathbf{fact}\ x,\ \mathbf{fact}\ y_0,\ \ldots,\ \mathbf{fact}\ y_{n-1})\boxed{\hookrightarrow}$$
$$[\![v_0]\!]$$
$$\ldots$$
$$[\![v_{p-1}]\!]\boxed{\hookleftarrow}$$
$$\mathbf{qed}$$
$$[\![\langle v| \cdot | \alpha\rangle]\!] :=$$
$$\mathbf{proof}\ (\mathbf{rule\ impI},\ \ldots cl(v)\ \text{times}\ldots,\ \mathbf{rule\ impI})\ \boxed{\hookrightarrow}$$
$$[\![v]\!]\boxed{\hookleftarrow}$$
$$\mathbf{qed}$$

In the rule for $[\![\mu\alpha : T_\alpha. \langle v|E|\tilde{\mu}x : T_x.c\rangle]\!]$, $\langle v|E|\alpha\rangle$ is not well-typed (the α may be unbound or of the wrong type), but that doesn't matter, because the 'name' α is never used in the Isar output (it translates to $\mathbf{qed}$). One can vary on this translation, producing different Isar proofs – that we claim have the same logical structure.

7 Disjunction

As an example to extension to propositional logic, we briefly treat disjunction; again these constructs are catalogued in [6]; the natural language rendering is ours:

$$v ::= \ldots \mid \iota_{1,2}(v) \mid [\alpha : A, \beta : B].c \mid \lambda_{1,2}\alpha : A.v$$
$$e ::= \ldots \mid [e, e]$$

$$\frac{\Gamma \vdash v : T \mid \Delta}{\Gamma \vdash \iota_1(v) : T \vee T' \mid \Delta} \qquad \frac{\Gamma \vdash v : T' \mid \Delta}{\Gamma \vdash \iota_2(v) : T \vee T' \mid \Delta}$$

$$\frac{c : (\Gamma \vdash \beta : T', \alpha : T, \Delta)}{\Gamma \vdash ([\alpha : T, \beta : T'].c) : T \vee T' \mid \Delta} \qquad \frac{\Gamma \mid e : T \vdash \Delta \qquad \Gamma \mid e' : T' \vdash \Delta}{\Gamma \mid [e, e'] : T \vee T' \vdash \Delta}$$

$$\frac{\Gamma \vdash v : T' \mid \alpha : T, \Delta}{\Gamma \vdash (\lambda_1\alpha : T.v) : T \vee T' \mid \Delta} \qquad \frac{\Gamma \vdash v : T \mid \alpha : T', \Delta}{\Gamma \vdash (\lambda_2\alpha : T'.v) : T \vee T' \mid \Delta}$$

$[\![\iota_1(v)]\!] :=$ it suffices to prove the left part of the disjunction $[\![v]\!]$

$[\![\iota_2(v)]\!] :=$ it suffices to prove the right part of the disjunction $[\![v]\!]$

$[\![\lambda_{1,2}\alpha : A.v]\!] :=$ keeping in mind that we may prove A (α), $[\![v]\!]$

$[\![[\alpha : A, \beta : B].c]\!] :=$ thesis A (α) or B (β) $[\![[e, e']]\!] :=$ either

$$\boxed{\hookrightarrow}\,[\![c]\!] \qquad\qquad\qquad \boxed{\hookrightarrow}\,[\![e]\!]$$

or

$$\boxed{\hookrightarrow}\,[\![e']\!]\,\boxed{\hookleftarrow}$$

Many of the improvements of section 4 apply to disjunction mutadis mutandis. This is further detailed in [7], but here are a few examples:

- $[\alpha : A, \beta : B]$ is a μ-like construct; the definitions and extended rules that deal with μ should thus deal also with it, suitably adapted. For example, the "backwards proofs" enhancement:

$$[\![([\alpha : A, \beta : B].c) \circ e]\!] :=$$ the thesis is reduced to T (α) or T' (β)
$$\boxed{\hookrightarrow}\,[\![c]\!]\,[\![e]\!]$$

Similarly, the definition of "α is used intuitionistically in c" has to be changed to "no path from the μ that binds α to an occurrence of α traverses a μ or a $[\cdot, \cdot]$". In particular, if α is bound by a $[\cdot, \cdot]$, it is not used intuitionistically; $[\cdot, \cdot]$ is an inherently non-intuitionistic construct.

- $\lambda_{1,2}$, $\iota_{1,2}$ and $[e, e']$ benefit particularly strongly from typing information:

$[\![\iota_{1,2}(v)]\!] :=$ it suffices to prove $t(v)$ $[\![[e, e']]\!] :=$ either $t(e)$

$$[\![v]\!] \qquad\qquad\qquad\qquad \boxed{\hookrightarrow}\,[\![e]\!]$$

$[\![\lambda_1\alpha : A.(v)]\!] :=$ keeping in mind we may prove A (α),   or $t(e')$

we proceed with the proof of $t(v)$ $\boxed{\hookrightarrow}\,[\![e']\!]\,\boxed{\hookleftarrow}$

- Just as sequences of $\circ$ can be collected in an emulation of n-ary implication, sequences of ι_1, ι_2 can be collected in an emulation of n-ary disjunction.

The introduction of $[\cdot, \cdot]$ has interesting consequences for the enhanced rendering: there is not anymore unicity of the terminal environment constructor (the first environment constructor that does not syntactically recurse back into category e). For example, $[\ldots \alpha, \ldots \tilde{\mu}x : T.c]$.

Definition 3. *If all terminal environment constructors of an environment e are the same (the same α or $\tilde{\mu}x : T.c$ with the same x and the same T), then e is said to terminate* uniformly *in that constructor. Similarly, e (ultimately)* uniformly concludes α *if all its branches (ultimately) conclude α.*

The adaptation to that situation is to then defer the decisions (on commands) that depend on the terminals of the environment, when these are not uniform. For example, the "announce thesis changes" enhancement:

$$[[e, e']]_\beta := \text{either } t(e)\boxed{\hookrightarrow}$$

> *if e does not conclude β and uniformly concludes α*
>
> > we now consider thesis α
> >
> > $[\![e]\!]_\alpha$
>
> *else if e terminates uniformly in $\tilde\mu x : T.c$*
>
> > we now prove $T\ (x)$
> >
> > $\boxed{\hookrightarrow}[\![e(\cdot)]\!]_\beta$
> >
> > $[\![c]\!]_\beta$
>
> *else*
>
> > $[\![e]\!]_\beta$
>
> *end if*
>
> or $t(e')\boxed{\hookrightarrow}$
>
> *the same for e'*
>
> $\boxed{\hookrightarrow}$

Similarly, in the alternative syntax of remark 1 page 413, a command now needs to take a list of es whose length matches the number of $\cdot$s in the E.

8   Future Work

There are several directions in which this can be taken further. The most obvious is extending, with the same concern for saturation, to some predicate logic. The second glaring need is that the problem of capturing automation in theorem provers needs to be addressed for the language to be really functional in practise as a proof interchange language. A natural idea is to store as part of the term a witness provided by the automation, which would be used to produce a proof in a system whose automation is weaker. Alas, it may not be practical or possible to get a witness from some provers' automation (no access to its source code, prover not structured in De Bruijn criterion conformant way (that is the automation is not already forced to provide a witness to a kernel), . . .).

Also, our calculus is not completely saturated in its treatment of classical logic; one can convincingly argue that e.g. De Morgan laws and classical decomposition of disjunction tend to be considered as valid atomic deduction steps in the vernacular; a saturated calculus should thus have constructs for them.

The proof of the pudding being in the eating, we will concretely implement the transformations from and to various proof languages (various proof assistants, but also e.g. a standard sequent calculus); this would find a natural place as part of a proof assistant.

On a more theoretical side, our natural language rendering has an underlying concept of structure of a natural language proof, which (when restricted to single-goal logic) is quite close to the structure of declarative proof languages like Mizar or Isar. But, in particular in its notion of active thesis, and when a change of it needs to be explicitly announced (i.e. always), it is not in complete agreement with the structure of proofs in $LK_{\mu\tilde{\mu}}$ (sequent calculus with 'stoup', basically just $\bar{\lambda}\mu\tilde{\mu}$ without term/expression/command information). One would like changes in the active thesis to be characterised as either a cut, or an explicit active thesis change. We will thus define a sequent calculus whose notion of cut matches the notion of introducing an arbitrary new thesis (a forward step) in our natural language, and that matches the natural language's need for an explicit switch action between the theses. Restricting the hypothesis-creating children of a left introduction rule to a left introduction rule or an axiom may be the main thing needed to achieve the former. It would then be interesting to study proof intent conserving transformations between that calculus, $LK_{\mu\tilde{\mu}}$, standard stoup-free sequent calculus and other proof formats.

References

1. Sacerdoti Coen, C.: Explanation in natural language of $\bar{\lambda}\mu\tilde{\mu}$ terms. In: Kohlhase, M. (ed.) MKM 2005. LNCS (LNAI), vol. 3863, pp. 234–249. Springer, Heidelberg (2006)
2. Sacerdoti Coen, C.: Declarative representation of proof terms. In: Prog. Lang. for Mechanized Math., University of Linz. RISC Reports, vol. 07-10, pp. 3–18 (2007)
3. Kirchner, F.: Interoperable proof systems. Ph.D thesis, École Polytechnique (2007)
4. Autexier, S., Sacerdoti Coen, C.: A formal correspondence between OMDoc with alternative proofs and the $\bar{\lambda}\mu\tilde{\mu}$-calculus. In: Borwein, J.M., Farmer, W.M. (eds.) MKM 2006. LNCS (LNAI), vol. 4108, pp. 67–81. Springer, Heidelberg (2006)
5. Curien, P.L., Herbelin, H.: The duality of computation. In: ICFP 2000, vol. 35(9), pp. 233–243. ACM, New York (2000)
6. Herbelin, H.: C'est maintenant qu'on calcule; au cœur de la dualité. Habilitation à diriger des recherches, Université Paris 11 (December 2005)
7. Mamane, L.E., Geuvers, H., McKinna, J.: A logically saturated extension of $\bar{\lambda}\mu\tilde{\mu}$ to propositional logic, http://www.mamane.lu/science/lbmmt_extension/
8. Corbineau, P.: A declarative proof language for the Coq proof assistant. In: Miculan, M., Scagnetto, I., Honsell, F. (eds.) TYPES 2007. LNCS, vol. 4941, pp. 69–84. Springer, Heidelberg (2008)
9. Wenzel, M.: Isar - a generic interpretative approach to readable formal proof documents. In: Bertot, Y., Dowek, G., Hirschowitz, A., Paulin, C., Théry, L. (eds.) TPHOLs 1999. LNCS, vol. 1690, p. 167. Springer, Heidelberg (1999)
10. Owre, S., Rushby, J.M., Shankar, N.: PVS: A prototype verification system. In: Kapur, D. (ed.) CADE 1992. LNCS (LNAI), vol. 607, pp. 748–752. Springer, Heidelberg (1992)
11. Trybulec, A.: Mizar language, http://mizar.org/language/
12. Asperti, A., Guidi, F., Padovani, L., Sacerdoti Coen, C., Schena, I.: Mathematical knowledge management in HELM. AMAI 38(1), 27–46 (2003)
13. Coscoy, Y.: Explication textuelles de preuves pour le calcul des constructions inductives. Ph.D thesis, Université de Nice-Sophia-Antipolis (September 2000)

A   Basic Propositional Logic

$$v ::= \dots |(v,v)|\lambda_\neg x : A.v| \ltimes \; | \, \mathrm{TE}(T)| \, \mathrm{DN}(v)$$

$$e ::= \dots |\pi_{1,2}[e]|(x : A, y : B).c|\lambda_{1,2}x : A.e|\neg[v]| \rtimes \; | \, \mathrm{DN}(e)$$

$$\frac{\Gamma \vdash v : T|\Delta \qquad \Gamma \vdash v' : T'|\Delta}{\Gamma \vdash (v,v') : T \wedge T'|\Delta} \qquad \frac{c : (\Gamma, x : T, x' : T' \vdash \Delta)}{\Gamma|((x : T, x' : T').c) : T \wedge T' \vdash \Delta}$$

$$\frac{\Gamma, x : T|e : T' \vdash \Delta}{\Gamma|(\lambda_1 x : T.e) : T \wedge T' \vdash \Delta} \qquad \frac{\Gamma, x : T'|e : T \vdash \Delta}{\Gamma|(\lambda_2 x : T'.e) : T \wedge T' \vdash \Delta}$$

$$\frac{\Gamma|e : T \vdash \Delta}{\Gamma|\pi_1[e] : T \wedge T' \vdash \Delta} \qquad \frac{\Gamma|e : T' \vdash \Delta}{\Gamma|\pi_2[e] : T \wedge T' \vdash \Delta} \qquad \frac{\Gamma, x : T \vdash v : \bot|\Delta}{\Gamma \vdash (\lambda_\neg x : T.v) : \neg T|\Delta}$$

$$\frac{}{\Gamma \vdash \ltimes : \top|\Delta} \qquad \frac{\Gamma \vdash v : T|\Delta}{\Gamma|\neg[v] : \neg T \vdash \Delta} \qquad \frac{}{\Gamma|\rtimes : \bot \vdash \Delta}$$

$$\frac{}{\Gamma \vdash \mathrm{TE}(P) : P \vee \neg P|\Delta} \qquad \frac{\Gamma \vdash v : \neg\neg T|\Delta}{\Gamma \vdash \mathrm{DN}(v) : T|\Delta} \qquad \frac{\Gamma|e : T \vdash \Delta}{\Gamma|\mathrm{DN}(e) : \neg\neg T \vdash \Delta}$$

$$[\![(x : A, y : B).c]\!] := \text{we have proven } A \; (x) \text{ and } B \; (y)$$
$$[\![c]\!]$$

$$[\![\lambda_{1,2}x : T.e]\!] := \text{we have proven } T \; (x) \text{ and } [\![e]\!]$$

$$[\![\mathrm{DN}(v)]\!] := \text{proof by contradiction}$$

$$[\![\mathrm{DN}(e)]\!] := \text{and by double negation elimination}$$

$$[\![(v,v')]\!] := \bullet [\![v]\!] \qquad\qquad\qquad [\![\pi_{1,2}[e]]\!] := \text{in particular } [\![e]\!]$$
$$\bullet [\![v']\!]$$

$$[\![(\alpha : A, \beta : B].c]\!] := \text{thesis } A \; (\alpha) \text{ or } B \; (\beta) \qquad [\![\neg[v]]\!] := \text{and } [\![v]\!]$$
$$\boxed{\hookrightarrow} [\![c]\!] \qquad\qquad\qquad\qquad \boxed{\hookleftarrow} \text{done } (ECQ)$$

$$[\![(e,e']]\!] := \text{either} \qquad\qquad\qquad\qquad [\![\ltimes]\!] := \text{true}$$
$$\boxed{\hookrightarrow} [\![e]\!] \qquad\qquad\qquad\qquad\qquad [\![\rtimes]\!] := \boxed{\hookleftarrow} \text{done } (EFQ)$$
$$\text{or} \qquad\qquad\qquad\qquad\qquad\qquad [\![\lambda_\neg x : A.v]\!] := \text{assume } A \; (x) \; [\![v]\!]$$
$$\boxed{\hookrightarrow} [\![e']\!] \boxed{\hookleftarrow} \qquad\qquad\qquad [\![\mathrm{TE}(T)]\!] := \text{by TE}$$

ECQ is an abbreviation for "ex contradictione (sequitur) quodlibet" (from a contradiction, anything follows), EFQ for "ex falso (sequitur) quodlibet" (from falsehood, anything follows) and TE for "(principium) tertii exclusi" (principle of excluded middle).

Again, some natural language rendering enhancements apply mutatis mutandis; others need to be adapted. For example, dually to what happens with disjunction, $(x : A, y : B).c$ is a $\tilde{\mu}$-like construct and the enhancements that deal with $\tilde{\mu}$ need to similarly deal with it. Also, the definition of "e concludes α" has to be changed so that e.g. $\rtimes$ and $\neg[v]$ conclude α for any α.

From Tessellations to Table Interpretation

Ramana C. Jandhyala[1], Mukkai Krishnamoorthy[1], George Nagy[1],
Raghav Padmanabhan[1], Sharad Seth[2], and William Silversmith[1]

[1] DocLab, Rensselaer Polytechnic Institute, Troy, NY 12180, USA
[2] Computer Science and Engineering, University of Nebraska-Lincoln,
Lincoln, NE 68502, USA
nagy@ecse.rpi.edu, seth@cse.unl.edu

Abstract. The extraction of the relations of nested table headers to content cells is automated with a view to constructing narrow domain ontologies of semi-structured web data. A taxonomy of tessellations for displaying tabular data is developed. *X-Y tessellations* that can be obtained by a divide-and-conquer method are asymptotically only an infinitesimal fraction of all partitions of a rectangle into rectangles. *Admissible* tessellations are the even smaller subset of all partitions that correspond to the structures of published tables and that contain only rectangles produced by successive guillotine cuts. Many of these can be processed automatically. Their structures can be conveniently represented by X-Y trees, which facilitate relating hierarchical row and column headings to content cells. A formal grammar is proposed for characterizing the X-Y trees of layout-equivalent admissible tessellations. Algorithms are presented for transforming a tessellation into an X-Y tree and hence into multidimensional, layout-independent Category Trees (Wang abstract data types).

Keywords: document understanding, tables, rectangular tilings, X-Y trees, table grammars, Wang notation.

1 Introduction

Most quantitative data available in electronic form appears in the form of tables. We study formal aspects of web tables with a view to extracting their content. Various configurations of rectilinear tessellations defined on a grid can convey information in tabular form to human readers. In order to simplify the development of algorithms that recover the information from frequently occurring configurations automatically we construct a taxonomy of tabular layouts that may be considered equivalent from the perspective of table analysis.

Our work differs from earlier work w.r.t. (1) focusing on computer-constructed web tables rather than tables from scanned documents, (2) making use of commercial software to import web tables into a spreadsheet, (3) describing tables by X-Y trees and, most importantly, (4) facilitating content analysis by extracting the relationship of headers to content cells rather than only the geometric cell structure. This research is part of a larger project [1] to generate narrow-domain ontologies (e.g., for automobiles, obituaries, geopolitics) from semi-structured web data, which is itself a step

J. Caretteet al. (Eds.): Calculemus/MKM 2009, LNAI 5625, pp. 422–437, 2009.

towards realization of the Semantic Web [2,3]. Concentrating on tabular sources of quantitative information avoids some difficulties of natural language processing.

Comprehensive reviews of two decades of research on table processing appear in [4,5]. Algorithms were first developed for specifying cell location in terms of rulings or, in the case of unruled tables, according to the geometric alignment and typographic similarity of cell content. A recent proposal for an end-to-end system divides the task into table detection, segmentation, function analysis, structural analysis and interpretation, but was not implemented and does not define which tables can and cannot be processed [6]. None of the methods that address web tables (e.g. [7]), carries the analysis to the layout-independent multi-category level.

This paper formalizes the methods we used in an experiment on 200 tables randomly chosen from eight large web sites. The 200 tables were imported into Excel and edited into a form that could be processed algorithmically. The average size of the tables was 587 cells, and editing required on average 104 seconds [8]. *Augmentations* such as aggregates, annotations, footnotes and titles that are important componenents of most tables were also processed, but they are not included in the formalism presented here.

1.1 Rectangular Tessellations

A *discrete rectilinear tessellation,* or a *rectangular tiling,* is the partition of an isothetic rectangle into rectangles defined on an m x n lattice. The geometry of such a construct can be uniquely represented by the locations and types of all its junction points, i.e., points at which two non-collinear lines meet or cross. The number of tilings, $N_{all}(m) \equiv N_{all}(m,m)$, increases exponentially with the size of the grid. A quick count reveals that even a 4x4 grid has 70,878 different partitions. Some of these, called X-Y-tessellations, can be obtained by a divide-and-conquer method based on successive horizontal and vertical guillotine cuts. Klarner and Magliveras proved that the number $N_{xy}(m)$ of X-Y-tilings decreases quickly with the size of the grid [9]. Although $N_{xy}(4) = 68,480$, which does not differ in order of magnitude from 70,878,

$$\lim_{m \to \infty} N_{xy}(m) / N_{all}(m) = 0 .$$

Figure 1 shows a simple X-Y-tessellation, and Figure 2 shows tilings that are not X-Y-tessellations. In the VLSI literature these are known as *nonslicing structures* [10]. It is known that horizontal and vertical polar graphs (that are duals of each other) can be drawn for any rectangular tiling, and that for a *slicing structure* (X-Y-tessellation) the polar graphs are series parallel. The concept of polar graph goes back to a 1940 paper on the dissection of rectangles into squares [11].

Polar graphs abstract away the geometry of rectangular tilings but preserve the adjacency relationship between the tiles in the horizontal and vertical directions. X-Y

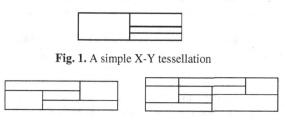

Fig. 1. A simple X-Y tessellation

Fig. 2. Two non-X-Y tessellations

trees similarly abstract the geometry X-Y of tessellations by providing a *structural* representation of the rectangles obtained by horizontal and vertical cuts at alternating levels. Such partitions can be represented by X-Y trees that we originally proposed for page layout analysis [12, 13]. They have been periodically rediscovered and are also known by other names like *puzzle tree* or *treemap* [14]. They transform a 2-D structure into two interlaced 1-D structures, thereby facilitating analysis. Figure 3 shows two X-Y-tessellations defined on a 4 x 4 lattice that are geometrically different but are both represented by the X-Y tree shown on the right. We don't know the number of structurally different X-Y tessellations, $N_{S,xy}(m)$, but it clearly is much smaller than the number of (geometrically) different X-Y tessellations $N_{xy}(m)$. The transformation of an X-Y-tessellation to an X-Y tree is discussed in Section 2.

Fig. 3. Two geometrically different but structurally identical tessellations

1.2   Web Tables

The layout of tables for the presentation of information is dictated by convention. The Chicago Manual of Style [15] and the US Government Printing Office Style Manual [16] both have lengthy chapters describing these conventions. All tables have a *stub, column headings, row headings*, and *data cells*. Several common layouts are illustrated in Figure 4. Tessellations that correspond to such layouts are called *admissible tessellations* or table candidates because the location of each data cell is specified by a set of hierarchical row and column headings.

Many tables that appear in the literature do not strictly follow conventions yet are readily understandable by their intended readers. For example, a common occurrence is

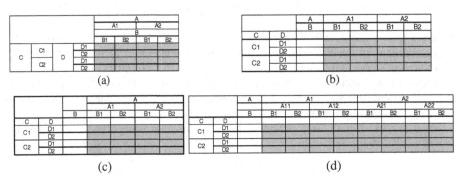

Fig. 4. Common table layouts. The blank top-left area is the *stub*. Only the *column* and *row* headings are labeled. The gray areas are content (*delta*) cells. Combinations of (a) for columns and (b) for rows are popular. (c) and (d) are more unusual hybrids.

the absence of a *root*, or spanning heading, for a category. Let us call the mathematically indefinable and unknown number of human-understandable tables $N_{T,S,xy}(m)$. We propose to process tables in this category by interactively transforming them into a smaller set of *admissible* tables that can be formally described and algorithmically analyzed. The number of admissible tables is $N_{A,S,xy}(m)$.

For the purpose of algorithmic analysis we need consider only *layout-equivalent* admissible table candidates that do not differ in the number of categories, but only with respect to the depth of their heading hierarchies, or the number of rows and columns, as do the examples in Figure 5.

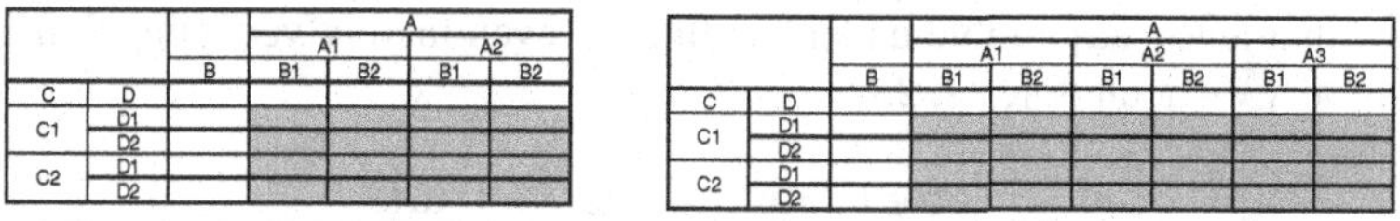

Fig. 5. Layout equivalent tables. The blank areas must be empty. Gray areas contain data.

Context-free grammars can help to characterize entire families of layout-equivalent admissible tessellations, as first demonstrated in [17, 18, 19] and revived here in Section 3. A few such families account for the vast majority of tables encountered in books, journals, and the web. The number of different layout-equivalent admissible table candidates is $N_{L,S,xy}(m)$. We cannot yet process automatically all structurally equivalent admissible tables, therefore $N_{L,S,xy}(m) < N_{A,S,xy}(m)$.

X-Y trees represent only the physical layout of a table, which can be modified to suit page size or column width, or display characteristics. The first step in *understanding* a table is to analyze its logical structure, which is independent of the presentation aspects. Interpretation requires understanding the relationship between *headings* and *content cells*. An abstract data structure for this purpose was proposed by Wang in 1996 [20]. It represents headings in terms of category trees (*labeled domains*), whose Cartesian product provides the paths to every content cell (called *delta cells*). The number of categories in a table is called its *dimensionality*. Figure 6 displays the category trees for a simple table. The *size* of the table is the product of the number of rows and columns of delta cells, and it is also equal to the product of the number of leaf nodes in the category trees. An algorithm for extracting the Wang Notation from the X-Y trees is presented in Section 4.

Labeled table candidates for which Wang Notation exists are called Well Formed Tables (WFT). They are only a subclass of tables encountered in practice. However,

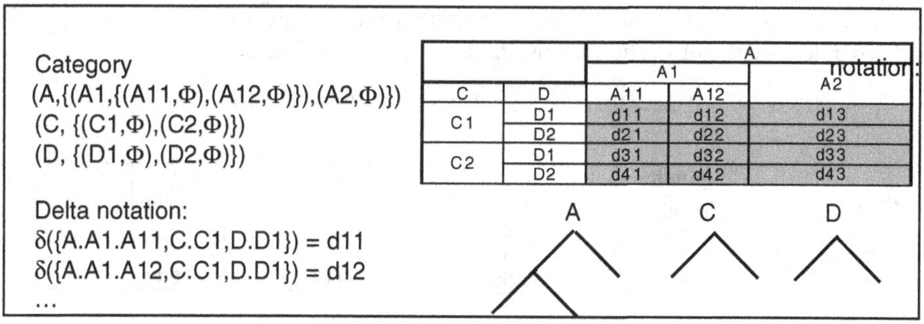

Fig. 6. Wang notation for the categories and data cells of a simple 3-category table

most such tables can be transformed to WFT format with little effort. Figure 7 shows a table that is not well formed, and its WFT equivalent, obtained by the addition of *virtual headings*. The headings shown are sensible, but any arbitrary labels would do for the Wang notation.

Analyzing the logical structure of a table is necessary but by no means sufficient for understanding it. Understanding most tables requires considerable context and knowledge that extends far beyond the table under consideration. There is ample evidence that automating table understanding, or even merely verifying claims to this effect, is very difficult [21, 22, 23].

Table I Maximum temperature

	2000		2001		2002	
	Summer	Winter	Summer	Winter	Summer	Winter
Montreal	35	11	36	2	37	13
Vancouver	28	18	29	19	30	20
James Bay	8	4	9	5	10	6

Table I Maximum temperature

	YEAR					
	2000		2001		2002	
	SEASON					
CITY	Summer	Winter	Summer	Winter	Summer	Winter
Montreal	35	11	36	2	37	13
Vancouver	28	18	29	19	30	20
James Bay	8	4	9	5	10	6

Fig. 7. Top: Rootless categories: not an admissible table. Bottom: Virtual headings added to obtain an admissible configuration that is also a WFT.

As mentioned, our project is the front end of a larger undertaking that endeavors to create narrow-domain ontologies by combining information from web tables [1, 24, 25]. Suppose, for instance, that we process the left-hand table in Figure 8 and include it into the ontology. Then when we encounter the right-hand table we hope to be able to learn that the *hepth* of *goldam* is *320 gd* [26]. Our current plans to build interactive software for harvesting web tables based on the formalisms described above are outlined in Section 5.

Our approach to the gradual automation of table processing is based on the following inequalities, which show that useful tessellations are only a very small fraction of all possible tessellations. The various classes of tables are illustrated in Fig. 9.

$$N_{L,S,xy}(m) < N_{A,S,xy}(m) < N_{T,S,xy}(m) \ll N_{S,xy}(m) \ll N_{xy}(m) \ll N_{all}(m).$$

fleck	gonsity (ld/gg)	hepth (gd)
burlam	1.2	120
falder	2.3	230
multon	2.5	350

goldam	1.3 ld/gg	320 gd
falder	2.3 ld/gg	230 gd
elmer	2.9 ld/gg	350 gd

Fig. 8. Two tables with overlapping information

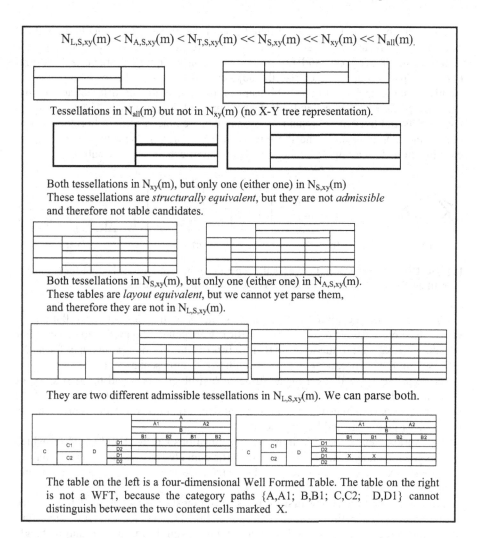

$$N_{L,S,xy}(m) < N_{A,S,xy}(m) < N_{T,S,xy}(m) \ll N_{S,xy}(m) \ll N_{xy}(m) \ll N_{all}(m).$$

Tessellations in $N_{all}(m)$ but not in $N_{xy}(m)$ (no X-Y tree representation).

Both tessellations in $N_{xy}(m)$, but only one (either one) in $N_{S,xy}(m)$
These tessellations are *structurally equivalent*, but they are not *admissible* and therefore not table candidates.

Both tessellations in $N_{S,xy}(m)$, but only one (either one) in $N_{A,S,xy}(m)$.
These tables are *layout equivalent*, but we cannot yet parse them, and therefore they are not in $N_{L,S,xy}(m)$.

They are two different admissible tessellations in $N_{L,S,xy}(m)$. We can parse both.

The table on the left is a four-dimensional Well Formed Table. The table on the right is not a WFT, because the category paths {A,A1; B,B1; C,C2; D,D1} cannot distinguish between the two content cells marked X.

Fig. 9. Discrete rectangular isothetic tessellations. Our taxonomy does not include human-readable tables to which Wang Notation is inapplicable. The top table without a row header in Fig. 7 is certainly in $N_{T,S,xy}(m)$, but we cannot formally define *all* human-readable tables.

2   Tessellations to X-Y Trees

As discussed above, the X-Y tree is an economical representation of layouts that are of interest in table processing. Similar table layouts yield X-Y trees with similar structures. We can identify tables from which we can algorithmically extract Wang Notation. We shall also attempt to characterize families of inadmissible table structures that can be converted into admissible structures by a few editing steps. We expect to be able to automate such frequently used editing protocols.

The horizontally and vertically ordered lists of the *indices* of the junction points of a tessellation are not sufficient to derive the corresponding X-Y tree, although the combination of pre-order and post-order traversals uniquely characterizes general trees. The lists do not characterize the adjacency topology of the tessellations sufficiently for table analysis. Figure 10 illustrates tilings that are not differentiated by the structure of their X-Y trees, and different tilings with identical lists. For table analysis, the lists and trees must contain additional data, i.e., the vertical or horizontal location of the junction points and the type (e.g. NE-corner, T-connection, crossing) of each junction. This allows checking the alignment of cuts in separate subtrees.

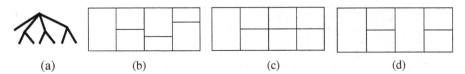

| (a) | (b) | (c) | (d) |

Fig. 10. The vertical cut X-Y tree (a) is the same for tilings (b) and (c), but not for (d). However, tilings (c) and (d) have the same lists of junction point coordinates.

The recursive algorithm EX2XY obtains the X-Y tree for any tessellation for which the tree exists. We use it to transform web tables imported into Excel. For portability, EX2XY produces an XML file. It takes the junction-points data for an X-Y tessellation and produces a fully-parenthesized representation for it, which can either be printed (saved) as a linear string of leaf-block labels and the two kinds of parentheses. It can also include geometric information attached to the labels and the left parentheses (of either type) in an internal data structure. The latter representation is useful for geometric and lexical checks.

The workhorses of the algorithm are two functions CutV(R) and CutH(R) which cut the given rectangle R, respectively, in vertical and horizontal directions. R may be specified as (x1,y1,x2,y2), where (x1,y1) are their top-left and (x2,y2) are the bottom-right junction points.

CutV and CutH return the first (leftmost or topmost) sub-rectangle of R, obtained by a guillotine cut. In the example rectangle of Fig. 11, CutV((1,1,4,4)) would return the sub-rectangle (1,1,3,4) and CutH((1,1,4,4)) would return the sub-rectangle (1,1,4,2). The cut may be *trivial* or *degenerate*, e.g for R = (2,1,3,2) in the example, CutV(R) = CutH(R) = R.

CutV and CutH are used in a pair of procedures, P1 and P2, which call each other recursively (Fig 12). P1 cuts a given rectangle vertically, submitting the leftmost sub-rectangle to P2 for horizontal cuts. Similarly, P2 cuts a given rectangle horizontally, submitting the topmost sub-rectangle to P1 for vertical cuts. The main procedure calls P1 with the outermost rectangle (1,1,4,4) for vertical-cut first, and P2 for horizontal-cut first.

Although most of our illustrations contain simple examples created directly in Excel, Figure 13 shows part of an actual web table, its Excel version created by the built-in IMPORT functionality, and its appearance after editing.

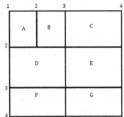

Fig. 11. A simple example to illustrate algorithm EX2XY

```
P1(R); {
     Declare S: rectangle

     S = CutV(R);
     if (S == R) then {if CutH(R) == R) then {print(label(R)); return} }
     /* Also, attach coordinates of R with the label of R */

     else /* have a non-trivial cut */
     {
               print("[");          /* Also, attach coordinates of  R with this "[" */

               Loop {
                         P2(S);   /* H-Cut S */
                         R = R-S;
                           S = CutV(R);
                     } until S==R;

               P2(S);                 /* H-Cut the last rectangle */

               print("]")
     }
}

P2(R); {
     Declare S: rectangle

     S = CutH(R);
     if (S == R) then {if CutV(R) == R) then {print(label(R)); return} }
     /* Also, attach coordinates of R with the label of R */

     else /* have a non-trivial cut */
     {
               print("{");          /* Also, attach coordinates of  R with this "{" */

               Loop {
                         P1(S);   /* V-Cut S */
                         R = R-S;
                           S = CutH(R);
                     } until S==R;

               P1(S);                 /* V-Cut the last rectangle */

               print("}")
     }
}
```

Fig. 12. Algorithm EX2XY

U.S. Coal Supply, Disposition, and Prices

Report No.: DOE/EIA 0584 (2007)
Report Released: September 2008
Next Release Date: September 2009

Table ES1.   xls   pdf   **Annual Coal Report**

Table ES1. U.S. Coal Supply, Disposition, and Prices, 2006-2007
(Million Short Tons and Dollars per Short Ton)

Item	2006	2007
Production by Region		
Appalachian	391.2	377.8
Interior	151.4	146.7
Western	619.4	621.0
Refuse Recovery	0.8	1.2
Total	1,162.8	1,146.6
Consumption by Sector		
Electric Power	1,026.6	1,045.1
Coke Plants	23.0	22.7
Other Industrial Plants	59.5	56.6
Residential/Commercial	3.2	3.5
Total	1,112.3	1,128.0

(a)

Table ES1. U.S. Coal Supply, Disposition, and Prices, 2006-2007
(Million Short Tons and Dollars per Short Ton)

Item	2006	2007
Production by Region		
Appalachian	391.2	377.8
Interior	151.4	146.7
Western	619.4	621.0
Refuse Recovery		
Total		
Consumption by Sector		
Electric Power		
Coke Plants		
Other Industrial Plants		
Residential/Commercial		
Total		
Year-End Coal Stocks		
Electric Power		

(b)

Table ES1. U.S. Coal Supply, Disposition, and Prices, 2006-2007
(Million Short Tons and Dollars per Short Ton)

			2006	2007
Production by Region		Appalachian	391.2	377.8
		Interior	151.4	146.7
		Western	619.4	621.0
		Refuse Recovery	0.8	1.2
		Total	1,162.8	1,146.6
		Electric Power	1,026.6	1,045.1
		Coke Plants	23.0	22.7
		Other Industrial Plants	59.5	56.6
		Residential/Commercial	3.2	3.5
		Total	1,112.3	1,128.0
Consumption by Sector		Year-End Coal Stocks		
		Electric Power	141.0	151.2
Item		Coke Plants	2.9	1.9

(c)

Fig. 13. Part of a US Energy Information Administration table. (a) As it appears on the web; (b) Imported into Excel; (c) After editing. http://www.eia.doe.gov/cneaf/coal/page/acr/tables1.html

3 A Grammar for Table Candidates

Although EX2XY produces X-Y trees as verbose XML files, in this section we represent the trees with nested parentheses notation. This notation has a 1:1 correspondence with general trees provided that the order of the symbols is preserved [27]. We present the notation and the *Look Ahead Left to Right* (LALR) grammar G_1 [28,29] constructed to parse the X-Y trees of table-like tessellations by means of an example. The grammar was implemented in yacc [30].

Consider the following column headings for two Wang categories of Employment Status and Education (Fig. 14) which result in the derivation of Fig. 15.

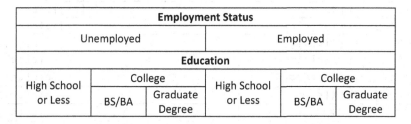

Employment Status					
Unemployed			Employed		
Education					
High School or Less	College		High School or Less	College	
	BS/BA	Graduate Degree		BS/BA	Graduate Degree

Fig. 14. Sample table row heading for grammar G_1

Textual labels (like Employment Status) have no bearing on the structure, so we will replace them by the generic symbol c. We alternate brackets and braces for ease of reading, but they are equivalent. The X-Y tree "sentence" S_{XY} for this partition of the tessellation is:

$$S_{XY} = \{ c [c c] c [c \{ c [c c] \} c \{ c [c c] \}] \}$$

Grammar G_1 for parsing all layout-equivalent tessellations of this kind is:

```
S := A
A := { B }
B := c[X]B | c[X]
X := cX | AX | A | c
```

This grammar can parse fully parenthesized input for column headers of tables with arbitrary dimensions and any number of levels in each dimension. It is a simple matter to add a mirror-image grammar to parse the row headings and delta cells. The non-terminals in G_1 serve the following functions.

S is the start symbol (eventually to generate all admissible strings for tables).
A is the nonterminal that generates all admissible strings for column headers.
B generates one or more instances of categories in the form "c[X]".

Each c becomes a root category and X generates its subcategory tree. X generates strings of length ≥ 1, with arbitrary occurrences of c and A.

We rewrite the grammar in the following equivalent form for ease of reference:

G₁ RULES:

1. S := A	3. B := c [X] B	5. X := cX	7. X := A
2. A := { B }	4. B := c [X]	6. X := AX	8. X := c

Action	Stack state	Remaining Input
-	Null	{ c [c c] c [c { c [c c] } c { c [c c] }] }
Shift⁵	{ c [c **c**	] c [c { c [c c] } c { c [c c] }] }
R 8	{ c [**c X**	] c [c { c [c c] } c { c [c c] }] }
R 5	{ c [X	] c [c { c [c c] } c { c [c c] }] }
Shift⁹	{ c [X] c [c { c [c **c**	] } c { c [c c] }] }
R 8	{ c [X] c [c { c [**c X**	] } c { c [c c] }] }
R 5	{ c [X] c [c { c [X	] } c { c [c c] }] }
Shift	{ c [X] c [c { **c [X]**	} c { c [c c] }] }
R 4	{ c [X] c [c { B	} c { c [c c] }] }
Shift	{ c [X] c [c { **B }**	c { c [c c] }] }
R 2	{ c [X] c [c A	c { c [c c] }] }
Shift⁶	{ c [X] c [c A c { c [c **c**	] }] }
R 8	{ c [X] c [c A c { c [**c X**	] }] }
R 5	{ c [X] c [c A c { c [X	] }] }
Shift	{ c [X] c [c A c { **c [X]**	}] }
R 4	{ c [X] c [c A c { B	}] }
Shift	{ c [X] c [c A c { **B }**	] }
R 2	{ c [X] c [c A c A	] }
R 7	{ c [X] c [c A **c X**	] }
R 5	{ c [X] c [c **A X**	] }
R 6	{ c [X] c [**c X**	] }
R 5	{ c [X] c [X	] }
Shift	{ c [X] **c [X]**	}
R 4	{ **c [X] B**	}
R 3	{ B	}
Shift	**{ B }**	
R 2	A	
R 1	S	

Fig. 15. Derivation for the example of Fig. 14

An LALR is a shift-reduce parser that at each step either *shifts* the next input on to the stack or *reduces* the symbols on top of the stack according to a rule of the grammar. It produces leftmost reductions as it scans the input from left to right, which yields a rightmost derivation in reverse order. The first column in Fig. 15 shows the action (shift or reduce); with Shift^n denoting n consecutive shifts and R m denoting reduction according to rule number m. The bold characters represent the *handle* (right-hand side) of the production that is reduced to the left-hand symbol by the rule listed on the next row.

This example demonstrates both the power and the limitations of using a grammatical approach to parsing: A grammar can be written to recognize a broad class of tilings. On the other hand, a context-free grammar is not powerful enough to check that the headings are labeled appropriately for a WFT. If a candidate structure is accepted by G_1, then we must conduct additional geometrical alignment and lexical checks to verify the Wang Notation.

4 X-Y Tree to Wang Notation

In this section we demonstrate XY2WANG an algorithm that converts an X-Y tree generated from a restricted family of *admissible tables* to Wang Notation. An example of this family can be seen in Figure 4(a). Figure 16 shows a simple example from this family that that XY2WANG can process. Although in Section 3 we used parenthesis notation for the trees, here we use an indented table-of-contents that reveals the underlying data structure (Figure 17). Figure 18 shows the top level pseudo-code for XY2WANG (which was implemented in Python and produces XML output). The algorithm accepts table trees with an arbitrary number of categories and levels of headings. For the selected example, the algorithm returns the Wang notation (in a verbose XML format) for a two-category table $T = (C, d)$:

Category Notation (Labeled Domains):

$$C = \{ \ (A, \{ \ (A_1, F), (A_2, \Phi) \ \}), \ (B, \{ \ (B_1, \Phi), (B_2, \Phi), (B_3, \Phi) \ \}) \ \}$$

Delta Mappings:

$$\delta(\{A.A_1, B.B_1\}) = d_{11}$$
$$\delta(\{A.A_1, B.B_2\}) = d_{12}$$
$$\delta(\{A.A_1, B.B_3\}) = d_{13}$$

...

The algorithm first locates the four principal regions of the table in the XY tree: the stub, row-headings, column-headings, and content cells. It then extracts the Wang labeled domains from the category regions under the assumption that each spanning cells in the row header is the parent category of smaller cells to its right, and each spanning cell in the column header is the parent of smaller cells below it. After the category notation is generated, the Cartesian product of the category paths is computed and each key is matched to the content of a delta cell.

		B		
		B1	B2	B3
A	A1	d11	d12	d13
	A2	d21	d22	d23

Fig. 16. Example table for XY2WANG

Index	Node	Parent	Children	H x W
1	Outer Frame	None	2,9	4x5
2	Left Side	1	3,4	2x5
3	Stub	2	None	2x2
4	A+(A1+A2)	2	5,6	2x2
5	A	4	None	2x1
6	(A1+A2)	4	7,8	2x1
7	A1	6	None	1x1
8	A2	6	None	1x1
9	Right Side	1	10,11,15,19	4x3
10	B	9	None	1x3
11	B1+B2+B3	9	12,13,14	1x3
12	B1	11	None	1x1
13	B2	11	None	1x1
14	B3	11	None	1x1
15	d11+d12+d13	9	16,17,18	1x3
16	d11	15	None	1x1
17	d12	15	None	1x1
18	d13	15	None	1x1
19	d21+d22+d23	9	20,21,22	1x3
20	d21	19	None	1x1
21	d22	19	None	1x1
22	d23	19	None	1x1

Fig. 17. Data structure created by XY2WANG for the X-Y tree of the table in Fig. 17. The index represents a depth-first traversal of the X-Y tree, which has 8 internal nodes (including the root) and 14 leaf nodes corresponding to the cells of the table. Children are listed top-to-bottom or left-to-right. Borders enclose the four principal regions of the table.

XY2WANG must be able to handle more complex scenarios than Figure 16, such as higher Wang dimensionality, deeper nesting of headers, repetitive headers, and the detection of not well-formed tables. Provisions for such scenarios are included in the Python program outlined by the pseudo-code of Figure 18.

Pseudo-code for XY2WANG
Divide Left Side into *Stub* and *Row-headings* // (Stub is first child of Left Side)
Divide Right Side into *Col-headings* and *Data Cells* // separate using *stub-height*
Separate row-category subtrees at nodes with spanning heights
Separate column-category subtrees at nodes with spanning widths
Traverse breadth-first each category tree while removing duplicate labels
Return Wang Category notation
Form Cartesian *product* of unique paths
Compare size of *product* to number of data cells
If it differs from number of leaf nodes in *Data Cells*, return "tree not well formed"
Else assign a data cell in *Data Cells* to each path
Return Wang Delta notation

Fig. 18. Top-level pseudo-code for algorithm XY2WANG

5   Conclusion

Web tables intended for human readers are generally laid out on a grid. The data cells are referenced by row and column headings which form labeled domains of categories. The hierarchical structure of categories and the flat structure of the data cells can be recovered by interleaved vertical and horizontal partitions represented as X-Y trees. An X-Y tree represents a generic rectangular tiling and indiscriminately makes all the cuts in each direction. The table grammar reorders the cuts so as to represent the structure of the table according to specific style(s) of tables.

We defined geometric and topological equivalence classes on tessellations and their X-Y trees. Many tables encountered in practice correspond to well-defined subsets of these equivalence classes. They can be identified by parsing the X-Y tree with a context-free grammar. If the labels of the headings are consistent, then the table is well formed, and we can algorithmically extract its Wang category notation.

The current formalism does not account for *augmentations* although our experimental system does process them and includes them in the XML output. Common augmentations are *aggregates* (sums, averages and weighted averages, medians), *footnotes, units, annotations, table titles* and *captions*. As Wang noted and our 200-table experiment confirms, these are essential components of most tables. Because they are not revealed by the tiling itself, so far we have not been able to treat them uniformly, but they must eventually be integrated into any practical table understanding system.

The precise representation of layout-invariant table syntax is the first step towards semantic interpretation of groups of conceptually overlapping tables. The approach we propose towards this goal is to import the web tables into a spreadsheet, interactively edit them as necessary, and then algorithmically transform the data into Wang Notation in a portable XML format. We believe that syntactic analysis of the X-Y trees will allow identifying tables requiring similar edit steps, so that these edit steps can be applied automatically. This will effectively expand the number of admissible layouts and thereby reduce the amount of necessary interaction.

Acknowledgments. This work was supported by the National Science Foundation under Grants# 044114854 and 0414644 and by the Rensselaer Center for Open Source Software. We gratefully acknowledge the influence of two decades of discussions about tables and ontologies with Prof. D.W. Embley of Brigham Young University.

References

1. Tijerino, Y.A., Embley, D.W., Lonsdale, D.W., Nagy, G.: Towards ontology generation from tables. World Wide Web Journal 6, #3 (2005)
2. Berners-Lee, T., Hendler, J., Lassila, O.: The Semantic Web. Scientific American (2001)
3. Halevy, A., Norvig, P., Pereira, F.: The Unreasonable Effectiveness of Data. IEEE Transactions on Intelligent Systems, 8–12 (March/April 2009)
4. Lopresti, D., Embley, D.W., Hurst, M., Nagy, G.: Table Processing Paradigms: A Research Survey. Int. J. Doc. Anal. Recognit. 8(2-3), 66–86 (2006)
5. Zanibbi, R., Blostein, D., Cordy, J.R.: A survey of table recognition: Models, observations, transformations, and inferences. Int. J. Doc. Anal. Recognit. 7(1), 1–16 (2004)
6. Silva, E.C., Jorge, A.M., Torgo, L.: Design of an end-to-end method to extract information from tables. Int. J. Doc. Anal. Recognit. 8(2), 144–171 (2006)
7. Gatterbauer, W., Bohunsky, P., Herzog, K.M., Pollak, B.: Towards Domain-Independent Information Extraction from Web Tables. In: Proceedings of World Wide Web, Banff, pp. 71–80 (2007)
8. Padmanabhan, R., Jandhyala, R.C., Krishnamoorhty, M., Nagy, G., Seth, S., Silversmith, W.: How many different kinds of tables are there. In: Procs. Eights Int'l. Workshop on Graphics Recognition (GREC 2009) (2009) (in press)
9. Klarner, D.A., Magliveras, S.S.: Tilings of a Block with Blocks. Europ. J. Combinatorics 9, 317–330 (1988)
10. Kuh, E.S., Ohtsuki, T.: Recent Advances in VLSI Layout. Proceedings of the IEEE 78(2) (1990)
11. Brooks, R.L., Smith, C.A.B., Stone, A.H., Tutte, W.T.: The dissection of rectangles into squares. Duke Math. J. 7, 312–340 (1940)
12. Nagy, G., Seth, S.: Hierarchical Image Representation with Application to Optically Scanned Documents. In: Procs. Int. Conf. Pat. Recog. VII, Montreal, pp. 347–349 (1984)
13. Krishnamoorthy, M., Nagy, G., Seth, S., Viswanathan, M.: Syntactic Segmentation and Labeling of Digitized Pages from Technical Journals. IEEE Transactions on Pattern Analysis and Machine Intelligence 15, #7, 737–747 (1993)
14. Samet, H.: Foundations of Multidimensional and Metric Data Structures. Morgan Kaufmann, San Francisco (2006)
15. The Chicago Manual of Style, 15th edn. Univ. of Chicago Press, Chicago (2003)
16. U.S. Government Style Manual, 29th edn. (2000)
17. Green, E.A., Krishnamoorthy, M.: Model-based analysis of printed tables. In: Procs. of Third International Conference on Document Analysis and Recognition (ICDAR 1995), Montreal, Canada, pp. 214–217 (1995)
18. Green, E.A., Krishnamoorthy, M.: Recognition of tables using table grammars. In: Procs. of Symposium on Document Analysis and Recognition (SDAIR 1995), Las Vegas, NV, pp. 261–277 (1995)

19. Green, E.A., Krishnamoorthy, M.: Model-based analysis of printed tables. In: Procs. Third Int'l. Workshop on Graphics Recognition (GREC 1995), pp. 234–242 (1995); in Graphics Recognition Methods and Applications. LNCS, vol. 1072, pp. 80–91. Springer, Heidelberg (1996)

20. Wang, X.: Tabular Abstraction, Editing, and Formatting. Ph.D Dissertation, University of Waterloo, Waterloo, ON, Canada (1996)

21. Hu, J., Kashi, R., Lopresti, D., Nagy, G., Wilfong, G.: Why table ground-truthing is hard. In: Procs. of Sixth International Conference on Document Analysis and Recognition, Seattle, WA, pp. 129–133 (2001)

22. Lopresti, D., Nagy, G.: A Tabular Survey of Table Processing. In: Chhabra, A.K., Dori, D. (eds.) GREC 1999. LNCS, vol. 1941, pp. 93–120. Springer, Heidelberg (2000)

23. Nagy, G., Lopresti, D.: Issues in ground-truthing graphic documents. In: Blostein, D., Kwon, Y.-B. (eds.) GREC 2001. LNCS, vol. 2390, pp. 46–66. Springer, Heidelberg (2002) (selected papers from the Fourth International Workshop on Graphics Recognition)

24. Tao, C., Embley, D.W.: Automatic Hidden-Web Table Interpretation by Sibling Page Comparison. In: Parent, C., Schewe, K.-D., Storey, V.C., Thalheim, B. (eds.) ER 2007. LNCS, vol. 4801, pp. 566–581. Springer, Heidelberg (2007)

25. Tao, C., Embley, D.W., Liddle, S.W.: Enabling a Web of Knowledge. Brigham Young University. manuscript submitted to the special issue about the web of data for the Journal of Web Semantics (2009)

26. Embley, D.W., Lopresti, D., Nagy, G.: Notes on Contemporary Table Recognition. In: Bunke, H., Spitz, A.L. (eds.) DAS 2006. LNCS, vol. 3872, pp. 164–175. Springer, Heidelberg (2006)

27. Horowitz, E., Sahni, S.: Fundamentals of Data Structures. W.H. Freeman & Co., New York (1983)

28. Aho, A.V., Sethi, R., Ullman, J.D.: Compilers: Principles, Techniques, and Tools. Addison-Wesley, Reading (1986)

29. DeRemer, F., Pennello, T.: Efficient Computation of LALR(1) Look-Ahead Sets. ACM Trans. Prog. Lang. and Sys. (TOPLAS) 4(4), 615–649 (1982)

30. Johnson, S.C.: YACC: Yet another Compiler-Compiler. Unix Programmer's Manual 2b (1979)

Finite Groups Representation Theory with CoQ*

Sidi Ould Biha

INRIA Sophia Antipolis,
INRIA Microsoft Research Joint Centre
Sidi.Biha@sophia.inria.fr

Abstract. Representation theory is a branch of algebra that allows the study of groups through linear applications, i.e. matrices. Thus problems in abstract groups can be reduced to problems on matrices. Representation theory is the basis of character theory. In this paper we present a formalization of finite groups representation theory in the CoQ system that includes a formalization of Maschke's theorem on reducible finite group algebra.

Keywords: Representation theory, Maschke's theorem, linear algebra, CoQ, SSReflect.

1 Introduction

The use of proof assistants for the formalization of mathematical theories has increased considerably in recent years. Success stories like the formal proofs of the Four Colour theorem [1] or the prime number theorem [2] have shown that formal proof systems have reached the age of maturity. After these successes, ambitious projects were launched, for example the Flyspeck [3] project which aims to develop a formal proof of Kepler's conjecture. Projects such as the C-CoRN [4] have developed large repositories of mathematical formal proof libraries, but the number of formal mathematics libraries remains low compared to the number of libraries developed in Computer Algebra System (CAS) like Mathematica [5] or GAP [6]. This is one of the reasons for the limited number of users of formal proof systems, especially among mathematicians.

This work is a part of the Mathematical Components project [7] which aims to develop a formal proof of the Feit-Thompson theorem [8]. Finite group representation theory is among the large variety of mathematical theories covered by the proof of the Feit-Thompson Theorem.

This paper presents a formalisation of finite group representation theory and generic libraries for linear algebra : theory of finitely generated modules over algebras and fields. A formal proof of the Maschke theorem was also developed. This is done using the SSReflect [9] extension of the CoQ proof assistant [10,11].

The paper is organized as follows. In Section 2 we give an introduction to finite group representation theory and show how it is linked to module theory. In Section 3, we present the Mathematical Components project, the CoQ extension

---

* This research work was funded by the Microsoft Research INRIA joint centre.

J. Carette et al. (Eds.): Calculemus/MKM 2009, LNAI 5625, pp. 438–452, 2009.
© Springer-Verlag Berlin Heidelberg 2009

SSReflect and the project libraries we reused in our development. Finally, in Section 4, we present the two components of this development : the linear algebra and representation libraries.

2   Representations Theory

Finite group representation theory studies the structure of a finite group by presenting it as a matrix. For example, the symmetric group of index three S_3 can be represented as the group of the isometries of an equilateral triangle. The symmetric group of index four S_4 can be represented as the group of rotations of a cube. This same technic can be used to study other algebraic structures likes associative algebra and Lie algebra. Historically the theory was introduced in the second half of the nineteenth century by Frobenius to solve problems from Galois theory. It was largely developed afterwards to be a basic tool for the classification of finite groups and an important part of the proof of the Feit-Thompson theorem. Representation theory is used in algebraic number theory through the class field theory. It is also used in the Langlands program [12], an active field of contemporary mathematical research.

Algebra representation : Given a field F, an integer n and an F-algebra A, a representation of A is an algebra homomorphism $\phi : A \rightarrow M_n(F)$, where $M_n(F)$ is the algebra of square matrix of size n and coefficients in F. Generally, A can be an associative algebra or Lie algebra.

In representation theory literature [13,14,15], a common way to study representations is to see them as modules. In Isaacs book [13], which is the reference for representation theory for the proof of Feit-Thompson theorem, a theory of finitely generated modules over algebra is developped to introduce representation theory. A module over a finite algebra has also a structure of finite F-vector space, since for any F-algebra A, $F.1 = \{c1|c \in F\}$ is a subalgebra of A. Thus a module over an F-algebra A is a F-vector space V with a right action of A on V such that for all $x, y \in A, v, w \in V$ and $c \in F$ the following properties hold :

- $(v + w)x = vx + wx$,
- $v(x + y) = vx + vy$,
- $(vx)y = v(xy)$,
- $(cv)x = c(vx) = v(cx)$,
- $v1 = v$

With this definition we have that for any F-algebra A, every representation of A has an A-module structure and conversely every A-module provides a representation of A. Thus we have an equivalence between representations and A-modules. The advantage of this approach is that many definitions and results on representations can be borrowed from module theory. With this equivalence we can introduce some definitions on representations. In the following, A is an F-algebra and V is a representation (in others words an A-module) :

- A *subrepresentation* of V is an A submodule W of V or an F subspace W of V which is stable under the action of A. A *subrepresentation* is also a representation.
- V is *irreducible* if its only submodules are 0 and V. It is *semisimple* if for every submodule $W \subseteq V$, there exists another submodule $U \subseteq V$ such that $V = W \oplus U$ in other words V is the direct sum of W and U.
- A representation or more generally a module is *semisimple* if it is the finite direct sum of *irreducible* submodules. These two last definitions are equivalent.

Finite group representation : Let G be a finite group, F a field and $GL(n, F)$ the multiplicative group of non-singular $n \times n$ matrices on F. An F-representation of G is a group homomorphism ρ from G to $GL(n, F)$. The integer n is called the degree of ρ. Thus a representation is a function $\rho : G \to GL(n, F)$ such that :

$$\rho(1_G) = I_n \quad and \quad \forall g\, h \in G\; \rho(gh) = (\rho g)(\rho h)$$

Group algebra is a key structure in representation theory. It links the definition above of group representation to that of algebra representation and modules theory. Given a finite group G and a field F, the group algebra $F[G]$ is the set $\{\sum_{g \in G} a_g g \mid a_g \in F\}$. This set has a structure of F-vector space and F-algebra. Indeed, the function that associates to any element g of G the element $\sum_{h \in G} a_h h$ with $a_g = 1$ and $a_h = 0\; if\; h \neq g$ embeds G into $F[G]$, so we can see G as a basis for $F[G]$. The addition and external multiplication are defined as follows :

$$\sum_{g \in G} a_g g + \sum_{g \in G} b_g g = \sum_{g \in G} (a_g + b_g)g \qquad \alpha \sum_{g \in G} a_g g = \sum_{g \in G} (\alpha * a_g)g$$

The $F[G]$ internal multiplication law is defined by considering the group multiplication :

$$(\sum_{g \in G} a_g g)(\sum_{h \in G} b_h h) = \sum_{k \in G} (\sum_{gh=k} a_g b_h)k$$

With this law and 1_G, $F[G]$ has a structure of an F-algebra. It follows that for any finite group G, an F-representation of G can be seen as the restriction on $G \subset F[G]$ of a representation of $F[G]$. Thus any group representation has an $F[G]$-module structure and vice versa.

An important result on finite group representation is the Maschke's theorem. It states that :

For any finite group G and a field F whose characteristic does not divide $\mid G \mid$, every representation ($F[G]$-module) is semisimple.

Maschke's theorem reduces the study of group representations into the study of *irreducible* representations. This is given by the fact that every group representation is the direct sum of *irreducible* representations. In order to know all the representations of a finite group, it suffices to know all its irreducible representations.

3   Mathematical Component Project

In the classification of finite groups, the Feit-Thompson theorem is a central result. His proof revolutionized group theory not only by the techniques it introduces but also by its length. The original paper [8] is more than 250 page long and remains roughly the same despite all the efforts to simplify it. The verification of the paper proof took about a year for a team of specialists in group theory. More generally, several results in the theory of classification have been published in papers whose length reaches hundreds of pages. The formalization of these proofs is a real challenge for proof assistants. Based on the experience gained in the proof of the Four Colour theorem, the Mathematical Components project aims to develop a formal proof of Feit-Thompson theorem.

SSREFLECT

In the Mathematical Components project, the development enviroment is SSREFLECT, a Coq extension who was developed by G. Gonthier for the formal proof of the Four Colour theorem. SSREFLECT (for *Small Scale Reflection*) introduces a new language for tactics that eases the development of proof scripts. It allows the user to write more concise proof scripts than those written using the standard Coq tactic language. Another main feature is the generic reflection mechanism. In the Coq proof system, the default logic is intuitionistic. In this logic, logical propositions and boolean values are distinct. Logical propositions are objects of type Prop which is the carrier of intuitionistic reasoning. The boolean type is an inductive type with two values : true and false. These two structures are complementary. The first one makes it possible to have structured proofs by using natural deduction whereas the second makes it possible to perform computation. SSREFLECT introduces a generic reflection mechanism that allows to combine the best of the two views and to switch from the propositional version of a decidable predicate to its boolean version. More details on the SSREFLECT tactics language and the view mechanism are presented in the SSREFLECT manual [9].

Libraries

In the project, we have a large variety of libraries that gives definitions and properties for a variety of mathematical structures. In all this development, libraries are independent from the excluded middle and the choice axiom. The logical requirements are internalized in the structure definitions.

SSREFLECT includes, among others, the following libraries :

- eqtype: type with a a decidable equality which is equivalent to the Leibniz one.
- choice: type with choice operator.
- fintype: type with finite elements.
- finfun: type of function of finite domain.
- bigops: generic indexed "big" operations, like $\sum_{i=0}^{n} f(i)$ or $\max_{i \in I} f(i)$.

- `groups`: finite groups theory.
- `ssralg`: algebraic structures from abelian group to algebraic closed field.
- `matrix`: determinant theory and matrix decomposition (LUP decomposition).

The libraries include also results on finite groups theory like the Sylow theorems and the Cauchy-Frobenius lemma. The Cayley-Hamilton theorem about matrix and polynomial was also formalised. For more precise details on theses libraries we refer to [16,17].

4   Formalization

The Feit-Thompson theorem covers a variety of different mathematical theories such as linear algebra and finite group theory. The work of formalization of such large theory requires an approach similar to software engineering. In this design process, the choice of which data structure to use to represent a mathematical concept is important. This choice must take into account the needs of genericity and reusability. The proof assistant COQ provides mechanisms such as dependent types and records, coercions or canonical structures that meet those needs.

COQ's dependent records [18] are useful to encode data types such as algebraic structures where we have a set of axioms and operations associated to the type of elements. They have been used in several algebraic hierarchy formalization such as [19] and [20]. Coercions provides a sub-typing mechanisms. They facilitate the development of generic theories and sharing of notations for abstract structures. This is very useful especially for the development of theory on algebraic structures where it is common to have inheritance : vector spaces are a sub-type of commutative groups and algebras are a sub-type of vector spaces and rings. Canonical structures allow the inference of a specific structure for a specific type. It works in the opposite direction to that of coercions. For example the matrix type can be equiped in a canonical way by a ring structure. With canonical structures, this structure can be implicitly infered by the system. When A and B are matrices, when writing the expression $A + B$, the system automatically infers that the + corresponds to the additive group law of matrices. When m and n are integers, when writing the expression $m + n$, it automatically infers that the + corresponds to the additive group law of integers. It is common in mathematical literature to let the reader infer from the context the corresponding operation.

The use of `ssralg` requires that the domain type on which the algebraic structure is defined has a decidable equality and a choice operator. This two requirements are internalized in the structures `eqType` and `choiceType`. This means that in order to use the theories implemented by this library for a certain type, this axioms should be valid on the elements of the type.

In an intuitionistic type theory base proof assitant like COQ, this design approch gives a general quotient construction, like ideal quotient or canonical basis for finitely generated modules. It also allows the use of COQ's powerfull rewriting system thanks to the inclusion of the decidable equality in the Leibniz's one (COQ's default equality). The alternative to this approach is the use of Setoids

[21] like what is done in [19] and [20]. The corresponding equivalence relation has then to be handled explicitly. This approach is costly especially when, as in our case, we deal with advanced mathematicals statements that involve several mathematical structures. Also, in the proof of the Feit-Thompson theorem, the algebraic structures handled are mainly finite groups or finite dimension vector spaces over finite fields or the algebraic number field. For these structures the decidable equality and the choice operator can be defined constructively.

4.1   Linear Algebra

As already said in Section 2, finite group representation theory inherits many definitions and results from linear algebra. Our main motivation is the formalization of finite group representation theory but the part on linear algebra can be used independently in other formalizations. In finite group representation theory, modules that are taken into consideration are all of finite dimension and are either defined over a field (finite or not) or on an algebra. We formalized a linear algebraic structures hierarchy that covers the theories of :

- finite dimension vector space
- finite dimension algebra
- finitely generated module over an algebra

The development is built on top of the algebraic hierarchy given by the **ssralg** library. It consists of three layers. The first layer defines the interfaces and

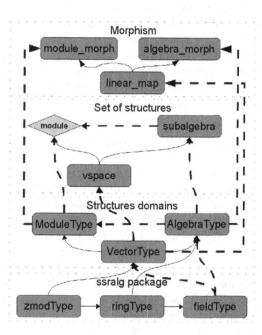

Fig. 1. The hierarchy for linear algebra

provides the generic theory for the domain of the considered structures. The second layer defines the interfaces and provides the generic theory for the special sets associated with the domain type structures of the first layer: sub-vector spaces, sub-algebras and sub-modules. The third layer is dedicated to morphisms of structures: linear applications, algebra and module morphisms.

Figure 1 provides an overview of the composition of these layers. In this figure, the plain arrows represent the sub-typing, for example `AlgebraType` is a subtype of `VectorType` and `ringType`. The dashed arrows represent type dependency, for example `VectorType` type depends on a `fieldType`.

Structures domains : An abstract algebraic structure is a combination of a representation type (domain), constants and operations on this type, and axioms satisfied by this constant and operations. For example, a group is a set G together with a constant e, an internal operation $*$ and a list of axioms (neutral element, associativity of $*$ ...). Also, algebraic structures are defined according to a hierarchical scheme. A vector space is a commutative group that has an external law which acts from a field. An algebra is the combination of a vector space and a ring structure with additional axioms. For the definition of our domains of algebraic structures, we apply the generic method introduced in the `ssralg` library. This method gives a design pattern for the definition of such domains. It is mainly motivated by problems of packaging (inheritance and sharing) and performances. For example, Figure 2 gives the interface of the finite dimension vector space.

We use the COQ `Module` system to create a separate name space for the F-vector space structure and avoid any clash of definitions, since the declaration of other algebraic structures follows the same scheme. In Figure 2, `zmodType` represents the type of commutative group that also has a decidable equality on its elements and a choice operator. The module `Equality` packages types with a decidable equality and the module `Choice` packages types with a choice operator. In this figure, different structures are defined :

- The `mixin_of` structure packages the additional operator (external law), the constants (basis) and the axioms needed by a commutative group to be an F-vector space.
- The `class_of` structure packages all the theories needed by a representation type (here V : Type) to be an F-vector space. The first projection `base` is a proposition that states that V has a structure of commutative group. The second projection `ext` is a proposition that states that the `zmodType` built on V (with the call of the function `Zmodule.Pack`) with the structure `base` has the additional structure to be an F-vector space.
- The `type` structure corresponds to the F-vector space interface. In this structure, the declaration `sort :> Type` makes the `sort` projection a *coercion* from `type` to `Type`. This form of explicit sub-typing allows any V : `vecType` to be used as a `Type`, e.g., the declaration x : V is understood as x : `sort` V. It is useful for getting generic theorems for abstracts structures.
- The declarations `eqType`, `choiceType` and `zmodType` at the end of the `Module` define the inheritance rules.

```
Module Vector.
Section VectorTypeDef.
Variable F : fieldType.

Structure mixin_of (V : zmodType) : Type := Mixin {
  mul : F -> V -> V;
  _ : forall a b u, mul a (mul b u) = mul (a * b) u;
  _ : forall u, mul 1 u = u;
  _ : forall a, {morph mul a : u v / u + v};
  _ : forall a b u, mul (a + b) u = (mul a u) + (mul b u);
  basis : seq V;
  _ : forall s, \sum_(i < size basis) mul s`_i basis`_i = 0 -> forall i
      , i < size basis -> s`_i = 0;
  _ : forall v, exists s, v = \sum_(i < size basis) mul s`_i basis`_i
}.
Structure class_of (V : Type) : Type := Class {
  base :> Zmodule.class_of V;
  ext :> mixin_of (Zmodule.Pack base V)
}.
Structure type : Type :=
 Pack {sort :> Type; _ : class_of sort}.
...
Definition eqType cT := Equality.Pack (class cT) cT.
Definition choiceType cT := Choice.Pack (class cT) cT.
Coercion zmodType cT := Zmodule.Pack (class cT) cT.

End VectorTypeDef.
End Vector.
...
Notation "a *: v" := (mulv a v) (at level 40) : ring_scope.
```

Fig. 2. Vector space interface

- The last line of the listing declares *: as a notation for the external law of the vector space.

The other structures of the hierarchy, algebra and finitely generated modules over algebra are defined following the same design pattern. The only difference comes from the composition of the structures mixin_of and class_of.

Set of Structures: In finite group representation theory, questions about relations between different representations of a given group are very frequent. Is it true that a given representation of a group is irreducible? Is it true that two representations are equivalent or complementary? As we have already said, these questions can be reduced to questions on relations between modules defined on an algebra. In the case of representations, the algebra will be the group algebra. We have thus formalized a theory of sub-vector spaces. The algebraic structures we are interested in are algebras and modules of algebras. They have a structure of vector space. Thus, the corresponding sub-structures are sub-vector spaces with an additional property on the internal multiplicative law (algebra) and the external law (module).

In a type theory framework, sub algebraic structure, likes sub-group or sub-space, are usually defined as a propositional or boolean predicate. For example, in the COQ system, they can be represented as a dependent structure with two elements : a propositional predicate and a closure property. The type of sub-groups of given group can be defined as follows :

```
Structure sub_group (G : group) : Type := SubGroup {
 set :> G -> Prop;
 is_sub_group : forall a b : G, set (a - b)
}.
```

The problem with this representation is that in order to have equality between sub-structures, we need an axiom of extensionality. In finite dimension vector space theory, there is no need for this axiom. A set of vectors of a finite dimension vector space always has a family of generators. If, in addition, this family is free then the set is a vector space. Conversely, every family of vectors defines a sub-space. Thus, a sub-space of finite dimension vector space can be represented by a list of vectors : the generators. Deciding the membership to a family of generators is equivalent to deciding if there is a solution for a linear system. To be able to view a set of vectors as a boolean predicate, we have added two axioms assuming that for all linear systems the problem of the existence of a solution is decidable :

```
Axiom system_dec : forall m n (F : fieldType),
  matrix F m n -> matrix F m 1 -> bool.
Axiom system_dec_ex : forall m n (F : fieldType) (A : matrix F m n) v,
(exists vs, A *m vs = v) <-> (system_dec A v).
```

These two axioms can be removed once a procedure for solving linear systems is formalized. In the project libraries, we are not far from having one since the matrix LUP decomposition has already been formalized.

We define the type of sub-spaces of a vector space as a list of vectors with a predicate specifing that it is the canonical family of generators. We use choose, the choice operator, to quotient with the spanning set equality relation and then identify with a Leibniz equality all the families that generate the same vector space.

```
Variable (K : fieldType) (vT : vecType K).

Structure vspace : Type := VSpace {
 gf :> seq vT;
 _ : gf == choose [pred x | free_gf x && gf_eq gf x ] (basis_for_gf gf)
}.
```

In this definition seq vT is the type of lists over vT. The notation == corresponds to the decidable equality associated with the type seq vT. Thanks to canonical structures, COQ will automatically infer the corresponding equality. The function choose is provided by the Choice interface. It takes two parameters: a predicate and an element. It returns a "canonical" element that satisfies the given predicate if the element given as parameter satisfies already the parameter predicate, i.e. it is the witness that the predicate is satisfiable. In the definition above, we require that the list of vectors is equal to the result of the application of choose to the predicate that checks if a given family of vector is free (free_gf) and the set its generates is equal to the one generated by gf (relation gf_eq). The function basis_for_gf returns a free basis for a given family of generators. It proceeds by removing the dependent vectors.

After that, and in order to be able to view sub-vector spaces as extensional sets (functions of type vT -> bool) which are more practical for proofs, we have declared a coercion from the vspace type to predPredType. It is a generic interface for the type of boolean predicates provided by the SSREFLECT library ssrbool :

```
Coercion pred_of_vs :=
 (fun F (vT : vecType F) (V : vspace vT) => mem_gf V : _ -> _).
```

We constructively define sub-space operations likes the sum and intersection of two sub-spaces and the complement of a given sub-spaces. We also prove some membership properties for sub-vector spaces :

```
Lemma vs_mul : forall c v, c *: v \in V = ((c == 0) || (v \in V)).
Lemma eq_vsP : forall V1 V2,
  (forall v, v \in V1 = (v \in V2)) <-> (V1 == V2).
```

In this statement, the equality = stands for COQ standard equality between boolean values. The first lemma provides a rewriting rule for the membership of a product. The second lemma gives an equivalence between the extensional equality of sub-spaces and their decidable equality. It allows switching between the two views.

We define sub-algebras and sub-modules as boolean predicates over vspace :

```
Definition is_sub_algebra (V : {vspace aT}) :=
 forallb i : 'I_(\dim_V), forallb j : 'I_(\dim_V), (V'_i * V'_j) \in V.
Definition module (A : salgebra aT) (V : {vspace mT}) :=
 forallb i : 'I_(\dim_A), forallb j : 'I_(\dim_V), (V'_j :* A'_i) \in V.
```

In these definitions, forallb is a notation for the boolean universal quantifier and 'I_(dim_V) is the finite type of all integers less than dim_V the dimension of the sub-space V.

In order to check if a sub-space of an algebra is a sub-algebra we only need to check that the multiplication of any two elements of the basis is included in the basis. The definition of the module predicate is parametrized by a sub-algebra and not the whole algebra domain. Indeed, any module on an algebra has also the structure of module on every sub-algebra of the original algebra. The genericity will be useful when defining sub-group representations.

Morphisms: A linear application between two vector spaces V and W is a function $f : V \to W$ such that :

$$\forall a\, u\, v, f(a * u + v) = a * fu + fv$$

If V and W are of finite dimension, then every linear application from V to W can be represented as a matrix. Conversely, every $n \times m$ matrix defines a linear application. The functional view is more practical to handle when doing proofs. The matrix view gives a finite description, which is suitable for encoding, and inherits Leibniz equality and choice operator from the corresponding field of the vector spaces.

We represent linear applications as singleton type that contains a matrix. In order to be able to see them as functions, we define a coercion from the type of linear applications to that of functions :

```
Variable (K : fieldType) (vT wT : vecType K).
```

```
Inductive linear_map : Type :=
LinearMap of (matrix K (size (basis wT)) (size (basis vT))).
```

```
Definition lmap_mx f := let: LinearMap M := f in M.
```

```
Definition fun_of_lmap := (fun K vT wT f v =>
 let fv := (@lmap_mx K vT wT f) *m (@mx_of_vec K vT [:: v]) in
 \sum_(i < size (basis wT)) fv i (Ordinal (ltnSn 0)) *: (basis wT)'_i).
```

The library also provides a constructor of linear applications (elements of type linear_map) from a function. It builds the matrix corresponding to the image of the basis of the domain according to the basis of the codomain.

```
Definition lmap_of_fun := (fun K (vT wT : vecType K) (f : vT -> wT) =>
 let m := size (basis wT) in let n := size (basis vT) in
 let M := \matrix_(i < m, j < n) (decomp (f (basis vT)'_j))'_i in
      LinearMap M).
```

In the library we have constructively defined the kernel of a linear application and the linear projection on a sub-space. The canonical vector space construction for the linear application type is also defined.

4.2   Representations

The main motivation of this work is the formalization of finite group representation theory. In this development, our main reference is the book by I. Martin Isaacs[13]. That is why we started our work by formalizing a theory of free modules on algebras and fields. We also found inspiration in the ideas presented in [15].

Definitions : The representation type is defined using the CoQ dependent type record.

```
Variables (gT : finGroupType) (F : fieldType) (n : pos_nat).
```

```
Structure representation (A : {set gT}) : Type := RepPack {
 ro_ :> gT -> (matrix F n n);
 _ : {in A &, forall g h, ro_ (g * h) = ro_ g * ro_ h};
 _ : ro_ 1 = 1
}.
```

The definition is parametrized by a finite group domain type gT, a field F and a positive integer n. The representation type is defined for a given set A of the finite group domain gT. The first component of the structure is a function from the finite group domain gT to the type of square matrix of size n on F. The two other components state that a representation is a group morphism.

In this definition and thanks to the use of Canonical Structures, the structure of ring for matrix is automatically infered by CoQ. This allow us to use the standard notations for ring structures : the ring multiplication * and neutral element 1 in the second part of the last two statements.

Group algebra: To link the above definition of representation to the module theory, the group algebra is defined. This is done using the SSREFLECT finfun library which contains a complete formalization of finite domain functions theory. For a finite group G and a field F, the group algebra $F[G]$ is defined as the type of functions of type G -> F.

```
Variables (F : fieldType) (G : finGroupType).
Notation Local "F ,[ G ]" := {ffun G -> F}.

Definition gA0 : F,[G] := [ffun g => 0].
Definition gA1 : F,[G] := [ffun g => if g == 1 then 1%R else 0].
Definition opprgA (v : F,[G]) : F,[G] := [ffun g => - (v g)].
Definition addrgA (v1 v2 : F,[G]) : F,[G] := [ffun g => (v1 g) + (v2 g)].
Definition mulvgA (a : F) (v : F,[G]) : F,[G] := [ffun g => (a * (v g))%R
    ].
Definition mulrgA (v1 v2 : F,[G]) : F,[G] :=
  [ffun g => \sum_(k : G) (v1 k) * (v2 ((k^-1) * g)%g) ].

Definition gAbasis : seq F,[G] :=
  map (fun g => [ffun k => if k == g then 1%R else 0]) (enum G).
```

The ring and vector space operations for this type are defined using the generic constructor [ffun g => E] which constructs the graph of the function that associates to g the expression E. We also define the associated sub-algebra structure for a sub-group of the original group domain. It is the sub-space of $F[G]$ generated by the elements of the sub-group.

Maschke's Theorem: Our library for representations theory includes a formal proof of Maschke's theorem. The CoQ's statement of the theorem is the following :

```
Section Maschke.
Variables (gT : finGroupType) (G : {group gT}) (F : fieldType).
Variable (mT : modType F,[gT]).
Notation Local "|G|" := (#|G| %:R : F).
Notation Local "[F/G]" := (groupSAlg F G).
Hypothesis (HcardG : |G| != 0).
Theorem Maschke :
 forall V : {vspace mT}, module [F/G] V -> semisimple [F/G] V.
Proof.
 ...
Qed.
```

In this statement [F/G] is the notation for the F[gT] sub-algebra associated to the sub-group G. The proposition semisimple expresses the fact that every sub-module W of V has a direct complement.

The idea of the proof [13] is to take, for a sub module W of V, an F-linear projection on it. From this projection, we build a new $F[G]$-projection on it. The kernel of this new projection is an $F[G]$ sub-module of V and also a complement of W. In our formalization, the proof is 34 line long, the doube of the standard paper proof.

5   Related Works

In the proof assistant community, there has been some developments on linear algebra that cover part of what we have formalized. To our knowledge, none has tackled representation theory.

The set-theoretic Mizar Mathematical Library (MML) [22], which has the largest library of formal mathematics, contains formalizations of algebraic structures such as groups, rings, modules and real vector spaces. This structures are defined in various articles and by various authors.

In the COQ system, there are essentially two constructive algebraic hierarchy that have been developed. The first is the seminal Algebra repository [20], which constructs algebraic structures from monoids to modules. The second is the C-CoRN hierarchy [19], mainly devoted to a constructive formalisation of real numbers and including a proof of the fundamental theorem of algebra. Both are setoid based and have been proved difficult to extend with theories like linear or multilinear algebra.

6   Conclusion

The work we present here provides a formalization of finite group representation theory in the COQ system. It also include also a formal proof of the Maschke's theorem. It shows that the different components of our development work well together.To facilitate the reuse of this development, we used a modular approach. This is an important point especially for large formalizations such as the one we work on in the Mathematical Component project. The formalization of algebraic theory is not an easy task especially in terms of packaging and reuse. COQ's dependent types, Coercions and Canonical Structure have contributed to the achievement of this work. They provide a powerful mechanisms for formalizing abstract algebraic structures. SSREFLECT and its large libraries of formal proofs and theories have also been very useful. We have used several of these libraries and especially ssrlag for basic algebraic structures (groups, rings, field ...) and bigops for indexed operations.

In this formalization, representations are defined as a subclass of algebras and modules. For the latter we have formalized a theory of linear algebra which covers the theories of finitely generated modules over algebras or fields. This development consists of four libraries. Together, they are about 2800 lines of code. In these libraries, approximately 95 % of lemmas are proved. The sources are available at the following address : `http://www-sop.inria.fr/marelle/Sidi.Biha/reptheo`

The first perspective of this work is the formalization of more advanced results of representation theory. The Artin-Wedderburn theorem is an example of such results. Once this achieved, a formalization of character theory is possible. Another interesting perspective is to link this work with work on representations theory that has already been done in computer algebra system like GAP. We can use GAP to compute the irreducible representations of a given group and import it in CoQ. We did some experiments to links the two systems by using XML to encode the data exchanged (a finite group generated in GAP), but work on the external communication interfaces of the two systems is still needed.

References

1. Gonthier, G.: Formal proof - The Four-Color Theorem. Notices of the American Mathematical Society 55(11) (2008)
2. Avigad, J., Donnelly, K., Gray, D., Raff, P.: A formally verified proof of the prime number theorem. CoRR abs/cs/0509025 (2005)
3. The Flyspeck project, `http://code.google.com/p/flyspeck/`
4. The C-CoRN project, `http://c-corn.cs.ru.nl/`
5. Stephen, W.: The Mathematica Book, 5th edn. Wolfram Media (2003)
6. The GAP Group: GAP – Groups, Algorithms, and Programming, Version 4.4.12 (2008)
7. Mathematical Components manifesto: `http://www.msr-inria.inria.fr/Projects/math-components/manifesto`
8. Feit, W., Thompson, J.G.: Solvability of groups of odd order. Pacific Journal of Mathematics 13(3), 775–1029 (1963)
9. Gonthier, G., Mahboubi, A.: A small scale reflection extension for the Coq system. INRIA Technical report, `http://hal.inria.fr/inria-00258384`
10. Bertot, Y., Castéran, P.: Interactive Theorem Proving and Program Development, Coq'Art: the Calculus of Inductive Constructions. Springer, Heidelberg (2004)
11. Coq development team: The Coq Proof Assistant Reference Manual, version 8.2 (2009)
12. Gelbart, S.: An Elementary Introduction to the Langlands Program. Bulletin of the American Mathematical Society 10(2) (1984)
13. Isaacs, I.M.: Character Theory of Finite Groups. American Mathematical Society (1994)
14. James, G., Liebeck, M.: Representations and Characters of Groups. Cambridge University Press, Cambridge (2001)
15. Webb, P.: Finite Group Representations for the Pure Mathematician. The manuscript of the book is available on the author web page, `http://www.math.umn.edu/~webb/RepBook/index.html`
16. Gonthier, G., Mahboubi, A., Rideau, L., Tassi, E., Théry, L.: A modular formalisation of finite group theory. In: Schneider, K., Brandt, J. (eds.) TPHOLs 2007. LNCS, vol. 4732, pp. 86–101. Springer, Heidelberg (2007)
17. Bertot, Y., Gonthier, G., Biha, S.O., Pasca, I.: Canonical big operators. In: Mohamed, O.A., Muñoz, C., Tahar, S. (eds.) TPHOLs 2008. LNCS, vol. 5170, pp. 86–101. Springer, Heidelberg (2008)

18. Pollack, R.: Dependently Typed Records for Representing Mathematical Structure. In: Aagaard, M.D., Harrison, J. (eds.) TPHOLs 2000. LNCS, vol. 1869, pp. 462–479. Springer, Heidelberg (2000)
19. Geuvers, H., Pollack, R., Wiedijk, F., Zwanenburg, J.: A constructive algebraic hierarchy in Coq. Journal of Symbolic Computation 34(4), 271–286 (2002)
20. Pottier, L.: User contributions in Coq, Algebra (1999), http://coq.inria.fr/contribs/Algebra.html
21. Barthe, G., Capretta, V., Pons, O.: Setoids in type theory. Journal of Functional Programming 13(2), 261–293 (2003)
22. Mizar Mathematical Library, http://mizar.org/library/

Collaborative Assistant to Handle MathML Expressions

Aslam Muhammad[1], Ana Maria Martinez Enriquez[2], and Gonzalo Escalada-Imaz[3]

[1] Department of CS & E., U.E.T. Lahore, Pakistan
{maslam@uet.edu.pk}
[2] Department of Computer Science, CINVESTAV-IPN, Mexico
{ammartin@cinvestav.mx}
[3] Artificial Intelligence Research Institute, IIIA-CSIC, Barcelona, Spain
{gonzalo@iiia.csic.es}

Abstract. The lack of an assistance support may result in disturbance of coauthors by beginners who ask for help when they are in trouble to produce or reuse shared resources. Additionally, collaborators may not be sure if their respective production is consistent with the collaborative common contribution. We tackle this issue by developing a group awareness knowledge based system that takes the responsibility to automatically evaluate and reuse mathematical formulae, and deduces which participant is a possible expert to help others.

Keywords: Knowledge based systems, awareness, collaborative writing, MathML.

1 Introduction

In order to coordinate the cooperative production, it is important that collaborators are aware of: What objects are being modified? What action is performed? Who is present? This information highlights the "Object" and "Changes" awareness elements that have been well identified in [5]. Hence, as a result of using inadequate notification services, users are not able to produce consistently. In addition, a groupware requires tools to evaluate collaborative production, without them, this task needs a lot of patience and skill. An editorial scenario depicts this situation: A new author Jorge, is confronted with producing MathML statements, in a large document, while one of the editors Sara, checks periodically whether expressions are well formed after getting confirmation from user's production updates.

Our research focused on how people interact within the workspace, how useful workspace information can be obtained, and the way in which that information can be presented, as awareness information is the key of coordination support for a groupware. Some collaborative writing applications are: Basic Support for Cooperative Work, BSCW [1], EquiText [7], REal-time Distributed, Unconstrained Cooperative Editing system, REDUCE [8]. However, none of the above cooperative writing application provides implicitly or explicitly assistance to users when they may have production problems. Thus, in order to improve the coauthoring production (Section 2) we have implemented a group awareness assistant as a knowledge based system (Section 3). One of the goals of our group awareness system is to help "beginners", allowing

J. Caretteet al. (Eds.): Calculemus/MKM 2009, LNAI 5625, pp. 453–459, 2009.
© Springer-Verlag Berlin Heidelberg 2009

them to be assisted by "experts" establishing a focused communication. Furthermore, our application provides the facility to reuse fragments of the document (figures, formulas, tables) as a resource (Section 4).

2 Cooperative Writing

In our studied test bed application [3], the writing actions (document handling, editing, presentation decoration/style) performed by authors on the produced document are captured in the form of events by means of a distributed event management service (DEMS) [4]. An event represents a state change of a shared entity. A shared entity may be a document, hardware/software, and participants. However, several actions, not necessarily cause a change within a resource, such as select and copy or highlight any part of the document, and they are also perceived as events. DEMS acts as a communication mechanism between the producer and consumer applications of events: - **Producer** generates events and can be configured for extending or restricting broadcasting of some events depending upon their scope. **Consumer** subscribes to DEMS to receive events. There are some rules to filter events by category and their sources. When an author wants to stay intensively focused on his production, he may allow those events to be received which are resulted from coauthors who annotate the part of document he produces, whereas, other event notifications are restricted.

In order to deliver events to consumers, DEMS maintains a dedicated storage space to memorize them. Events are maintained as long as user is present in the session, in such case the event storage space is volatile. Whereas, non-volatile storage space maintains events in the form of a log file which can be used at any later stage. This space is also updated periodically until the manager of the shared production decides it, for instance, when the production is considered as completed and accepted.

Producers and consumers are uniquely identified to control the event broadcasting, and different meta-data is associated to events: - **Entity user** who produces events (*login*, his *ID*, the *working site* from where the user works, as well as the cooperative/non-cooperative application from which events are generated). - **Entity resource** within the cooperative environment (document fragment, figure, MathML statement) on which writing actions are performed. - **Entity action** is any performed action within the environment using cooperative/non cooperative application.

In the following scenario: "tito" user works from the "galaxy" site, and selects a MathML statement. Due to the "select" action DEMS identifies who is on line and his working site. The produced event is: galaxy_tito_writing-editor_MathML_select.

3 Group Awareness Knowledge Based System

Our designed and developed group awareness knowledge based system [6] uses rules written in the first order predicate logic. A rule is composed of the premise and the action part. The Inference Engine (IE) deduces new knowledge or trigger actions to provide a dynamic user environment. When a user starts a session IE is automatically launched until session is closed. The steps followed by IE are: - **Information catching**: The retrieval of facts like opening/closing session, data consultation, writing actions, user's role on each fragment, nature of the produced section, working and storage sites, communication interest of users. - **The deduction of new knowledge**: Using caught facts, IE verifies which rules are satisfied and deduces other facts like the number of well structured formulas, their complexity, users in trouble, requirement of communication services. - **The proposition of actions**: IE suggests some tools to be used or characterizes user's production (multimedia, mathematical, annotation) to detect experts (mathematical or multimedia producer).

Our authoring application uses MathML[2] to write, represent, and interpret mathematical statements. Specific events are generated during the production or selection of an expression. As a result of selecting, the author is notified about whether a formula is well-structured (wsf), implicit contextual functions related with (keyboard shortcuts), rewriting it as infix, prefix, or postfix form, transformation from one to another MathML constructor, as observed below. A formula is a wsf if all elements composing a MathML pattern are completed and each expression, in its turn is a wsf too. IE notifies possible errors: missing operator, bad symbol. Once a formula is updated, IE informs this fact to users who makes reference to it.

> **Startrule** "Selecting Mathematical object"
> **If** author(fragment)= X
> nature(fragment)= "Formula"   /*the nature of the fragment is a formula*/
> action(X)= "select_element"   /* X selects an element of the formula */
> **Then**   /* evaluation and related contextual function are notified */
> announce(X)←evaluate_wsf(fragment)
>   */the symbol "←" stands for a function of 1:1 assignment */
> announce(X)←list_tools(Math_handling)
> **Endrule**

Many tools have been introduced to check/spell/evaluate textual documents but as far as we know mathematical statements are not included in the evaluation. In our approach, a criterion is established to evaluate the complexity of a formula that can be modified by modifying the specific rule. A weight to the following steps is assigned: a) the way by which it is created: by producing a sequence of symbols or by using the Math Menu option, b) steps taken to produce it: sequence of characters from the standard input (keyboard) or using menus dialog boxes or mathematical symbol palette, c) elapsed time to complete it, d) computation involved.

4   Online Support Awareness Service: Expert Assistance

The dedicated online support service for coauthors provides technical assistance for writing, production questions, and diagnostic advice that may include quick reference, task specific help, full explanation, and tutorial. This support does not require that users have knowledge of the collaborative writing application. Users are assisted step by step to complete their task. A user who produces many (resp few) elaborated well-structured formulas of high complexity (resp elementary) in the same session and continuously adds different kind of elements, can be considered as an expert (resp beginner). Experts avoid loss of recent changes by saving the document after a gradual period of time, or certain number of characters, or after addition of a figure, or concluding a section. By contrast, beginners take a lot of time to perform a simple task doing irrelevant and unneeded actions because they do not know how to proceed. User characterization is stored in the database (DB), including working site and availability to communicate with, as we see in the rule.

Startrule "Writing formula expertise"
If author(fragment) = X
role(X)="Writer"
nature(fragment)="Formula"   /*X authors a formula */
update(fragment) ="True"   /* fragment is brought up to date */
computed_complexity(fragment)="High"/*formula complexity calculation*/
summary_wsf(formula) = "Many"   /* number of produced wsf */
status_open(authors_definition_DB)="True" /* DB characterization exists */
Then   /*multiple assigned values for author definition */
author_definition(X)<=="expert_MathML"
                        /* the symbol <== stands for a function of 1:N assignment*/
Endrule

A context-based communication is characterized by synchronous exchange of messages between collaborators focused on selected objects (a formula, table, figure). The object on which coauthors center their discussion is named work focus. Thanks to the unique identifier (ID) associated to each element of the structured document, our awareness functionality has the possibility to send, display, and highlight it within the environment of all concerned coauthors distributed on the Internet (Fig. 1).

Suppose, Jorge is reviewing a document mainly composed of formulas and he wants to suggest the producer Carmin to write an alternative expression. Jorge starts a synchronous communication with Carmin, so she asks him to make precise the topic of discussion. He selects the expression (Reader focus). This event is recovered by DEMS and sent to IE. IE retrieves the ID and displays its content in the Carmin's environment. The "Work focus communication Reader-Writer" rule is applied. Collaborators must have access to the same document. Once the communication is established (talking session) and the work focus is selected (1), it is highlighted in both environments (Reader focus and Writer perception). When a user performs modifications on a section that affects the current contextual work focus, the highlighted selection is updated and concerned users are notified about this fact.

Startrule "Work focus communication Reader-Writer"
If author(fragment) = X
author(fragment) = Y
role(X)= "Writer"
role(Y)= "Reader"   /* Y has the reader role on fragment */
action(Y)= "select_element"   /* reader Y selects an object */
sync_comm(X)="True"   /* communication is enabled */
sync_comm(Y)="True"
Then   /* the unique ID of the selected object is transmitted */
display(Y)←send(id_fragment)
announce(Y)←"successfully sent Information"
Endrule

The assistant observes sequence of user actions on the basis of which suggestions or hints may be useful to complete a specific task. As this kind of help may become intrusive, users can avoid it at any time. However, whenever a user starts to produce a new object and IE deduces that many irrelevant actions are performed (do/undo, textual based actions while a formula has been started) IE informs about experts present (open session) who may solve problems. Synchronous communication focused on particular elements gives to collaborators the possibility to coordinate their activities and to be more efficient.

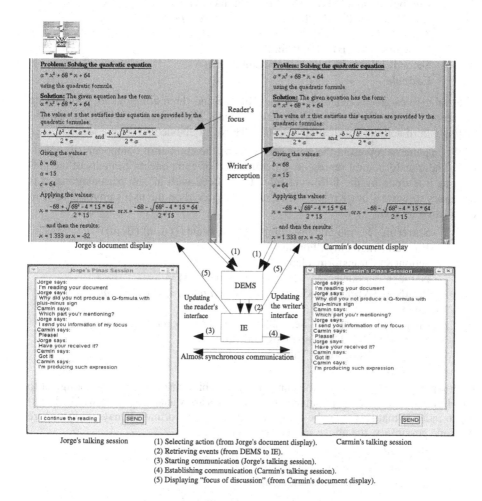

(1) Selecting action (from Jorge's document display).
(2) Retrieving events (from DEMS to IE).
(3) Starting communication (Jorge's talking session).
(4) Establishing communication (Carmin's talking session).
(5) Displaying "focus of discussion" (from Carmin's document display).

Fig. 1. Context-based Communication

Mathematical Expression as Resource. MathML offers constructors characterized by the number of components to be filled as arguments (one, two, and three). The structure of a constructor is represented by a tree, thus, it is feasible to reuse these expressions by transforming one pattern (A, n) to the targeted pattern (B, m), where **n**

and **m** represent respectively the number of components of each one: Transformation ((A, n) (B, m)). Some cases are presented:

a) When patterns have equal number of components and different functions, the transformation only moves the content of A to the content of B:

> **Startrule** "Exchanging patterns 2:2 ((A, n) (B, m))"
> **If** nature(fragment) = "Formula"
> action(X) = "transform_element"
> Number_parameter(A) = Number_parameter(B)=2
> RootNode(A)≠RootNode(B)   /* *different patterns* */
> **Then**
> FirstChildNode(B)←FirstChildNode(A)
> SecondChildNode(B) ←SecondChildNode(A)
> **Endrule**

Similar rules are included for the case of one and three components, eliminating the second action (SecondChildNode(B)←SecondChildNode(A)) for the former and adding a third action (ThirdChildNode(B)←ThirdChildNode(A)) for the second case.

b) When patterns represent the same function and the number of components is equals to 3. The transformation represents a permutation of components contents of the pattern, as it is done by the following rule. To proceed with all the permutations, it is necessary to write 4 other rules. All cases are displayed on the user's environment, so users can select one.

> **Startrule** "Permutation of components contents ((A, n) (B, m))"
> **If** nature(fragment)="Formula"
> action(X)="transform_element"
> Number_parameter(A) = Number_parameter(B)=3
> RootNode(A)=RootNode(B)   /* *both patterns are the same* */
> **Then**
> FirstChildNode(B))← SecondChildNode(A) /* *exchange contents* */
> SecondChildNode(B))←FirstChildNode(A)
> ThirdChildNode(A))←ThirdChildNode(B)
> **Endrule**

c) When **n** is smaller than **m**, the contents of the children nodes of A are assigned to the children of B whereas, remained children nodes are vacant, and they will be blinking to indicate that they must be completed. As this transformation may produce more than one possibility, the above rule can be later applied.

> **Startrule** "Completing components ((A, n) (B, m))"
> **If** nature(fragment) = "Formula"
> action(X) = "transform_element"
>
> Number_parameter(A)←Number_parameter(B)
>
> **Then**
> FirstChildNode(B))←FirstChildNode(A)
> SecondChildNode(B))←SecondChildNode(A)
>   /* *whether it exists or Nil otherwise* */
> Blinking (Vacant_Child_Node(B))   /* *blinking vacant/empty nodes* */
> **Endrule**

One of possible application of transformations is to produce reusable expressions to construct other more complex than the former. Depending upon the current produced object, the assistant informs whether this kind of resource exists in the database.

5 Conclusions

There are some applications that support the single user environment to produce mathematical expressions like the Open Math Editor [9], MathType [10], WIRIS [11], Mathematica [12]. Some ones detect syntactical errors, others have friendly interface, but none of them provides the facility to transform from one pattern to other like our cooperative writing application. Our approach takes advantage of produced atomic or composite events to enhance group awareness among collaborators. It means, from events taken as facts, it is possible to infer other knowledge and adapt the user work space. The awareness knowledge based system provides functionalities to assist users during their collaboration. Users have the possibility to establish a synchronous contextual communication to coordinate their actions and produce consistent documents. Thus, collaborators are concentrated in their production and the group awareness system informs them about all what is going on the shared environment.

One of the future extensions of our platform is to include the management of different versions of the shared document in order to evaluate, for instance, the graphical production which is produced in more than one session.

References

1. Bentley, R., Horstmann, T., Trevor, J.: The World Wide Web as Enabling Technology for CSCW: The Case Study of BSCW. Computer Supported Cooperative Work. The Journal of Collaborative Computing 6(2-3), 111–134 (1997)
2. Carlisle, C., Ion, P., Miner, R., Poppelier, N.: Mathematical Markup Language MathML Version 2.0, 2nd edn., W3C Recommendation, October 21 (2003)
3. Decouchant, D., Favela, J., Martínez, A.: PIÑAS: A Middleware for Web Distributed Cooperative Authoring. In: SAINT 2001, San Diego, CA, USA, pp. 187–194 (2001)
4. Decouchant, D., Martínez, A.M., Favela, J., Morán, A.L., Mendoza, S., Jafar, S.: A Distributed Event Service for Adaptive Group Awareness. In: Coello Coello, C.A., de Albornoz, Á., Sucar, L.E., Battistutti, O.C. (eds.) MICAI 2002. LNCS (LNAI), vol. 2313, pp. 22–26. Springer, Heidelberg (2002)
5. Gutwin, C., Greenberg, S.: A Descriptive Framework of Workspace Awareness for Real-Time Groupware. CSCW, pp. 411–446. Kluwer Academic Publishers, Dordrecht (2002)
6. Martínez, A.M., Muhammad, A., Decouchant, D., Favela, J.: An Inference Engine for Web Adaptive Cooperative Work. In: Coello Coello, C.A., de Albornoz, Á., Sucar, L.E., Battistutti, O.C. (eds.) MICAI 2002. LNCS (LNAI), vol. 2313, pp. 526–535. Springer, Heidelberg (2002)
7. Rizzi, C.B., Alonso, M.C., Seixas, D.E., Costa, J.S., Tamusiunas, F.R., DA Rosa Martins, A.: Collaborative Writing via Web - EquiText. Informatic Education (2000)
8. Yang, Y., Sun, C., Zhang, Y., Jia, X.: Real-time Cooperative Editing on the Internet. IEEE Internet Computing 4(3), 18–25 (2000)
9. http://www.openmath.org/
10. http://www.dessci.com/en/products/mathtype/sysreqs.htm
11. http://www.wiris.com
12. http://www.Wolfram.com/Mathematica

Confidence Measures in Recognizing
Handwritten Mathematical Symbols

Oleg Golubitsky and Stephen M. Watt

University of Western Ontario, London, Ontario, CANADA N6A 5B7
http://publish.uwo.ca/~ogolubit
http://www.csd.uwo.ca/~watt

Abstract. Recent work on computer recognition of handwritten mathematical symbols has reached the state where geometric analysis of isolated characters can correctly identify individual characters about 96%
of the time. This paper presents confidence measures for two classification methods applied to the recognition of handwritten mathematical
symbols. We show how the distance to the nearest convex hull of nearest neighbors relates to the classification accuracy. For multi-classifiers
based on support vector machine ensembles, we show how the outcomes
of the binary classifiers can be combined into an overall confidence value.

1   Introduction

Recognition of handwritten mathematics is a substantially different problem
from natural language text recognition. Because mathematical formulae use a
larger variety symbols, which are better segmented, and because the applicability of dictionary-based classification methods is limited in the mathematical
context, the problem of recognition of individual mathematical symbols is of
special importance. In case of online recognition, it can be thought of as the
problem of classification of parametric plane curves.

Previous work has proposed a model for classification of curves based on the
representation of the curves in a finite-dimensional vector space by the coefficient vectors of their coordinate functions in an orthogonal functional basis. It
has been shown that truncated Legendre-Sobolev series of order about 10 approximate most handwritten character curves to the extent that the approximation
is visually indistinguishable from the original curve [1,2,3]. Furthermore, robust
classification methods based on linear support vector machines and distance to
the convex hull of nearest neighbors can be applied to this representation. These
methods achieve a correct retrieval rate of over 96% for about 230 symbol classes
and at least 9 training samples per class [4,5,6].

Our next goal is to incorporate individual symbol recognition into the classification of entire mathematical expressions. Earlier work [7] has shown that,
depending on the mathematical area, statistically, there is a strong preference
towards certain symbols or their combinations. This statistical information provides a measure of likelihood of a given symbol within its context, which can be

J. Carette et al. (Eds.): Calculemus/MKM 2009, LNAI 5625, pp. 460–466, 2009.

used to improve the classification results. In order to combine this information with the outcome of the individual symbol recognizer, the latter must have a similar format: namely, together with the suggested class or list of classes, it must supply confidence values associated to each choice. These values represent the likelihood that the choice made by the character recognizer is correct.

For nearest-neighbor-based classification, it is natural to use the distance to the nearest neighbor(s) to produce a confidence measure. In this paper, we show that the error rate increases with the distance following a cubic law, which becomes nearly quadratic for large distances. For support vector machines, the distance to the separating hyperplane can be used. For binary linear classifiers, we show that, independent of the choice of the class pair, the error rate decreases exponentially with the distance to the hyperplane. For an ensemble of binary linear classifiers, we give a formula to combine the confidence values of the individual binary classifiers into a final confidence value, which reflects the likelihood of correctness of the majority vote. Finally, we compare the nearest-neighbor-based and SVM-based confidence measures.

2   Representation and Classification

Initially, handwritten symbols are usually represented as a sequence of points, which is sampled in real time by a digital pen. Given the sequences of X and Y coordinates of the points, we compute the moments of the coordinate functions, that is, approximations of the integrals $\int_0^T x(t)t^k\,dt$ and, similarly, for $y(t)$. From the moment integrals, we obtain the Legendre-Sobolev coefficients of the coordinate functions through a linear transformation of the moment vector [1,2,3]. By translating and normalizing the Legendre-Sobolev coefficient vector, we center and normalize the curve with respect to size. We obtain a representation of the symbol curve as a point in a 20–30 dimensional vector space, which is device-independent and invariant with respect to variations in the speed of writing. Then, vector-space-based classification techniques can be applied to this representation.

Among such techniques, linear support vector machines and the nearest convex hull of nearest neighbors have been considered. These techniques yield high correct retrieval rates (about 95–96%) and allow fast classification among multiple classes [4,5,6]. Moreover, as will be shown in the next two sections, the decisions produced by these classifiers can be accompanied by reliable confidence measures, without incurring any significant computational overhead.

3   Confidence of SVM Classification

As classes of handwritten symbol curves are highly linearly separable [5], it is natural to apply linear support vector machines for classification. It has been observed previously [8] that the distance to the separating hyperplane can be used to produce a reliable measure of confidence in the classifier's outcome. Our experiments with various pairs of handwritten symbol classes confirm that the error rates decrease exponentially with the distance to the separating hyperplane,

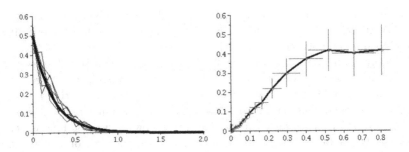

Fig. 1. *Left:* Error rate *vs* distance to hyperplane. Thin curves are for different class pairs, thick curve for exponential fit. *Right:* Ensemble uncertainty (horizontal/vertical error bars correspond to 95th percentiles of the normal/Bernoulli distributions.)

see Fig. 1 (left). The thick line, which fits in the envelope of the frequency curves, is $y = 0.5 \exp(-4.4x)$, which we take for the confidence measure of the binary linear classifier. Note that, when the distance to the hyperplane approaches zero, the error rate tends to 50%, which agrees with the intuition that points on the hyperplane should equally likely belong to either class.

In a multi-class setting, we use a majority voting scheme, with each binary linear classifier casting one vote for the winning class in the pair. If more than one class gets the maximal number number of votes, the tie is broken randomly. The confidence values for the individual classifiers can be combined into an ensemble confidence value using the following observation. Each individual binary classifier makes a decision with a certain confidence, which approximates the likelihood of this decision being correct. On the other hand, with a certain probability, the decision is incorrect, in which case the vote will go to the opposite class. As a result, the winner of the election may lose enough votes, and another class gain enough votes, so that the outcome of the election changes. The probability of this event is the uncertainty of the ensemble classifier.

An exact computation of this uncertainty would incur exponential complexity. We therefore compute its approximation using the following assumption. Let C_1 be the class that has won the election, and let C_i be another class, for which we are going to compute the probability of winning the election instead of C_1. We assume that this different outcome can occur as a result of (some of) the following events:

1. The vote between C_1 and C_i is reversed.
2. C_1 loses a vote to another class C_j, $j \neq i$.
3. C_1 wins a vote from another class C_j, $j \neq i$.
4. C_i loses a vote to another class C_j, $j \neq 1$.
5. C_i wins a vote from another class C_j, $j \neq 1$.

In other words, we assume that the probability that C_1 or C_i wins/loses more than one vote from/to another class can be neglected.

Let ξ_{ij} be the probability that the vote between classes i and j is correct (approximated by the confidence value of the binary classifier between C_i and

C_j). Then the probability that the vote between C_1 and C_i is reversed equals $1 - \xi_{1i}$. If W_1 denotes the set of classes C_j, $j \neq i$, from which C_1 has won a vote, then $1 - \prod_{j \in W_1} \xi_{1j}$ is the probability of the second event in the above list. The probabilities of the remaining events are given by similar formulae.

Given the current numbers of votes collected by C_1 and C_i, we select those combinations of the events 1–5 that would result in C_i taking over C_1, and compute the sum of the probabilities of these combinations (for combinations that lead to a tie between C_1 and C_i, we divide the corresponding probability by 2). This sum, denoted η_{1i}, represents the probability that C_1 has wrongly defeated C_i in the election, because of possible errors made by the binary classifiers. Then, $\prod_{i \neq 1}(1 - \eta_{1i})$ is the probability that C_1 is the correct winner of the election.

The error rate versus the resulting measure of uncertainty (equal to one minus confidence) is shown in Fig. 1 (right). Apart from the uncertainty values that are very close to or very far from zero, we can see that the error rates are closely approximated by the uncertainties, the latter being slightly higher. This small difference is due to the fact that, in our setting, the classes may overlap, so more than one class can be considered as correct winner of the election.

4   Confidence of Nearest Neighbor Classification

The distance to the convex hull of nearest neighbors [9] is the technique that has so far yielded the highest correct retrieval rates for classes of handwritten symbol curves [6]. Since this technique is much slower than SVM classification, we apply it only at the last stage, to distinguish among the top few classes that have received many votes. In each of the top S classes, we find k nearest neighbors to the test sample and compute the distance from the sample to their convex hull. The class with the closest convex hull is then chosen.

In Figure 2 (left), the dependence of the error rate on this distance is shown (computed for $S = 10$ and $k = 11$). However, the error bars, corresponding to the 95% confidence intervals, are too wide to allow a definite conclusion about the dependence of the error rate on the distance. The outcome is also influenced by the choice of the bins used to compute frequencies. This especially applies to distances near zero, where the error bars may cross an axis, rendering the corresponding points meaningless, as well as far away from zero, where few data points are available.

A more accurate estimate, which avoids the direct calculation of frequencies in subintervals, can be obtained as follows. Let $e(\rho)$ and $N(\rho)$ be the percentages of misclassified and all samples, respectively, whose distance to the nearest convex hull does not exceed ρ. These cumulative distributions are smooth functions, for which a good fit can be found in the family $f_{a,b,c,d}(t) = (at^b + c)^{-1} + d$. The values of the parameters that provide the lowest root mean square approximation errors are summarized in Table 4. Given the analytic formulae for $e(\rho)$ and $N(\rho)$, we can calculate the error rate as $e'(\rho)/N'(\rho)$. The graphs of this quotient, for dimensions 12, 16, 20, and 24, are shown in Figure 2 (right). The lowest curve (for dimension 24) models the direct error measurement shown in Figure 2 (left).

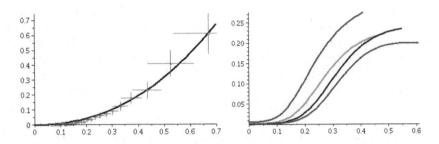

Fig. 2. Error rate *vs* distance to the nearest convex hull of nearest neighbors

5 Comparison of Confidence Measures

A good confidence measure should yield a high value for most correctly classified samples and a low value for most misclassified samples. Let X be the set of all samples, and let X^+ and X^- be the subsets of correctly classified and misclassified samples, respectively. As a measure of quality of a confidence measure $f(x)$ on X, we propose the function

$$q(\xi) = (\#\{x \in X^+ \mid f(x) \geq \xi\} + \#\{x \in X^- \mid f(x) \leq \xi\}) / \#X,$$

where ξ ranges over all possible confidence values, that is, over the interval $[0, 1]$. When deciding between the outputs of the character recognizer and another independent classifier (such as the statistical character predictor described in the introduction), we will always choose the more confident one (it is easy to show that this choice is optimal). Then, if ξ is the confidence of the character predictor, then the greater $q(\xi)$, the more likely we will make a correct choice.

The qualities of the two proposed confidence measures are shown in Fig. 3 (left), and their difference in Fig. 3 (right). We can see that the SVM confidence measure is better at accepting correct classification results and should be used for high confidence values, while the confidence measure based on the distance to the convex hull of nearest neighbors is better at rejecting incorrect results and should be used for low confidence values. The dividing line between the two is at about 96%, which is the mean correct retrieval rate.

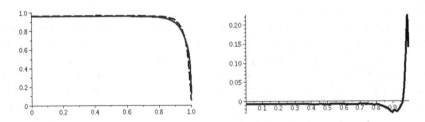

Fig. 3. *Left:* Qualities of confidence measures. Solid is SVM, dashed is convex hull of nearest neighbors. *Right:* Their difference.

Dim	a_e	b_e	c_e	d_e	error	a_N	b_N	c_N	d_N	error
12	0.18	-2.98	22.5	-0.00023	0.00021	0.0010	-3.00	1.0037	0.024	0.0038
14	0.21	-3.31	28.9	-0.00004	0.00013	0.0012	-3.08	1.0043	0.022	0.0030
16	0.29	-3.39	33.3	-0.00014	0.00013	0.0013	-3.16	1.0058	0.022	0.0027
18	0.39	-3.49	35.3	-0.00014	0.00012	0.0014	-3.22	1.0065	0.021	0.0025
20	0.42	-3.63	36.4	-0.00014	0.00016	0.0015	-3.29	1.0074	0.021	0.0027
22	0.44	-3.77	37.5	-0.00014	0.00012	0.0017	-3.33	1.0078	0.021	0.0026
24	0.40	-3.95	38.9	-0.00010	0.00011	0.0019	-3.36	1.0079	0.021	0.0026

Fig. 4. Parameters of the best fits to cumulative distributions

6   Combining Prediction and Recognition

When human readers interpret a handwritten mathematical formula, they recognize some symbols and infer the others from context. A handwriting recognition system can also use both approaches in order to achieve high retrieval rates. One way to take into account a symbol's context is by looking at the frequencies of the n-grams involving it and neighboring symbols. This approach assumes that at least some neighbors have been recognized with a high confidence and that there are only a few high-frequency choices for the symbol under consideration. In such a setting, we would have several choices proposed by the n-gram predictor, with a probability associated to each choice. It would be convenient to have the character recognizer's output in a similar format. Then, assuming that these two classifiers are independent (indeed, their decisions are based on very different considerations), we can combine their outputs by maximizing the posterior probability. This implies that we choose the class for which the product of the probabilities associated by the character recognizer and the n-gram predictor is maximal. We may also let the value of n (the size of n-grams) vary in order to maximize this product.

Using the confidence values presented in this paper, we can obtain a distribution on the set of all classes as follows. Let the confidence value be the probability associated to the winning class (denote it p_1). Then, discard the winning class from consideration and repeat the classification process. Associate to the new winner the resulting confidence value, multiplied by $(1 - p_1)$. Our experiments show that these probabilities decrease very rapidly. In fact it takes on average 7 and at most 26 iterations for the probabilities to become less than 10^{-10}.

Moreover, in the case of SVM classification, we do not need to collect all the votes again, but instead we discard only the ones that involve the winning class and recalculate the ensemble confidence values. This will incur only a mild computational overhead. Indeed, let

$$W_i = \{j \mid C_i \text{ won the vote } C_i - C_j\}, \quad L_i = \{j \mid C_i \text{ lost the vote } C_i - C_j\},$$

where i and j range over the indices of classes still under consideration. Assume that the products $\prod_{j \in W_i} \xi_{ij}$, $\prod_{j \in L_i} \xi_{ij}$ have been computed. When the winning class is discarded, exactly one element will be removed from W_i or L_i, for each

i; call it j_i. Then the above products for the set of remaining classes can be obtained using a single division by ξ_{ij_i}. Using the new values of the products, the probabilities of events 1–5 in Section 3 can be obtained in time proportional to the number of classes. Since the computation of ξ_{ij} is quadratic in the number of classes, and since the number of times we need to discard the winner and calculate the new confidence values is small, the complexity of computing the probabilities is by an order of magnitude lower than the complexity of computing the initial confidence values.

In the case of distance-based classification, very little additional computation is needed to obtain the probabilities, since the confidence values for all classes are derived directly from the distances.

7   Conclusions

We have derived confidence measures for two classifiers, one based on SVM and one based on nearest neighbor geometry. We have demonstrated quantitatively that the SVM ensemble confidence measure performs better than the distance to the convex hull of nearest neighbors at samples classified with high confidence. Future work will be to combine the character recognizer with statistical frequency data using the proposed confidence measures.

References

1. Char, B., Watt, S.M.: Representing and Characterizing Handwritten Mathematical Symbols through Succinct Functional Approximation. In: Proc. Intl. Conf. on Docum. Anal. and Rec. (ICDAR), pp. 1198–1202 (2007)
2. Golubitsky, O., Watt, S.M.: Online Stroke Modeling for Handwriting Recognition. In: Proc. 18th Intl. Conf. on Comp. Sci. and Soft. Eng. (CASCON), pp. 72–80 (2008)
3. Golubitsky, O., Watt, S.M.: Online Computation of Similarity between Handwritten Characters. In: Proc. Document Recognition and Retrieval (DRR XVI), pp. C1–C10 (2009)
4. Golubitsky, O., Watt, S.M.: Online Recognition of Multi-Stroke Symbols with Orthogonal Series. In: ICDAR (accepted, 2009)
5. Golubitsky, O., Watt, S.M.: Improved Character Recognition through Subclassing and Runoff Elections. Ontario Research Center for Computer Algebra Tecnical Report TR-09-01
6. Golubitsky, O., Watt, S.M.: Tie Breaking for Curve Multiclassifiers. Ontario Research Center for Computer Algebra Tecnical Report TR-09-02
7. Watt, S.M.: Mathematical Document Classification via Symbol Frequency Analysis. In: Proc. Towards Digital Mathematics Library (DML 2008), pp. 29–40 (2008)
8. Li, M., Sethi, I.: Confidence-Based Classifier Design. Pattern Recognition 39(7), 1230–1240 (2006)
9. Vincent, P., Bengio, Y.: K-local Hyperplane and Convex Distance Nearest Neighbor Algorithms. In: Adv. in Neural Inform. Proc. Systems, pp. 985–992. MIT Press, Cambridge (2002)

Using Open Mathematical Documents to Interface Computer Algebra and Proof Assistant Systems[*]

Jónathan Heras, Vico Pascual, and Julio Rubio

Departamento de Matemáticas y Computación, Universidad de La Rioja,
Edificio Vives, Luis de Ulloa s/n, E-26004 Logroño (La Rioja, Spain)
{jonathan.heras,vico.pascual,julio.rubio}@unirioja.es

Abstract. Mathematical Knowledge can be encoded by means of Open
Mathematical Documents (OMDoc) to interface both Computer Alge-
bra and Proof Assistant systems. In this paper, we show how a unique
OMDoc structure can be used to dynamically generate, both a Graphical
User Interface for a Computer Algebra system and a script for a Proof
Assistant. So, the OMDoc format can be used for representing different aspects. This generic approach has been made concrete through a
first prototype interfacing the Kenzo Computer Algebra system and the
ACL2 Theorem Prover, both based on the Common Lisp programming
language. An OMDoc repository has been developed allowing the user
to customize the application in an easy way.

1 Introduction

OpenMath [2] is an XML standard widely adopted to express mathematical
knowledge. In [9] we presented an architecture based on OpenMath and allowing,
in principle, to dynamically load new modules in a Graphical User Interface (GUI)
for the Kenzo system [3] (a Common Lisp system devoted to Symbolic Computation in Algebraic Topology). The main obstacle to plug-in dynamically new modules, was that OpenMath content dictionaries (the OpenMath technology used in
that paper) are not designed to store code parts. So, in our new development we
have moved from OpenMath content dictionaries to Open Mathematical Documents, OMDoc [11], which allow us to encode information on both the user interface specification and the code, carrying out the functionality of the GUI. This
fulfills our objective of dynamically loading new modules into the GUI.

In addition, we realized that it was possible to take advantage of the OM-
Doc technology to represent different kinds of information allowing us to extend
our system by including deduction capabilities. In order to do this we have exploited the content dictionaries' capacities for representing richer mathematical
knowledge. To be precise, each mathematical structure used in Kenzo has been

---

[*] Partially supported by Universidad de La Rioja, project API08/08, and Ministerio
de Educación y Ciencia, project MTM2006-06513.

J. Carette et al. (Eds.): Calculemus/MKM 2009, LNAI 5625, pp. 467–473, 2009.

represented by means of an algebraic specification which has been embedded in OMDoc documents. These OMDoc documents are the basis to construct some *encapsulates* in the ACL2 theorem prover [10]. ACL2 is a system for proving properties of programs written in (a subset of) Common Lisp.

Thus, from some OMDoc documents all the pieces needed to dynamically customize a GUI and to integrate a Computer Algebra system, Kenzo [3], with a Proof Assistant, namely ACL2 [10], can be generated. Therefore, the definitions and examples included in the OMDoc documents can be formally validated.

2   Specifying with Open Mathematical Documents

The Kenzo system works with the main mathematical categories used in Combinatorial Algebraic Topology, [12]. In [9], a framework wrapping Kenzo with a Phrasebook as external interface was developed, so the unique possible interaction with it is by means of OpenMath. In addition, an OpenMath content dictionary [2] was defined for each mathematical category which the Kenzo system works with.

In order to make the development of clients of our framework easier, we have intended to supplement it with information related to both the user interaction and functionality of the user interface. For this task, we have used OMDoc.

The OMDoc [11] format is an open markup language for mathematical documents and the knowledge encapsulated in them which allows for the representation of three levels of information in mathematical knowledge: formulæ, mathematical statements and mathematical theories. Different sub-languages including only part of the OMDoc functionality have been specified, and the respective modules developed. We have focused on three of them, namely the *Basic OMDoc*, *OMDoc content dictionaries* and *MathWeb OMDoc* sub-languages; the complete list of sub-languages and their descriptions can be found in [11].

The first task we have had to deal with has been to specify with OMDoc documents the basic mathematical structures used in Kenzo as well as the Kenzo functionality itself. On the one hand, the OpenMath content dictionaries developed for the Kenzo system provide not only the different objects used in the Kenzo system (spheres, Moore spaces, loop spaces and so on) but also a specification of the mathematical structures that they represent. So, the signature (which consists of the headers of the functions) and their formal properties (this will be useful to interact with theorem provers) are included in the content dictionaries. The sub-language for OMDoc content dictionaries allows us to specify the meaning of basic mathematical objects (symbols) by axioms and definitions; and grouping them, it is possible to refer to the symbols defined via their theory. In general, OMDoc content dictionaries can add some functionalities with respect to the OpenMath content dictionaries, so our previous OpenMath content dictionaries can be embedded into OMDoc content dictionaries. On the other hand, an OMDoc document including the functionality of our framework (related to that mathematical structure) has been written. This functionality has been included into the OMDoc document by using the `<code>` tag of the

EXT OMDoc Module, which is aimed at embedding pieces of program code into an OMDoc document. This OMDoc document can be interpreted as a Kenzo wrapper.

Starting from these OMDoc documents, we have specified both the interaction and functionality of a user interface for Kenzo and the integration with the ACL2 Theorem Prover.

2.1 Generating Dynamically a GUI for the Kenzo System

Thinking about increasing the usability of Kenzo, a GUI has been developed to interact with the system, making the interaction with the user easier. The first GUI presented, detailed in [8], allowed the user to build different spaces (like spheres, Moore spaces, loop spaces and so on) and compute homology and some homotopy groups, using the Kenzo system as its kernel. Although it worked in a correct way, it had an extensibility problem. At this moment the Kenzo system keeps on growing. So it would be desirable that our GUI could evolve with Kenzo. To add new functionality to the Kenzo system, a file with the new functions must be included; the Kenzo main code is not modified. It is worthwhile having an extensibility system for the GUI that consists in including only a file with the new functionality, as it is done in Kenzo itself.

The first approach consisted in extracting all the functionality included in the first version of the GUI, changing from a static to a dynamic GUI. So the starting point was a meaningless GUI, and OpenMath content dictionaries dealt with the evolution of the interface itself: when loading a content dictionary, the interface changed, with new options appearing in the toolbar. Each content dictionary had a module associated with it, including the extension of the system, both the GUI and the functionality. Even if this extensibility way worked in a correct way, it had a drawback: adding the necessary modules to our GUI in order to extend its functionality had to be programmed in Allegro Common Lisp [4].

In [6], a proposal for the declarative programming of user interfaces (UI) was presented with the aim of abstracting the ingredients for high-level UI programming. To be precise, three constituents are distinguished: *structure*, *functionality* and *layout*. To be based on the previous proposal, the structure of our GUI is provided by XUL (XUL, [7], is Mozilla's XML-based user interface language that lets us build feature rich cross-platform applications defining all the elements of a UI), functionality has been programmed in Allegro Common Lisp and the default layout has been used, although we could have used a style sheet to customize our application. Thus, we have all the ingredients to extend our meaningless GUI.

Some OMDoc documents being based on the MathWeb sub-language and containing all the information needed to dynamically add the Kenzo functionality to the GUI have been defined. For each Kenzo mathematical structure, an OMDoc document including the definition of the GUI corresponding to the specific mathematical structure has been written. This OMDoc document contains the structure, functionality and layout of our GUI. In order to include the XUL containing the structure and the layout, an OpenMath Foreign object (`<OMForeign>`), which allows us to associate NON-OpenMath objects with OpenMath objects, has been used. With respect to the functionality, that is, the event handlers, it has been included into a `<code>` tag again.

Note that the `<code>` tag would be able to include code for different applications allowing us to build several UIs. This last aspect is related to transform our desktop application into a Web User Interface.

2.2 An Interpreter from OpenMath to the ACL2 Theorem Prover

With the aim of including some deductive capabilities in our system, we have added, in our content dictionaries, the properties which the mathematical structures encoded in our OMDoc documents must really satisfy. This opens the possibility of interfacing our system with the Common Lisp theorem prover ACL2 (some aspects of Simplicial Topology has already been formalized in [1]).

ACL2 [10] supports the constrained introduction of new function symbols by means of the *encapsulate* notion. An *encapsulate* in ACL2 is composed of a set of function signatures, a set of properties of these functions and a "witness" for each one of the functions, where a witness is an existing function that can be proved to have the required properties.

We have included signatures in OpenMath object definitions. In addition, we have specified their properties in two different ways (by means of `<FMP>` and `<CMP>` tags) and we have associated an instance example with them. Gathering together all the previous aspects, it is possible to include in the content dictionaries all the needed information to generate an ACL2 encapsulate from an OMDoc content dictionary.

By using the OMDoc content dictionaries sub-language to define the objects, an interpreter which transforms each OMDoc content dictionary into an executable encapsulate in ACL2 has been developed.

The necessary functions to transform the OMDoc content dictionaries into the respective ACL2 encapsulates are stored in an OMDoc document which is based on the MathWeb sub-language. By collecting all OMDoc documents of this kind a specific purpose interpreter from OpenMath to ACL2 is obtained.

3   Integrating All the Pieces

Finally, the Basic OMDoc sub-language is sufficient for mathematical documents that do not introduce new symbols or concepts. In our system, a Basic OMDoc document glues the different documents associated with a specific mathematical structure together. To be precise, for a specific mathematical structure three different OMDoc documents are provided: the first one gives an algebraic specification of the mathematical structure using the OMDoc content dictionaries, the second one supplies the functions to build these mathematical structures in our system (abstracting the ones of Kenzo), and the last one defines the GUI that can be loaded as a new module of our main GUI. These documents try to make the interaction with Kenzo easier. In addition, some OMDoc documents represent the integration with other systems. These documents can be considered as interpreters from Kenzo (by means of OMDoc) to the specific system. To sum up, both mathematical knowledge and different kinds of interactions have been specified by means of OMDoc documents.

From these different kinds of documents, *templates* customizing the system can be generated. If a user wants to work with a specific structure of the Kenzo system, he must create a document that links to the documents that provide the corresponding content dictionaries, the necessary Kenzo functionality and the part of the GUI which must be added to the meaningless GUI. Besides, if he wanted to interact with another system, he must supply an interpreter from Kenzo OMDoc documents to the system and a client to interact (for instance, a GUI, a web service, and so on). If some of the OMDoc documents is not available, the user can develop them and in this way the system grows up. In addition the different templates, that is, an OMDoc grouping all the OMDoc documents needed for an specific interaction, can also be added to the repository to be used by any other clients.

To define these templates the Basic OMDoc sub-language is used. Namely, the *DOC* module provides the document infrastructure (in particular, the <omgroup> tag allows us to group the references to other documents) and the *DC* module supplies the metadata.

Now, we can present a concrete example which integrates all the pieces explained in the previous sections. On the one hand, we want to be able to work with the simplicial sets using the Kenzo functionalities, for instance, compute their homology and homotopy groups. On the other hand, we want to use ACL2 to prove that the simplicial sets built by Kenzo (spheres, Moore Space, cartesian product and so on), and used in our system, are really simplicial sets. In this way, representation, computation and deduction will be integrated in the same system.

In our OMDoc repository, we can find almost all the ingredients to customize our application to achieve our objective. The relations among the components and their role in the workflow of our system customization in this example is shown in Figure 1. For the simplicial sets, an OMDoc content dictionary defining their mathematical structure (SS-definition), the logic to interact with Kenzo (SS-Kenzo-functionality) and the presentation for the GUI (SS-GUI) are available. With respect to ACL2, an interpreter (ACL2-interpreter) which is able to translate from an OMDoc content dictionary (in particular, simplicial sets content dictionary) into an ACL2 encapsulate can be found. An OMDoc document to customize the GUI (ACL2-GUI) allowing the ACL2 system to interact with our system has been developed.

As can be seen in Figure 1, the only interaction made by the user consists in loading the Basic OMDoc document. The numbers in the diagram indicate the execution workflow.

Figure 2 shows a customized Kenzo GUI for simplicial sets integrating and ACL2 GUI.

All this work is made without any changes in the code of our previous framework. The only thing that must be done consists in loading an OMDoc document,

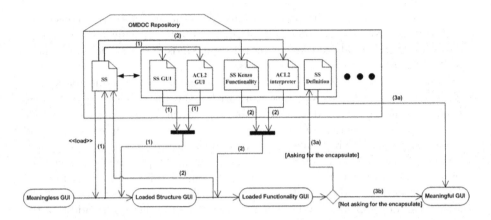

Fig. 1. Workflow diagram

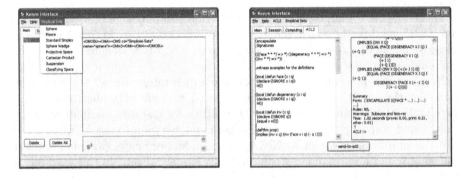

Fig. 2. Screen-shots of our customized Kenzo GUI

containing all the necessary information to customize the application adding the new functionality.

4   Conclusions and Further Work

In this paper we have reported on an OMDoc documents repository. This repository is composed of several OMDoc documents which have been defined using different OMDoc sub-languages to reach different goals. On the one hand, some OMDoc documents, based on the OMDoc content dictionaries sub-language, supply the mathematical structures of our system. On the other hand, OMDoc documents in the MathWeb sub-language provide us with the necessary tools to specify user interfaces, the functionality of these interfaces, the functionality of the system itself and also the interaction with other systems.

As an example of the use of this repository we have described a first prototype that allows an integration of the Kenzo computer algebra system and the ACL2 theorem proving system. The interaction part of our OMDoc documents generates new modules in the GUI, and the axiomatic part generates an *encapsulate* in ACL2, allowing us to check, in an automated way, that the properties are consistent. This allows us, to a limited extent, to integrate, in a same system interaction (i.e. representation), computation (through the Kenzo kernel) and deduction (by means of ACL2). Other prototypes can be developed in order to integrate our system with different mathematical systems (for instance, GAP [5] has already been connected with Kenzo through OpenMath in [13]).

Once the ACL2 and the Kenzo systems are integrated in a same GUI, much more work is needed to implement more interesting interactions. For instance, the encapsulates should be the basis for more complex theorem proving inside the system. As an example, let us consider the construction of a sphere in Kenzo. The GUI should prepare an ACL2 script stating that this concrete (Common Lisp) object is a (functional) instance of the encapsulate `simplicial-set`. ACL2 very likely will not be able to prove those statements automatically, and some user interaction will be needed. Then, both the interface and the OMDoc documents should be enriched to cope with the user actions, allowing the system to recover, in further sessions, the full proof script, and then automating the verification of each construction generated in the system.

References

1. Andrés, M., Lambán, L., Rubio, J., Ruiz Reina, J.L.: Formalizing simplicial topology in ACL2. In: Proceedings Workshop ACL2, pp. 34–39. Austin University (2007)
2. Buswell, S., Caprotti, O., Carlisle, D.P., Dewar, M.C., Gaëtano, M., Kohlhase, M.: OpenMath Version 2.0 (2004), http://www.openmath.org/
3. Dousson, X., Sergeraert, F., Siret, Y.: The Kenzo program, Institut Fourier, Grenoble (1999), http://www-fourier.ujf-grenoble.fr/~sergerar/Kenzo/
4. Franz Inc. Allegro Common Lisp, http://www.franz.com
5. GAP - Groups, Algorithms, Programming - a System for Computational Discrete Algebra, http://www.gap-system.org/
6. Hanus, M., Kluß, C.: Declarative Programming of User Interfaces. In: PADL 2009. Lectures Notes in Computer Science, vol. 5418, pp. 16–30. Springer, Heidelberg (2009)
7. Hyatt, D., et al.: XML User Interface Language (XUL) 1.0, http://www.mozilla.org/projects/xul/
8. Heras, J., Pascual, V., Rubio, J.: Mediated Access to Symbolic Computation Systems. In: Autexier, S., Campbell, J., Rubio, J., Sorge, V., Suzuki, M., Wiedijk, F. (eds.) AISC 2008, Calculemus 2008, and MKM 2008. LNCS (LNAI), vol. 5144, pp. 446–461. Springer, Heidelberg (2008)
9. Heras, J., Pascual, V., Rubio, J.: Mediated Access to Symbolic Computation Systems: An OpenMath Approach (preprint)
10. Kaufmann, M., Manolios, P., Moore, J.: Computer-Aided Reasoning: An Approach. Kluwer Academic Press, Boston (2000)
11. Kohlhase, M.: OMDoc – An open markup format for mathematical documents [Version 1.2]. Springer, Heidelberg (2006)
12. May, J.P.: Simplicial objects in Algebraic Topology. Van Nostrand Mathematical Studies (11) (1967)
13. Romero, A., Ellis, G., Rubio, J.: Interoperating between Computer Algebra systems: computing homology of groups with Kenzo and GAP. In: Proceedings of ISSAC 2009 (to appear, 2009)

OpenMath in SCIEnce: SCSCP and POPCORN

Peter Horn[1] and Dan Roozemond[2]

[1] Universität Kassel, Heinrich Plett Straße 40, 34132 Kassel,
horn@math.uni-kassel.de,
http://www.mathematik.uni-kassel.de/ hornp
[2] Technical Universiteit Eindhoven, Den Dolech 2, Postbus 513, 5600 MB Eindhoven,
d.a.roozemond@tue.nl,
http://www.win.tue.nl/ droozemo

Abstract. In this short communication we want to give an overview of how OpenMath is used in the European project "SCIEnce" [12]. The main aim of this project is to allow unified communication between different computer algebra systems (CASes) or different instances of one CAS. This may involve one or more computers, clusters, and even grids.

The main topics are the use of OpenMath to marshal mathematical objects for transport between different CASes, an alternative textual OpenMath representation more suitable for human reading and writing, and finally the publicly released Java Library developed for the project.

1 Marshaling Mathematics in SCSCP

When designing a uniform communication interface for Computer Algebra Systems, the first problem that needs to be solved is how to transport the mathematical objects from one system to another. Here, the obvious choice for us was OpenMath [6], since it is a widely used standard with a long history of assisting communication between CASes [1,2]. In this section we briefly comment on the problems faced and the choices made.

To simplify the communication between the various CASes, we have developed a protocol called "Symbolic Computation Software Composability Protocol", abbreviated SCSCP [9,13]. This protocol does not only enable the computation of simple commands in a different system or on a different machine, but it will also serve as a means of conveying constituents of larger, more complex, computations.

The protocol is XML-based; in particular, the protocol messages are in the OpenMath language, and its TCP-sockets based implementation uses XML processing instructions to delimit these messages and convey small pieces of information on a higher level. Communication takes place using port 26133, reserved for SCSCP by the Internet Assigned Numbers Authority (IANA). At the moment of writing the protocol has reached version 1.3 and both client and server implementations exist in GAP, KANT, Maple, and MuPAD. The TRIP system also supports the protocol, using their own publicly available implementation of SCSCP [3]. Moreover, we have developed a Java library `org.symcomp.scscp`

J. Carette et al. (Eds.): Calculemus/MKM 2009, LNAI 5625, pp. 474–479, 2009.
© Springer-Verlag Berlin Heidelberg 2009

[14] to facilitate third party developers in exposing their own applications using SCSCP.

Apart from two OpenMath Content Dictionaries accompanying the SCSCP protocol [10,11] several other Content Dictionaries were developed in the project, concerning for example efficient matrix representations or polynomial factorization. We expect to submit these to the OpenMath community for consideration in the Summer of 2009.

2   POPCORN – A Tasty OpenMath Representation

When handling OpenMath objects, one frequently finds oneself typing and reading lots of OMAs, OMSs, and so on. This may lead one to the conclusion that humans were not designed to parse XML. Therefore, whenever people discuss their experiences with OpenMath, they tend to use more human-readable shortcuts, often inspired by LATEXor a Maple-like syntax.

That is why we decided to produce an OpenMath representation taking this into account, and created POPCORN, which is an acronym standing for "Possibly Only Practical Convenient OpenMath Replacement Notation". For the sake of typographic beauty, we write it as "Popcorn".

We emphasize that Popcorn is merely an OpenMath representation that we consider convenient for humans, similar to the XML representation that is obviously more convenient for machines. Furthermore, if a two-dimensional environment such as a web browser is available, more sophisticated editors such as the MathDox formula editor [5] are even better. However, we still think Popcorn is a valuable addition, e.g. for quick tests, command line applications, etc.

Parsing of Popcorn is sufficiently fast for small examples, but for larger OpenMath trees the XML representation can be parsed more efficiently, if only because when parsing Popcorn code an intermediate abstract syntax tree has to be constructed, while such a tree is inherent to the XML representation.

The Popcorn language itself not easily user-extensible, but because the Popcorn grammar is included in the libraries, an advanced user may change the grammar, e.g. add infix operators, special symbols, etc, and use it in his or her own application.

2.1   Elementary OpenMath in Popcorn

We first look at the notation used for the elementary OpenMath objects.

Integers are typed just as one expects: as decimal numbers without whitespace inside or prefixed with 0x in hexadecimal representation;

Floats are typed either as, e. g. 2.34e12 or 0f### where the # represent hexcharacters as in the hex attribute of OMF;

Strings are wrapped in " or ' ;

References are given either in the simple form #name (for local references <OMR href="#name"/>) or the more complex form ##http://somewhere/ something/## (for non-local references);

Variables are whitespace-free strings prefixed with $;
Symbols are written as cdname.name.

To add an id value to any object, simply postpone it with :theid.

2.2 Compound OpenMath in Popcorn

Application is encoded by postponing the parenthesized arguments to the applied object, e.g. arith1.plus(1,2,3);
Binding is done by typing square brackets behind the bound object. Within the brackets the comma-separated bound variables are separated from the expression by ->, e.g. quant1.forall[$x, $y -> ...];
Attribution is done by adding key/value-pairs as a comma-separated list in braces to the attributed object, e.g. 1.2{aa.bb -> "cc", "dd" -> 3}

2.3 Syntactically Sugared/Salted Popcorn

To allow for more intuitive notation, we added some shortcuts:

For the arith1 symbols plus, times, and the relation1 symbols we added the obvious infix symbols +, *, =, <, <=, and so on. The same is true for the logic1 symbols, all with a well-defined operator precedence.

For a reasonably large number of frequently used symbols, we decided to get the cdname-free name into the global context, e. g., sum, sin, true, lambda, pi, i, etc.

To construct a list1.list, one may simply use square brackets, and to construct set1.set, braces can be used (The Popcorn parser automatically checks whether for example an expression in square brackets matches the binding pattern, so that no confusion arises). Constructing nums1.rational can be done by separating numerator and denominator with //. Similarly nums1.complex_cartesian can be constructed by separating the real and imaginary part with |.

Also, some of the functionality of the experimental prog1 Content Dictionary is exposed in a Maple-like syntax.

2.4 Popcorn Examples

```
sin(3)                    <OMA>
                            <OMS cd="arith1" name="sin">
                            <OMI>3</OMI>
                          </OMA>

lambda[$x->1+$x]          <OMBIND>
                            <OMS cd="fns1" name="lambda" />
                            <OMBVAR>
                              <OMV name="x" />
                            </OMBVAR>
                            <OMA>
                              <OMS cd="arith1" name="plus" />
                              <OMI>1</OMI>
                              <OMV name="x" />
                            </OMA>
                          </OMBIND>
```

```
$a := [1//2, (2|8):x]        <OMA>
                               <OMS cd="prog1" name="assign" />
                               <OMV name="a" />
                               <OMA>
                                 <OMS cd="list1" name="list" />
                                 <OMA>
                                   <OMS cd="nums1" name="rational" />
                                   <OMI>1</OMI><OMI>2</OMI>
                                 </OMA>
                                 <OMA id="x">
                                   <OMS cd="nums1" name="complex_cartesian" />
                                   <OMI>2</OMI><OMI>8</OMI>
                                 </OMA>
                               </OMA>
                             </OMA>

1.2{aa.bb -> "cc"}           <OMATTR>
                               <OMATP>
                                 <OMS cd="aa" name="bb" />
                                 <OMSTR>cc</OMSTR>
                               </OMATP>
                               <OMF dec="1.2" />
                             </OMATTR>
```

Here another motivation for the name Popcorn can be seen – it turns something pretty small into something rather giant.

3   org.symcomp.openmath – Convenient Handling of OpenMath with Java

For the development of tools and applications within SCIEnce, Java seemed a natural choice because of its portability and the availability of many libraries. Although there are some Java OpenMath Libraries available [7,8], these are older (last update in 2000 and 2004, respectively) and we disagreed with some of the design choices made.

We therefore created a new library that takes advantage of the recent developments in Java, such as annotations and generics, and we designed it from the ground up to be as easily extensible as possible. It provides many convenience classes and handy methods to traverse, construct, and analyze OpenMath trees. Furthermore, it has completely transparent support for OpenMath Attributions, eliminating the need to handle these objects in any special way.

Import and export to OpenMath 2 XML, OpenMath 2 Binary, and Popcorn are included. Moreover, we have implemented export to LaTeX to demonstrate the great extensibility of the library, and because it easily enables rendering of OpenMath in browsers using the well known jsMath package. We expect to include Strict Content MathML 3 soon as well, in view of the recent developments with respect to OpenMath 3.

We hope to clarify the simplicity and elegance this library offers by means of the small example in Listing 1.1 (the usual Java preliminaries have been omitted).

3.1   Custom Renderers

To feed OpenMath data into other applications, it is often necessary (or at least convenient) to produce a specific format. This is wired into

```
1  //Creating an OpenMath application
   OMSymbol s = new OMSymbol("somecd", "add");
   assert s.isSymbol("somecd", "add");
   OpenMathBase[] params = new OpenMathBase[] {
       new OMInteger(1),
6      new OMString("lala")
   };
   OMApply oma = s.apply(params);
   assert oma.isApplication("somecd", "add");

11 //Creating the same OpenMath object from Popcorn
   OpenMathBase oma2 = OpenMathBase.parse("somecd.add(1, 'lala')");
   assert oma.equals(oma2);

   //Creating an OpenMath object from the XML representation
16 OpenMathBase omi = OpenMathBase.parse("<OMOBJ><OMI>42</OMI></OMOBJ>");
   assert omi.deOMObject().isInteger(42);

   //Creating an OpenMath Binding, using a combination of pure Java
   //  creations and Popcorn.
21 OMVariable[] omvs = new OMVariable[] { new OMVariable("x") };
   OMBind ombind = s.bind(omvs, OpenMathBase.parse("$x + 1"));
   assert ombind.isBinding(s);

   //Convenient equality testing (note that it is a literal comparison,
26 //  e.g. alpha-conversion is not included)
   assert ombind.toPopcorn().equals("cdname.name[$x -> $x + 1]");
```

Listing 1.1. Using the org.symcomp.openmath library

org.symcomp.openmath as *custom renderers*. We designed these classes in such a way that producing e. g. a renderer for the Magma language took only a few lines of code.

The LaTeX- and Popcorn-renderer are made using the same mechanism. These also give the user a great starting point for developing his/her own custom renderer.

4   Conclusion

In this short communication we have given an overview of the current developments in the European project "SCIEnce," in particular the implementation of the OpenMath based SCSCP protocol. We presented two Java libraries assisting this implementation, one for conveniently handling OpenMath objects, the other for executing the SCSCP protocol itself. These libraries enable a developer to expose his own application to other systems using OpenMath and SCSCP, requiring nothing but the strictly necessary from that developer.

Future activities include extending OpenMath support in the participating systems, porting the libraries to C++, both for developers using C or C++ as well as for improved performance, and adding MathML support to the library once OpenMath3 and MathML3 have been finalized.

5   License and Availability

The org.symcomp.openmath library and the SCSCP library org.symcomp. scscp are released under the Apache 2 License. In February 2009 the first public release was made [14]. The libraries are available as binaries, source packages or they may be used as Maven [4] dependencies. Available on the website is also a comprehensive (and continuously improving) API documentation.

Part of this release is an extensive example that uses both the OpenMath and SCSCP library, and shows how little is needed to use the libraries to set up SCSCP clients and servers.

References

1. Caprotti, O., Cohen, A.: Connecting proof checkers and computer algebra using OpenMath. In: Bertot, Y., Dowek, G., Hirschowitz, A., Paulin, C., Théry, L. (eds.) TPHOLs 1999. LNCS, vol. 1690, p. 109. Springer, Heidelberg (1999)
2. Caprotti, O., Cohen, A.M., Riem, M.: Java Phrasebooks for Computer Algebra and Automated Deduction. SIGSAM Bulletin (2000) (Special Issue on OpenMath)
3. Gastineau, M.: SCSCP C Library - A C/C++ library for Symbolic Computation Software Composibility Protocol, IMCCE (2009), http://www.imcce.fr/Equipes/ASD/trip/scscp/
4. Maven: A software project management and comprehension tool, http://maven.apache.org/
5. MathDox OpenMath Formula Editor, http://mathdox.org/formulaeditor/
6. OpenMath, http://www.openmath.org/
7. PolyMath/OpenMath, http://pdg.cecm.sfu.ca/openmath/
8. RIACA OpenMath Library, http://www.mathdox.org/new-web/openmath.html
9. Freundt, S., Horn, P., Konovalov, A., Linton, S., Roozemond, D.: Symbolic Computation Software Composability Protocol (SCSCP) specification, Version 1.2 (2008), http://www.symbolic-computation.org/scscp/
10. Roozemond, D.: OpenMath Content Dictionary: scscp1, http://www.win.tue.nl/SCIEnce/cds/scscp1.html
11. Roozemond, D.: OpenMath Content Dictionary: scscp2, http://www.win.tue.nl/SCIEnce/cds/scscp2.html
12. Symbolic Computation Infrastructure for Europe, http://www.symbolic-computation.org/
13. Freundt, S., Horn, P., Konovalov, A., Linton, S., Roozemond, D.: Symbolic Computation Software Composability. In: Autexier, S., Campbell, J., Rubio, J., Sorge, V., Suzuki, M., Wiedijk, F. (eds.) AISC 2008, Calculemus 2008, and MKM 2008. LNCS, vol. 5144, pp. 285–295. Springer, Heidelberg (2008)
14. Homepage of the org.symcomp.openmath and org.symcomp.scscp libraries, http://java.symcomp.org/

A Knowledge Repository for Indefinite Integration Based on Transformation Rules

A.D. Rich[1] and D.J. Jeffrey[2]

[1] 62-3614 Loli'i Way, Kamuela, Hawaii, USA
[2] Department of Applied Mathematics
The University of Western Ontario
London, Ontario, Canada N6A 5B7

Abstract. Taking the specific problem domain of indefinite integration, we describe the on-going development of a repository of mathematical knowledge based on transformation rules. It is important that the repository be not confused with a look-up table. The database of transformation rules is at present encoded in Mathematica, but this is only one convenient form of the repository, and it could be readily translated into other formats. The principles upon which the set of rules is compiled is described. One important principle is minimality. The benefits of the approach are illustrated with examples, and with the results of comparisons with other approaches.

1 Introduction

Ever since the automating of the simplification of mathematical expressions was first attempted, there has been a lively discussion as to whether a rule-based approach should complement algorithmic methods, and if so which should be tried first. We contend that a computer algebra system should try rules first, and turn to general purpose algorithms only if no rules apply. This frees developers of algorithms from having to worry about the annoying and trivial problems and the special cases, and instead focus on the genuinely hard and interesting problems.

Unfortunately, rule-based systems, owing to poor implementations, have a reputation for being inefficient and plagued by endless loops. For the problem area of indefinite integration, this paper describes the development of a rule-based repository of knowledge that is compact, efficient, transparent and modular. For the purposes of this preliminary discussion, we shall not address questions of combining our approach with algorithmic approaches, in order to arrive at a full integration system, but concentrate on the questions of constructing a database of knowledge and show examples of how it performs in practice.

It must be emphasized that what is not being described is a scheme for table look-up. Such schemes were described, for example, in [2]. Their approach was to consider data structures and search techniques which would allow them to encode all the entries is reference books such as [1]. Adopting this approach for integration — or *a fortiori* for all simplification — would result in huge databases

J. Carette et al. (Eds.): Calculemus/MKM 2009, LNAI 5625, pp. 480–485, 2009.

which would be unwieldy to maintain, debug and utilize. The set of rules whose development is described here is relatively compact, verifiable and efficient.

2   Proper Definition of Transformation Rules

Many of the problems common to rule-based systems can be avoided by defining transformation rules according to the principles below. A transformation rule will be written as $A \rightarrow B$, where A and B are mathematical expressions.

- Define Functionally. The right side of a properly defined rule consists of a mathematical expression followed by any required restrictions on the domains of the variables in the rule. Procedural programming constructs such as loops or conditionals are not allowed, and nor are assignments to global or fluid variables. The application of rules defined this way results in a single, comprehensible step in the simplification of an expression.
- Restrict to domains of validity. Many rules are valid only if their variables are restricted to a certain domain. Conditions on a properly defined rule must restrict its application to the domain over which it is valid. For example, the transformation $\sqrt{z^2} \rightarrow z$ should only be applied if z is known to be purely imaginary or in the right half of the complex plane (unrestricted versions of this transformation caused the well remembered 'square-root bug' in Maple).
- Restrict to simplification. To avoid infinite loops, applications of rules must eventually result in an expression that can be made no simpler (i.e. an expression to which no rules applies). The conditions attached to a rule must limit its application to those expressions for which its application results in a simpler expression. For example, if F stands for any trigonometric function and n for a rational number, the goal of transformations of $F(n\pi)$ is to reduce the magnitude of the angle. Thus, specifically, although $\sin(n\pi) \rightarrow \cos((n-1/2)\pi)$ is valid for all real and complex n, it should only be applied if n is in the interval $(\pi/2, \pi)$, thereby reducing the angle to the interval $(0, \pi/2)$. Note that if n is negative, application of this rule would actually increase the magnitude of the angle.
- Provision for local variables. Sometimes it is convenient to assign a value to a local variable so it can be used multiple times in a rule's conditions or body without having to recompute it. To provide for this need while preserving the functional nature of rules, assignments to local variables are allowed in properly define rules. However, assignments to fluid and global variables are not allowed.
- Mutually exclusive For a collection of transformation rules to be properly defined, at most one of the rules can be applicable to any given expression. Mutual exclusivity is critical to ensuring that rules can be added, removed or modified without affecting the other rules. Such stand-alone, order-independent rules make it possible to build a rule-based repository of knowledge incrementally and as a collaborative effort.

3   Transformation Rules versus Mathematical Identities

There exist numerous collections of mathematical formulas and identities available in books and on the Internet (e.g. the Wolfram Functions website `functions.wolfram.com` lists over 300,000 formulas). Superficially, properly defined transformation rules and mathematical identities look the same, since both have left and right sides which are mathematical expressions that are equivalent. The obvious question is then why build a repository of knowledge based on rules, when huge libraries of formulas and identities already exist? The answer to that question requires that we understand the difference between collections of rules and collections of identities; they are fundamentally different in nature.

- Rules are active; formulas are passive. Rules are precise instructions on when and how to transform expressions of a particular form into equivalent, but simpler, ones. Identities, on the other hand, are statements of the fact that their left and right sides are mathematically equivalent.
- Rules include application restrictions. Both rules and identities specify the domains of their variables over which they are valid. Properly defined rules include additional conditions so that they are applied only if a simplification actually results and so that collections of rules are mutually exclusive.
- Rule collections are minimized; formula collections are maximized. When crafting a rule-based repository of knowledge, one goal is to minimize the number of rules, by making them mutually exclusive while at the same time maximizing their generality. However, for a library of formulas, one goal is to include all commonly occurring cases of more general formulas, so readers are not required to derive special cases. For example, a library may have dozens of identities giving the algebraic equivalents of trigonometric functions of special angles. However, a repository would have just the handful of rules required to transform trigonometric functions of special angles into algebraic form.
- Integration formulas give final results; Rule collections may not. The ideal entry in an integration table gives an algebraic expression for an integral, and expressing one integral in terms of another is a less satisfactory formula. However, many rules in a repository will express one integral in terms of another, and even if it is possible to express an integral directly in algebraic terms, such a transformation may be excluded in order to keep the repository compact. Thus although ideally an integration table could be used in one pass, a transformation repository will necessarily be recursive.

4   Integration Examples

One conspicuous benefit of rule-based integration is the greater simplicity of its results. Simplicity can include not just one integral, but consistent behavior over families of integrals. For example, the following integrals show symmetry between trigonometric and hyperbolic functions:

$$\int \frac{dx}{\sqrt{a+x}\sqrt{b+x}} = 2\operatorname{arctanh}\frac{\sqrt{a+x}}{\sqrt{b+x}} \; , \tag{1}$$

$$\int \frac{dx}{\sqrt{a+x}\sqrt{b-x}} = -2\arctan\frac{\sqrt{b-x}}{\sqrt{a+x}} \; . \tag{2}$$

In contrast, Mathematica and Maple express integral (2) using arctangent as shown, but use logarithm for the integral (1):

$$\int \frac{dx}{\sqrt{a+x}\sqrt{b+x}} = \ln\left(a+b+2x+2\sqrt{a+x}\sqrt{b+x}\right) \; . \tag{3}$$

The current version of the repository is being tested on a database of over 5000 integrals, and the results compared with other major computer algebra systems. The comparisons are based on a variety of metrics; for example, simplicity is measured by counting the leaves of the tree structures used to represent the expressions. These metrics, and the results of the comparisons, will be detailed in a future publication.

5   Platform Requirements

An efficient and reliable software platform is required to build a rule-based repository of knowledge. As a minimum the support platform needs to provide the following services:

- Transformation rules. The platform must make it possible to define and recursively apply transformation rules to expressions of a specified form. This requires a flexible and natural syntax for the patterns used to specify the form of expressions.
- Efficient pattern matching. A rule-based system may have thousand of rules, and hundreds of rule applications may be required to simplify an expression. Thus when given an expression to simplify, it is essential that the pattern matcher quickly find the applicable rule, if any. Thoughts on how to implement an efficient pattern matcher is discussed below.
- Exact and arbitrary precision arithmetic. Numerical routines are required for built-in functions and operators since a rule-based approach is usually not appropriate for numerical computations.
- Programming environment. The platform must provide the ability to input, evaluate and display expressions, as well as provide a suitable environment for testing and debugging the repository.
  Since most modern computer algebra systems provide the above capabilities, they are suitable platforms for crafting a rule-based repository of knowledge.

6 Efficient Pattern Matching

A general purpose repository might require thousands or even tens of thousands of rules. Obviously sequentially searching a list of that many rules to find a match would be unacceptably slow. Even having a separate list of rules for each built-in function or operator is insufficient, since some functions may have a large number of rules associated with it (e.g. our integrator requires over a 1000 rules).

Therefore instead of a list, the software platform supporting a repository should store the rules in the form of a discrimination net based on the tree structure of expressions. Then, for example, all rules applicable to expressions of the form $\sin(u)$ will be collected in one branch of the tree, all differentiation rules in another, all integration rules in another, etc. Then the rules in each branch will be recursively subdivided based on the form of its arguments, etc.

With the rules stored in such a discrimination net, the rule applicable to a given expression can be quickly found by a simple tree walk in $\log(n)$ time, where n is the number of rules in the repository.

7 Advantages

The following summarizes the advantages of storing mathematical knowledge in the form of a repository based on properly defined transformation rules:

- Human and machine readable. Since rules are defined using mathematical formulas rather than procedural programming constructs, they express a self-contained mathematical fact that can be attractively displayed in standard two-dimensional mathematical notation.
- Able to show simplification steps. The successive application of rules exactly corresponds to the steps required to simplify an expression. Thus when a rule is applied, it can display itself in standard mathematical notation as justification for the step, and then suspend further simplification so the partially simplified result is returned.
- Mechanical rule verification. Since the right side of a properly defined rule is just a mathematical expression, the rules validity can often be mechanically verified. For example, the right side of integration rules can be differentiated to see if they equal the integrand on the left.
- Facilitates program development. The fact that properly defined rules are inherently self-contained and free of side-effects makes it easy to test the effect on the system of selectively adding, modifying or deleting rules. Although collections of rules may be highly recursive, each individual rule must be able to stand on its own, thus making it possible to test it on examples before adding it to the collection.
- Platform independent. Since properly defined transformation rules consists only of mathematical expressions and pattern matching specifications, the translation of rules from the syntax of one computer algebra system to another is relatively straight-forward.

- White box transparency. For the most part, computer algebra systems appear as mysterious black boxes to users with little or no explanation given as to how results are obtained. However, if the source file of rules on which a CAS is based were included with the system, it becomes a transparent white box, making it possible for users to modify existing rules and even add new ones.
- Fosters community development. The open source nature of a rule-based repository of knowledge would foster an active community of users. A website blog dedicated to a repository could provide developers the ability to propose new rules and improvements to existing ones. Developers would vie with one another to get credit for adding new rules to the repository. Others would shoot down defective ones. Thus the system would grow and evolve in Darwinian fashion much the same way Wikipedia does.
- An active repository. Encyclopedias and reference manuals, even on-line ones, are inherently passive repositories in the sense that users have to find the knowledge required to solve a given problem, and then manually apply it. However, given a problem a rule-based repository actively finds and applies the knowledge required to solve it. Thus the knowledge in such repositories is in a much more useful form.

References

1. Abramowitz, M., Stegun, I.: Handbook of Mathematical Functions with Formulas, Graphs, and Mathematical Tables. US Government Printing Office (1964) (10th Printing December 1972)
2. Einwohner, T.H., Fateman, R.J.: Searching techniques for integral tables. In: Proceedings ISSAC 1995, pp. 133–139. ACM Press, New York (1995)

Natural Deduction Environment for Matita

Claudio Sacerdoti Coen* and Enrico Tassi*

Department of Computer Science, University of Bologna
Mura Anteo Zamboni, 7 — 40127 Bologna, ITALY
{sacerdot,tassi}@cs.unibo.it

Abstract. Matita is a proof assistant characterised by a rich, user extensible, output facility based on a widget for the rendering of MathML Presentation, and by the automatic handling of overloading by means of a flexible disambiguation mechanism. We show how to use these features to obtain a simple learning environment for natural deduction, without modifying the source code or Matita.

1 Introduction

There is at least one good reason for pushing the adoption of Interactive Theorem Provers as teaching instruments that go beyond any pedagogical consideration: make students familiar with Interactive Theorem Provers, grasping their interest for future project works or thesis assignments.

To do that, Interactive Theorem Provers have to be turned into learning platforms, for example tools for learning induction or other fundamental concepts like formal proofs. Learning logics, students meet formal proofs as derivation trees, for example following the natural deduction calculus.

In this paper we present our effort in implementing on top of the Matita Interactive Theorem Prover [2] a learning environment for Natural Deduction, exploiting the flexibility of the notational mechanism the tool offers and its peculiar ambiguity management. Our requirements were:

- allow students to input possibly incorrect derivation trees
- force the user to input exactly the same information he would write on paper, even if redundant or inferable by the system
- notify the user highlighting erroneously applied derivation rules, but allow him to complete the tree
- graphically display the derivation tree, facilitating its navigation in the frequent case of huge derivations not fitting the screen
- allow a quick (batch) correction of exercises to the teacher
- introduce the user to a textual syntax for the procedural construction of derivation trees in order to smooth the transition from derivation trees to procedural/declarative scripts
- help the user in learning such syntax and haste the input phase

* Partially supported by the Strategic Project "DAMA: Dimostrazia Assistita per la Matematica e l'Apprendimento" of the University of Bologna.

J. Carette et al. (Eds.): Calculemus/MKM 2009, LNAI 5625, pp. 486–491, 2009.
© Springer-Verlag Berlin Heidelberg 2009

D Introduction rules
▽ Elimination rules

Implication (⇒_e)

Conjunction left (∧_e_l)

Conjunction right (∧_e_r)

Disjunction (v_e)

Negation (¬_e)

Bottom (⊥_e)

Universal (∀_e)

Existential (∃_e)

▽ Misc rules

Reduction to Absurdity (RAA)

Use lemma (lem)

Discharge (discharge)

lemma ex1: $\neg\,(\exists x.P\ x)\ \Rightarrow \forall x.\neg\ P\ x$.
apply rule (**prove** $(\neg\,(\exists x.P\ x)\ \Rightarrow \forall x.\neg\ P\ x)$);
apply rule ($\Rightarrow_i$ [h1] $(\forall x.\neg\ P\ x)$);
apply rule ($\forall_i$ {l} $(\neg P\ l)$);
apply rule ($\neg_i$ [h2] $(\bot)$);
apply rule ($\neg_e$ $(\neg\,(\exists x.P\ x))$ $(\exists x.P\ x)$);
[**apply rule** (**discharge** [h1]);
| **apply rule** ($\exists_i$ {l} (P l));
  apply rule (**discharge** [h2]);
]
qed.

Fig. 1. Input palette and proof script

We presents the input and output interfaces in Section 2. The real contribution of the paper is in Section 3 where we show how to achieve the aforementioned goals without modifying the system and exploiting the MathML Presentation based notational system [4] and the peculiar management of notational overloading [6] Matita provides.

2   User Interface

Derivation trees are described using the following subset of the procedural language of Matita:

apply rule (*rule name* arguments...); [*subproof* | *subproof*...]

The apply-rule tactic is the standard application of a proof-term (which is usually, but not in our case, the name of a lemma), while square brackets are standard tacticals used to structure the proof script. Unlike other systems like LCF, Coq or Isabelle, the execution of a structured script is performed in Matita one tactic at a time [5], so that the incremental build of a structured proof is comfortable.

To apply a derivation rule the user is asked to list after the rule name all the information he would write as a side condition and above the inference line. For example the partial derivation tree of Figure 2 is obtained by executing the script of Figure 1 up to the end of the blue region in Figure 2.

The choice of annotating rules with the name of the hypotheses they discharge was a teacher choice, and it reflects the way the teacher wants the students to write the tree even on paper. Similarly, hypotheses are discharged by labelling them with their names (e.g. [·]^{h1}) and witnesses are explicitly provided for existential elimination. Missing sub-derivations are represented by numbered question marks; clicking on them the user is reminded of what assumptions are in scope.

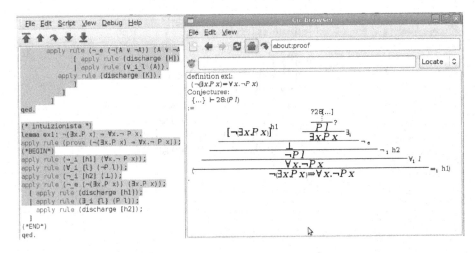

Fig. 2. Rendering of a partial derivation tree

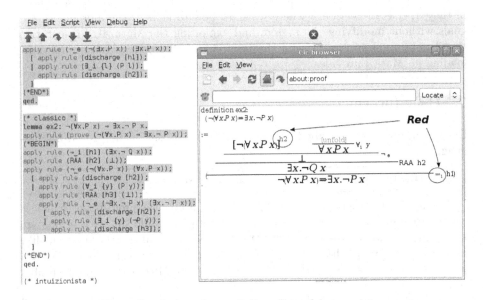

Fig. 3. Rendering of a partially collapsed incorrect tree

The concrete syntax consistently uses round brackets for formulae, square brackets to name hypotheses and curly brackets to name terms/variables.

A palette (Figure 1) can be used to insert in the script a template for every derivation rule, where no formula or term is inserted for the user. It is thus possible that the user incorrectly applies a rule. In such a case, the rule name is coloured in red as in Figure 3.

To improve readability, correct sub-trees can be collapsed to "[unfold]" by clicking on their root formula (see Figure 3).

3   Implementation

The notational system of Matita is based on three different languages for the representation of formulae and proofs: semantics, which is logic dependent, content, where we use MathML Content and OMDoc, and presentation, where we use MathML Presentation within a simple layout language called BoxML [3]. The user may define two sets of rules to map back and forth semantic objects to content objects, and to map content objects to presentation objects [1].

The two mappings need not to be one-to-one, and disambiguation is employed in the parsing phase to resolve overloading in favour of the interpretation that is meaningful (i.e. the well typed semantics objects).

Thanks to MathML Presentation we can render derivation trees in a partially satisfactory way as fractions (<mfrac>) and we can use <maction> nodes to collapse sub-trees and to show hypotheses in scope; finally we can use <mstyle> to highlight errors in red.

Disambiguation is used to associate to the same rule name two different semantics objects. The first one is the constant corresponding to the introduction/elimination rule in the logic of Matita. Thus, the object is well typed (i.e. meaningful for the disambiguation engine) only if its conclusion matches the current goal and its arguments are the expected ones. The second interpretation, that is always meaningful and is only used when the former fails, uses an (axiomatic, non admissible) cast operator to fix the incorrect rule application. Another notational rule associates <mstyle>, that produces the rendering in red, to occurrences of that particular cast constant.

The second interpretation can be deactivated by the teacher to make the system reject incorrect proofs. This is useful for batch correction of exercises.

We show now all the definitions, interpretations (from semantics to content) and notations (from content to presentation) that deal with logical conjunction and its elimination rule.

```
inductive And (A,B:CProp) : CProp :=And_intro: A → B → And A B.

definition And_elim_l : ∀A,B.And A B →A :=
  λ A,B,f. match f with [ And_intro l r ⇒l].

axiom any : CProp.
```

Note that the definition of conjunction is the standard one used in Matita, and not an embedding only useful for natural deduction. This allows the reuse of the standard machinery of Matita (e.g. automation, type checking) and of its library (e.g. in order to propose proofs of arithmetical statements). It also allows to progressively drop the natural deduction trees when the student is ready to embrace the full procedural language, or the declarative or to mix the three modes. The any proposition is necessary in the interpretation of a badly-applied conjunction elimination rule.

Show is used to make statements of sub-proofs explicit (so that notation can label the root of every subtree with the formula it proves). Cast is used to accept

> (* Used to replace a proof 'p' of 'P' with '(show P p)' to use 'P' in rendering *)
> **definition** show: ∀ A:CProp.A→ A := λ A,a.a.
> **axiom** cast: ∀ A,B:CProp.B → A.

a formula erroneously typed by the user by "casting" a proof of any proposition B to any proposition A.

We present now the output notations and relative interpretations used to render the conjunction elimination rule.

> **notation** < "\infrule ab a mstyle color #ff0000 (∧ $_{e\_l}$)"
>   **with** precedence 19 for @{ 'And_elim_l_ko ab a }.
> **interpretation** "And_elim_l_ko" 'And_elim_l_ko ab a =
>   (show a (cast _ _ (And_elim_l _ _ (cast _ _ ab)))).
>
> **notation** < "maction (\infrule ab a (∧ $_{e\_l}$)) [**unfold**]"
>   **with** precedence 19 for @{ 'And_elim_l_ok ab a }.
> **interpretation** "And_elim_l_ok" 'And_elim_l_ok ab a = (show a (And_elim_l _ _ ab)).

The first output notation displays the content symbol 'And_elim_l_ko with an \infrule layout (mapped to MathML <mfraction> plus <mstyle> directions to avoid font shrinking in fractions of fractions) colouring the rule name in red. Its corresponding interpretation is associated to a term containing the cast constant. The second output rule displays a correct rule application, adding the possibility to fold the tree clicking on it (<maction> node, whose default behaviour is to toggle between its children).

Dually, we introduce an input notation and two corresponding interpretations, respectively for a correct and a wrong application of conjunction elimination.

> **notation** > "∧ _e_l term 90 ab" **with** precedence 19 for @{ 'And_elim_l (show ab ?) }.
> **interpretation** "And_elim_l KO" 'And_elim_l ab =
>   (cast _ _ (And_elim_l _ _ (cast (And any any) _ ab))).
> **interpretation** "And_elim_l OK" 'And_elim_l ab = (And_elim_l _ _ ab).

The input notation is associated to the rule name. The second interpretation is preferred and inserts no cast constant, while the first one, used only if the second fails, applies the rule casting both its conclusion and premise. In particular, the premise is cast to the conjunction of two occurrences of the previously declared dummy proposition any in order to apply elimination of conjunction.

4   Conclusion

Matita is a proof assistant based on the Curry-Howard isomorphism: proofs are internally represented at the semantics level with proof terms. Thus the machinery used to associate the familiar mathematical notation to formulae can as well be used to render proofs. In particular, in place of the usual rendering as natural language text with embedded MathML formulae, we can output derivation trees as interactive MathML formulae, exploiting <mfrac>, <maction> (to collapse and expand sub-trees) and <mstyle> (to highlight incorrect parts of the proof). Thanks to the typical MKM layered representation of knowledge (semantics, content and presentation [1]) that is at the core of the notational machinery of Matita, we have been able to achieve the latter result without changing the source code, and with around 800 lines of notation and interpretation commands.

We then proceeded to fulfil the minimal requirements for a learning system, that we identified in the possibility for the student to input exactly the same information he would write on paper and to be allowed to commit errors and continue. We achieved this easily thanks to the disambiguation engine of Matita [6] that allows to resolve overloading in favour of the interpretations that yield meaningful formulae: by introducing an axiomatic, non admissible cast rule, any derivation tree becomes legal, but incorrect trees can be easily distinguished and displayed accordingly.

This way of implementing the learning system using only notational devices allows to progressively abandon the derivation trees in favour of the standard declarative or procedural language of Matita, when the students are ready. The system was adopted in a first course of logic for computer scientists at the University of Bologna in the academic year 2008/2009.

The only current limitation is the impossibility to develop derivation trees in a bottom-up fashion, from the leaves to the root, by working with hypotheses that will be discharged only later. This reflects the way proofs are developed in Fitch style and in informal mathematics, and in class we carefully avoid making students work with un-assumed hypotheses anyway. On the other hand, sub-proofs (containing no free assumptions) can be proved in advance and plugged in the main proof.

References

1. Adams, A.A.: Digitisation, representation and formalisation: Digital libraries of mathematics. In: Davenport, J.H., Asperti, A., Buchberger, B. (eds.) MKM 2003. LNCS, vol. 2594, pp. 1–16. Springer, Heidelberg (2003)
2. The Matita interactive theorem prover, http://matita.cs.unibo.it
3. Padovani, L.: A math canvas for the GNOME desktop. In: 5th Annual GNOME User and Developer European Conference (GUADEC 2004), vol. 107. Agder University College (2004)
4. Padovani, L., Zacchiroli, S.: From notation to semantics: There and back again. In: Borwein, J.M., Farmer, W.M. (eds.) MKM 2006. LNCS (LNAI), vol. 4108, pp. 194–207. Springer, Heidelberg (2006)
5. Sacerdoti Coen, C., Tassi, E., Zacchiroli, S.: Tinycals: step by step tacticals. In: Proceedings of User Interface for Theorem Provers 2006. Electronic Notes in Theoretical Computer Science, vol. 174, pp. 125–142. Elsevier Science, Amsterdam (2006)
6. Sacerdoti Coen, C., Zacchiroli, S.: Spurious disambiguation errors and how to get rid of them. Journal of Mathematics in Computer Science, special issue on Management of Mathematical Knowledge 2, 355–378 (2008)

Author Index